# DIMENSION THEORY OF
# GENERAL SPACES

# DIMENSION THEORY
# OF GENERAL
# SPACES

**A. R. PEARS**

*Lecturer in Mathematics*
*Queen Elizabeth College, University of London*

**CAMBRIDGE UNIVERSITY PRESS**

CAMBRIDGE

LONDON · NEW YORK · MELBOURNE

CAMBRIDGE UNIVERSITY PRESS
Cambridge, New York, Melbourne, Madrid, Cape Town, Singapore, São Paulo, Delhi

Cambridge University Press
The Edinburgh Building, Cambridge CB2 8RU, UK

Published in the United States of America by Cambridge University Press, New York

www.cambridge.org
Information on this title: www.cambridge.org/9780521205153

First published 1975
This digitally printed version 2008

A catalogue record for this publication is available from the British Library

Library of Congress Catalogue Card Number: 74–12955

ISBN 978-0-521-20515-3 hardback
ISBN 978-0-521-09302-6 paperback

# CONTENTS

# INTRODUCTION

This book is concerned with the dimension theory of general topological spaces. It provides a complete and self-contained account of the theory laying particular emphasis on the dimensional properties of non-metrizable spaces. It is intended to serve as a reference work for mathematicians with an interest in general topology. It is written in such a way as to be accessible to beginning postgraduate students and might be used as a textbook.

A theory of dimension starts with a 'dimension function', which is a function $d$ defined on the class of topological spaces such that $d(X)$ is an integer or $\infty$, with the properties that $d(X) = d(Y)$ if $X$ and $Y$ are homeomorphic and $d(\mathbf{R}^n) = n$ for each positive integer $n$, where $\mathbf{R}^n$ denotes Euclidean $n$-space. Dimension theory reveals the properties of dimension functions. For example we wish to know under what circumstances it can be asserted that $d(A) \leqslant d(X)$ if $A$ is a subspace of a topological space $X$. Assertions of this form are known as subset theorems for the dimension function $d$. An affirmation that $d(X) = \sup d(A_\lambda)$ for certain coverings $\{A_\lambda\}$ of some spaces $X$ is called a sum theorem for the dimension function $d$. In dimension theory we also examine the relationships between different dimension functions. An objective is to ascertain circumstances under which 'reasonable' dimension functions coincide.

The principal dimension functions are the covering dimension function which is studied in Chapter 3 of this book and the small and large inductive dimension functions which are studied in Chapter 4. Chapters 5 and 6 are devoted to a study of related dimension functions. There is a substantial theory of covering dimension for normal spaces and the covering dimension function satisfies a general subset theorem for the class of totally normal spaces. The large inductive dimension function satisfies sum and subset theorems for totally normal spaces. The small inductive dimension function has the greatest intuitive appeal and satisfies the subset theorem for arbitrary spaces. There is no theory of small inductive dimension but, being easier to calculate, the small inductive dimension function gives information about the other dimension functions. The coincidence of all three principal dimension functions for separable metrizable spaces explains the elegance of classical dimension theory, so well exposed by Hurewicz and Wallman [1941]. Chapter 7 shows that a very satisfactory theory of dimension can be constructed for metrizable spaces, but also includes P. Roy's example of a metrizable space for which small inductive dimension is different from covering dimension and large inductive dimension. Most of Chapter 8 is occupied with the pathological dimension theory of bicompact spaces and includes an example due to V. V. Filippov which shows that the small and large inductive dimensions can differ for such spaces. There are many other examples throughout the book. In Chapter 10 the covering dimension function is modified for non-normal spaces. Dimension-theoretic applications of the algebra of continuous bounded real-valued functions on a topological space are made in this chapter and Katětov's beautiful algebraic characterization of the dimension of a Tihonov space is given. Information about the dimension of bicompactifications is obtained here and in Chapter 6. Universal spaces for various classes of spaces of given dimension are constructed during the course of the work, which concludes with Pasynkov's universal spaces for Tihonov spaces of given dimension.

In 1968 I gave postgraduate lectures on dimension theory at the University of London. The notes of those lectures provided a starting point for the writing of this book. My objective was to write an account of the now mature theory of dimension within point–set topology. I hope this book will be read by postgraduate students at the beginning of their careers. The reader is assumed to be familiar with naive set theory and its standard notation and to know as much general topology as one might expect an undergraduate course to cover. The first section of the first chapter contains a brief résumé of elementary topology. The rest of Chapter 1 is concerned with standard general topology and will be used for reference only by many readers. Chapter 2 covers material on paracompactness and metrization which may be less familiar. Both chapters contain some topics which arose in dimension theory and these may be unfamiliar to readers well acquainted with general topology. Expanded treatments of many of the topics in Chapters 1 and 2, and the examples necessary to complement the theory, will be found in the books of Dugundji [1966], Engelking [1968] and Kelley [1955]. I am glad to acknowledge my debt to these works.

Throughout the book proofs are given in detail except for those of some standard results on normality and compactness in Chapter 1. The symbol * follows the statement of these propositions, with which it is assumed the reader will be familiar. Their proofs are given in many textbooks. The Brouwer theorem, that the unit sphere in Euclidean space is not a retract of the closed unit ball which it bounds, is used in the determination of the dimension of Euclidean space. An *ad hoc* proof of this theorem might have been given but it seems that the 'right' proof is by means of homology theory. I hope the omission of its proof will be acceptable since its truth is easy to take on trust. An effort has been made to employ separation properties of topological spaces only where they are appropriate. In particular normal and paracompact spaces are not assumed to satisfy the Hausdorff separation axiom. In places where a known theorem and its proof have been modified in accordance with this principle, then the new proof usually seems to be more natural.

The book is written in the 'definition–proposition–proof–remark' style. I hope this will allow the logical structure of the subject to show clearly. Much of the informal and motivational material has been collected into notes at the end of each chapter. Each chapter is divided into sections. The items (definitions, theorems, propositions, corollaries, examples, remarks) in a section are numbered by pairs $n.p$ of integers, where $n$ is the section number within the chapter and $p$ is the item number within the section. Elsewhere in the book, item $n.p$ of Chapter $m$ is referred to as item $m.n.p$ except in Chapter $m$ itself where the reference is abbreviated to $n.p$. A Halmos symbol $\square$ marks the end of a proof. The appearance of $\square$ immediately after the statement of a proposition or corollary signifies that its truth is obvious or occasionally that its easy proof is left to the reader.

There are notes at the end of every chapter except Chapter 1. These contain the references to original sources, listed in the bibliography. There are also comments on the historical development of the subject. These do not form a complete record and there has been no attempt to make a definitive assignment of credit. I hope some impression is given of the subject as a human activity, which will add to the reader's enjoyment. The notes also survey some recent developments which have not been included in the book. I made much use of the survey articles of Aleksandrov [1951, 1960, 1964] and Nagata [1966, 1971] whilst writing the notes.

I have been helped by several people whom I wish to thank. Professor C. H. Dowker of Birkbeck College, London, helped me to plan the course of lectures from which the book has grown. He has answered many questions and advised me most generously as the work has developed. I was sustained during the labour of transforming my lecture notes into a book by the interest shown in the work by Drs A. G. R. Calder and E. H. Kronheimer of Birkbeck College. Mr A. T. Al-Ani, whilst a postgraduate student at Queen Elizabeth College, London, read several versions of the manuscript and made many helpful suggestions. Dr A. J. Ward of Emmanuel College, Cambridge, and Dr D. J. White of the University of Reading each read a chapter of the final manuscript and commented in detail and most helpfully. In 1972, I enjoyed some months of collaboration with Professor J. E. Mack of the University of Kentucky and I am also grateful to him for a conversation which gave me the courage to study Prabir Roy's example. I am grateful to Professor W. B. Bonnor and my colleagues at Queen Elizabeth College for providing a pleasant environment in which to work.

I wish to thank the Syndics of the Cambridge University Press for accepting the book for publication and their staff for their helpfulness and understanding.

A. R. PEARS

# 1

## TOPOLOGICAL SPACES, NORMALITY AND COMPACTNESS

### 1 Topological spaces

In this section the definitions and results from point-set topology with which we shall assume familiarity are listed. In addition to making clear the prerequisites for reading the book, the purpose is to establish what forms of concepts, about which there is no general agreement, will be used, and to introduce notation and terminology. In addition to the topics covered in the brief discussion of this section, some acquaintance will be assumed with the conditions of normality and compactness. Later sections of this chapter are devoted to these concepts. We begin by recalling some definitions from the theory of ordered sets.

A *quasi-order* on a set $X$ is a binary relation $\leqslant$ on $X$ which is reflexive and transitive. Thus $x \leqslant x$ for every $x$, and if $x \leqslant y$ and $y \leqslant z$ then $x \leqslant z$. A *quasi-ordered set* is a set together with a quasi-order on the set. A quasi-ordered set $X$ is said to be a *directed set* if for each pair $x, y$ of elements of $X$ there exists $z$ in $X$ such that $x \leqslant z$ and $y \leqslant z$. A *partial order* on a set is a quasi-order on the set which is anti-symmetric in the sense that $x = y$ if $x \leqslant y$ and $y \leqslant x$. A *partially ordered set* is a set together with a partial order on the set. Let $X$ be a partially ordered set with partial order $\leqslant$. An element $x$ of $X$ is said to be *maximal* if there exists no element $y$ distinct from $x$ such that $x \leqslant y$. Similarly $x$ is said to be *minimal* if there exists no $y$ distinct from $x$ such that $y \leqslant x$. If $A$ is a subset of $X$, then an *upper bound* of $A$ is an element $x$ such that $a \leqslant x$ for every $a$ in $A$. The *least upper bound* or *supremum* of $A$ is an upper bound $x$ of $A$ such that $x \leqslant y$ for every upper bound $y$ of $A$. We shall denote the least upper bound of $A$ by sup $A$ when it exists. The definitions of *lower bound* and *greatest lower bound* or *infimum* are similar. The notation inf $A$ is used for the greatest lower bound of $A$ when it exists. If $X$ has a lower bound it is called the least element of $X$ and an upper bound for $X$, if it exists, is called the greatest element of $X$. The least and greatest elements of $X$, when they exist, are called *universal bounds* of $X$. A partially ordered set in which each finite subset has a least upper bound and a greatest lower bound is called a *lattice*. If $x$ and $y$ are elements of a lattice, we write

$$x \vee y = \sup\{x, y\} \quad \text{and} \quad x \wedge y = \inf\{x, y\}.$$

A lattice is called a *distributive lattice* if the identitites

$$x \wedge (y \vee z) = (x \wedge y) \vee (x \wedge z),$$

$$x \vee (y \wedge z) = (x \vee y) \wedge (x \vee z),$$

hold. A partial order $\leqslant$ on a set $X$ is called a *linear order* if for each pair $x, y$ of elements of $X$ either $x \leqslant y$ or $y \leqslant x$ holds. A *linearly ordered set* is a set together with a linear order on the set. Each subset of a partially ordered set has an induced order; a subset of a partially ordered set in which the induced order is a linear order is called a *chain*. A linearly ordered set is said to be *order-complete* if each non-empty subset has a least upper bound and a greatest lower bound. If $\leqslant$ is a linear order on a set $X$ let us introduce a relation $<$ on $X$ by putting $x < y$ if $x \leqslant y$ and $x \neq y$. If $a, b \in X$ and $a < b$ let us put

$$(a, b) = \{x \in X \mid a < x < b\},$$

$$(a, b] = \{x \in X \mid a < x \leqslant b\},$$

$$[a, b) = \{x \in X \mid a \leqslant x < b\},$$

$$[a, b] = \{x \in X \mid a \leqslant x \leqslant b\}.$$

This notation will always be used for linearly ordered sets. A subset of a linearly ordered set of any one of the above four forms is called an *interval*. A linearly ordered set is *well-ordered* if each non-empty subset has a least element. Thus if $A$ is a non-empty subset of a well-ordered set, there exists $x$ in $A$ such that $x \leqslant y$ if $y \in A$. We shall frequently use the equivalent form of the axiom of choice that any set can be well-ordered. It follows that any set can be well-ordered so that it has a greatest element. [For if $X$ is a well-ordered set with order $\leqslant$, and $x$ is the least element of $X$, a new linear order $\leqslant'$ can be defined on $X$ so that $y \leqslant' z$ if and only if $y \leqslant z$ if $y$ and $z$ are distinct from $x$, and $y \leqslant' x$ for all $y$. With the linear order $\leqslant'$, $X$ is well-ordered and $x$ is the greatest element of $X$.] We shall also use *Zorn's lemma* which states that if each chain in a partially ordered set has an upper bound then the set contains a maximal element, and the Kuratowski lemma which states that each chain in a partially ordered set is contained in a maximal chain. Familiarity will be assumed with ordinal numbers and with proof and construction by transfinite induction. Some knowledge of cardinal numbers and their arithmetic will be assumed. The cardinal number of a set $X$ will be denoted by $|X|$. The set of positive integers will be denoted by $\mathbf{N}$. The cardinal number $|\mathbf{N}|$ is the first infinite cardinal, denoted by $\aleph_0$. The first uncountable cardinal is denoted by $\aleph_1$. The cardinal number of the set $\mathbf{R}$ of real numbers is denoted by $\mathbf{c}$.

A *topological space* is a set $X$ and a set $\mathscr{T}$ of subsets of $X$, called the *open sets* of $X$, such that $\mathscr{T}$ contains $\varnothing$ and $X$, and is closed under the formation of finite intersections and arbitrary unions. The set $\mathscr{T}$ of open sets is called the *topology* of the topological space. Thus a topological space is a pair $(X, \mathscr{T})$. If no confusion can arise, the topology $\mathscr{T}$ is not mentioned and we speak of the 'topological space $X$'. If $X$ is a set and $\mathscr{T}$ is the set of all subsets of $X$ then $\mathscr{T}$ is a topology called the *discrete topology*. The set of subsets of $X$ which consists of $\varnothing$ and $X$ only is a topology called the *trivial topology*. A subset $N$ of a topological space $X$ is a *neighbourhood* of a point $x$ of $X$ if there exists an open set $U$ such that $x \in U \subset N$. Similarly $N$ is a neighbourhood of a subset $A$ of $X$ if there exists an open set $U$ such that $A \subset U \subset N$. A set $U$ is open if and only if $U$ contains a neighbourhood of each of its points. A *base* for the topology of a space $X$ is a set $\mathscr{B}$ of open sets such that for every point $x$ of $X$ and every neighbourhood $N$ of $x$ there exists $B$ in $\mathscr{B}$ such that $x \in B \subset N$. Equivalently $\mathscr{B}$ is a base if each open set is a union of members of $\mathscr{B}$. A *subbase* for a topology is a set $\mathscr{S}$ of subsets such that the set of all finite intersections of members of $\mathscr{S}$ is a base for the topology. The *interior* $A^\circ$ of a subset $A$ of a topological space is the largest open set contained in $A$. A set $U$ is open if and only if $U = U^\circ$. For subsets $A$ and $B$ we have

$$(A \cap B)^\circ = A^\circ \cap B^\circ.$$

A subset $A$ of a topological space $X$ is said to be *closed* if its complement $X \backslash A$ is open. Clearly any intersection of closed sets is a closed set and any finite union of closed sets is a closed set. In a topological space $X$, the sets $\varnothing, X$ are closed, and so are open-and-closed sets. The topological space $X$ is said to be *connected* if there are no open-and-closed sets other than $\varnothing$ and $X$. The *closure* $\bar{A}$ of a subset $A$ of a topological space is the smallest closed set containing $A$. A point $x$ belongs to $\bar{A}$ if and only if $N \cap A \neq \varnothing$ for every neighbourhood $N$ of $x$. For typographical convenience, the closure of $A$ will sometimes be denoted by $(A)^-$. A set $A$ is closed if and only if $A = \bar{A}$. If $A$ is a subset of a topological space $X$ then

$$X \setminus \bar{A} = (X \setminus A)^\circ.$$

For subsets $A$ and $B$ we have

$$(A \cup B)^- = \bar{A} \cup \bar{B}.$$

If $A$ is a subset of a topological space $X$, then a point $x$ of $X$ is said to be an *accumulation point* of $A$ if every neighbourhood of $x$ contains a point of $A$ distinct from $x$. Clearly $\bar{A}$ is the union of $A$ and the set of accumulation points of $A$. A subset $A$ is *dense* in a topological space $X$ if $\bar{A} = X$. If $A$ is a subset of a topological space $X$ and $V$ is an open set of $X$, then $V \cap \bar{A} \subset (V \cap A)^-$. [For if $x \in V \cap \bar{A}$ and $N$ is a neighbourhood of $x$, then $N \cap V$ is a neighbourhood of $x$ so that

$$N \cap V \cap A \neq \varnothing$$

and hence $x \in (V \cap A)^-$.] It follows that if $A$ is a dense subset and $V$ is an open set then $V \subset (V \cap A)^-$, so that $\bar{V} \subset (V \cap A)^-$ and hence $\bar{V} = (V \cap A)^-$. A *covering* of a topological space is a family $\mathscr{A} = \{A_\lambda\}_{\lambda \in \Lambda}$ of subsets such that

$$\bigcup_{\lambda \in \Lambda} A_\lambda = X.$$

If each set $A_\lambda$ is open, then $\mathscr{A}$ is called an *open covering*, and if each set $A_\lambda$ is closed, then $\mathscr{A}$ is called a *closed covering*. A covering $\{B_\gamma\}_{\gamma \in \Gamma}$ is said to be a *refinement* of a covering $\{A_\lambda\}_{\lambda \in \Lambda}$ if for each $\gamma$ in $\Gamma$ there exists some $\lambda$ in $\Lambda$ such that $B_\gamma \subset A_\lambda$. If $Y$ is a subset of a topological space $X$, the induced topology for $Y$ consists of the sets of the form $U \cap Y$, where $U$ is an open set of $X$. The subset $Y$ with the induced topology is a *subspace* of $X$. If $Y$ is a subspace of $X$ and $B$ is a subset of $Y$, then the closure of $B$ in $Y$ is $\bar{B} \cap Y$, where $\bar{B}$ is the closure of $B$ in $X$. A subset $Y$ of a topological space is said to be a connected set if the subspace $Y$ is connected. Since the union of connected sets with a common point is connected, it follows that each point of a topological space is contained in a largest connected set which is called the *component* of the point. Since the closure of a connected set is connected, the components are closed sets. The set of all distinct components in a space forms a partition of the space.

Many examples in dimension theory involve linearly ordered spaces. The interval topology on a linearly ordered set $X$ has a subbase consisting of all sets of the form

$$\{x \in X \mid x > a\} \quad \text{and} \quad \{x \in X \mid x < b\}$$

for $a,b$ in $X$. A linearly ordered set with the interval topology is called a *linearly ordered space*. If $X$ has neither a least element nor a greatest element, then a base for the interval topology consists of all 'open' intervals $(a,b)$, where $a,b$ are elements of $X$ such that $a < b$. If $X$ has a least element $x_0$, all intervals $[x_0, b)$ for $b$ in $X\backslash\{x_0\}$ must be added, and if $X$ has a greatest element $x_1$, all intervals $(a, x_1]$ for $a$ in $X\backslash(\{x_1\}$ must be added. The interval topology on the set $\mathbf{R}$ of real numbers with respect to its natural linear order is the usual topology for $\mathbf{R}$. The closed subspace $[0, 1]$ is the *unit interval* and will usually be denoted by $I$.

If $X$ and $Y$ are sets and $f\colon X \to Y$ is a function, then the inverse image of a subset $B$ of $Y$ is the set

$$f^{-1}(B) = \{x \in X \mid f(x) \in B\}.$$

In the case of one-point subsets we change the notation slightly, writing

$$f^{-1}(y) = \{x \in X \mid f(x) = y\}$$

if $y \in Y$. If $X$ and $Y$ are topological spaces, then $f\colon X \to Y$ is said to be a *continuous function* (mapping or continuous mapping) if $f^{-1}(U)$ is an open set of $X$ for each open set $U$ of $Y$. We list a number of conditions equivalent to continuity: $(a)$ the inverse image of each member of a subbase for the topology of $Y$ is an open set of $X$; $(b)$ the inverse image of each closed set of $Y$ is a closed set of $X$; $(c)$ for every point $x$ of $X$, the inverse image of every neighbourhood of $f(x)$ is a neighbourhood of $x$; $(d)$ $f(\bar{A}) \subset (f(A))^{-}$ for each subset $A$ of $X$; $(e)$ $(f^{-1}(B))^{-} \subset f^{-1}(\bar{B})$ for each subset $B$ of $Y$. The composite of continuous functions is continuous. The continuous real-valued functions on a topological space are of particular importance. Let $X$ be a topological space and let $f, g\colon X \to \mathbf{R}$ be continuous functions, where $\mathbf{R}$ has the usual topology. The real-valued functions $|f|, f+g, fg$, given by

$$|f|\,(x) = |f(x)|,$$

$$(f+g)\,(x) = f(x) + g(x),$$

$$(fg)\,(x) = f(x)\,g(x),$$

if $x \in X$, are continuous. If $\lambda \in \mathbf{R}$, then the function $\lambda f$, given by $(\lambda f)\,(x) = \lambda f(x)$ if $x \in X$, is continuous. The function $h = \max\{f, g\}$ is continuous, where

$$h(x) = \max\{f(x), g(x)\}$$

if $x \in X$. Similarly the function $\min \{f, g\}$ is continuous. If $g(x) \neq 0$ for all $x$ in $X$ we can define a real-valued function $f/g$ by putting

$$(f/g)(x) = f(x)/g(x)$$

if $x \in X$, and $f/g$ is continuous. If $\{f_i\}_{i \in \mathbf{N}}$ is a sequence of continuous real-valued functions on a topological space $X$, and for each $i$, $|f_i(x)| \leqslant M_i$ for all $x$, where the series $\Sigma_{i=1}^{\infty} M_i$ is convergent, then the series $\Sigma_{i=1}^{\infty} f_i(x)$ is absolutely convergent for each $x$, and we can define a real-valued function $f$ on $X$ by putting

$$f(x) = \sum_{i=1}^{\infty} f_i(x)$$

if $x \in X$, and $f$ is continuous. If $X$ and $Y$ are topological spaces, a bijection $h: X \to Y$ is said to be a *homeomorphism* if both $h$ and $h^{-1}$ are continuous. Equivalently, a continuous function $h: X \to Y$ is a homeomorphism if there exists a continuous function $g: Y \to X$ such that

$$g \circ h = 1_X \quad \text{and} \quad h \circ g = 1_Y,$$

where $1_X$ is the identity mapping on $X$ given by $1_X(x) = x$ for all $x$, and similarly $1_Y$ is the identity mapping on $Y$. If $h: X \to Y$ is a homeomorphism then $h$ induces a bijective correspondence between the topologies of $X$ and $Y$. If there exists a homeomorphism between $X$ and $Y$ then $X$ and $Y$ are said to be homeomorphic. The relation of being homeomorphic is an equivalence relation in the class of all topological spaces. A property which, when possessed by a given space, is possessed by all homeomorphic spaces is called a *topological property* or a topological invariant. Clearly each property of a topological space which is defined in terms of open sets and concepts derived from open sets and the notions of set theory is a topological property. A continuous function $f: X \to Y$ is said to be *open* if $f(U)$ is an open set of $Y$ for each open set $U$ of $X$, and $f$ is said to be *closed* if $f(E)$ is a closed set of $Y$ for each closed set $E$ of $X$. The following statements about a function $h$ are equivalent: (a) $h$ is a homeomorphism; (b) $h$ is a continuous open bijection; (c) $h$ is a continuous closed bijection. A mapping $f: X \to Y$ is closed if and only if $f(\bar{A}) = (f(A))^-$ for every subset $A$ of $X$. If $X$ and $Y$ are topological spaces, $f: X \to Y$ is a continuous function and $A$ and $B$ are subspaces of $X$ and $Y$ respectively such that $f(A) \subset B$, then the function $g: A \to B$ given by $g(x) = f(x)$ for each $x$ in $A$ is continuous. The mapping $g$ is said to be given by restriction of $f$. If $f$ is open and $A$ is open then $g$ is open, and similarly $g$ is closed if $f$ is closed and $A$ is closed. We shall write $f|A$ to denote the mapping of

$A$ into $Y$ given by restriction of $f$. If $B$ is a subset of $Y$, then for each subset $H$ of $X$

$$f\big(f^{-1}(B) \cap H\big) = B \cap f(H).$$

Thus the mapping of $f^{-1}(B)$ into $B$ given by restriction of $f$ is open if $f$ is open and closed if $f$ is closed. A continuous function $f: X \to Y$ is said to be an *embedding* of $X$ in $Y$ if the mapping of $X$ onto $f(X)$, given by restriction of $f$, is a homeomorphism. If there exists an embedding of $X$ in $Y$, then $X$ is homeomorphic with a subspace of $Y$ and thus $X$ has any topological property which is inherited by subspaces of $Y$. If $A$ is a subspace of a topological space $X$, then a continuous function $r: X \to A$ such that $r(x) = x$ for each $x$ in $A$ is called a *retraction* of $X$ onto $A$. If there exists a retraction of $X$ onto $A$, then $A$ is said to be a *retract* of $X$.

Let $\{X_\lambda\}_{\lambda \in \Lambda}$ be a family of topological spaces and let $X$ be the cartesian product $\Pi_{\lambda \in \Lambda} X_\lambda$. The points of $X$ are indexed families $\{x_\lambda\}_{\lambda \in \Lambda}$, where $x_\lambda \in X_\lambda$ for each $\lambda$. For $\lambda_0$ in $\Lambda$, the projection $\pi_{\lambda_0}: X \to X_{\lambda_0}$ is given by $\pi_{\lambda_0}(\{x_\lambda\}) = x_{\lambda_0}$. The set $X$ with the smallest topology such that every projection $\pi_\lambda$ is continuous is the *topological product* of the family $\{X_\lambda\}_{\lambda \in \Lambda}$. A base for the product topology consists of all sets of the form $\bigcap_{\lambda \in M} \pi_\lambda^{-1}(V_\lambda)$, where $M$ is a finite subset of $\Lambda$ and $V_\lambda$ is an open set of $X_\lambda$ for $\lambda$ in $M$. In fact $X$ has a subbase for its topology which consists of all sets of the form $\pi_\lambda^{-1}(V_\lambda)$, where $\lambda \in \Lambda$ and $V_\lambda$ is a member of a given subbase for the topology of $X_\lambda$. If for each $\lambda$, $g_\lambda: Z \to X_\lambda$ is a continuous function, then the unique function $g: Z \to X$ such that $\pi_\lambda \circ g = g_\lambda$, is continuous. It follows that if $\{X_\lambda\}_{\lambda \in \Lambda}$ and $\{Y_\lambda\}_{\lambda \in \Lambda}$ are families of topological spaces with topological products $X$ and $Y$ respectively and projections $\pi_\lambda$ and $\rho_\lambda$ for each $\lambda$, and $f_\lambda: X_\lambda \to Y_\lambda$ is a continuous function for each $\lambda$, then there exists a unique continuous function $f: X \to Y$ such that $\rho_\lambda \circ f = f_\lambda \circ \pi_\lambda$ for every $\lambda$. We shall write $f = \Pi_{\lambda \in \Lambda} f_\lambda$. If $\{X_\lambda\}_{\lambda \in \Lambda}$ is a family of topological spaces with topological product $X$ and $A_\lambda$ is a subspace of $X_\lambda$ for each $\lambda$, then the topology induced on $\Pi_{\lambda \in \Lambda} A_\lambda$ as a subspace of $X$ coincides with the product topology. The *topological sum* of a family $\{X_\lambda\}_{\lambda \in \Lambda}$ of topological spaces is the disjoint union $X$ of the sets $X_\lambda$ with the largest topology such that the inclusion of each $X_\lambda$ in $X$ is continuous. The open sets of the topological sum are the disjoint unions of families $\{U_\lambda\}_{\lambda \in \Lambda}$, where $U_\lambda$ is an open set of $X_\lambda$ for each $\lambda$. Each space $X_\lambda$ is embedded in $X$ as an open-and-closed subspace. If $X$ is a topological space, $Y$ is a set and $f: X \to Y$ is a surjection, then the *identification topology* on $Y$ with respect to $f$ is the largest topology on $Y$ such that

$f$ is continuous. Thus a subset $V$ of $Y$ is an open set if and only if $f^{-1}(V)$ is an open set of $X$. If $X$ and $Y$ are topological spaces and $f\colon X \to Y$ is a continuous surjection, then $f$ is called an *identification mapping* if $Y$ has the identification topology with respect to $f$. Any continuous open surjection and any continuous closed surjection is an identification mapping. If $f\colon X \to Y$ is an identification mapping and $h\colon X \to Z$ is a continuous function such that $h$ is constant on $f^{-1}(y)$ for each $y$ in $Y$, then the function $g\colon Y \to Z$ such that $g \circ f = h$ is continuous. If $R$ is an equivalence relation in a topological space $X$, then the *quotient space* of $X$ with respect to $R$ is the set $Y$ of equivalence classes of $X$ with respect to $R$ with the identification topology with respect to the surjection $f\colon X \to Y$ such that $x \in f(x)$ for each $x$ in $X$. If $f\colon X \to Y$ is an identification mapping, then $Y$ is homeomorphic with the quotient space of $X$ with respect to the equivalence relation $R$ defined by $xRx'$ if and only if $f(x) = f(x')$ for $x, x'$ in $X$.

Next we introduce some 'separation axioms'. In this work, a topological space will not be assumed to satisfy any separation axiom unless it is explicitly stated. The most important separation property for dimension theory is normality which will be studied in §3. Consider the following conditions on a topological space:

$(T_0)$ For every pair of distinct points there exists a neighbourhood of one of them which does not contain the other.

$(T_1)$ For every pair of distinct points there exists a neighbourhood of each of them which does not contain the other.

$(T_2)$ For every pair of distinct points $x$ and $y$ there exist a neighbourhood of $x$ and a neighbourhood of $y$ which are disjoint.

A topological space is said to be a $T_i$-*space*, where $i = 0, 1, 2$, if it satisfies the axiom $(T_i)$ above. Topological spaces which satisfy $(T_2)$ are also called *Hausdorff spaces*. Clearly every $T_2$-space is a $T_1$-space and every $T_1$-space is a $T_0$-space. The reverse implications do not hold and there exist spaces which do not satisfy the axiom $(T_0)$. A topological space is a $T_1$-space if and only if each set which consists of a single point is closed. A topological space is said to be a *regular space* if for each point $x$ and each closed set $F$ not containing $x$ there exist disjoint open sets $U$ and $V$ such that $x \in U$ and $F \subset V$. Equivalently, a topological space is a regular space if each neighbourhood of each point contains a closed neighbourhood of the point. Hausdorff spaces are not necessarily regular and regular spaces are not necessarily $T_0$-spaces. A regular $T_0$-space, however, is a Hausdorff space. A regular

$T_0$-space is said to be a $T_3$-space. The separation properties so far considered are preserved under the operations of taking subspaces and forming topological products.

A *pseudo-metric* on a set $X$ is a function $d: X \times X \to \mathbf{R}$ such that for all $x, y, z$ in $X$: (i) $d(x, y) = d(y, x)$; (ii) $d(x, z) \leqslant d(x, y) + d(y, z)$; (iii) $d(x, x) = 0$. If $d$ is a pseudo-metric on a set $X$, then $d(x, y) \geqslant 0$ for all $x, y$ in $X$. If $d$ is a pseudo-metric on a set $X$ and $d$ satisfies in addition the condition that $d(x, y) = 0$ implies $x = y$, then $d$ is a *metric* on $X$. A *pseudo-metric space* is a pair $(X, d)$, where $X$ is a set and $d$ is a pseudo-metric on $X$. If $d$ is a metric then $(X, d)$ is a *metric space*. If $(X, d)$ is a pseudo-metric space then there is a topology induced by $d$ on $X$. For each point $x$ of $X$ and each positive real number $r$ let

$$B_r(X) = \{y \in X \mid d(x, y) < r\}.$$

The family of sets $B_r(x)$ for $r > 0$ and $x$ in $X$ is the base for a topology on $X$. This topology is called the pseudo-metric topology on $X$ or the topology on $X$ induced by $d$. We shall call $B_r(x)$ the *open ball* with centre $x$ and radius $r$. The notation $B_r(x)$ for an open ball will be used without further explanation whenever a pseudo-metric or metric space is discussed. If $(X, d)$ is a pseudo-metric space, $x \in X$ and $A$ is a non-empty subset of $X$, then the set of real numbers $\{d(x, y) \mid y \in A\}$ is non-empty and bounded below. We define

$$d(x, A) = \inf_{y \in A} d(x, y).$$

The real-valued function on $X$ which associates with each point $x$ the real number $d(x, A)$ is continuous, and

$$\bar{A} = \{x \in X \mid d(x, A) = 0\},$$

$$A^0 = \{x \in X \mid d(x, X \backslash A) > 0\}.$$

For each positive real number the set

$$B_r(A) = \{x \in X \mid d(x, A) < r\}$$

is open. If $E$ and $F$ are non-empty subsets of a pseudo-metric space, then the set of real numbers $\{d(x, y) \mid x \in E, y \in F\}$ is bounded below. We define

$$d(E, F) = \inf\{d(x, y) \mid x \in E, y \in F\}.$$

If $d(E, F) > 0$, then $\bar{E}$ and $\bar{F}$ are disjoint. A pseudo-metric space $(X, d)$ is a regular space and $(X, d)$ is a $T_3$-space if and only if $d$ is a metric. A topological space $X$ is said to be *pseudo-metrizable* if there exists a pseudo-metric $d$ on $X$ such that the topology induced by $d$ is the topology of $X$. Similar $X$ is *metrizable* if its topology is induced by a metric. A pseudo-metrizable space is metrizable if it is a $T_0$-space. The pseudo-metrizable and metrizable spaces will be characterized in several ways in Chapter 2. The space $\mathbf{R}$ of real numbers is metrizable, its topology being induced by the metric $d$ given by $d(x,y) = |x-y|$ for $x, y$ in $\mathbf{R}$.

Let $n$ be a positive integer and consider the cartesian product $\mathbf{R}^n$ of $n$ copies of the set $\mathbf{R}$ of real numbers. If $x = (x_1, \ldots, x_n)$ is an element of $\mathbf{R}^n$, let

$$\|x\| = \sqrt{\left(\sum_{i=1}^{n} x_i^2\right)}.$$

In the case $n = 1$, the usual notation $|x|$ instead of $\|x\|$ will be used. Euclidean $n$-dimensional space is the set $\mathbf{R}^n$ with the topology induced by the metric $d$, where

$$d(x,y) = \|x-y\|$$

for $x, y$ in $\mathbf{R}^n$. Thus 1-dimensional Euclidean space is the usual space for real numbers. For each positive integer $n$ let

$$S^{n-1} = \{x \in \mathbf{R}^n \mid \|x\| = 1\},$$

$$E^n = \{x \in \mathbf{R}^n \mid \|x\| \leqslant 1\}.$$

We shall call $S^{n-1}$ and $E^n$ the *unit sphere* and the *closed unit ball* respectively. The origin $(0, \ldots, 0)$ in $\mathbf{R}^n$ will be denoted by $\mathbf{0}$. The following result is an important part of the proof that Euclidean $n$-dimensional space is indeed $n$-dimensional in the sense of the dimension theories to be studied in this book.

**Brouwer Theorem.** *For each positive integer $n$, $S^{n-1}$ is not a retract of $E^n$.*

## 2  Local finiteness, the weak topology and the weight of topological spaces

If $\{A_\lambda\}_{\lambda \in \Lambda}$ is a family of closed sets of a space $X$, it is not necessarily true that $\bigcup_{\lambda \in \Lambda} A_\lambda$ is a closed set. It is, of course, true if $\Lambda$ is finite. We can extend this result to families which are 'locally finite'. The concept of local finiteness proves to be important throughout general topology.

**2.1 Definition.** *A family* $\{A_\lambda\}_{\lambda \in \Lambda}$ *of subsets of a space* $X$ *is said to be* point-finite *if for each point* $x$ *of* $X$ *the set*

$$\{\lambda \in \Lambda \mid x \in A_\lambda\}$$

*is finite. The family* $\{A_\lambda\}_{\lambda \in \Lambda}$ *is said to be* locally finite *if for each point* $x$ *of* $X$ *there exists a neighbourhood* $N_x$ *of* $x$ *such that the set*

$$\{\lambda \in \Lambda \mid N_x \cap A_\lambda \neq \varnothing\}$$

*is finite. The family* $\{A_\lambda\}_{\lambda \in \Lambda}$ *is said to be* discrete *if for each point* $x$ *of* $X$ *there exists a neighbourhood* $N_x$ *of* $x$ *such that the set*

$$\{\lambda \in \Lambda \mid N_x \cap A_\lambda \neq \varnothing\}$$

*has at most one member.*

Each of these definitions applies to a set $\mathscr{A}$ of subsets of a space $X$ by considering $\mathscr{A}$ to be a family indexed by itself. Thus for example $\mathscr{A}$ is locally finite if each point $x$ of $X$ has a neighbourhood $N_x$ such that

$$\{A \in \mathscr{A} \mid N_x \cap A \neq \varnothing\}$$

is finite.

There is a related concept, the importance of which was first realized in connection with the metrization problem. A family $\{A_\lambda\}_{\lambda \in \Lambda}$ of subsets of a space $X$ is said to be *σ-locally finite* if $\Lambda = \bigcup_{i \in \mathbf{N}} \Lambda_i$ and for each $i$, the family $\{A_\lambda\}_{\lambda \in \Lambda_i}$ is locally finite. If we put $A_{i\lambda} = A_\lambda$ if $\lambda \in \Lambda_i$, and $A_{i\lambda} = \varnothing$ if $\lambda \notin \Lambda_i$, then we obtain a family $\{A_{i\lambda}\}_{i \in \mathbf{N}, \lambda \in \Lambda}$ such that for each integer $i$ the family $\{A_{i\lambda}\}_{\lambda \in \Lambda}$ is locally finite. It is often convenient to index a σ-locally finite family in this way. The definition of a *σ-discrete* family of subsets of a space is similar.

**2.2 Proposition.** *If* $\{A_\lambda\}_{\lambda \in \Lambda}$ *is a locally finite family of subsets of a space* $X$, *then* $\{\bar{A}_\lambda\}_{\lambda \in \Lambda}$ *is a locally finite family of subsets of* $X$, *and*

$$(\textstyle\bigcup_{\lambda \in \Lambda} A_\lambda)^- = \bigcup_{\lambda \in \Lambda} \bar{A}_\lambda.$$

*Proof.* Let $x$ be a point of $X$ and let $V_x$ be an open neighbourhood of $x$ such that $\{\lambda \in \Lambda \mid V_x \cap A_\lambda \neq \varnothing\}$ is finite. If $V_x \cap A_\lambda = \varnothing$, then

$$V_x \cap \bar{A}_\lambda = \varnothing,$$

so that

$$\{\lambda \in \Lambda \mid V_x \cap \bar{A}_\lambda \neq \varnothing\}$$

is finite. Thus $\{\bar{A}_\lambda\}_{\lambda \in \Lambda}$ is locally finite. [This argument also shows that if $\{A_\lambda\}_{\lambda \in \Lambda}$ is a discrete family, then $\{\bar{A}_\lambda\}_{\lambda \in \Lambda}$ is discrete.] Since

$$\textstyle\bigcup \bar{A}_\lambda \subset (\bigcup A_\lambda)^-,$$

for the second part of the proposition it is only necessary to prove that the union of a locally finite family of closed sets is closed. Let $\{A_\lambda\}_{\lambda \in \Lambda}$ be a locally finite family of closed sets, let $A = \bigcup_{\lambda \in \Lambda} A_\lambda$ and suppose that $x \notin A$. There exists a neighbourhood $N$ of $x$ such that $N \cap A_\lambda \neq \varnothing$ if and only if $\lambda \in M$, where $M$ is a finite subset of $\Lambda$. If

$$G = N \cap \bigcap_{\lambda \in M}(X \backslash A_\lambda),$$

then $G$ is a neighbourhood of $x$ and $G \cap A = \varnothing$. $\square$

The following result will be useful later.

**2.3 Proposition.** *Let $\{E_\lambda\}_{\lambda \in \Lambda}$ be a family of subsets of a space $X$ and let $\{B_\gamma\}_{\gamma \in \Gamma}$ be a locally finite closed covering of $X$ such that for each $\gamma$ in $\Gamma$, the set*

$$\{\lambda \in \Lambda \mid B_\lambda \cap E_\lambda \neq \varnothing\}$$

*is finite. Then there exists a locally finite family $\{U_\lambda\}_{\lambda \in \Lambda}$ of open sets of $X$ such that $E_\lambda \subset U_\lambda$ for each $\lambda$ in $\Lambda$.*

*Proof.* For each $\lambda$ let

$$U_\lambda = X \backslash \bigcup \{B_\gamma \mid B_\gamma \cap E_\lambda = \varnothing\}.$$

Clearly $E_\lambda \subset U_\lambda$ and since $\{B_\gamma\}_{\gamma \in \Gamma}$ is locally finite it follows from Proposition 2.2 that $U_\lambda$ is open. Let $x$ be a point of $X$. There exists a neighbourhood $N$ of $x$ and a finite subset $K$ of $\Gamma$ such that $N \cap B_\gamma = \varnothing$ if $\gamma \notin K$. Hence $N \subset \bigcup_{\gamma \in K} B_\gamma$. Now $B_\gamma \cap U_\lambda \neq \varnothing$ if and only if

$$B_\gamma \cap E_\lambda \neq \varnothing.$$

For each $\gamma$ in $K$, the set $\{\lambda \in \Lambda \mid B_\gamma \cap E_\lambda \neq \varnothing\}$ is finite. Hence the set $\{\lambda \in \Lambda \mid N \cap U_\lambda \neq \varnothing\}$ is finite. $\square$

The concept of the boundary of a set is important in dimension theory. The *boundary* of a subset $A$ of a space $X$ is the set of points which belong neither to the interior of $A$ nor to the interior of $X \backslash A$. The boundary of $A$ is denoted by $\operatorname{bd}(A)$. Clearly

$$\operatorname{bd}(A) = \operatorname{bd}(X \backslash A) = \bar{A} \cap (X \backslash A)^- = \bar{A} \backslash A^\circ,$$

and hence if $V$ is an open set then

$$\operatorname{bd}(V) = \bar{V} \backslash V.$$

In particular a subset $W$ is open-and-closed if and only if $\operatorname{bd}(W)$ is empty. For each subset $A$ of a space $X$, the sets $A^\circ$, $\operatorname{bd}(A)$ and $(X \backslash A)^\circ$

form a partition of $X$. If $B$ is a subspace of a space $X$ and $A \subset B$, then the boundary of $A$ in $B$ will be denoted by $\mathrm{bd}_B(A)$. Since

$$\mathrm{bd}_B(A) = B \cap \bar{A} \cap (B \backslash A)^-,$$

it is clear that

$$\mathrm{bd}_B(A) \subset B \cap \mathrm{bd}(A).$$

**2.4  Proposition.** (a) If $\{A_\lambda\}_{\lambda \in \Lambda}$ is a locally finite family of subsets of a topological space $X$, then

$$\mathrm{bd}(\bigcup_{\lambda \in \Lambda} A_\lambda) \subset \bigcup_{\lambda \in \Lambda} \mathrm{bd}(A_\lambda).$$

(b) If $A$ and $B$ are subsets of a topological space, then

$$\mathrm{bd}(A \cap B) \subset \mathrm{bd}(A) \cup \mathrm{bd}(B).$$

*Proof.* (a) Let $\{A_\lambda\}_{\lambda \in \Lambda}$ be a locally finite family of subsets of a space $X$ and let $A = \bigcup_{\lambda \in \Lambda} A_\lambda$. Then

$$\mathrm{bd}(A) = \bar{A} \cap (X \backslash A)^- = (\bigcup_{\lambda \in \Lambda} \bar{A}_\lambda) \cap (X \backslash A)^-$$
$$= \bigcup_{\lambda \in \Lambda} (\bar{A}_\lambda \cap (X \backslash A)^-).$$

But $X \backslash A \subset X \backslash A_\lambda$ for every $\lambda$, so that

$$\mathrm{bd}(A) \subset \bigcup_{\lambda \in \Lambda} (\bar{A}_\lambda \cap (X \backslash A_\lambda)^-) = \bigcup_{\lambda \in \Lambda} \mathrm{bd}(A_\lambda).$$

(b) If $A$ and $B$ are subsets of a space $X$, then

$$\mathrm{bd}(A \cap B) = (A \cap B)^- \cap (X \backslash A \cap B)^-$$
$$= (A \cap B)^- \cap ((X \backslash A)^- \cup (X \backslash B)^-)$$
$$\subset (\bar{A} \cap (X \backslash A)^-) \cup (\bar{B} \cap (X \backslash B)^-)$$
$$= \mathrm{bd}(A) \cup \mathrm{bd}(B). \square$$

If $f$ is a function with domain a topological space $X$ and $\{U_\lambda\}_{\lambda \in \Lambda}$ is an open covering of $X$ such that $f|U_\lambda$ is continuous for every $\lambda$, then $f$ is continuous. For let $V$ be an open set of the range of $f$ and let $x \in f^{-1}(V)$. Then $x \in U_\lambda$ for some $\lambda$ and $(f|U_\lambda)^{-1}(V)$ is an open neighbourhood of $x$ in $U_\lambda$, and hence in $X$, which is contained in $f^{-1}(V)$. Hence $f^{-1}(V)$ is open in $X$ so that $f$ is continuous. We shall obtain a similar result for certain closed coverings. First we make a definition.

Let $X$ be a set which has a covering $\{A_\lambda\}_{\lambda \in \Lambda}$, where for each $\lambda$ in $\Lambda$ there is a topology defined on $A_\lambda$ such that the following condition is satisfied: if $\lambda, \mu \in \Lambda$, then the set $A_\lambda \cap A_\mu$ is closed in $A_\lambda$ and in $A_\mu$, and the topologies induced on $A_\lambda \cap A_\mu$ by the topologies of $A_\lambda$ and $A_\mu$ coincide. Let $\mathscr{F}$ be the set of subsets $F$ of $X$ such that $F \cap A_\lambda$ is closed in $A_\lambda$ for every $\lambda$. It is easily verified that $\mathscr{F}$ is the set of closed sets for a topology on $X$. In this topology each set $A_\lambda$ is closed. Furthermore the induced topology on $A_\lambda$ as a subspace of $X$ is the given topology on $A_\lambda$ for every $\lambda$, for the induced topology is clearly not larger than the given topology of $A_\lambda$. If $B$ is a closed set in $A_\lambda$ with the given topology, then $B \cap A_\mu = B \cap A_\lambda \cap A_\mu$ so that $B \cap A_\mu$ is closed in $A_\lambda \cap A_\mu$ and hence $B \cap A_\mu$ is closed in $A_\mu$ for every $\mu$. The topology thus defined on $X$ is called the *weak topology* with respect to the covering $\{A_\lambda\}_{\lambda \in \Lambda}$. It is the largest topology on $X$ inducing the given topology on $A_\lambda$ for every $\lambda$.

**2.5  Proposition.** *If $f$ is a function with domain a topological space $X$ which has the weak topology with respect to a covering $\{A_\lambda\}_{\lambda \in \Lambda}$, and $f|A_\lambda$ is continuous for every $\lambda$, then $f$ is continuous.*

*Proof.* Let $E$ be a closed set of the range of $f$. Then

$$\left(f^{-1}(E)\right) \cap A_\lambda = (f|A_\lambda)^{-1}(E)$$

which is closed in $A_\lambda$ since $f|A_\lambda$ is continuous. Since $X$ has the weak topology with respect to $\{A_\lambda\}_{x \in \Lambda}$, it follows that $f^{-1}(E)$ is closed in $X$. Hence $f$ is continuous. $\square$

**2.6  Definition.** *A topological space $X$ is said to be dominated by a covering $\{A_\lambda\}_{\lambda \in \Lambda}$ if for each subset $\Lambda'$ of $\Lambda$, $\bigcup_{\lambda \in \Lambda'} A_\lambda$ is closed and the subspace $\bigcup_{\lambda \in \Lambda'} A_\lambda$ of $X$ has the weak topology with respect to the covering $\{A_\lambda\}_{\lambda \in \Lambda'}$.*

If a space $X$ is dominated by a covering $\{A_\lambda\}_{\lambda \in \Lambda}$, then $X$ has the weak topology with respect to $\{A_\lambda\}_{\lambda \in \Lambda}$. The statement that a space $X$ is dominated by a covering $\{A_\lambda\}_{\lambda \in \Lambda}$ is, however, much stronger. A space $X$ can have the weak topology with respect to a covering $\{A_\lambda\}_{\lambda \in \Lambda}$ which has the property that $\bigcup_{\lambda \in \Lambda'} A_\lambda$ is closed for each subset $\Lambda'$ of $\Lambda$ and yet $X$ is not dominated by $\{A_\lambda\}_{\lambda \in \Lambda}$. We shall prove a number of propositions which assert that if a space $X$ is dominated by a covering $\{A_\lambda\}$ and each $A_\lambda$ has some topological property $P$ then $X$ has the property $P$. An important case of domination is the following:

**2.7  Proposition.** *A space is dominated by a locally finite closed covering.*

*Proof.* Let $\{A_\lambda\}_{\lambda \in \Lambda}$ be a locally finite closed covering of a space $X$. If $\Lambda' \subset \Lambda$, then by Proposition 2.2, $\bigcup_{\lambda \in \Lambda'} A_\lambda$ is a closed set. If $B$ is a subset of $\bigcup_{\lambda\in\Lambda'}A_\lambda$ such that $B\cap A_\lambda$ is closed if $\lambda\in\Lambda'$, then $\{B\cap A_\lambda\}_{\lambda\in\Lambda'}$ is a locally finite family of closed sets so that $B=\bigcup_{\lambda\in\Lambda'}B\cap A_\lambda$ is a closed set of $X$, by Proposition 2.2, and thus is closed in $\bigcup_{\lambda\in\Lambda'}A_\lambda$. $\square$

We conclude this section by introducing a cardinal topological invariant which is essential for the theorems to be proved on the classification of topological spaces. Every base for a topological space $X$ has cardinality not exceeding the cardinal number of the topology of $X$. Since the set of cardinal numbers not exceeding some given cardinal number is well-ordered, it follows that there exists a base for the topology of $X$ of least cardinality.

**2.8  Definition.** *The* weight *of a topological space $X$ is the least cardinal number of a base for the topology of $X$. The weight of a space $X$ is denoted by $w(X)$.*

It will be shown that for spaces of infinite weight, every base contains a base of cardinality equal to the weight of the space. We note that if $\tau$ is an infinite cardinal number and there exists a subbase $\mathscr{S}$ for the topology of a space $X$ such that $|\mathscr{S}|=\tau$, then $w(X)\leqslant\tau$.

**2.9  Lemma.** *Let $\{U_\lambda\}_{\lambda\in\Lambda}$ be a family of open sets of a topological space $X$ such that $w(X)=\tau$. Then there exists a subset of $M$ of $\Lambda$ such that $\bigcup_{\lambda\in M}U_\lambda=\bigcup_{\lambda\in\Lambda}U_\lambda$ and $|M|\leqslant\tau$.*

*Proof.* Let $\mathscr{B}$ be a base for the topology of $X$ such that $|\mathscr{B}|=\tau$, and let $\mathscr{B}_1$ be the subset of $\mathscr{B}$ consisting of those sets $B$ such that $B\subset U_\lambda$ for some $\lambda$. For each $B$ in $\mathscr{B}_1$ choose $\lambda(B)$ in $\Lambda$ such that $B\subset U_{\lambda(B)}$ and let

$$M=\{\lambda\in\Lambda\mid\lambda=\lambda(B)\quad\text{for some}\quad B\text{ in }\mathscr{B}_1\}.$$

Then $|M|\leqslant\tau$. Furthermore if $\lambda\in\Lambda$ and $x\in U_\lambda$, then there exists $B$ in $\mathscr{B}$ such that $x\in B\subset U_\lambda$. Thus $B\in\mathscr{B}_1$ and $x\in U_{\lambda(B)}$. Hence

$$\bigcup_{\lambda\in\Lambda}U_\lambda\subset\bigcup_{\lambda\in M}U_\lambda.$$

The reverse inclusion is obvious. $\square$

**2.10  Proposition.** *Let $X$ be a topological space of weight $\tau$, where $\tau$ is an infinite cardinal number and let $\mathscr{B}$ be a base for the topology of $X$. Then there exists a subset $\mathscr{B}_1$ of $\mathscr{B}$ such that $\mathscr{B}_1$ is a base for the topology of $X$ and $|\mathscr{B}_1|=\tau$.*

*Proof.* Let $\mathscr{W}$ be a base for the topology of $X$ such that $|\mathscr{W}| = \tau$. If $W \in \mathscr{W}$ then $W$ is the union of members of $\mathscr{B}$. Thus by Lemma 2.9 there exists a subset $\mathscr{B}(W)$ of $\mathscr{B}$ such that $|\mathscr{B}(W)| = \tau$ and

$$W = \bigcup_{B \in \mathscr{B}(W)} B.$$

If $\mathscr{B}_1 = \bigcup_{W \in \mathscr{W}} \mathscr{B}(W)$, then $\mathscr{B}_1$ is a subset of $\mathscr{B}$ of cardinality equal to $\tau$ and a base for the topology of $X$.$\square$

It is evident that the weight of a subspace does not exceed the weight of the space. For the classification theorems we shall need to be able to determine the weight of a topological product.

**2.11  Proposition.** *Let $X$ be the topological product of a family $\{X_\lambda\}_{\lambda \in \Lambda}$ of spaces of infinite weight and let $\tau$ be a cardinal number such that $w(X_\lambda) \leqslant \tau$ for every $\lambda$ and $w(X_{\lambda_0}) = \tau$ for some $\lambda_0$ in $\Lambda$. Then the weight of $X$ is the larger of the cardinal numbers $|\Lambda|$ and $\tau$.*

*Proof.* For each $\lambda$, let $\pi_\lambda$ be the projection of $X$ onto $X_\lambda$. Let $|\Lambda| = \nu$ and for each $\lambda$ let $\mathscr{B}_\lambda$ be a base for the topology of $X_\lambda$ such that $|\mathscr{B}_\lambda| \leqslant \tau$. A base $\mathscr{B}$ for the topology of $X$ consists of all sets

$$\bigcap_{\lambda \in M} \pi_\lambda^{-1}(V_\lambda),$$

where $M$ is a finite subset of $\Lambda$ and $V_\lambda \in \mathscr{B}_\lambda$ if $\lambda \in M$. For each non-empty finite subset $M$ of $\Lambda$ the cardinality of the set of subsets of $X$ of the form $\bigcap_{\lambda \in M} \pi_\lambda^{-1}(V_\lambda)$, where $V_\lambda \in \mathscr{B}_\lambda$, does not exceed $\tau$. Thus if $\nu$ is finite, $|\mathscr{B}| \leqslant \tau = \max \{\nu, \tau\}$. If $\nu$ is infinite, then the cardinality of the set of non-empty finite subsets of $\Lambda$ is $\nu$. Thus $|\mathscr{B}| \leqslant \nu.\tau = \max \{\nu, \tau\}$. Thus $w(X) \leqslant \max \{\nu, \tau\}$. Suppose that $\max \{\nu, \tau\} = \tau$. Since $X$ has a subspace which is homeomorphic with $X_{\lambda_0}$ it follows that $w(X) = \tau$. Now suppose that $\tau < \nu$ so that $\max \{\nu, \tau\} = \nu$. If $w(X) < \nu$ then by Proposition 2.10, there exists a subset $\mathscr{B}_1$ of $\mathscr{B}$ such that $|\mathscr{B}_1| < \nu$ and $\mathscr{B}_1$ is a base for the topology of $X$. Since each space $X_\lambda$ is of infinite weight, for each $\lambda$ in $\Lambda$ we an choose a point $x_\lambda$ which belongs to a proper open subset of $X_\lambda$. Let $x = \{x_\lambda\}_{\lambda \in \Lambda}$ and let $\mathscr{B}_2$ be the subset of $\mathscr{B}_1$ consisting of those elements which contain $x$. For each $B$ in $\mathscr{B}_2$ let $M_B$ be a finite subset of $\Lambda$ such that

$$B = \bigcap_{\lambda \in M_B} \pi_\lambda^{-1}(V_\lambda),$$

where $V_\lambda \in \mathscr{B}_\lambda$ and $V_\lambda \neq X_\lambda$ if $\lambda \in M_B$. Then since $|\mathscr{B}_2| < \nu$, the set $\bigcup_{B \in \mathscr{B}_2} M_B$ has cardinality less than $\nu$ so that there exists $\mu$ belonging to $\Lambda \backslash (\bigcup_{B \in \mathscr{B}_2} M_B)$. Let $W_\mu$ be a proper open subset of $X_\mu$ containing $x_\mu$.

Then $\pi_\mu^{-1}(W_\mu)$ is an open set of $X$ and $x \in \pi_\mu^{-1}(W_\mu)$. But there exists no member $B$ of $\mathscr{B}_2$ such that $x \in B \subset \pi_\mu^{-1}(W_\mu)$, which is absurd. Hence $w(X) = \nu$. $\square$

## 3 Normal Spaces

Normal spaces form the most important class of topological spaces for dimension theory.

**3.1 Definition.** *A topological space $X$ is said to be* normal *if for each pair of disjoint closed sets $A$ and $B$ of $X$ there exist disjoint open sets $U$ and $V$ such that $A \subset U$ and $B \subset V$.*

It is easily seen that a topological space $X$ is normal if and only if for each closed set $A$ and each open set $U$ such that $A \subset U$, there exists an open set $G$ such that $A \subset G \subset \bar{G} \subset U$. Thus if $A$ and $B$ are disjoint closed sets in a normal space $X$, there exist open sets $G$ and $H$ such that $A \subset G$, $B \subset H$ and $\bar{G}$ and $\bar{H}$ are disjoint. This result will be extended later.

A normal space need be neither regular nor a $T_0$-space. A normal $T_0$-space need be neither regular nor a $T_1$-space. But, since one-point sets are closed in a $T_1$-space, a normal $T_1$-space is a $T_3$-space. A normal $T_1$-space will be called a $T_4$-*space*.

It is clear that each closed subspace of a normal space is normal. It is not true that every subspace of a normal space is normal. The question of which subspaces of a normal space are normal will be taken up in the next section.

The following result is usually known as *Urysohn's lemma*.

**3.2 Proposition.** *A topological space $X$ is normal if and only if for each pair $A$, $B$ of disjoint closed sets of $X$, there exists a continuous function $f: X \to I$ such that $f(x) = 0$ if $x \in A$ and $f(x) = 1$ if $x \in B$.* *

**3.3 Remark.** Since $I$ and $[a, b]$ are homeomorphic, where $a$ and $b$ are real numbers such that $a < b$, Urysohn's lemma implies that if $X$ is a normal space and $A, B$ are disjoint closed sets of $X$, then there exists a continuous function $f: X \to [a, b]$ such that $f(x) = a$ if $x \in A$ and $f(x) = b$ if $x \in B$.

**3.4  Definition.** *A subset of $A$ of a topological space $X$ is called a zero-set of $X$ if there exists a continuous real-valued function $f$ on $X$ such that*

$$A = \{x \in X \mid f(x) = 0\}.$$

*A subset is called a* cozero-set *if its complement is a zero-set.*

If $f$ is a continuous real-valued function on a space $X$, let us put

$$Z(f) = \{x \in X \mid f(x) = 0\},$$

and call $Z(f)$ the zero-set of $f$. If $A$ is a zero-set of a space $X$, then there exists a continuous real-valued function $g$ on $X$ such that $g(x) \geqslant 0$ for all $x$ in $X$ and $A = Z(g)$. For if $A = Z(f)$, we can take $g = |f|$. Similarly there exists a continuous function $h : X \to I$ such that $A = Z(h)$, for we can take $h(x) = \min\{g(x), 1\}$ if $x \in X$. Any finite union or any finite intersection of zero-sets of a space $X$ is a zero-set of $X$. For

$$Z(f) \cup Z(g) = Z(fg)$$

and
$$Z(f) \cap Z(g) = Z(h),$$

where $h(x) = |f(x)| + |g(x)|$ if $x \in X$. If $a$ is a real number and $f$ is a continuous real-valued function on a space $X$, then the set

$$A = \{x \in X \mid f(x) \geqslant a\}$$

is a zero-set of $X$. For $A = Z(h)$, where $h(x) = \min\{f(x) - a, 0\}$ if $x \in X$. Similarly the set $\{x \in X \mid f(x) \leqslant a\}$ is a zero-set of $X$. Thus the sets

$$\{x \in X \mid f(x) < a\} \quad \text{and} \quad \{x \in X \mid f(x) > a\}$$

are cozero-sets of $X$.

It follows from Urysohn's lemma that if $A, B$ are disjoint closed sets in a normal space $X$, then there exist zero-sets $E, F$ and cozero-sets $U, V$ such that

$$A \subset E \subset U, \quad B \subset F \subset V$$

and $U$ and $V$ are disjoint. The following result can sometimes be used to extend results about normal spaces to more general spaces.

**3.5  Proposition.** *If $A$ and $B$ are disjoint zero-sets in a space $X$, then there exist disjoint cozero-sets $U$ and $V$ in $X$ such that $A \subset U$ and $B \subset V$.*

*Proof.* Let $A = Z(f)$ and $B = Z(g)$. Since $A$ and $B$ are disjoint, it follows that $|f(x)| + |g(x)| > 0$ for all $x$. Thus we can define a continuous function $h : X \to I$ by

$$h(x) = |f(x)| / \big( |f(x)| + |g(x)| \big) \quad \text{if} \quad x \in X.$$

Then $h(x) = 0$ if $x \in A$ and $h(x) = 1$ if $x \in B$. Let

$$U = \{ x \in X \mid h(x) < \tfrac{1}{2} \} \quad \text{and} \quad V = \{ x \in X \mid h(x) > \tfrac{1}{2} \}.$$

Then $U$ and $V$ are disjoint cozero-sets, $A \subset U$ and $B \subset V$.$\square$

A subset $A$ of a space $X$ is said to be an $F_\sigma$-*set* if $A = \bigcup_{i \in \mathbf{N}} A_i$, where $A_i$ is a closed set for each $i$. Dually, a subset $H$ is said to be a $G_\delta$-*set* if $H = \bigcap_{i \in \mathbf{N}} H_i$, where $H_i$ is an open set for each $i$.

**3.6   Proposition.** *A zero-set of a topological space is a closed $G_\delta$-set. A closed $G_\delta$-set in a normal space is a zero-set.*

*Proof.* Let $A$ be a zero-set in a topological space $X$. Then $A$ is evidently closed and if $A = Z(g)$, where $g$ is a continuous real-valued function such that $g(x) \geqslant 0$ if $x \in X$, then $A = \bigcap_{n \in \mathbf{N}} G_n$, where

$$G_n = \{ x \in X \mid g(x) < 1/n \},$$

so that $A$ is a $G_\delta$-set.

Now let $A$ be a closed set of a normal space $X$ such that $A = \bigcap_{n \in \mathbf{N}} G_n$, where each $G_n$ is an open set. For each $n$ there exists a continuous function $f_n : X \to [0, 1/2^n]$ such that $f_n(x) = 0$ if $x \in A$ and $f_n(x) = 1/2^n$ if $x \notin G_n$. Then the function $f : X \to I$ defined by

$$f(x) = \sum_{n=1}^{\infty} f_n(x) \quad \text{if} \quad x \in X,$$

is continuous, and it is clear that $A = Z(f)$.$\square$

If $X$ is a topological space and $A$ is a subspace of $X$, then an *extension* of a continuous function $f : A \to Y$ is a continuous function $g : X \to Y$ such that $g | A = f$. The following result is known as the *Tietze–Urysohn extension theorem*.

**3.7   Proposition.** *A topological space $X$ is normal if and only if for every closed subspace $A$ of $X$, each continuous function $f : A \to I$ has an extension $g : X \to I$.*∗

It is easily established that the Tietze–Urysohn theorem holds with any interval (not necessarily closed) replacing the unit interval $I$.

In particular, if $X$ is a normal space and $A$ is a closed subspace of $X$, then each continuous real-valued function defined on $A$ has an extension to $X$. The following property of normal spaces has numerous applications in dimension theory.

**3.8  Proposition.** *Let $X$ be a normal space, let $A$ be a closed subspace of $X$, and let $f: A \to S^n$ be a continuous function. Then there exist an open set $U$ and a continuous function $g: U \to S^n$ such that $A \subset U$ and $g|A = f$.*

*Proof.* For $x$ in $A$, let
$$f(x) = \big(f_1(x), \ldots, f_{n+1}(x)\big),$$
where $f_i(x) \in \mathbf{R}$. By the preceding remark, each of the continuous functions $f_i: A \to \mathbf{R}$ has a continuous extension $h_i: X \to \mathbf{R}$. Let
$$h: X \to \mathbf{R}^{n+1}$$
be the continuous function defined by
$$h(x) = \big(h_1(x), \ldots, h_{n+1}(x)\big) \quad \text{if} \quad x \in X.$$
If $U = \{x \in X \mid h(x) \neq \mathbf{0}\}$, then $U$ is open and $A \subset U$. If $x \in U$, let
$$g(x) = h(x)/\|h(x)\|.$$
Then $g: U \to S^n$ is a continuous function, and $g|A = f$. $\square$

An open covering $\{U_\lambda\}_{\lambda \in \Lambda}$ of a space $X$ is said to be *shrinkable* if there exists an open covering $\{V_\lambda\}_{\lambda \in \Lambda}$ of $X$ such that $\overline{V}_\lambda \subset U_\lambda$ for each $\lambda$ in $\Lambda$.

**3.9  Proposition.** *The following statements about a topological space $X$ are equivalent:*

(a) *$X$ is normal;*

(b) *each point-finite open covering of $X$ is shrinkable;*

(c) *each finite open covering of $X$ has a locally finite closed refinement.*

*Proof.* (a) $\Rightarrow$ (b). Let $\{U_\lambda\}_{\lambda \in \Lambda}$ be a point-finite open covering of a normal space $X$ and let $\Lambda$ be well-ordered. We shall construct a shrinking of $\{U_\lambda\}_{\lambda \in \Lambda}$ by transfinite induction. Let $\mu$ be an element of $\Lambda$ and suppose that for each $\lambda < \mu$ we have an open set $V_\lambda$ such that $\overline{V}_\lambda \subset U_\lambda$ and for each $\nu < \mu$
$$\bigcup_{\lambda \leqslant \nu} V_\lambda \cup \bigcup_{\lambda > \nu} U_\lambda = X.$$
Let $x$ be a point of $X$. Then since $\{U_\lambda\}_{\lambda \in \Lambda}$ is point-finite there exists a

largest element $\xi$, say, of $\Lambda$ such that $x \in U_\xi$. If $\xi \geqslant \mu$ then $x \in \bigcup_{\lambda \geqslant \mu} U_\lambda$, whilst if $\xi < \mu$ then $x \in \bigcup_{\lambda \leqslant \xi} V_\lambda \subset \bigcup_{\lambda < \mu} V_\lambda$. Hence

$$\bigcup_{\lambda < \mu} V_\lambda \cup \bigcup_{\lambda \geqslant \mu} U_\lambda = X.$$

Thus $U_\mu$ contains the complement of $\bigcup_{\lambda < \mu} V_\lambda \cup \bigcup_{\lambda > \mu} U_\lambda$. Since $X$ is normal there exists an open set $V_\mu$ such that

$$X \backslash (\bigcup_{\lambda < \mu} V_\lambda \cup \bigcup_{\lambda > \mu} U_\lambda) \subset V_\mu \subset \overline{V}_\mu \subset U_\mu.$$

Thus $\overline{V}_\mu \subset U_\mu$ and

$$\bigcup_{\lambda \leqslant \mu} V_\lambda \cup \bigcup_{\lambda > \mu} U_\lambda = X.$$

The construction of a shrinking of $\{U_\lambda\}_{\lambda \in \Lambda}$ is completed by transfinite induction.

(b) $\Rightarrow$ (c). Obvious.

(c) $\Rightarrow$ (a). Let $X$ be a space each finite open covering of which has a locally finite closed refinement and let $A$ and $B$ be disjoint closed sets of $X$. The covering $\{X \backslash A, X \backslash B\}$ of $X$ has a locally finite closed refinement $\mathscr{F}$. Let $E$ be the union of the members of $\mathscr{F}$ disjoint from $A$ and let $F$ be the union of the members of $\mathscr{F}$ disjoint from $B$. Then $E$ and $F$ are closed sets and $E \cup F = X$. Thus if $U = X \backslash E$ and $V = X \backslash F$, then $U$ and $V$ are disjoint open sets such that $A \subset U$ and $B \subset V$. Hence $X$ is a normal space.$\square$

If $f: X \to Y$ is a continuous surjection and $X$ is a normal space, it is not necessarily true that $Y$ is a normal space. Normality is however preserved by closed mappings.

**3.10**   *Proposition. If $f: X \to Y$ is a continuous closed surjection and $X$ is a normal space, then $Y$ is a normal space.*

*Proof.* Let $\{U_1, \ldots, U_k\}$ be an open covering of $Y$. Then

$$\{f^{-1}(U_1), \ldots, f^{-1}(U_k)\}$$

is an open covering of $X$. By Proposition 3.9 there exists a closed covering $\{E_1, \ldots, E_k\}$ of $X$ such that $E_i \subset f^{-1}(U_i)$ for each $i$. Then $\{f(E_1), \ldots, f(E_k)\}$ is a closed covering of $Y$ and $f(E_i) \subset U_i$ for each $i$. It follows from Proposition 3.9 that $Y$ is a normal space.$\square$

Let $\{f_\lambda\}_{\lambda \in \Lambda}$ be a family of continuous real-valued functions on a space $X$ and for each $\lambda$ let

$$S_\lambda = \{x \in X \mid f_\lambda(x) \neq 0\}.$$

If $\{S_\lambda\}_{\lambda \in \Lambda}$ is point-finite then for each $x$ in $X$ we can define

$$f(x) = \Sigma_{\lambda \in \Lambda} f_\lambda(x),$$

for the sum has only finitely many non-zero terms. If $\{S_\lambda\}_{\lambda \in \Lambda}$ is locally finite, the function $f$ thus obtained is continuous. For if $x \in X$, then there exists an open neighbourhood $V_x$ of $x$ such that $M = \{\lambda \in \Lambda \mid V_x \cap S_\lambda \neq \varnothing\}$ is finite. And if $y \in V_x$, then

$$f(y) = \Sigma_{\lambda \in M} f_\lambda(y).$$

Thus $\{V_x\}_{x \in X}$ is an open covering of $X$ such that $f \mid V_x$ is continuous for each $x$ in $X$.

**3.11**   **Definition.** *A covering $\{G_\lambda\}_{\lambda \in \Lambda}$ of a space $X$ is said to be* numerable *if for each $\lambda$ in $\Lambda$ there exists a continuous function $\phi_\lambda \colon X \to I$ such that*

$$\{x \in X \mid \phi_\lambda(x) > 0\} \subset G_\lambda,$$

*the family of open sets $\{x \in X \mid \phi_\lambda(x) > 0\}$ is locally finite and*

$$\Sigma_{\lambda \in \Lambda} \phi_\lambda(x) = 1$$

*if $x \in X$.*

**3.12**   **Proposition.** *Each locally finite open covering of a normal space is numerable.*

*Proof.* Let $\{G_\lambda\}_{\lambda \in \Lambda}$ be a locally finite open covering of a normal space $X$. A locally finite covering is point-finite, so that by Proposition 3.9 there exists an open covering $\{V_\lambda\}_{\lambda \in \Lambda}$ of $X$ such that $\overline{V}_\lambda \subset G_\lambda$ for each $\lambda$. And by Urysohn's lemma, for each $\lambda$ there exists a continuous function $f_\lambda \colon X \to I$ such that $f_\lambda(x) = 0$ if $x \notin G_\lambda$ and $f_\lambda(x) = 1$ if $x \in \overline{V}_\lambda$. By the remark above we can define a continuous real-valued function $f$ on $X$ by putting

$$f(x) = \Sigma_{\lambda \in \Lambda} f_\lambda(x) \quad \text{if} \quad x \in X.$$

And $f(x) \geqslant 1$ if $x \in X$. Hence for each $\lambda$, we can define a continuous function $\phi_\lambda \colon X \to I$ by putting

$$\phi_\lambda(x) = f_\lambda(x)/f(x) \quad \text{if} \quad x \in X.$$

Thus the covering $\{G_\lambda\}_{\lambda \in \Lambda}$ is numerable. $\square$

The following result will have an important application in the dimension theory of normal spaces.

**3.13  Proposition.** *If $\mathcal{G} = \{G_\lambda\}_{\lambda \in \Lambda}$ is a locally finite open covering of a normal space $X$, then there exists a locally finite open refinement $\{H_\gamma\}_{\gamma \in \Gamma}$ of $\mathcal{G}$ and a locally finite open covering $\{U_\gamma\}_{\gamma \in \Gamma}$ of $X$ such that for each $\beta$ in $\Gamma$ the set*

$$\{\gamma \in \Gamma \mid U_\beta \cap H_\gamma \neq \varnothing\}$$

*is finite.*

*Proof.* By Proposition 3.12 there exist continuous real-valued functions $\phi_\lambda$ for $\lambda$ in $\Lambda$, such that $\phi_\lambda(x) \geqslant 0$ for all $x$, $\phi_\lambda(x) = 0$ if $x \notin G_\lambda$ and $\Sigma_{\lambda \in \Lambda} \phi_\lambda(x) = 1$ for all $x$. Let $\Gamma$ denote the set of finite subsets of $\Lambda$ and for each $\gamma$ in $\Gamma$ and each $\epsilon > 0$ let

$$M(\gamma; \epsilon) = \{x \in X \mid \Sigma_{\lambda \in \gamma} \phi_\lambda(x) > 1 - \epsilon\}$$

if $\gamma \neq \varnothing$, whilst $M(\varnothing; \epsilon) = \varnothing$ if $0 < \epsilon \leqslant 1$ and $M(\varnothing; \epsilon) = X$ if $\epsilon > 1$. Then $M(\gamma; \epsilon)$ is an open set for all $\gamma, \epsilon$. It is clear that if $\beta \subset \gamma$ then $M(\beta; \epsilon) \subset M(\gamma; \epsilon)$, and it is easily verified that

$$M(\beta; \epsilon) \cap M(\gamma; \epsilon') \subset M(\beta \cap \gamma; \epsilon + \epsilon').$$

Now if $\gamma \in \Gamma$ and $\gamma$ has $m$ elements we put

$$V_\gamma = M(\gamma; (\tfrac{1}{4})^{m+1}), \quad W_\gamma = M(\gamma; \tfrac{1}{2}(\tfrac{1}{4})^{m+1}),$$

and provided $m > 0$ we put

$$K_\gamma = \{x \in X \mid \Sigma_{\lambda \in \gamma} \phi_\lambda(x) \geqslant 1 - \tfrac{1}{2}(\tfrac{1}{4})^{m+1}\},$$

whilst if $m = 0$, $K_\gamma = \varnothing$. Then the sets $V_\gamma$ and $W_\gamma$ are open, $K_\gamma$ is closed and $\overline{W}_\gamma \subset K_\gamma \subset V_\gamma$; also $\{W_\gamma\}_{\gamma \in \Gamma}$ is a covering of $X$. For if $x \in X$, there exists $\gamma$ in $\Gamma$ such that $x \notin G_\lambda$ if $\lambda \notin \gamma$ so that $\Sigma_{\lambda \in \gamma} \phi_\lambda(x) = 1$ and hence $x \in W_\gamma$. Clearly $\{K_\gamma\}$ and $\{V_\gamma\}$ are also coverings of $X$. And if $\beta, \gamma \in \Gamma$ and $\beta \neq \beta \cap \gamma \neq \gamma$ then

$$V_\beta \cap V_\gamma \subset W_{\beta \cap \gamma}.$$

For if $\beta$ has $m$ elements and $\gamma$ has $n$ elements, then

$$V_\beta \cap V_\gamma \subset M(\beta \cap \gamma; (\tfrac{1}{4})^{m+1} + (\tfrac{1}{4})^{n+1}).$$

But if $r$ is the number of elements in $\beta \cap \gamma$, then $r < \min\{m, n\}$ so that

$$(\tfrac{1}{4})^{m+1} + (\tfrac{1}{4})^{n+1} \leqslant (\tfrac{1}{4})^{r+2} + (\tfrac{1}{4})^{r+2} = \tfrac{1}{2}(\tfrac{1}{4})^{r+1}.$$

Thus

$$V_\beta \cap V_\gamma \subset M(\beta \cap \gamma; \tfrac{1}{2}(\tfrac{1}{4})^{r+1}) = W_{\beta \cap \gamma}.$$

Now if $\gamma \in \Gamma$, let

$$H_\gamma = V_\gamma \backslash \cup \{K_\beta \mid \beta \subset \gamma, \beta \neq \gamma\}.$$

Then $H_\gamma$ is open. If $x \in X$, there exists $\gamma$ in $\Gamma$ such that $x \in V_\gamma$ and $x \notin V_\beta$ if $\beta \subset \gamma$ and $\beta \neq \gamma$, so that $x \in H_\gamma$. Thus $\{H_\gamma\}_{\gamma \in \Gamma}$ is an open covering of $X$. If $x \in H_\gamma$, then $x \in V_\gamma$ so that $\Sigma_{\lambda \in \gamma} \phi_\lambda(x) > 1 - (\frac{1}{4})^{m+1}$, where $m$ is the number of elements in $\gamma$. If $\lambda_0 \in \gamma$ and $x \notin G_{\lambda_0}$ then $\phi_{\lambda_0}(x) = 0$. Thus $\beta = \gamma \backslash \{\lambda_0\}$ is non-empty and

$$\Sigma_{\lambda \in \beta} \phi_\lambda(x) > 1 - (\tfrac{1}{4})^{m+1} > 1 - \tfrac{1}{2}(\tfrac{1}{4})^m,$$

so that $x \in K_\beta$, which contradicts the fact that $x \in H_\gamma$. Hence $H_\gamma \subset G_\lambda$ if $\lambda \in \gamma$. Thus $\{H_\gamma\}_{\gamma \in \Gamma}$ is a refinement of $\mathcal{G}$.

If $\beta, \gamma \in \Gamma$ and $\beta \neq \beta \cap \gamma \neq \gamma$, then $V_\beta \cap V_\gamma \subset W_{\beta \cap \gamma} \subset K_{\beta \cap \gamma}$, so that $V_\beta \cap H_\gamma = \varnothing$ and hence $W_\beta \cap H_\gamma = \varnothing$. If $\beta \subset \gamma$ and $\beta \neq \gamma$, then $W_\beta \cap H_\gamma = \varnothing$ since $W_\beta \subset K_\beta$. Thus if $W_\beta \cap H_\gamma \neq \varnothing$, then $\gamma \subset \beta$ so that for given $\beta$ in $\Gamma$ the set $\{\gamma \in \Gamma \mid W_\beta \cap H_\gamma \neq \varnothing\}$ is finite. Since $\{W_\gamma\}_{\gamma \in \Gamma}$ is an open covering of $X$, it follows that $\{H_\gamma\}_{\gamma \in \Gamma}$ is a locally finite open covering. Finally if $\gamma \in \Gamma$ and $\gamma$ has $m$ elements let

$$L_\gamma = \{x \in X \mid \Sigma_{\lambda \in \gamma} \phi_\lambda(x) \geqslant 1 - (\tfrac{1}{4})^{m+2}\}$$

if $m > 0$, whilst $L_\gamma = \varnothing$ if $\gamma = \varnothing$. Also, for each $\gamma$ let

$$U_\gamma = W_\gamma \backslash \cup \{L_\beta \mid \beta \subset \gamma, \beta \neq \gamma\}.$$

Then by arguments similar to those above, $\{U_\gamma\}_{\gamma \in \Gamma}$ is a locally finite open covering of $X$. If $\beta \in \Gamma$, then $U_\beta \subset W_\beta$ and hence for given $\beta$ in $\Gamma$ the set

$$\{\gamma \in \Gamma \mid U_\beta \cap H_\gamma \neq \varnothing\}$$

is finite. $\square$

The families $\{A_\lambda\}_{\lambda \in \Lambda}$ and $\{B_\lambda\}_{\lambda \in \Lambda}$ of subsets of a set $X$ are said to be *similar* if for each finite subset $\gamma$ of $\Lambda$ the sets $\cap_{\lambda \in \gamma} A_\lambda$ and $\cap_{\lambda \in \gamma} B_\lambda$ are either both empty or both non-empty.

**3.14**   **Proposition.** *Let $\{U_\lambda\}_{\lambda \in \Lambda}$ be a locally finite family of open sets of a normal space $X$ and let $\{F_\lambda\}_{\lambda \in \Lambda}$ be a family of closed sets such that $F_\lambda \subset U_\lambda$ for each $\lambda$. Then there exists a family $\{G_\lambda\}_{\lambda \in \Lambda}$ of open sets such that*

$$F_\lambda \subset G_\lambda \subset \bar{G}_\lambda \subset U_\lambda$$

*and the families $\{F_\lambda\}_{\lambda \in \Lambda}$ and $\{\bar{G}_\lambda\}_{\lambda \in \Lambda}$ are similar.*

*Proof.* Let $\Lambda$ be well-ordered with a last element. By transfinite induction we shall construct a family $\{G_\lambda\}_{\lambda \in \Lambda}$ of open sets such that

$$F_\lambda \subset G_\lambda \subset \bar{G}_\lambda \subset U_\lambda,$$

and for each element $\nu$ of $\Lambda$ the family $\{K_\lambda^\nu\}_{\lambda \in \Lambda}$, given by

$$K_\lambda^\nu = \begin{cases} \bar{G}_\lambda & \text{if} \quad \lambda \leqslant \nu, \\ F_\lambda & \text{if} \quad \lambda > \nu, \end{cases}$$

is similar to $\{F_\lambda\}_{\lambda \in \Lambda}$.

Suppose that $\mu \in \Lambda$ and that $G_\lambda$ has been defined for $\lambda < \mu$ such that for each $\nu < \mu$ the family $\{K_\lambda^\nu\}_{\lambda \in \Lambda}$ is similar to $\{F_\lambda\}_{\lambda \in \Lambda}$. Let $\{L_\lambda\}_{\lambda \in \Lambda}$ be the family given by

$$L_\lambda = \begin{cases} \bar{G}_\lambda & \text{if} \quad \lambda < \mu, \\ F_\lambda & \text{if} \quad \lambda \geqslant \mu. \end{cases}$$

Then $\{L_\lambda\}_{\lambda \in \Lambda}$ is similar to $\{F_\lambda\}_{\lambda \in \Lambda}$. For suppose that $\lambda_1, \ldots, \lambda_r \in \Lambda$ and $\lambda_1 < \ldots < \lambda_j < \mu \leqslant \lambda_{j+1} < \ldots < \lambda_r$. Then

$$\bigcap_{i=1}^{r} L_{\lambda_i} = \bigcap_{i=1}^{r} K_{\lambda_i}^{\lambda_j},$$

so that $\bigcap_{i=1}^r L_{\lambda_i} = \varnothing$ if and only if $\bigcap_{i=1}^r F_{\lambda_i} = \varnothing$, since $\{K_\lambda^{\lambda_j}\}_{\lambda \in \Lambda}$ is similar to $\{F_\lambda\}_{\lambda \in \Lambda}$. Since $L_\lambda \subset U_\lambda$ for each $\lambda$ the family $\{L_\lambda\}_{\lambda \in \Lambda}$ is locally finite. Thus if $\Gamma$ is the set of finite subsets of $\Lambda$ and for each $\gamma$ in $\Gamma$, $E_\gamma = \bigcap_{\lambda \in \gamma} L_\lambda$, then $\{E_\gamma\}_{\gamma \in \Gamma}$ is a locally finite family of closed sets. Hence

$$E = \bigcup \{E_\gamma \mid E_\gamma \cap F_\mu = \varnothing\}$$

is a closed set which is disjoint from $F_\mu$. Therefore there exists an open set $G_\mu$ such that

$$F_\mu \subset G_\mu \subset \bar{G}_\mu \subset U_\mu \quad \text{and} \quad \bar{G}_\mu \cap E = \varnothing.$$

Now the open sets $G_\lambda$ are defined for $\lambda \leqslant \mu$ and to complete the proof it remains to show that the family $\{K_\lambda^\mu\}_{\lambda \in \Lambda}$ is similar to $\{F_\lambda\}_{\lambda \in \Lambda}$. It is sufficient to show that the families $\{K_\lambda^\mu\}_{\lambda \in \Lambda}$ and $\{L_\lambda\}_{\lambda \in \Lambda}$ are similar. Suppose that $\lambda_1, \ldots, \lambda_r \in \Lambda$ and that $\bigcap_{i=1}^r L_{\lambda_i} = \varnothing$. It must be shown that $\bigcap_{i=1}^r K_{\lambda_i}^\mu = \varnothing$. Suppose that $\lambda_1 < \ldots < \lambda_j \leqslant \mu < \lambda_{j+1} < \ldots < \lambda_r$. If $\lambda_j \neq \mu$, there is nothing to prove. If $\lambda_j = \mu$, then

$$L_{\lambda_1} \cap \ldots \cap L_{\lambda_{j-1}} \cap F_\mu \cap L_{\lambda_{j+1}} \cap \ldots \cap L_{\lambda_r} = \varnothing.$$

Hence by the construction

$$L_{\lambda_1} \cap \ldots \cap L_{\lambda_{j-1}} \cap \bar{G}_\mu \cap L_{\lambda_{j+1}} \cap \ldots \cap L_{\lambda_r} = \varnothing.$$

Thus $\bigcap_{i=1}^r K_{\lambda_i}^\mu = \varnothing$ as required. $\square$

**3.15 Corollary.** *The following statements about a space $X$ are equivalent:*

(a) *$X$ is normal;*

(b) *for each finite family* $\{F_1, \ldots, F_k\}$ *of closed sets of* $X$ *there exists a family* $\{G_1, \ldots, G_k\}$ *of open sets of* $X$ *such that each* $F_i \subset G_i$ *and the families* $\{F_1, \ldots, F_k\}$ *and* $\{\bar{G}_1, \ldots, \bar{G}_k\}$ *are similar;*

(c) *for each pair* $F_1, F_2$ *of disjoint closed sets of* $X$ *there exists a pair* $G_1, G_2$ *of open sets such that* $F_i \subset G_i$ *for* $i = 1, 2$ *and the sets* $\bar{G}_1$ *and* $\bar{G}_2$ *are disjoint.*

*Proof.* The implication $(a) \Rightarrow (b)$ holds, for if $\{F_1, \ldots, F_k\}$ is a family of closed sets in a normal space $X$, then we take $U_i = X$ for $i = 1, \ldots, k$ and apply the preceding proposition. The implications $(b) \Rightarrow (c)$ and $(c) \Rightarrow (a)$ are obvious. $\square$

Finally we prove a result for normal regular spaces which should be compared with Proposition 3.9.

**3.16** **Proposition.** *Let* $X$ *be a normal regular space, let* $A$ *be a closed set of* $X$ *and let* $\{U_\lambda\}_{\lambda \in \Lambda}$ *be a locally finite family of open sets of* $X$ *such that* $A \subset \bigcup_{\lambda \in \Lambda} U_\lambda$. *Then there exists a family* $\{V_\lambda\}_{\lambda \in \Lambda}$ *of open sets of* $X$ *such that* $\bar{V}_\lambda \subset U_\lambda$ *for each* $\lambda$, $A \subset \bigcup_{\lambda \in \Lambda} V_\lambda$, *and the families* $\{\bar{V}_\lambda\}_{\lambda \in \Lambda}$ *and* $\{U_\lambda \cap A\}_{\lambda \in \Lambda}$ *are similar.*

*Proof.* Let $\Gamma$ be the set of finite subsets $\gamma$ of $\Lambda$ such that

$$A \cap \bigcap_{\lambda \in \gamma} U_\lambda \neq \varnothing.$$

The family $\{A \cap \bigcap_{\lambda \in \gamma} U_\lambda\}_{\gamma \in \Gamma}$ of non-empty open sets of $A$ is locally finite. Since $A$ is regular, for each $\gamma$ in $\Gamma$ there exists a non-empty closed set $D_\gamma$ of $A$ such that $D_\gamma \subset A \cap \bigcap_{\lambda \in \gamma} U_\lambda$. The family $\{D_\gamma\}_{\gamma \in \Gamma}$ consists of closed sets of $X$ and is locally finite in $X$. Since $A$ is normal it follows from Proposition 3.9 that there exists a locally finite closed covering $\{E_\lambda\}_{\lambda \in \Lambda}$ of $A$ such that $E_\lambda \subset U_\lambda \cap A$ for each $\lambda$. For each $\lambda$ in $\Lambda$ let

$$F_\lambda = E_\lambda \cup \bigcup_{\lambda \in \gamma} D_\gamma.$$

Then for each $\lambda$, $F_\lambda$ is a closed set of $X$ and $F_\lambda \subset U_\lambda \cap A$. And $\{F_\lambda\}_{\lambda \in \Lambda}$ is a closed covering of $A$. Furthermore $\{F_\lambda\}_{\lambda \in \Lambda}$ is similar to

$$\{U_\lambda \cap A\}_{\lambda \in \Lambda}.$$

For if $\gamma \in \Gamma$, then $D_\gamma \subset \bigcap_{\lambda \in \gamma} F_\lambda$ so that $\bigcap_{\lambda \in \gamma} F_\lambda \neq \varnothing$. Now since $X$ is normal, by Proposition 3.14 there exists a family $\{V_\lambda\}_{\lambda \in \Lambda}$ of open sets of $X$ such that

$$F_\lambda \subset V_\lambda \subset \bar{V}_\lambda \subset U_\lambda$$

for each $\lambda$, and the family $\{\bar{V}_\lambda\}_{\lambda \in \Lambda}$ is similar to $\{F_\lambda\}_{\lambda \in \Lambda}$. Clearly $A \subset \bigcup_{\lambda \in \Lambda} V_\lambda$ and the family $\{\bar{V}_\lambda\}_{\lambda \in \Lambda}$ is similar to $\{U_\lambda \cap A\}_{\lambda \in \Lambda}$. $\square$

## 4 Complete, total and perfect normality

A subspace of a normal space need not be normal. Each closed subspace of a normal space is normal and in this section we shall obtain a more general sufficient condition for a subspace of a normal space to be normal.

**4.1 Definition.** *A topological space is said to be* completely normal *if every subspace is normal.*

**4.2 Proposition.** *The following statements about a topological space $X$ are equivalent:*

(a) *$X$ is completely normal;*

(b) *every open subspace of $X$ is normal;*

(c) *if $A$ and $B$ are subsets of $X$ such that $A \cap \bar{B} = \varnothing$ and $\bar{A} \cap B = \varnothing$, then there exist disjoint open sets $U$ and $V$ such that $A \subset U$ and $B \subset V$.*

*Proof.* (a) $\Rightarrow$ (b). Obvious.

(b) $\Rightarrow$ (c). Let $X$ be a space of which every open subspace is normal, let $A$ and $B$ be subsets of $X$ such that $A \cap \bar{B} = \varnothing$ and $\bar{A} \cap B = \varnothing$, and let $C = \bar{A} \cap \bar{B}$. Then $\bar{A} \backslash C$ and $\bar{B} \backslash C$ are disjoint closed sets of the open subspace $X \backslash C$. But by hypothesis, $X \backslash C$ is normal so that there exist disjoint sets $U$ and $V$, open in $X \backslash C$, such that $\bar{A} \backslash C \subset U$ and $\bar{B} \backslash C \subset V$. Since $A \cap \bar{B} = \varnothing$, it follows that $A \cap C = \varnothing$ so that $A \subset X \backslash C$ and hence $A \subset \bar{A} \backslash C$. Similarly $B \subset \bar{B} \backslash C$. Thus $A \subset U$ and $B \subset V$. Furthermore $U$ and $V$ are open in $X$ since $X \backslash C$ is open.

(c) $\Rightarrow$ (a). Let $X$ be a space which satisfies condition (c), let $A$ be a subspace of $X$ and let $E$ and $F$ be disjoint closed sets of $A$. If $\bar{E}$ is the closure of $E$ in $X$ then $\bar{E} \cap A = E$ so that $\bar{E} \cap F = \varnothing$. Similarly $E \cap \bar{F} = \varnothing$. Hence there exist disjoint open sets $U$ and $V$ such that $E \subset U$ and $F \subset V$. Then $U \cap A$ and $V \cap A$ are disjoint open sets of $A$, containing $E$ and $F$ respectively. Thus $A$ is normal. $\square$

Next we obtain a result which will be useful in the study of subspaces of a normal space and which is of independent interest.

**4.3 Proposition.** *A topological space which is dominated by a covering $\{A_\lambda\}_{\lambda \in \Lambda}$ is normal if each subspace $A_\lambda$ is normal.*

*Proof.* Let $X$ be a space dominated by the covering $\{A_\lambda\}_{\lambda \in \Lambda}$, where each subspace $A_\lambda$ is normal. Let $B$ be a closed set of $X$ and let $f: B \to I$ be a continuous function. The normality of $X$ will be established if we find an extension of $f$ to $X$. Let $\mathscr{G}$ be the set of continuous functions $g$ which are extensions of $f$ and have domains of the form $B \cup \bigcup_{\lambda \in \Lambda'} A_\lambda$ for some subset $\Lambda'$ of $\Lambda$. By the Tietze–Urysohn extension theorem, $\mathscr{G}$ is a non-empty set. Let $\mathscr{G}$ be partially ordered by the relation: if $g, h \in \mathscr{G}$ then $g \leqslant h$ if $h$ is an extension of $g$. It follows from Proposition 2.5 that every chain in $\mathscr{G}$ has an upper bound in $\mathscr{G}$. Thus by Zorn's lemma, $\mathscr{G}$ contains a maximal element, $g$ say. Let $D$ be the domain of $g$ and suppose that $D \neq X$. Choose $\lambda$ such that $A_\lambda$ is not contained in $D$. By the Tietze–Urysohn extension theorem, $g|A_\lambda \cap D$ has an extension $g'$ to $A_\lambda$. By Proposition 2.5 the function $h$ with domain $D \cup A_\lambda$ such that $h|D = g$ and $h|A_\lambda = g'$ is continuous. This is a contradiction of the maximality of $g$ in $\mathscr{G}$. Thus $D = X$ and $g$ is the required extension of $f$. $\square$

If $A$ is a closed set in a normal space $X$ and $U$ is an open set such that $A \subset U$, then by Urysohn's lemma, there exists an open $F_\sigma$-set $H$, such that $A \subset H \subset U$. We consider two generalizations of this situation. Firstly a subset $M$ of a space $X$ is called a *generalized $F_\sigma$-set* if for each open set $U$ such that $M \subset U$ there exists an $F_\sigma$-set $F$ of $X$ such that $M \subset F \subset U$. Secondly we have the following concept which in the case of normal spaces is more general.

**4.4 Definition.** *A subset $M$ of a space $X$ is said to be* normally situated *in $X$ if for each open set $U$ such that $M \subset U$ there exists an open set $G$ such that $M \subset G \subset U$ and $G = \bigcup_{\lambda \in \Lambda} G_\lambda$, where $\{G_\lambda\}_{\lambda \in \Lambda}$ is a family, locally finite in $G$, of open $F_\sigma$-sets of $X$.*

**4.5 Proposition.** *If $M$ is a generalized $F_\sigma$-set in a normal space $X$, then $M$ is normally situated in $X$.*

*Proof.* If $M$ is a generalized $F_\sigma$-set in a normal space $X$ and $M \subset U$, where $U$ is an open set, then there exists a family $\{A_i\}_{i \in \mathbf{N}}$ of closed sets such that

$$M \subset \bigcup_{i \in \mathbf{N}} A_i \subset U.$$

But since $X$ is a normal space, $A_i \subset H_i \subset U$ for each $i$, where $H_i$ is an open $F_\sigma$-set and $M \subset \bigcup H_i \subset U$. Hence $M$ is normally situated in $X$. $\square$

We shall see that a normally situated set in a normal space is normal. Two lemmas are required.

**4.6  Lemma.** *If an open set $G$ is normally situated in a normal space $X$, then for each positive integer $i$ there is a family $\{W_{i\lambda}\}_{\lambda \in \Lambda}$, locally finite in $G$, of disjoint open sets and a family $\{F_{i\lambda}\}_{\lambda \in \Lambda}$ of closed sets such that $F_{i\lambda} \subset W_{i\lambda} \subset G$ and $\bigcup_{i \in \mathbf{N}} \bigcup_{\lambda \in \Lambda} F_{i\lambda} = G$.*

*Proof.* If $G$ is a normally situated open set of a normal space $X$ then $G = \bigcup_{\lambda \in \Lambda} G_\lambda$, where $\{G_\lambda\}$ is locally finite in $G$ and each $G_\lambda$ is an open $F_\sigma$-set of $X$. By Proposition 3.6, $G_\lambda$ is a cozero-set so that for each $\lambda$ there exists a continuous function $f_\lambda : X \to I$ such that $f_\lambda(x) > 0$ if and only if $x \in G_\lambda$. Let $\Lambda$ be well-ordered and for each positive integer $i$ and each $\lambda$ in $\Lambda$ let

$$W_{i\lambda} = \{x \in X \mid f_\lambda(x) > 1/(i+1), \quad f_\mu(x) < 1/(i+1) \quad \text{for all} \quad \mu < \lambda\}.$$

If $x_0 \in W_{i\lambda}$, then $x_0 \in G$ so that there is an open neighbourhood $N$ of $x_0$ such that the set $\gamma = \{\mu \in \Lambda \mid \mu < \lambda \text{ and } N \cap G_\mu \neq \varnothing\}$ is finite. If $\mu \in \gamma$, let $N_\mu = \{x \in X \mid f_\mu(x) < 1/(i+1)\}$ and let

$$M = \{x \in X \mid f_\lambda(x) > 1/(i+1)\}.$$

If $P = N \cap M \cap \bigcap_{\mu \in \gamma} N_\mu$, then $P$ is an open set and $x_0 \in P \subset W_{i\lambda}$. Thus $W_{i\lambda}$ is open and since $W_{i\lambda} \subset G_\lambda$, for each positive integer $i$ the family $\{W_{i\lambda}\}_{\lambda \in \Lambda}$ is locally finite in $G$. Clearly if $\lambda \neq \mu$ then

$$W_{i\lambda} \cap W_{i\mu} = \varnothing.$$

For each integer $i$ and each $\lambda$ in $\Lambda$ let

$$F_{i\lambda} = \{x \in X \mid x \notin \bigcup_{\mu < \lambda} G_\mu \quad \text{and} \quad f_\lambda(x) \geqslant 1/i\}.$$

Then $F_{i\lambda}$ is closed and $F_{i\lambda} \subset W_{i\lambda}$. If $x \in G$ there exists $\lambda_0$ in $\Lambda$ such that $x \in G_{\lambda_0}$ and $x \notin G_\mu$ for $\mu < \lambda_0$. Then $f_{\lambda_0}(x) > 0$ and hence $f_{\lambda_0}(x) \geqslant 1/i$ for some $i$. Since $x \notin \bigcup_{\mu < \lambda_0} G_\mu$, it follows that $x \in F_{i\lambda_0}$. Thus

$$\bigcup_{i \in \mathbf{N}} \bigcup_{\lambda \in \Lambda} F_{i\lambda} = G. \quad \square$$

**4.7  Lemma.** *If a space $X$ has a $\sigma$-locally finite covering consisting of open sets whose closures are normal subspaces of $X$, then $X$ is a normal space.*

*Proof.* Let $\{V_{i\lambda}\}_{i \in \mathbf{N}, \lambda \in \Lambda}$ be an open covering of a space $X$ such that each $\overline{V}_{i\lambda}$ is normal and for each positive integer $i$ the family $\{V_{i\lambda}\}_{\lambda \in \Lambda}$ is locally finite. If $i \in \mathbf{N}$, let $C_i = \bigcup_{\lambda \in \Lambda} \overline{V}_{i\lambda}$. Since $\{\overline{V}_{i\lambda}\}_{\lambda \in \Lambda}$ is a locally finite closed covering of $C_i$, it follows from Proposition 4.3 that $C_i$ is normal. Let $E_0$ and $F_0$ be any two disjoint closed sets of $X$. We shall construct increasing sequences $\{E_i\}_{i \in \mathbf{N}}$ and $\{F_i\}_{i \in \mathbf{N}}$ of closed sets such that $E_i$ and $F_i$ are disjoint if $i \in \mathbf{N}$ and

$$E_{i-1} \cap C_i \subset G_i \subset E_i, \quad F_{i-1} \cap C_i \subset H_i \subset F_i,$$

where $G_i$ and $H_i$ are disjoint and open in $C_i$. Let us suppose that we have found $E_{i-1}$ and $F_{i-1}$. Then $E_{i-1} \cap C_i$ and $F_{i-1} \cap C_i$ are disjoint closed sets of the normal space $C_i$. Hence there exist open sets $G_i$ and $H_i$ of $C_i$ such that $E_{i-1} \cap C_i \subset G_i$, $F_{i-1} \cap C_i \subset H_i$ and $\bar{G}_i \cap \bar{H}_i = \varnothing$. Let $E_i = E_{i-1} \cup \bar{G}_i$ and $F_i = F_{i-1} \cup \bar{H}_i$. Then $E_i$ and $F_i$ are disjoint closed sets of $X$, $G_i \subset E_i$ and $H_i \subset F_i$. It follows by induction that we can construct the required sets. Now let $V_i = \bigcup_{\lambda \in \Lambda} V_{i\lambda}$ and let

$$U = \bigcup_{i \in \mathbf{N}} G_i \cap V_i, \quad V = \bigcup_{i \in \mathbf{N}} H_i \cap V_i.$$

Since $G_i$ is open in $C_i$, $G_i \cap V_i$ is open in $V_i$ and hence open in $X$. Therefore $U$ is open in $X$ and similarly $V$ is open in $X$. Since

$$G_i \cap V_i \subset G_i \subset E_i,$$

it follows that $U \subset \bigcup_{i \in \mathbf{N}} E_i$ and similarly $V \subset \bigcup_{i \in \mathbf{N}} F_i$. And if $i, j \in \mathbf{N}$ and $k = \max\{i, j\}$, then $E_i \cap F_j \subset E_k \cap F_k = \varnothing$. Hence $U$ and $V$ are disjoint open sets. Since $X = \bigcup_{i \in \mathbf{N}} V_i$ it follows that

$$E_0 = \bigcup_{i \in \mathbf{N}} (E_0 \cap V_i) \subset \bigcup_{i \in \mathbf{N}} (E_{i-1} \cap V_i) \subset \bigcup_{i \in \mathbf{N}} (G_i \cap V_i) = U.$$

Similarly $F_0 \subset V$. Therefore $X$ is normal. $\square$

**4.8  Proposition.** *If $M$ is normally situated in a normal space, then the subspace $M$ is normal.*

*Proof.* Let $M$ be normally situated in a normal space $X$ and let $A$ and $B$ be disjoint closed sets of $M$. Then $A = M \cap E$ and $B = M \cap F$, where $E$ and $F$ are closed sets in $X$. Hence $M \subset X \backslash (E \cap F)$ which is open in $X$ so that

$$M \subset G \subset X \backslash (E \cap F),$$

where $G = \bigcup_{\lambda \in \Lambda} G_\lambda$, the family $\{G_\lambda\}_{\lambda \in \Lambda}$ being locally finite in $G$ and each $G_\lambda$ being an open $F_\sigma$-set of $X$. Thus by Lemma 4.6, for each positive integer $i$ there exist a family $\{W_{i\lambda}\}_{\lambda \in \Lambda}$ locally finite in $G$, of disjoint open sets of $X$ and a family $\{F_{i\lambda}\}_{\lambda \in \Lambda}$ of closed sets of $X$ such that $F_{i\lambda} \subset W_{i\lambda} \subset G$ for each $i$ and $\lambda$ and $\bigcup_{i \in \mathbf{N}} \bigcup_{\lambda \in \Lambda} F_{i\lambda} = G$. Since $X$ is normal, for each positive integer $i$ and each $\lambda$ in $\Lambda$ there exists an open set $V_{i\lambda}$ such that

$$F_{i\lambda} \subset V_{i\lambda} \subset \bar{V}_{i\lambda} \subset W_{i\lambda}.$$

The family $\{V_{i\lambda}\}_{\lambda \in \mathbf{N}, \lambda \in \Lambda}$ is a $\sigma$-locally finite open covering of $G$. The closure of $V_{i\lambda}$ in $G$ is $\bar{V}_{i\lambda}$ which is normal since it is closed in the normal space $X$. It follows from Lemma 4.7 that $G$ is normal. Now $G \cap E$ and $G \cap F$ are disjoint closed sets of $G$ so that there exist disjoint sets $U$ and $V$ which are open in $G$ and hence open in $X$ such that $G \cap E \subset U$ and $G \cap F \subset V$. Then $M \cap U$ and $M \cap V$ are disjoint open sets of $M$ such that $A \subset M \cap U$ and $B \subset M \cap V$. Hence $M$ is normal. $\square$

**4.9  Corollary.** *A generalized $F_\sigma$-set in a normal space is normal.* $\square$

**4.10  Definition.** *A normal space $X$ is said to be* totally normal *if every subset of $X$ is normally situated in $X$.*

Equivalently, a normal space $X$ is totally normal if each open set $G$ is the union of a family $\{G_\lambda\}_{\lambda \in \Lambda}$, locally finite in $G$, of open $F_\sigma$-sets of $X$.

**4.11  Proposition.** *A totally normal space is completely normal and regular.*

*Proof.* If $X$ is a totally normal space and $M$ is a subspace of $X$, then $M$ is normally situated in $X$ so that $M$ is normal by Proposition 4.8. Thus $X$ is completely normal. And if $G$ is an open set and $x \in G$, then $x \in U \subset G$, where $U$ is an open $F_\sigma$-set. Hence there exists a closed set $F$ such that $x \in F \subset G$. Since $X$ is normal there exists an open set $V$ such that $F \subset V \subset \bar{V} \subset G$. Then $x \in V \subset \bar{V} \subset G$ and hence $X$ is a regular space. $\square$

**4.12  Remark.** If $X$ is a totally normal space then every subspace of $X$ is *totally* normal. For let $A$ be a subspace of $X$ and suppose that $M \subset U \subset A$, where $U$ is open in $A$. Then there exists $V$ open in $X$ such that $U = V \cap A$. Since $M$ is normally situated in $X$ and $M \subset V$, there exists a family $\{G_\lambda\}_{\lambda \in \Lambda}$, locally finite in $G = \bigcup_{\lambda \in \Lambda} G_\lambda$, of open $F_\sigma$-sets of $X$ such that $M \subset G \subset V$. Then $G \cap A = \bigcup_{\lambda \in \Lambda} G_\lambda \cap A$ and for each $\lambda$ the set $G_\lambda \cap A$ is an open $F_\sigma$-set of $A$. Since $M \subset G \cap A \subset U$ it follows that $M$ is normally situated in $A$. Hence $A$ is totally normal.

Next we consider an example which shows that there exist completely normal regular spaces which are not totally normal. This example will also be used in work on dimension.

Let $P$ be the set of ordinals not exceeding the first uncountable ordinal $\omega_1$. Then $P$ is an uncountable well-ordered set. Let

$$T = P \backslash \{\omega_1\} = [0, \omega_1).$$

Then $T$ is an uncountable well-ordered set with the property that any countable subset of $T$ has a least upper bound in $T$. For let $A$ be a countable subset of $T$. For each $\alpha$ in $A$, the set $[0, \alpha)$ is countable so that if $B = \bigcup_{\alpha \in A}[0, \alpha)$, then $B$ is countable and hence $B \neq T$. Let $\gamma$ be the first element of $T$ such that $\gamma \notin B$. Then $\gamma \geqslant \alpha$ for every $\alpha$ in $A$, and if $\gamma' < \gamma$, then $\gamma' \in B$ so that $\gamma' < \alpha$ for some $\alpha$ in $A$. Thus $\gamma$ is the least upper bound of $A$. If $P$ has the interval topology, then the sets $(\alpha, \beta)$ and $(\alpha, \beta]$ are open for all $\alpha, \beta$ in $P$ such that $\alpha < \beta$, and if $\alpha$ has an immediate predecessor (or $\alpha = 0$) then the one-point subset $\{\alpha\}$ is open, together with the sets $[\alpha, \beta)$ and $[\alpha, \beta]$ for $\beta > \alpha$. The subspace $T$ of $P$ is also a linearly ordered space. We establish a property of $T$ which we shall exploit here, and several times in later examples, involving the spaces $T$ and $P$. Let us say that a covering $\{U_\lambda\}_{\lambda \in \Lambda}$ of a space is point-countable if for each point $x$, the set $\{\lambda \in \Lambda \mid x \in U_\lambda\}$ is countable.

**4.13   Lemma.** *If $\mathcal{U}$ is a point-countable open covering of the linearly ordered space $T$ of ordinals less than the first uncountable ordinal $\omega_1$, then there exists $\alpha$ in $T$ such that $(\alpha, \omega_1)$ is contained in some member of $\mathcal{U}$.*

*Proof.* Consider first a countable open covering $\{U_i\}_{i \in \mathbf{N}}$ of $T$ and suppose that for each $\alpha$ in $T$ and $j$ in $\mathbf{N}$ there exists some $\beta$ in $T$ such that $\beta > \alpha$ and $\beta \notin U_j$. Let $\beta_j(\alpha)$ be the least such $\beta$ and let $\gamma(\alpha)$ be the least upper bound of the sequence of ordinals $\{\beta_j(\alpha)\}_{j \in \mathbf{N}}$. Then $\alpha < \gamma(\alpha)$. Now let us construct inductively a sequence $\{\alpha_r\}_{r \in \mathbf{N}}$ of ordinals by putting $\alpha_1 = \gamma(0)$ and $\alpha_{r+1} = \gamma(\alpha_r)$ for all $r$. And let $\delta$ be the least upper bound of this sequence. Since $\{\alpha_r\}$ is strictly increasing, it follows that $\delta > \alpha_r$ for all $r$. But $\delta \in U_j$ for some $j$, so that there exist $\alpha, \beta$ in $T$ such that $\delta \in (\alpha, \beta) \subset U_j$. Then $\alpha < \delta$ and hence $\alpha < \alpha_r$ for some $r$ and

$$\alpha_{r+1} = \gamma(\alpha_r) \geqslant \beta_j(\alpha_r) \geqslant \beta > \delta,$$

which is absurd. Hence $(\alpha, \omega_1) \subset U_j$ for some $\alpha$ in $T$ and some $j$.

Now let $\mathcal{U}$ be a point-countable open covering of $T$. Then there exists some $\beta$ in $T$ such that $[\beta, \omega_1)$ is contained in the union of the members of $\mathcal{U}$ which contain $\beta$. For suppose this is not the case. We take $\alpha_1 = 0$ and define a sequence $\{\alpha_n\}$ of ordinals in $T$ by taking $\alpha_{n+1}$ to be the first ordinal not in any member of $\mathcal{U}$ which contains $\alpha_n$. Let $\delta$ be the least upper bound of this sequence. Since $\{\alpha_n\}$ is strictly increasing, it follows that $\alpha_n < \delta$ for all $n$. Now $\delta \in U$ for some $U$ in $\mathcal{U}$ so that there exists $\gamma < \delta$ such that $(\gamma, \delta] \subset U$. For some $m$, $\gamma < \alpha_m < \delta$ and it follows that $\alpha_n \in U$ for all $n \geqslant m$, which is absurd.

Let $\beta$ be an element of $T$ such that $[\beta, \omega_1)$ is contained in the union of the members of $\mathscr{U}$ which contain $\beta$. Then $[0, \beta)$, together with the members of $\mathscr{U}$ which contain $\beta$, form a countable open covering of $T$. It follows from the first part of the proof that some member of this covering contains $(\alpha, \omega_1)$ for some $\alpha$ in $T$. Thus there exists $\alpha$ in $T$ such that $(\alpha, \omega_1)$ is contained in some member of $\mathscr{U}$.☐

**4.14** *Example.* The linearly ordered space $P$ of ordinals not exceeding $\omega_1$ is a completely normal Hausdorff space which is not totally normal.

Let $A$ and $B$ be subsets of $P$ such that $A \cap \bar{B} = \varnothing$ and $\bar{A} \cap B = \varnothing$. For each $\alpha$ in $A$, there exists $a(\alpha) < \alpha$ such that $(a(\alpha), \alpha]$ contains no point of $B$, and for $\beta$ in $B$ there exists $b(\beta) < \beta$ such that $(b(\beta), \beta]$ contains no point of $A$. If $U = \bigcup_{\alpha \in A}(a(\alpha), \alpha]$ and $V = \bigcup_{\beta \in B}(b(\beta), \beta]$, then $U$ and $V$ are disjoint open sets such that $A \subset U$ and $B \subset V$. Thus $P$ is a completely normal space.

Now suppose that the open set $T$ of $P$ is normally situated in $P$. Then $T$ is covered by a family $\mathscr{G}$, locally finite in $T$, of open $F_\sigma$-sets of $P$. It follows from Lemma 4.13 that there exists $\alpha$ in $T$ such that $(\alpha, \omega_1)$ is contained in some member $G$ of $\mathscr{G}$. But $G = \bigcup_{i \in \mathbf{N}} G_i$, where each set $G_i$ is closed in $P$ and $\omega_1 \notin G_i$. Thus for each $i$ there exists $\gamma_i$ in $T$ such that $(\gamma_i, \omega_1]$ is disjoint from $G_i$. If $\beta$ is the least upper bound of the sequence $\{\gamma_i\}$ then $(\beta, \omega_1]$ is disjoint from $G$. This is absurd. Thus $T$ is not normally situated in $P$ and it follows that $P$ is not a totally normal space.

Finally we consider a further strengthening of normality.

**4.15** *Definition.* A topological space is said to be perfectly normal if each closed set is a zero-set.

**4.16** *Proposition.* The following statements about a topological space $X$ are equivalent:
  (a) $X$ is perfectly normal;
  (b) $X$ is normal and each open set of $X$ is an $F_\sigma$-set;
  (c) $X$ is normal and each subset of $X$ is a generalized $F_\sigma$-set.

*Proof.* (a) $\Rightarrow$ (b). Let $X$ be a perfectly normal space and let $A$ and $B$ be disjoint closed sets of $X$. Then $A$ and $B$ are disjoint zero-sets so that by Proposition 3.5 there exist disjoint open sets $U$ and $V$ such that $A \subset U$ and $B \subset V$. Thus $X$ is normal. Furthermore each open set of $X$ is a cozero-set and hence an $F_\sigma$-set.

$(b) \Leftrightarrow (c)$. Obvious.

$(b) \Rightarrow (a)$. This follows immediately from Proposition 3.6.$\square$

**4.17 Corollary.** *A perfectly normal space is totally normal.*

*Proof.* This follows from Propositions 4.16 and 4.5.$\square$

Not every totally normal space is perfectly normal. An example of a totally normal space which is not perfectly normal will be given in the next chapter.

**4.18 Proposition.** *A pseudo-metric space is perfectly normal.*

*Proof.* Let $(X, d)$ be a pseudo-metric space and let $A$ be a closed set of $X$. The real-valued function $d_A$ defined on $X$ by $d_A(x) = d(x, A)$ is continuous and $A = \{x \in X \mid d_A(x) = 0\}$.$\square$

A space dominated by perfectly normal subspaces is perfectly normal. To prove this we use the following lemma.

**4.19 Lemma.** *Let $X$ be a perfectly normal space, let $A, B$ be closed sets of $X$ and let $f : A \to I$ be a continuous function such that*

$$A \cap B = \{x \in A \mid f(x) = 0\}.$$

*Then $f$ has an extension $g : X \to I$ such that*

$$B = \{x \in X \mid g(x) = 0\}.$$

*Proof.* Let $f' : A \cup B \to I$ be the extension of $f$ defined by putting $f'(x) = 0$ if $x \in B$, and let $g' : X \to I$ be an extension of $f'$ given by Proposition 3.7. Since $X$ is perfectly normal, there exists a continuous function $g'' : X \to I$ such that

$$A \cup B = \{x \in X \mid g''(x) = 0\}.$$

Now let $g = \max\{g', g''\}$. Clearly $g$ is a continuous function with the required properties.$\square$

**4.20 Proposition.** *A topological space which is dominated by a covering $\{A_\lambda\}_{\lambda \in \Lambda}$ is perfectly normal if each subspace $A_\lambda$ is perfectly normal.*

*Proof.* Let $B$ be a closed set of a space $X$ which is dominated by a covering $\{A_\lambda\}_{\lambda \in \Lambda}$, where each subspace $A_\lambda$ is perfectly normal. Let $\Lambda$ be well-ordered with last element $\xi$. If $\mu \in \Lambda$ let $C_\mu = \bigcup_{\lambda \leqslant \mu} A_\lambda$ and

$D_\mu = \bigcup_{\lambda < \mu} A_\lambda$. We shall construct continuous functions $f_\mu : C_\mu \to I$ such that $f_\mu | C_\nu = f_\nu$ if $\nu < \mu$ and

$$\{x \in C_\mu \,|\, f_\mu(x) = 0\} = B \cap C_\mu.$$

Let $\mu$ be an element of $\Lambda$ and suppose that the mappings $f_\lambda$ have been defined for $\lambda < \mu$. Let $g_\mu : D_\mu \to I$ be the function such that

$$g_\mu | A_\lambda = f_\lambda | A_\lambda$$

for each $\lambda < \mu$. Since $D_\mu$ has the weak topology with respect to $\{A_\lambda\}_{\lambda < \mu}$, it follows that $g_\mu$ is continuous. The set $A_\mu \cap D_\mu$ is closed in $A_\mu$, and if $g'_\mu = g_\mu | A_\mu \cap D_\mu$, then $g'_\mu$ is continuous and

$$\{x \in A_\mu \cap D_\mu \,|\, g'_\mu(x) = 0\} = B \cap A_\mu \cap D_\mu.$$

The space $A_\mu$ is perfectly normal so that by Lemma 4.19 there exists $h_\mu : A_\mu \to I$ which is an extension of $g'_\mu$ such that

$$\{x \in A_\mu \,|\, h_\mu(x) = 0\} = B \cap A_\mu.$$

If $f_\mu : C_\mu \to I$ is the function such that $f_\mu | D_\mu = g_\mu$ and $f_\mu | A_\mu = h_\mu$, then we have the required mappings $f_\lambda$ for $\lambda \leqslant \mu$. Hence there exists a continuous function $f = f_\xi : X \to I$ such that

$$B = \{x \in X \,|\, f(x) = 0\}.$$

Thus $X$ is perfectly normal. □

## 5   Compact and completely regular spaces

We recall that a topological space $X$ is said to be *compact* if each open covering of $X$ has a finite subcovering. Thus a space $X$ is compact if for each open covering $\{U_\lambda\}_{\lambda \in \Lambda}$ of $X$ there exists a finite subset $M$ of $\Lambda$ such that $\bigcup_{\lambda \in M} U_\lambda = X$. In accordance with the usual convention, a subset $A$ of a topological space $X$ is said to be compact if it is compact in the induced topology. Equivalently, $A$ is a compact subset if for each family $\{U_\lambda\}_{\lambda \in \Lambda}$ of open sets of $X$ such that $A \subset \bigcup_{\lambda \in \Lambda} U_\lambda$, there exists a finite subset $M$ of $\Lambda$ such that $A \subset \bigcup_{\lambda \in M} U_\lambda$. In any space, all finite subsets are compact.

Some familiar results about compactness are now listed. A family $\{B_\lambda\}_{\lambda \in \Lambda}$ of sets is said to have the *finite intersection property* if for each finite subset $M$ of $\Lambda$, the intersection $\bigcap_{\lambda \in M} B_\lambda$ is non-empty.

**5.1   Proposition.** (a) *A topological space is compact if and only if each family of closed sets with the finite intersection property has a non-empty intersection.*

   (b)　*A closed set of a compact space is compact.*

   (c)　*A compact subset of a Hausdorff space is closed.*

   (d)　*A finite union of compact subsets is a compact subset.*

   (e)　*The continuous image of a compact space is compact.*

   (f)　*Disjoint compact subsets in a Hausdorff space have disjoint open neighbourhoods.*

   (g)　*A subset of Euclidean space is compact if and only if it is closed and bounded.*∗

**5.2**　*Remarks.* (1) It follows from parts (b), (e) and (c) of Proposition 5.1 that if $X$ is a compact space, $Y$ is a Hausdorff space and $f: X \to Y$ is continuous, then $f$ is a closed mapping. Thus if $X$ is a compact space and $Y$ is a Hausdorff space, then a continuous bijection of $X$ onto $Y$ is a homeomorphism. It follows that if $(X, \mathcal{T})$ is a compact Hausdorff space then there is no topology $\mathcal{U}$ on $X$ strictly smaller than $\mathcal{T}$ such that $(X, \mathcal{U})$ is a Hausdorff space and no topology $\mathcal{V}$ on $X$ strictly larger than $\mathcal{T}$ such that $(X, \mathcal{V})$ is a compact space.

   (2) Since the continuous image of a connected space is connected and a connected subset of the space **R** of real numbers is an interval, it follows from parts (e) and (g) of Proposition 5.1 that if $X$ is a compact connected space and $f$ is a real-valued continuous function on $X$, then $f(X)$ is a bounded closed interval. A continuous real-valued function on a compact space is bounded and attains its bounds.

   (3) It follows from Proposition 5.1 (a) that if $\{A_\lambda\}_{\lambda \in \Lambda}$ is a family of closed sets in a compact space and $U$ is an open set such that

$$\bigcap_{\lambda \in \Lambda} A_\lambda \subset U,$$

then there exists a finite subset $M$ of $\Lambda$ such that $\bigcap_{\lambda \in M} A_\lambda \subset U$. In particular, if $\{H_\lambda\}_{\lambda \in \Lambda}$ is a family of open-and-closed sets in a compact space and $U$ is an open set such that $\bigcap_{\lambda \in \Lambda} H_\lambda \subset U$, then there exists an open-and-closed set $H$ such that $\bigcap_{\lambda \in \Lambda} H_\lambda \subset H \subset U$.

If $X$ is the topological product of a family $\{X_\lambda\}_{\lambda \in \Lambda}$ of topological spaces and $X$ is compact, then since each projection of $X$ onto $X_\lambda$ is continuous, it follows from Proposition 5.1(e) that each space $X_\lambda$ is compact. *Tihonov's theorem* which follows is the converse result.

**5.3**　**Proposition.** *The topological product of a family of compact spaces is a compact space.*∗

The class of compact Hausdorff spaces occupies an important position in general topology.

**5.4  Definition.** *A compact Hausdorff space will be called a bi-compact space.*

It follows from parts (*b*) and (*f*) of Proposition 5.1 that a bicompact space is a $T_4$-space. This result will be generalized in Chapter 2. We establish two properties of bicompact spaces which will be needed later.

**5.5  Proposition.** *The component of a point $x$ in a bicompact space $X$ is the intersection of the open-and-closed sets of $X$ which contain $x$.*

*Proof.* Let $C$ be the component of $x$ and let $H$ be an open-and-closed set containing $x$. Then $C \cap H$ is a non-empty open-and-closed set of $C$. It follows that $C \cap H = C$ so that $C \subset H$. Thus if $\mathscr{H}(x)$ is the set of open-and-closed sets of $X$ containing $x$ and $K = \bigcap_{H \in \mathscr{H}(x)} H$, then $C \subset K$. Now let $K_1$, $K_2$ be disjoint open sets of $K$ such that

$$K = K_1 \cup K_2$$

and suppose that $x \in K_1$. Since $K_1$ and $K_2$ are closed sets of $K$ it follows that $K_1$ and $K_2$ are disjoint closed sets of $X$. Since $X$ is a $T_4$-space, there exist disjoint open sets $G_1$, $G_2$ such that $K_1 \subset G_1$ and $K_2 \subset G_2$. Since $K \subset G_1 \cup G_2$, it follows from Remark 5.2 (3) that there exists an open-and-closed set $H$ such that $K \subset H \subset G_1 \cup G_2$. Since $G_1$ is closed in $G_1 \cup G_2$, $H \cap G_1$ is closed in $H$ so that $H \cap G_1$ is a closed set of $X$. Since $H \cap G_1$ is clearly open it follows that $H \cap G_1$ is an open-and-closed set of $X$ which contains $x$. Thus $K \subset H \cap G_1$ and since $K_2$ is disjoint from $G_1$ it follows that $K_2$ is empty. Thus $K$ is a connected set and hence $C = K$. $\square$

**5.6  Proposition.** *If $X$ is a topological space of infinite weight, $Y$ is a bicompact space and $f: X \to Y$ is a continuous surjection, then*

$$w(Y) \leqslant w(X).$$

*Proof.* Let $\mathscr{B}$ be a base for the topology of $X$ such that $|\mathscr{B}| = w(X)$. Let $\Lambda$ be the set of subsets $\lambda = \{B_1, B_2\}$ of $\mathscr{B}$ with two elements such that there exist disjoint members $U_1$, $U_2$ of the topology $\mathscr{T}$ of $Y$ containing $f(B_1)$ and $f(B_2)$ respectively. For each such subset $\lambda = \{B_1, B_2\}$ choose disjoint members $U_1^\lambda$ and $U_2^\lambda$ of $\mathscr{T}$ such that $f(B_1) \subset U_1^\lambda$ and $f(B_2) \subset U_2^\lambda$. Let $\mathscr{S}$ be the subset of $\mathscr{T}$ consisting of those sets $U$ such that $U = U_i^\lambda$ for $i = 1$ or $2$ and some $\lambda$ in $\Lambda$. Then $\mathscr{S}$ is a subbase for a topology $\mathscr{U}$ on $Y$ and it is clear that $\mathscr{U} \subset \mathscr{T}$. But $(Y, \mathscr{U})$ is a Hausdorff space. For if $y_1, y_2$ are distinct points of $Y$ there exist disjoint members $V_1, V_2$ of $\mathscr{T}$ such that $y_1 \in V_1$ and $y_2 \in V_2$. For $i = 1, 2$ there exists $x_i$ in $X$ such that $f(x_i) = y_i$ and there exists $B_i$ in $\mathscr{B}$ such that $x_i \in B_i \subset f^{-1}(V_i)$ so that $y_i \in f(B_i) \subset V_i$. Thus $\{B_1, B_2\}$ is an element $\lambda$ of $\Lambda$ and $U_1^\lambda$, $U_2^\lambda$ are disjoint members of $\mathscr{U}$ containing $y_1, y_2$ respectively. It follows from Remark 5.2 (1) that $\mathscr{T} = \mathscr{U}$. Thus $\mathscr{S}$ is a subbase for the given topology $\mathscr{T}$ of $Y$. Since the cardinality of $\mathscr{S}$ does not exceed $w(X)$ it follows that $w(Y) \leqslant w(X)$.□

The next result is a source of examples of bicompact spaces. The condition given for bicompactness of a linearly ordered space is also a necessary condition.

**5.7  Proposition.** *An order-complete linearly ordered space is bicompact.*

*Proof.* Let $X$ be an order-complete linearly ordered space. Then $X$ is a Hausdorff space. Since $X$ is order-complete, $X$ has a least element $x_0$ and a greatest element $x_1$. Let $\mathscr{U}$ be an open covering of $X$ and let $A$ be the subset of $X$ consisting of those elements $x$ such that $x_0 < x$ and $[x_0, x]$ is contained in the union of finitely many members of $\mathscr{U}$. The set $A$ is non-empty. For there exists $U_0$ in $\mathscr{U}$ such that $x_0 \in U_0$ and since $U_0$ is open there exists $x$ such that $x_0 < x$ and $[x_0, x) \subset U_0$. There exists $U$ in $\mathscr{U}$ such that $x \in U$ so that $[x_0, x] \subset U_0 \cup U$ which shows that $x \in A$. Let $a$ be the least upper bound of $A$ so that $x_0 < a$ and suppose first that $a < x_1$. There exists $V$ in $\mathscr{U}$ such that $a \in V$ and since $V$ is open there exist $a_1, b_1$ such that $a_1 < b_1$ and

$$a \in (a_1, b_1) \subset V.$$

Since $a$ is the least upper bound of $A$, there exists $c$ in $A$ such that $a_1 < c \leqslant a$. There exists $W$ in $\mathscr{U}$ such that $b_1 \in W$, and so we see that $[x_0, c]$ is contained in the union of finitely many members of $\mathscr{U}$ whilst $[c, b_1] \subset V \cup W$. Thus $b_1 \in A$, which is absurd since $a < b_1$. It follows that $a = x_1$. Now there exists $U_1$ in $\mathscr{U}$ such that $x_1 \in U_1$ and there exists $d$ such that $d < x_1$ and $(d, x_1] \subset U_1$. Since $x_1$ is the least upper bound of $A$, there exists $y$ in $A$ such that $d < y \leqslant x_1$. Thus $[x_0, y]$ is contained in the union of finitely many members of $\mathscr{U}$ and

$$[y, x_1] \subset U_1 \in \mathscr{U}.$$

Thus $X = [x_0, x_1]$ is the union of finitely many members of $\mathscr{U}$.□

**5.8** *Examples.* (i) The set $P$ of ordinals not exceeding $\omega_1$, the first uncountable ordinal, is well-ordered and has a greatest element, and is therefore order-complete. Thus the space $P$ discussed in Example 4.14 is bicompact.

(ii) It follows from Proposition 5.1 $(g)$ that the unit interval $I$ is bicompact (this might also be deduced from Proposition 5.7). Now let $\tau$ be any cardinal number and let $\Lambda$ be a set such that $|\Lambda| = \tau$. Consider the family of spaces $\{X_\lambda\}_{\lambda \in \Lambda}$, where $X_\lambda = I$ for each $\lambda$. Since $I$ is bicompact, the topological product $\Pi_{\lambda \in \Lambda} X_\lambda$ is a bicompact space by Tihonov's theorem. Up to homeomorphism this product depends only on the cardinal number $\tau$ of the index set $\Lambda$. We shall denote this space by $I^\tau$. More precisely it is the copy of $I^\tau$ indexed by $\Lambda$. If $\tau$ is infinite then it follows from Proposition 2.11 that the weight of $I^\tau$ is $\tau$. We call $I^\tau$ the *Tihonov cube* of weight $\tau$.

Next a class of spaces on which there is a plentiful supply of continuous real-valued functions is introduced. There exist regular spaces on which every continuous real-valued function is constant.

**5.9** *Definition. A topological space $X$ is said to be* completely regular *if for each point $x$ and each closed set $F$ such that $x \notin F$ there exists a continuous function $f: X \to I$ such that $f(x) = 1$ and $f(y) = 0$ if $y \in F$. A completely regular $T_0$-space is called a* Tihonov space.

Evidently a completely regular space is regular so that a Tihonov space is a $T_3$-space. A normal regular space is a completely regular space. In fact if $X$ is a normal regular space and $A$ is a subspace of $X$, then $A$ is a completely regular space. For let $x_0$ be a point of $A$ and let $K$ be a closed set of $A$ such that $x_0 \notin K$. There exists a closed set $F$ of $X$ such that $F \cap A = K$. Since $x_0 \notin F$ and $X$ is a regular space, there exists a closed set $E$ of $X$ such that $x_0 \in E$ and $E \cap F = \varnothing$. By Urysohn's lemma, there exists a continuous function $g: X \to I$ such that $g(x) = 1$ if $x \in E$, and $g(x) = 0$ if $x \in F$. If $f = g|A$, then $f: A \to I$ is continuous and $f(x_0) = 1$, whilst $f(x) = 0$ if $x \in K$. It follows that every subspace of a $T_4$-space is a Tihonov space.

**5.10** *Proposition. Every neighbourhood of a point in a completely regular space contains a neighbourhood of the point which is a zero-set.*

*Proof.* Let $X$ be a completely regular space, let $x_0$ be a point of $X$ and let $U$ be an open set such that $x_0 \in U$. Then there exists a continuous function $f: X \to I$ such that $f(x_0) = 1$ and $f(x) = 0$ if $x \notin U$.

If $V = \{x \in X \,|\, f(x) \geqslant \tfrac{1}{2}\}$, then $V$ is a zero-set such that $x_0 \in V \subset U$, and $V$ is a neighbourhood of $x$ since $\{x \in X \,|\, f(x) > \tfrac{1}{2}\}$ is a cozero-set containing $x_0$. $\square$

We can characterize the completely regular spaces as follows:

**5.11  Proposition.** *A space $X$ is completely regular if and only if the set of cozero-sets of $X$ is a base for its topology.*

*Proof.* We saw in the preceding proof that if $X$ is a completely regular space and $U$ is an open set containing a point $x_0$, then $U$ contains a cozero-set of $X$ which contains $x_0$. Thus the cozero-sets form a base for the topology of $X$.

Conversely, let the set of cozero-sets be a base for the topology of a space $X$, let $x_0$ be a point of $X$ and let $F$ be a closed set such that $x_0 \notin F$. Then there exists a continuous real-valued function $g$ such that $x_0 \in X \backslash Z(g) \subset X \backslash F$, where $Z(g)$ is the zero-set of $g$. If

$$|g(x_0)| = r > 0,$$

then let us define a continuous function $f : X \to I$ by putting

$$f(x) = \min\{1, (1/r)\,|g(x)|\} \quad \text{if} \quad x \in X.$$

Then $f(x_0) = 1$ and if $x \in F$ then $x \in Z(g)$ so that $f(x) = 0$. Thus $X$ is a completely regular space. $\square$

It follows from a remark above that every subspace of a bicompact space is a Tihonov space. We shall prove the converse that every Tihonov space is homeomorphic with a subspace of a bicompact space by showing that any Tihonov space can be embedded in a Tihonov cube. The next lemma will have many subsequent applications.

**5.12  Definition.** *Let $X$ be a space, let $\{Y_\lambda\}_{\lambda \in \Lambda}$ be a family of spaces and let $\mathscr{F} = \{f_\lambda\}_{\lambda \in \Lambda}$ be a family of continuous functions, where $f_\lambda$ has domain $X$ and range $Y_\lambda$. The family $\mathscr{F}$ is said to separate points of $X$ if for each pair $x, y$ of distinct points of $X$ there exists $\lambda$ in $\Lambda$ such that $f_\lambda(x) \neq f_\lambda(y)$. Also $\mathscr{F}$ is said to separate points of $X$ from closed sets if for each $x$ in $X$ and each closed set $F$ of $X$ such that $x \notin F$ there exists $\lambda$ in $\Lambda$ such that $f_\lambda(x) \notin (f_\lambda(F))^-$.*

**5.13  Lemma.** *Let $X$ be a space, let $\{Y_\lambda\}_{\lambda \in \Lambda}$ be a family of spaces and let $\mathscr{F} = \{f_\lambda\}_{\lambda \in \Lambda}$ be a family of continuous functions, where $f_\lambda$ has domain*

$X$ and range $Y_\lambda$. If $\mathscr{F}$ separates points of $X$ and separates points of $X$ from closed sets, then $X$ can be embedded in the topological product of the family $\{Y_\lambda\}_{\lambda \in \Lambda}$.

*Proof.* Let $Y$ be the topological product of the family $\{Y_\lambda\}_{\lambda \in \Lambda}$, for each $\lambda$ in $\Lambda$ let $\pi_\lambda : Y \to Y_\lambda$ be the projection and let $f : X \to Y$ be the unique continuous function such that $\pi_\lambda \circ f = f_\lambda$ for every $\lambda$. We shall prove that $f$ is an embedding. Since $\mathscr{F}$ separates points, $f$ is injective. To show that $f$ is an embedding it will be enough to prove that $f(U)$ is open in $f(X)$ for every open set $U$ of $X$. Thus let $U$ be open in $X$ and let $y \in f(U)$ so that $y = f(x)$ for some $x$ in $U$. Since $\mathscr{F}$ separates points from closed sets $f_\lambda(x) \notin (f_\lambda(X \backslash U))^-$ for some $\lambda$ in $\Lambda$. Let

$$ V = \pi_\lambda^{-1}\big(Y_\lambda \backslash (f_\lambda(X \backslash U))^-\big) $$

so that $V$ is open in $Y$ and $f(x) \in V$. If $x' \in X$ and $f(x') \in V$, then $f_\lambda(x') \in Y_\lambda \backslash (f_\lambda(X \backslash U))^-$ so that $x' \notin X \backslash U$. Thus $V \cap f(X) \subset f(U)$. It follows that $f(U)$ is open in $f(X)$. $\square$

**5.14**    *Proposition. A topological space is a Tihonov space if and only if it is a subspace of a bicompact space.*

*Proof.* We have already observed that each subspace of a $T_4$-space and thus each subspace of a bicompact space is a Tihonov space. Now let $X$ be a Tihonov space and let $\mathscr{F}$ be the set of continuous functions $f : X \to I$. The family $\mathscr{F}$ separates points from closed sets for if $F$ is closed and $x \notin F$, then there exists $f$ in $\mathscr{F}$ such that $f(x) = 1$ and $f(F) = \{0\} = (f(F))^-$. Since $X$ is a $T_1$-space, $\mathscr{F}$ also separates the points of $X$. It follows from Lemma 5.13 that $X$ can be embedded in the Tihonov cube indexed by $\mathscr{F}$. $\square$

It follows from Proposition 5.14 that every subspace of a Tihonov space is a Tihonov space and that the topological product of a family of Tihonov spaces is a Tihonov space. We show finally that each Tihonov space $X$ can be embedded in a Tihonov cube of weight equal to the weight of $X$.

**5.15**    *Definition. A* universal space *for a class $\mathscr{P}$ of topological spaces is a space $Z$ which is a member of the class $\mathscr{P}$ and has the property that each member of $\mathscr{P}$ can be embedded in $Z$.*

**5.16**    *Proposition. If $\tau$ is an infinite cardinal number, then the Tihonov cube $I^\tau$ is a universal space for Tihonov spaces of weight not exceeding $\tau$.*

*Proof.* The Tihonov cube $I^\tau$ is a Tihonov space and $w(I^\tau) = \tau$. To complete the proof we shall show that if $X$ is a Tihonov space such that $w(X) = \tau$, then $X$ can be embedded in $I^\tau$. Let $\mathscr{B}$ be a base for the topology of $X$ of cardinality equal to $\tau$. A pair $(U_1, U_2)$ of elements of $\mathscr{B}$ will be called distinguished if there exists a continuous function $f: X \to I$ such that $f(x) = 1$ if $x \in X \backslash U_2$ and $f(x) < \frac{1}{2}$ if $x \in U_1$. We note that for each $x$ in $X$ and each $U_2$ in $\mathscr{B}$ such that $x \in U_2$ there exists $U_1$ in $\mathscr{B}$ such that $x \in U_1$ and the pair $(U_1, U_2)$ is distinguished. Indeed since $X$ is a Tihonov space there exists a continuous function $f: X \to I$ such that $f(x) = 0$ and $f(y) = 1$ if $y \in X \backslash U_2$. If $V = f^{-1}([0, \frac{1}{2}))$, then $V$ is an open neighbourhood of $x$ so that there exists $U_1$ in $\mathscr{B}$ such that $x \in U_1 \subset V \subset U_2$. The pair $(U_1, U_2)$ is distinguished. Let us assign to each distinguished pair $(U_1, U_2)$ a particular continuous function $f$ such that $f(x) = 1$ if $x \in X \backslash U_2$, and $f(x) < \frac{1}{2}$ if $x \in U_1$, and let $\mathscr{F}$ be the family of functions so obtained. We note that $|\mathscr{F}| = \tau$. Since $X$ is a $T_1$-space, the proof will be complete if we show that $\mathscr{F}$ separates points from closed sets. Let $x$ be a point of $X$ and let $F$ be a closed set such that $x \notin F$. There exists $U_2$ in $\mathscr{B}$ such that $x \in U_2 \subset X \backslash F$, and there exists $U_1$ such that $x \in U_1$ and $(U_1, U_2)$ is distinguished. The member $f$ of $\mathscr{F}$ associated with $(U_1, U_2)$ separates $x$ and $F$ because $f(x) < \frac{1}{2}$ and $f(F) = \{1\}$. $\square$

**5.17** *Remark.* Since $I^\tau$ is a bicompact space it follows that $I^\tau$ is a universal space for the class of bicompact spaces of weight not exceeding $\tau$.

## 6 Bicompactifications

Let $X$ be a Tihonov space. It follows from Proposition 5.14 that $X$ is homeomorphic with a subspace of a bicompact space $Y$. If $Z$ is the closure of the image of $X$ in $Y$, then $Z$ is a bicompact space in which $X$ can be embedded as a dense subspace. In this section we shall study those bicompact spaces in which a given Tihonov space can be embedded as a dense subspace.

**6.1** *Definition. A bicompactification of a Tihonov space $X$ is a pair $(Z, h)$, where $Z$ is a bicompact space and $h: X \to Z$ is an embedding such that $h(X)$ is dense in $Z$.*

It follows from Proposition 5.16 that if $X$ is a Tihonov space of infinite weight, then there exists a bicompactification $(Z, h)$ of $X$ such that $w(Z) = w(X)$. For there exists an embedding $h: X \to I^{w(X)}$, and we can take $Z$ to be the closure of $h(X)$ in $I^{w(X)}$. The following lemma shows that if $(Z, h)$ is a bicompactification of $X$ then $w(Z) \leqslant 2^{w(X)}$.

**6.2   Lemma.** *If $Y$ is a regular space and $X$ is a dense subspace of $Y$ such that $w(X) = \tau$, then $w(Y) \leqslant 2^\tau$.*

*Proof.* Let $\mathscr{B}$ be a base for the topology of the subspace $X$ such that $|\mathscr{B}| = \tau$. Let $A$ be a subset of $X$ consisting of one point from each member of $\mathscr{B}$. Then $|A| \leqslant \tau$. Now let $U$ be open in $Y$. Then since $X$ is dense in $Y$, there exists a point $x$ of $X$ such that $x \in U$. Hence there exists a member $B$ of $\mathscr{B}$ such that $x \in B \subset U \cap X$. Thus $U \cap A$ is non-empty. Hence $A$ is dense in $Y$. Let $\mathscr{W}$ be a base for the topology of $Y$ and let $\mathscr{V}$ consist of those subsets $V$ of $Y$ such that $V = (W \cap A)^{-\circ}$ for some $W$ in $\mathscr{W}$. If $W \in \mathscr{W}$, then $W \subset (W \cap A)^{-\circ} \subset \overline{W}$ since $A$ is dense in $Y$. Now if $G$ is open in $Y$ and $y \in G$, then since $Y$ is a regular space there exists $W$ in $\mathscr{W}$ such that $y \in W \subset \overline{W} \subset G$ and hence there exists $V$ in $\mathscr{V}$ such that $y \in V \subset G$. Thus $\mathscr{V}$ is a base for the topology of $Y$. But the cardinality of $\mathscr{V}$ does not exceed the cardinality of the set of subsets of $A$. Thus $w(Y) \leqslant 2^\tau$.□

It follows from Lemma 6.2 that there is a set of homeomorphism classes of bicompact spaces $Z$ such that $(Z, h)$ is a bicompactification of $X$. We introduce a finer classification than homeomorphism into the class of bicompactifications of a space $X$. This classification takes account of the way $X$ is embedded in its bicompactification. The bicompactifications $(Z_1, h_1)$ and $(Z_2, h_2)$ of a Tihonov space $X$ are said to be *equivalent* if there exists a homeomorphism $\phi: Z_2 \to Z_1$ such that $\phi \circ h_2 = h_1$. It is clear that the equivalence classes of bicompactifications form a set. We shall often fail to distinguish between equivalent bicompactifications. In this spirit, we speak of the set of bicompactifications of $X$. We introduce a partial order of precedence on the set of bicompactifications of $X$. A bicompactification $(Z_1, h_1)$ is said to *precede* a bicompactification $(Z_2, h_2)$ if there exists a continuous function $\phi: Z_2 \to Z_1$ such that $\phi \circ h_2 = h_1$. Moreover, $(Z_2, h_2)$ follows $(Z_1, h_1)$ if $(Z_1, h_1)$ precedes $(Z_2, h_2)$. To see that precedence is a partial order we need the following fact.

**6.3   Proposition.** *If $f, g: X \to Y$ are continuous functions, where $Y$ is a Hausdorff space, then the set*

$$\{x \in X \mid f(x) = g(x)\}$$

*is closed in $X$.*

*Proof.* Let $A = \{x \in X \mid f(x) = g(x)\}$ and suppose that $x_0 \notin A$. Then $f(x_0) \neq g(x_0)$ so that since $Y$ is a Hausdorff space there exist disjoint open sets $U$ and $V$ of $Y$ such that $f(x_0) \in U$ and $g(x_0) \in V$. If

$$W = f^{-1}(U) \cap g^{-1}(V),$$

then $W$ is open in $X$, $x_0 \in W$ and $W \cap A = \varnothing$. Thus $A$ is closed in $X$. $\square$

**6.4  Corollary.** *Let $Y$ be a Hausdorff space and let $A$ be a dense subset of a space $X$. If $f, g : X \to Y$ are continuous functions such that*

$$f \mid A = g \mid A,$$

*then $f = g$.* $\square$

Now let $(Z_1, h_1)$ and $(Z_2, h_2)$ be bicompactifications of a Tihonov space $X$, each of which precedes the other. Since $(Z_1, h_1) \leqslant (Z_2, h_2)$ there exists a continuous function $\phi : Z_2 \to Z_1$ such that $\phi \circ h_2 = h_1$ and since $(Z_2, h_2) \leqslant (Z_1, h_1)$ there exists a continuous function

$$\theta : Z_1 \to Z_2$$

such that $\theta \circ h_1 = h_2$. Thus $\phi \circ \theta \circ h_1 = h_1$ so that

$$\phi \circ \theta \mid h_1(X) = 1_{Z_1} \mid h_1(X),$$

where $1_{Z_1}$ is the identity mapping of $Z_1$. Since $h_1(X)$ is dense in $Z_1$ it follows that $\phi \circ \theta = 1_{Z_1}$. Similarly $\theta \circ \phi$ is the identity mapping of $Z_2$. Thus $\theta$ and $\phi$ are inverse homeomorphisms. It follows that $(Z_1, h_1)$ and $(Z_2, h_2)$ are equivalent bicompactifications. It also follows from Corollary 6.4 that if $(Z_1, h_1)$ and $(Z_2, h_2)$ are bicompactifications of a Tihonov space such that $(Z_1, h_1)$ precedes $(Z_2, h_2)$, then the continuous surjection $\phi : Z_2 \to Z_1$ such that $\phi \circ h_2 = h_1$ is unique. We call $\phi$ the *natural identification mapping* of the bicompactifications.

We shall usually suppose that a space $X$ is contained as a subspace in its bicompactification and with this understanding we speak of the bicompactification $Z$ of a space $X$. The bicompactification $Z_1$ of $X$ precedes the bicompactification $Z_2$ of $X$ if there exists a continuous surjection $\phi : Z_2 \to Z_1$ such that $\phi(x) = x$ for all $x$ in $X$. If the bicompactification $Z_1$ of $X$ precedes the bicompactification $Z_2$ and $\phi$ is the natural identification mapping, then

$$\phi(Z_2 \backslash X) = Z_1 \backslash X.$$

For let $Y$ be the subspace $\phi^{-1}(X)$ of $Z_2$. The continuous mapping $\theta : Y \to Y$, which is the composite of the mapping of $Y$ onto $X$ given by restriction of $\phi$ with the inclusion of $X$ in $Y$, satisfies $\theta(x) = x$ for $x$ in $X$. It follows from Corollary 6.4 that $\theta$ is the identity mapping of $Y$ so that $Y = X$. Thus $\phi(Z_2 \backslash X) = Z_1 \backslash X$.

**6.5 Proposition.** *Let $Z_1$ and $Z_2$ be bicompactifications of a space $X$ such that $Z_1$ precedes $Z_2$ and let $\phi: Z_2 \to Z_1$ be the natural identification mapping. If there exist continuous surjections $\theta: Z_2 \to Z$ and $\psi: Z \to Z_1$ such that $\psi \circ \theta = \phi$ and $Z$ is a Hausdorff space, then $Z$ is a bicompatification of $X$ which follows $Z_1$ and precedes $Z_2$.*

*Proof.* Since $\theta$ is a continuous surjection, $Z$ is a bicompact space. Since $\psi\theta(x) = x$ for $x$ in $X$, the continuous mapping $h: X \to Z$ given by restriction of $\theta$ is an injection. And $h(X) = \theta(X)$ is dense in $Z$. Since $\phi(Z_2 \backslash X) = Z_1 \backslash X$ we see that $\theta(Z_2 \backslash X)$ is disjoint from $\theta(X)$. Thus if $E$ is a closed set of $X$ and $E = F \cap X$, where $F$ is closed in $Z$, then

$$h(E) = h(X) \cap \theta(F).$$

Since $\theta(F)$ is closed in $Z$, $h(E)$ is closed in $h(X)$. Thus $h$ is an embedding. It follows that $(Z, h)$ is a bicompactification of $X$. And $\theta, \psi$ are natural identification mappings. □

Let $\mathscr{B}(X)$ be the partially ordered set of bicompactifications of a Tihonov space $X$. If $\mathscr{C}$ is any subset of $\mathscr{B}(X)$, then $\mathscr{C}$ has a least upper bound. We shall not prove this fact but we establish that $\mathscr{B}(X)$ has a greatest element.

**6.6 Proposition.** *A Tihonov space $X$ has a bicompactification $\beta X$ with the property that each continuous function $\phi: X \to Y$, where $Y$ is a bicompact space, has an extension $\psi: \beta X \to Y$. Any two bicompactifications with this property are equivalent.*

*Proof.* Let $\mathscr{F}$ be the set of continuous functions $f: X \to I$. It was shown in Proposition 5.14 that $X$ can be embedded in the Tihonov cube $\hat{X}$ indexed by $\mathscr{F}$. We identify $X$ with its image in $\hat{X}$ under thi embedding. If $\pi_f: \hat{X} \to I$ is the projection, where $f \in \mathscr{F}$, then $\pi_f$ is an extension of $f$. Let $\beta X$ be the closure of $X$ in $\hat{X}$. It is clear that $\beta X$ is a bicompactification of $X$.

Now let $Y$ be a bicompact space and let $\phi: X \to Y$ be a continuous function. Let $\mathscr{G}$ be the set of continuous functions $g: Y \to I$. Then $Y$ is homeomorphic with a closed subspace of the Tihonov cube $\hat{Y}$ indexed by $\mathscr{G}$ and we identify $Y$ with this subspace. If $\rho_g: \hat{Y} \to I$ is the projection, where $g \in \mathscr{G}$, then $\rho_g$ extends $g$. There is a unique continuous function $\theta \colon \hat{X} \to \hat{Y}$ such that $\rho_g \circ \theta = \pi_{g \circ \phi}$ if $g \in \mathscr{G}$. If $x \in X$, then for each $g$ in $\mathscr{G}$ we have

$$\rho_g \theta(x) = \pi_{g \circ \phi}(x) = g(\phi(x)) = \rho_g(\phi(x)),$$

so that $\theta(x) = \phi(x)$. Thus $\theta$ extends $\phi$. Since $\theta$ is continuous and $Y$ is closed in $\hat{Y}$, it follows that $\theta(\beta X) \subset Y$. It follows that we can define a continuous function $\psi \colon \beta X \to Y$ by restriction of $\theta$, and $\psi$ is the required extension of $\phi$.

Finally, we note that a bicompactification of $X$ with the property described in this proposition follows every bicompactification of $X$. Hence any two bicompactifications with this property are equivalent. $\square$

The bicompactification $\beta X$ of a Tihonov space $X$ is called the *Stone–Čech bicompactification* of $X$. The next two results illustrate the usefulness of the 'universal' property by which the Stone–Čech bicompactification has been characterized.

**6.7   Proposition.** *If $X$ and $Y$ are Tihonov spaces and $f \colon X \to Y$ is a continuous function, then there exists a continuous function*

$$\beta f \colon \beta X \to \beta Y$$

*which extends $f$ in the sense that $(\beta f)(x) = f(x)$ if $x \in X$.*

*Proof.* By Proposition 6.6, the mapping of $X$ into $\beta Y$, which is the composite of $f$ and the inclusion of $Y$ in $\beta Y$, has an extension

$$\beta f \colon \beta X \to \beta Y. \square$$

**6.8   Proposition.** *If $X$ is a Tihonov space and $Y$ is a subspace of $\beta X$ such that $X \subset Y \subset \beta X$, then $\beta Y = \beta X$.*

*Proof.* Let $Z$ be a bicompact space. Each continuous function $\phi \colon X \to Z$ has an extension to $\beta X$ and so to $Y$. Thus since $X$ is dense in $Y$, $X$ is dense in $\beta Y$ and every continuous function $\phi \colon X \to Z$ extends to $\beta Y$. It follows that $\beta Y = \beta X$. $\square$

A number of properties are equivalent to the property by which the Stone–Čech bicompactification has been characterized.

**6.9   Proposition.** *Let $X$ be a dense subspace of a completely regular space $Z$. The following statements are equivalent:*

   *(a) every continuous function $\phi \colon X \to Y$, where $Y$ is a bicompact space, has an extension $\psi \colon Z \to Y$;*

   (b) *every continuous function* $f\colon X \to I$ *has an extension* $g\colon Z \to I$;

   (c) *each two disjoint zero-sets in* $X$ *have disjoint closures in* $Z$;

   (d) *for each pair* $A, B$ *of zero-sets of* $X$ *the closure of* $A \cap B$ *in* $Z$ *is* $\bar{A} \cap \bar{B}$, *where* $\bar{A}, \bar{B}$ *are the closures of* $A, B$ *respectively in* $Z$.

*Proof.* Throughout this proof, the closure in $Z$ of a subset $A$ of $X$ will be denoted by $\bar{A}$.

It is clear that $(a) \Rightarrow (b)$.

$(b) \Rightarrow (c)$. Suppose that $(b)$ holds and let $A = Z(h_1)$ and $B = Z(h_2)$ be disjoint zero-sets in $X$, where $h_1$ and $h_2$ are continuous real-valued functions on $X$. Let us define $f\colon X \to I$ as follows:

$$f(x) = |h_1(x)| / \big( |h_1(x)| + |h_2(x)| \big) \quad \text{if} \quad x \in X.$$

Then $f$ is continuous, $f(x) = 0$ if $x \in A$ and $f(x) = 1$ if $x \in B$. By hypothesis, $f$ has an extension $g\colon Z \to I$. If $z \in \bar{A}$ then $g(z) = 0$ and if $z \in \bar{B}$ then $g(z) = 1$. Thus $\bar{A}$ and $\bar{B}$ are disjoint.

$(c) \Rightarrow (d)$. Suppose that $(c)$ holds and let $A$ and $B$ be zero-sets in $X$. Suppose that $z \in \bar{A} \cap \bar{B}$ and let $V$ be a neighbourhood of $z$ which is a zero-set of $Z$. Then $z \in (V \cap A)^- \cap (V \cap B)^-$. But $V \cap A$ and $V \cap B$ are zero-sets of $X$ so that $V \cap A$ and $V \cap B$ are not disjoint since $(c)$ holds. Thus $V \cap (A \cap B) \neq \varnothing$. Since by Proposition 5.10 every neighbourhood of $z$ contains a neighbourhood of $z$ which is a zero-set, it follows that $z \in (A \cap B)^-$. Thus $\bar{A} \cap \bar{B} \subset (A \cap B)^-$. The reverse inclusion evidently holds.

It is clear that $(d) \Rightarrow (c)$ and we complete the proof by showing that $(c) \Rightarrow (a)$. Let $(c)$ hold and let $\phi\colon X \to Y$ be a continuous function, where $Y$ is a bicompact space. Let $z_0$ be a point of $Z$ and let $\mathscr{B}$ be the set of closed sets $B$ in $Y$ such that $\phi^{-1}(B)$ is a zero-set in $X$ such that $z_0 \in \big( \phi^{-1}(B) \big)^-$. The set $\mathscr{B}$ is non-empty since $Y \in \mathscr{B}$. Since $(c)$ holds, $\mathscr{B}$ has the finite intersection property. Thus $\bigcap_{B \in \mathscr{B}} B$ is non-empty since $Y$ is compact. Let $y_0 \in \bigcap_{B \in \mathscr{B}} B$. If $V$ is a neighbourhood of $y_0$ which is a zero-set, then $V \in \mathscr{B}$. For by Proposition 5.11 there exists a zero-set $H$ such that $y_0 \in Y \backslash H \subset V$. Thus $V \cup H = Y$ so that

$$\big( \phi^{-1}(V) \big)^- \cup \big( \phi^{-1}(H) \big)^- = Z.$$

It follows that either $V \in \mathscr{B}$ or $H \in \mathscr{B}$. But $H \notin \mathscr{B}$ since $y_0 \notin H$. Thus $V \in \mathscr{B}$. Now suppose that $y_1 \in \bigcap_{B \in \mathscr{B}} B$ and $y_1 \neq y_0$. Then there exists a neighbourhood of $y_0$ which does not contain $y_1$. Since $V \in \mathscr{B}$, this is absurd. Thus $\bigcap_{B \in \mathscr{B}} B = \{y_0\}$. Let us define $\psi(z_0) = y_0$. The function $\psi\colon Z \to Y$ thus defined is an extension of $\phi$. For if $z_0 \in X$ and $W$ is a zero-set of $X$ such that $z_0 \in \bar{W}$, then $z_0 \in W$ since $\bar{W} \cap X = W$. Thus

$z_0 \in \phi^{-1}(B)$ for each $B$ in $\mathscr{B}$ so that $\psi(z_0) = \phi(z_0)$. To complete the proof we must show that $\psi$ is continuous. Suppose that $U$ is open in $Y$ and $z_0 \in \psi^{-1}(U)$. By Propositions 5.10 and 5.11 there exist zero-sets $H$ and $V$ of $Y$ such that

$$\psi(z_0) \in Y \backslash H \subset V \subset U.$$

Since $\psi(z_0) \notin H$, it follows that $z_0 \notin (\phi^{-1}(H))^-$ and hence

$$N = Z \backslash (\phi^{-1}(H))^-$$

is a neighbourhood of $z_0$ in $Z$. But $V \cup H = Y$ so that

$$(\phi^{-1}(V))^- \cup (\phi^{-1}(H))^- = Z.$$

Thus if $z \in N$ then $z \in (\phi^{-1}(V))^-$ and it follows that $\psi(z) \in V$. Thus $z_0 \in N \subset \psi^{-1}(U)$ and we see that $\psi$ is continuous as required.$\square$

If $A$ and $B$ are disjoint closed sets of a $T_4$-space $X$, then by Urysohn's lemma there exist disjoint zero-sets $E$ and $F$ of $X$ such that $A \subset E$ and $B \subset F$. It follows from Proposition 6.9 that $E$ and $F$ have disjoint closures in $\beta X$ and hence $A$ and $B$ have disjoint closures in $\beta X$. We shall need the following more general result. We denote the closure in $\beta X$ of a subset $A$ of $X$ by $\mathrm{cl}_{\beta X}(A)$.

**6.10   Proposition.** *If $B_1, \ldots, B_n$ are closed sets of a $T_4$-space $X$, then*

$$\bigcap_{i=1}^{n} \mathrm{cl}_{\beta X}(B_i) = \mathrm{cl}_{\beta X}\left(\bigcap_{i=1}^{n} B_i\right).$$

*Proof.* It is enough to show that $\bigcap_{i=1}^{n} \mathrm{cl}_{\beta X}(B_i) \subset \mathrm{cl}_{\beta X}(\bigcap_{i=1}^{n} B_i)$. Evidently this result follows by an easy induction from the case $n = 2$. Thus let $A$ and $B$ be closed sets of a $T_4$-space $X$ and let

$$z \in \mathrm{cl}_{\beta X}(A) \cap \mathrm{cl}_{\beta X}(B).$$

If $G$ is an open set of $\beta X$ such that $z \in G$, then there exists an open set $H$ of $\beta X$ such that $z \in H \subset K \subset G$, where $K = \mathrm{cl}_{\beta X}(H)$. Since $H$ is an open set, it follows that $\mathrm{cl}_{\beta X}(A) \cap H \subset \mathrm{cl}_{\beta X}(A \cap H)$ so that

$$z \in \mathrm{cl}_{\beta X}(A \cap H) \subset \mathrm{cl}_{\beta X}(A \cap K).$$

And similarly $z \in \mathrm{cl}_{\beta X}(B \cap K)$. Since $A \cap K$ and $B \cap K$ are closed sets of the $T_4$-space $X$, it follows from the remark above that $A \cap K$ and $B \cap K$ are not disjoint. It follows that $G \cap (A \cap B) \neq \varnothing$. Thus

$$z \in \mathrm{cl}_{\beta X}(A \cap B).$$

Hence        $$\mathrm{cl}_{\beta X}(A) \cap \mathrm{cl}_{\beta X}(B) \subset \mathrm{cl}_{\beta X}(A \cap B).\square$$

We shall say that a subspace $A$ of a Tihonov space $X$ is *$\beta$-closed* in $X$ if every continuous function $f: A \to I$ has an extension $g: X \to I$. For example each closed subspace of a $T_4$-space is $\beta$-closed. The following corollary of Proposition 6.9 shows the significance of the $\beta$-closed subspaces of a Tihonov space.

**6.11** **Proposition.** *If $A$ is a $\beta$-closed subspace of a Tihonov space $X$, then the closure of $A$ in $\beta X$ is $\beta A$.*

*Proof.* The subspace $\mathrm{cl}_{\beta X}(A)$ of $\beta X$ is a bicompactification of $A$. Since every continuous function $g: X \to I$ has an extension to $\beta X$ it follows that every continuous function $f: A \to I$ has an extension to $\beta X$ and thus to $\mathrm{cl}_{\beta X}(A)$. Hence from Proposition 6.9 and the uniqueness of the Stone–Čech bicompactification it follows that

$$\mathrm{cl}_{\beta X}(A) = \beta A. \square$$

A second construction of the Stone–Čech bicompactification is obtained by considering maximal ideals in a certain lattice. Let $L$ be a distributive lattice with universal bounds 0 and 1. A non-empty subset $J$ of $L$ is called an *ideal* if the following two conditions hold:

(i) if $u \in L$, $v \in J$ and $u \leqslant v$, then $u \in J$;

(ii) if $u \in J$ and $v \in J$, then $u \vee v \in J$.

Let $W$ be a non-empty subset of $L$ and let $J$ consist of all elements $u$ of $L$ which have the property that $u \leqslant w_1 \vee \ldots \vee w_n$ for some finite subset $\{w_1, \ldots, w_n\}$ of $W$. Then $J$ is an ideal containing $W$, called the *ideal generated by $W$*, and $J$ is contained in every ideal which contains $W$. An ideal $J$ is *proper* if $J \neq L$. Clearly an ideal $J$ is proper if and only if $1 \notin J$. A proper ideal $J$ is called a *maximal ideal* if there exists no proper ideal $J'$ such that $J \subset J'$ and $J \neq J'$. It is easily established using Zorn's lemma that each proper ideal is contained in a maximal ideal. A proper ideal $M$ is maximal if and only if the following condition holds: if $u \notin M$ then there exists $v$ in $M$ such that $u \vee v = 1$. For let $M$ be a proper ideal and for each $u$ in $L$ such that $u \notin M$, let $\langle M, u \rangle$ be the ideal generated by $M \cup \{u\}$. Since $\langle M, u \rangle$ is an ideal which contains $M$ as a proper subset, we see that $M$ is maximal if and only if for each $u$ such that $u \notin M$ we have $\langle M, u \rangle = L$. But $\langle M, u \rangle = L$ if and only if there exists $v$ in $M$ such that $u \vee v = 1$. A maximal ideal $M$ has the property that if $u \wedge v \in M$, then either $u \in M$ or $v \in M$. For suppose $u \wedge v \in M$ and $u \notin M$. Then there exists $z$ in $M$ such that $u \vee z = 1$, and

$$v = (u \vee z) \wedge v = (u \wedge v) \vee (z \wedge v).$$

By hypothesis $u \wedge v \in M$, and $z \wedge v \in M$ since $z \in M$, so that $v \in M$ as asserted.

Let $\mathscr{M}$ be the set of maximal ideals in the distributive lattice $L$. For each $u$ in $L$, let
$$\mathscr{M}_u = \{M \in \mathscr{M} \mid u \notin M\}.$$
Then we have
$$\mathscr{M}_{u \vee v} = \mathscr{M}_u \cup \mathscr{M}_v, \quad \mathscr{M}_{u \wedge v} = \mathscr{M}_u \cap \mathscr{M}_v.$$

The first equality is obvious, and it is clear that $\mathscr{M}_{u \wedge v} \subset \mathscr{M}_u \cap \mathscr{M}_v$. But if $u \notin M$ and $v \notin M$, where $M \in \mathscr{M}$, then the maximality of $M$ implies that $u \wedge v \notin M$ and we have the reverse inclusion. Let

$$\mathscr{L} = \{\mathscr{M}_u \mid u \in L\}.$$

Then $\mathscr{L}$ is a set of subsets of $\mathscr{M}$ with the property that each finite intersection of members of $\mathscr{L}$ is a member of $\mathscr{L}$. Thus we can take $\mathscr{L}$ as a base for a topology on $\mathscr{M}$. With this topology, $\mathscr{M}$ is called the *maximal ideal space* of the distributive lattice $L$.

**6.12 Proposition.** *The maximal ideal space of a distributive lattice is compact.*

*Proof.* We use the notation introduced above. Each closed set of $\mathscr{M}$ is the intersection of sets of the form $\mathscr{F}_u = \mathscr{M} \backslash \mathscr{M}_u$. Thus it will be enough to show that if $W \subset L$ and $\{\mathscr{F}_w\}_{w \in W}$ has the finite intersection property, then $\bigcap_{w \in W} \mathscr{F}_w$ is non-empty. Suppose that this is not the case. Then since
$$\bigcap_{w \in W} \mathscr{F}_w = \{M \in \mathscr{M} \mid M \supset W\},$$
there is no maximal ideal which contains $W$. Since each proper ideal is contained in a maximal ideal, it follows that the ideal generated by $W$ is $L$. Thus there exist $w_1, \ldots, w_n$ in $W$ such that $w_1 \vee \ldots \vee w_n = 1$. Hence $\mathscr{M}_{w_1} \cup \ldots \cup \mathscr{M}_{w_n} = \mathscr{M}_1 = \mathscr{M}$. Thus $\mathscr{F}_{w_1} \cap \ldots \cap \mathscr{F}_{w_n} = \varnothing$, which is a contradiction. Hence $\bigcap_{w \in W} \mathscr{F}_w$ is non-empty and it follows that $\mathscr{M}$ is compact. $\square$

Let $X$ be a Tihonov space and let $\mathscr{L}$ be the set of cozero-sets of $X$. Then by Proposition 5.11, $\mathscr{L}$ is a base for the topology of $X$. Furthermore finite unions and finite intersections of members of $\mathscr{L}$ belong to $\mathscr{L}$. Thus $\mathscr{L}$ is a distributive lattice with respect to inclusion with least element $\varnothing$ and greatest element $X$.

**6.13 Proposition.** *If $\mathscr{L}$ is the set of cozero-sets of a Tihonov space $X$. then $\mathscr{L}$ is a distributive lattice and the maximal ideal space is the Stone–Čech bicompactification $\beta X$.*

*Proof.* For each $x$ in $X$, let

$$M_x = \{U \in \mathscr{L} \mid x \notin U\}.$$

It is clear that $M_x$ is an ideal in $\mathscr{L}$. In fact $M_x$ is a maximal ideal. For if $U \notin M_x$, then $x \in U$ so that by Proposition 5.10, $U$ contains a neighbourhood $W$ of $x$ which is a zero-set. If $V = X \backslash W$, then $V \in M_x$ and $U \cup V = X$. Furthermore if $x \neq y$, then $M_x \neq M_y$. For since $\mathscr{L}$ is a base for the topology of the Tihonov space $X$, there exists some $U$ in $\mathscr{L}$ such that $x \in U$ and $y \notin U$.

Let $\mathscr{M}$ be the set of maximal ideals of $\mathscr{L}$ and let $\mathscr{M}$ be indexed by a set $Z$ such that $X \subset Z$ and $\mathscr{M} = \{M_z\}_{z \in Z}$, where $M_z$ is the ideal defined above if $z \in X$, and $M_z \neq M_{z'}$ if $z, z' \in Z$ and $z \neq z'$. For each $U$ in $\mathscr{L}$, let

$$G_U = \{z \in Z \mid U \notin M_z\}.$$

Then

$$G_U \cup G_V = G_{U \cup V} \quad \text{and} \quad G_U \cap G_V = G_{U \cap V}$$

if $U, V \in \mathscr{L}$, and if the sets $G_U$ for $U$ in $\mathscr{L}$ are taken as a base for the topology of $Z$, then $Z$ is homeomorphic with the maximal ideal space of $\mathscr{L}$. Thus $Z$ is compact. Furthermore $Z$ is a Hausdorff space. For if $z, z' \in Z$ and $z \neq z'$ then there exists $V$ in $M_z$, say, such that $V \notin M_{z'}$. Hence there exists $V'$ in $M_{z'}$ such that $V \cup V' = X$. Thus $X \backslash V$ and $X \backslash V'$ are disjoint zero-sets of $X$, so that by Proposition 3.5 there exist disjoint cozero-sets $U$ and $U'$ such that $X \backslash V \subset U$ and

$$X \backslash V' \subset U'.$$

Since $V \cup U = X$, it follows that $U \notin M_z$ so that $z \in G_U$. Similarly $z' \in G_{U'}$. But $G_U$ and $G_{U'}$ are disjoint since $U$ and $U'$ are disjoint. Thus $Z$ is a bicompact space. The space $X$ is a subspace of $Z$. For if $U \in \mathscr{L}$, then

$$X \cap G_U = \{x \in X \mid U \notin M_x\} = U.$$

Thus a base for the induced topology of $X$ as a subspace of $Z$ coincides with a base for the given topology of $X$. Moreover each member of a base for the topology of $Z$ has a non-empty intersection with $X$ so that $X$ is dense in $Z$.

To complete the proof we show that $Z$ is the Stone–Čech bicompactification of $X$ by appealing to Proposition 6.9. If $A$ is a zero-set in $X$, let

$$\tilde{A} = \{z \in Z \mid X \backslash A \in M_z\}.$$

Then $\tilde{A}$ is closed in $Z$ since it is the complement of a member of a base for the topology of $Z$. And $A \subset \tilde{A}$ so that if $\bar{A}$ is the closure of

$A$ in $Z$, then $\bar{A} \subset \tilde{A}$. But if $z \in \tilde{A}$, then $X \backslash A \in M_z$. Thus if $V \in \mathscr{L}$ and $V$ is disjoint from $A$, then $V \in M_z$. Hence if $V \in \mathscr{L}$ and $z \in G_V$, then $G_V \cap A = V \cap A \neq \varnothing$ since $V \notin M_z$. Thus $z \in \bar{A}$. It follows that the closure of $A$ in $Z$ is

$$\bar{A} = \tilde{A} = \{z \in Z \mid X \backslash A \in M_z\}.$$

It is now clear that if $A$ and $B$ are disjoint zero-sets of $X$ then their closures $\bar{A}$ and $\bar{B}$ in $Z$ are disjoint. For if $z \in \bar{A} \cap \bar{B}$, then $X \backslash A \in M_z$ and $X \backslash B \in M_z$, so that $X = (X \backslash A) \cup (X \backslash B) \in M_z$ which is absurd. It follows from Propositions 6.6 and 6.9 that $Z = \beta X$. $\square$

## 7 Inverse limits

An *inverse system* of topological spaces over a directed set $\Omega$ is a family $\{X_\alpha\}_{\alpha \in \Omega}$ of spaces together with continuous functions

$$\pi_{\alpha\beta} : X_\beta \to X_\alpha$$

for each pair $\alpha, \beta$ of elements of $\Omega$ such that $\alpha \leqslant \beta$; it is required furthermore that $\pi_{\alpha\alpha}$ is the identity mapping on $X_\alpha$ for every $\alpha$ and that if $\alpha \leqslant \beta \leqslant \gamma$, then

$$\pi_{\alpha\beta} \circ \pi_{\beta\gamma} = \pi_{\alpha\gamma}.$$

The *inverse limit* of the inverse system is the subspace $X$ of the topological product $\Pi_{\alpha \in \Omega} X_\alpha$ consisting of those elements $x$ such that

$$\pi_{\alpha\beta} \circ p_\beta(x) = p_\alpha(x)$$

for each pair $\alpha, \beta$ of elements of $\Omega$ such that $\alpha \leqslant \beta$, where for each $\alpha$ in $\Omega$, $p_\alpha$ denotes the projection of the topological product onto $X_\alpha$. If $\alpha \in \Omega$, let $\pi_\alpha : X \to X_\alpha$ be the restriction of $p_\alpha$. If $\alpha, \beta \in \Omega$ and $\alpha \leqslant \beta$ then

$$\pi_{\alpha\beta} \circ \pi_\beta = \pi_\alpha.$$

We shall denote this inverse system by $\mathbf{X} = \{X_\alpha, \pi_{\alpha\beta}\}_{\alpha, \beta \in \Omega}$. The continuous functions $\pi_{\alpha\beta}$ are called the *connecting mappings* of $\mathbf{X}$ and the continuous functions $\pi_\alpha$ are called the *canonical mappings* of $\mathbf{X}$.

A subset $M$ of a directed set $\Omega$ is said to be *cofinal* if for each $\alpha$ in $\Omega$ there exists $\beta$ in $M$ such that $\alpha \leqslant \beta$.

**7.1 Proposition.** *Let* $\mathbf{X} = \{X_\alpha, \pi_{\alpha\beta}\}_{\alpha, \beta \in \Omega}$ *be an inverse system of spaces and for each $\alpha$ in $\Omega$ let $\mathscr{B}_\alpha$ be a base for the topology of $X_\alpha$. If $X$ is the inverse limit of $\mathbf{X}$ with canonical mappings $\pi_\alpha$ and $M$ is a cofinal* *subset of $\Omega$, then a base for the topology of $X$ consists of all sets of the form $\pi_\alpha^{-1}(U_\alpha)$, where $U_\alpha \in \mathcal{B}_\alpha$ and $\alpha \in M$.*

*Proof.* For each $\alpha$ in $\Omega$ let $p_\alpha$ be the projection of the topological product $\Pi_{\gamma \in \Omega} X_\gamma$ onto $X_\alpha$. Let $V$ be open in $X$ and let $x$ be a point of $V$. There exists a finite subset $B$ of $\Omega$ and an open set $V_\beta$ of $X_\beta$ for each $\beta$ in $B$ such that $x \in X \cap \left( \bigcap_{\beta \in B} p_\beta^{-1}(V_\beta) \right) \subset V$. Thus

$$x \in \bigcap_{\beta \in B} \pi_\beta^{-1}(V_\beta) \subset V.$$

Since $\Omega$ is a directed set and $M$ is a cofinal subset, there exists $\alpha$ in $M$ such that $\beta \leqslant \alpha$ if $\beta \in B$. If $V_\alpha = \bigcap_{\beta \in B} \pi_{\beta\alpha}^{-1}(V_\beta)$, then $V_\alpha$ is open in $X_\alpha$ and $x \in \pi_\alpha^{-1}(V_\alpha) \subset V$. Since $\pi_\alpha(x) \in V_\alpha$, there exists a member $U_\alpha$ of $\mathcal{B}_\alpha$ such that $\pi_\alpha(x) \in U_\alpha \subset V_\alpha$. Thus $x \in \pi_\alpha^{-1}(U_\alpha) \subset V.\ \square$

Since the inverse limit $X$ of an inverse system $\mathbf{X} = \{X_\alpha, \pi_{\alpha\beta}\}_{\alpha,\beta \in \Omega}$ of topological spaces is a subspace of the topological product of the family $\{X_\alpha\}_{\alpha \in \Omega}$, it follows that if each $X_\alpha$ has some topological property $P$ which is productive and hereditary, then $X$ also has the property $P$. For example, if each $X_\alpha$ is a Hausdorff space then $X$ is a Hausdorff space. In this case $X$ is a closed subspace of $\Pi_{\gamma \in \Omega} X_\gamma$. For suppose that $x_0 \in \Pi_{\gamma \in \Omega} X_\gamma$ and $x_0 \notin X$. Then there exist $\alpha, \beta$ in $\Omega$ such that $\alpha \leqslant \beta$ and $\pi_{\alpha\beta} p_\beta(x_0) \neq p_\alpha(x_0)$, where $p_\alpha, p_\beta$ are the projections. By Proposition 6.3 the set

$$U = \{x \in \Pi X_\gamma \mid \pi_{\alpha\beta} p_\beta(x) \neq p_\alpha(x)\}$$

is an open set of the product. Since $x_0 \in U$ and $U \cap X = \varnothing$, it follows that $X$ is closed.

The inverse limit of an inverse system $\{X_\alpha, \pi_{\alpha\beta}\}$ may be empty even if each $X_\alpha$ is non-empty and the connecting mappings $\pi_{\alpha\beta}$ are surjective. The next proposition gives a useful condition under which the inverse limit is non-empty.

**7.2 Proposition.** *The inverse limit of an inverse system of non-empty bicompact spaces is a non-empty bicompact space.*

*Proof.* Let $\mathbf{X} = \{X_\alpha, \pi_{\alpha\beta}\}_{\alpha,\beta \in \Omega}$ be an inverse system, where $X_\alpha$ is a non-empty bicompact space for each $\alpha$ in $\Omega$. Let $X$ and $Z$ be the inverse limit of $\mathbf{X}$ and the topological product of the family of spaces $\{X_\alpha\}_{\alpha \in \Omega}$ respectively. And if $\alpha \in \Omega$, let $p_\alpha : Z \to X_\alpha$ be the projection. It follows from the remark above that $X$ is a closed subspace of $Z$

and $Z$ is bicompact by Tihonov's theorem. Thus $X$ is a bicompact space and it remains to show that $X$ is non-empty. If $\beta \in \Omega$ let

$$F_\beta = \{x \in Z \mid p_\alpha(x) = \pi_{\alpha\beta} p_\beta(x) \quad \text{if} \quad \alpha \in \Omega \quad \text{and} \quad \alpha \leqslant \beta\}.$$

It follows from Proposition 6.3 that $F_\beta$ is the intersection of closed sets of $Z$ so that $F_\beta$ is a closed set of $Z$. Furthermore $F_\beta$ is non-empty. For let $x_\beta^\circ$ be a point of $X_\beta$ and let $x = \{x_\alpha\}_{\alpha \in \Omega}$ be a point of $Z$ such that $x_\alpha = \pi_{\alpha\beta}(x_\beta^\circ)$ if $\alpha \leqslant \beta$, and $x_\alpha$ is an arbitrary point of $X_\alpha$ otherwise. Then $x \in F_\beta$. It is clear that if $\alpha, \beta \in \Omega$ and $\alpha \leqslant \beta$ then $F_\beta \subset F_\alpha$. It follows that the family $\{F_\alpha\}_{\alpha \in \Omega}$ of closed sets of $Z$ has the finite intersection property, for if $B$ is a finite subset of $\Omega$ there exists $\gamma$ in $\Omega$ such that $\beta \leqslant \gamma$ if $\beta \in B$, and hence $F_\gamma \subset \bigcap_{\beta \in B} F_\beta$. Since $Z$ is compact it follows that $\bigcap_{\alpha \in \Omega} F_\alpha$ is non-empty. It is clear that if $x \in \bigcap_{\alpha \in \Omega} F_\alpha$ then $x \in X$. Thus $X$ is non-empty. $\square$

The inverse limit of an inverse system can be characterized by a 'universal property'.

**7.3 Proposition.** *Let* $\mathbf{X} = \{X_\alpha, \pi_{\alpha\beta}\}_{\alpha, \beta \in \Omega}$ *be an inverse system of topological spaces and let $X$ be the inverse limit of $\mathbf{X}$ with canonical mappings* $\pi_\alpha : X \to X_\alpha$ *for $\alpha$ in $\Omega$. Let $Z$ be a topological space and for each $\alpha$ in $\Omega$ let* $\phi_\alpha : Z \to X_\alpha$ *be a continuous function such that $\pi_{\alpha\beta} \circ \phi_\beta = \phi_\alpha$ if $\alpha, \beta \in \Omega$ and $\alpha \leqslant \beta$. Then there exists a unique continuous function* $\phi : Z \to X$ *such that $\pi_\alpha \circ \phi = \phi_\alpha$ if $\alpha \in \Omega$.*

*Proof.* There exists a unique continuous function $\psi : Z \to \Pi X_\alpha$ such that $p_\alpha \circ \psi = \phi_\alpha$ if $\alpha \in \Omega$, where $p_\alpha$ is the projection of the product onto $X_\alpha$. And if $z \in Z$, $\alpha, \beta \in \Omega$ and $\alpha \leqslant \beta$ then

$$\pi_{\alpha\beta} p_\beta \psi(z) = \pi_{\alpha\beta} \phi_\beta(z) = \phi_\alpha(z) = p_\alpha \psi(z).$$

Thus $\psi(z) \in X$ if $z \in Z$ and the function $\phi : Z \to X$ is given by restriction of $\psi$. $\square$

**7.4** *Remark.* The above property essentially determines the inverse limit $X$ of $\mathbf{X}$. For let $Y$ be a topological space and for each $\alpha$ in $\Omega$ let $\rho_\alpha : Y \to X_\alpha$ be a continuous function such that $\pi_{\alpha\beta} \circ \rho_\beta = \rho_\alpha$ if $\alpha, \beta \in \Omega$ and $\alpha \leqslant \beta$. Suppose that for each space $Z$ and each family $\psi_\alpha : Z \to X_\alpha$, $\alpha \in \Omega$, of continuous functions such that $\pi_{\alpha\beta} \circ \psi_\beta = \psi_\alpha$ if $\alpha, \beta \in \Omega$ and $\alpha \leqslant \beta$, there exists a unique continuous function $\psi : Z \to Y$ such that $\rho_\alpha \circ \psi = \psi_\alpha$ if $\alpha \in \Omega$. Then by Proposition 7.3 there exists a unique continuous function $h : Y \to X$ such that $\pi_\alpha \circ h = \rho_\alpha$ if $\alpha \in \Omega$. And by hypothesis there exists a unique continuous function $g : X \to Y$

such that $\rho_\alpha \circ g = \pi_\alpha$ if $\alpha \in \Omega$. If $\alpha \in \Omega$ then $\pi_\alpha \circ h \circ g = \rho_\alpha \circ g = \pi_\alpha$, so that $h \circ g$ is the identity mapping on $X$ by the uniqueness property. Similarly $g \circ h$ is the identity mapping on $Y$. Thus there exists a unique homeomorphism $h : Y \to X$ such that $\pi_\alpha \circ h = \rho_\alpha$ if $\alpha \in \Omega$.

Let $\mathbf{X} = \{X_\alpha, \pi_{\alpha\beta}\}_{\alpha,\beta\in\Omega}$ be an inverse system of topological spaces. If $M$ is a cofinal subset of $\Omega$, then it is clear that $M$ with its induced order is a directed set. Suppose that for each $\alpha$ in $M$, $B_\alpha$ is a subspace of $X_\alpha$ such that $\pi_{\alpha\beta}(B_\beta) \subset B_\alpha$ if $\alpha, \beta \in M$ and $\alpha \leqslant \beta$. If $\alpha, \beta \in M$ and $\alpha \leqslant \beta$, then restriction of $\pi_{\alpha\beta}$ gives a continuous mapping $\mu_{\alpha\beta} : B_\beta \to B_\alpha$. It is clear that we have an inverse system $\mathbf{B} = \{B_\alpha, \mu_{\alpha\beta}\}_{\alpha,\beta\in M}$ of topological spaces over the directed set $M$. The inverse system $\mathbf{B}$ is called a subsystem over $M$ of $\mathbf{X}$. The inverse limit of $\mathbf{B}$ is homeomorphic to a subspace of the inverse limit of $\mathbf{X}$. For let $X$ be the inverse limit of $\mathbf{X}$ and let $\pi_\alpha : X \to X_\alpha$ be the canonical mapping for each $\alpha$ in $\Omega$. Consider the subspace $B = \bigcap_{\alpha \in M} \pi_\alpha^{-1}(B_\alpha)$ and let $\mu_\alpha : B \to B_\alpha$ be given by restriction of $\pi_\alpha$ if $\alpha \in M$. Let $Z$ be a topological space and for each $\alpha$ in $M$ let $\phi_\alpha : Z \to B_\alpha$ be a continuous function. If $\alpha \in \Omega$, choose $\beta$ in $M$ such that $\alpha \leqslant \beta$ and let $\psi_\alpha : Z \to X_\alpha$ be the continuous function given by putting

$$\psi_\alpha(z) = \pi_{\alpha\beta}\phi_\beta(z) \quad \text{if} \quad z \in Z.$$

The function $\psi_\alpha$ is independent of the choice of $\beta$ in $M$. For suppose that $\gamma \in M$ and $\alpha \leqslant \gamma$. Choose $\xi$ in $M$ such that $\beta \leqslant \xi$ and $\gamma \leqslant \xi$. If $z \in Z$ then

$$\pi_{\alpha\beta}\phi_\beta(z) = \pi_{\alpha\xi}\phi_\xi(z) = \pi_{\alpha\gamma}\phi_\gamma(z).$$

If $\alpha, \beta \in \Omega$ and $\alpha \leqslant \beta$, choose $\gamma$ in $M$ such that $\beta \leqslant \gamma$. If $z \in Z$ then

$$\pi_{\alpha\beta}\psi_\beta(z) = \pi_{\alpha\gamma}\phi_\gamma(z) = \psi_\alpha(z),$$

so that $\pi_{\alpha\beta} \circ \psi_\beta = \psi_\alpha$. It follows that there exists a unique continuous function $\psi : Z \to X$ such that $\pi_\alpha \circ \psi = \psi_\alpha$ if $\alpha \in \Omega$. And $\psi(Z) \subset B$, for if $z \in Z$ and $\alpha \in M$, then $\pi_\alpha \psi(z) = \phi_\alpha(z) \in B_\alpha$. If $\phi : Z \to B$ is given by restriction of $\psi$, then $\mu_\alpha \circ \phi = \phi_\alpha$ if $\alpha \in M$. It follows from the uniqueness of $\psi$ that $\phi$ is the unique function with this property. It follows from Remark 7.4 that the subspace $B$ of $X$ can be taken to be the inverse limit of the subsystem $\mathbf{B}$ of $\mathbf{X}$. Two particular cases will be used frequently. (i) Suppose that $\xi \in \Omega$, $B_\xi$ is a subspace of $X_\xi$ and $M = \{\alpha \in \Omega \mid \xi \leqslant \alpha\}$. If $B_\alpha = \pi_{\xi\alpha}^{-1}(B_\xi)$ for $\alpha$ in $M$, then $\{B_\alpha\}_{\alpha \in M}$ is a subsystem of $\mathbf{X}$ with inverse limit the subspace $\pi_\xi^{-1}(B_\xi)$ of $X$. (ii) Suppose that $\xi, \eta \in \Omega$, $B_\xi$ and $B_\eta$ are subspaces of $X_\xi$ and $X_\eta$ respectively and $M = \{\alpha \in \Omega \mid \xi \leqslant \alpha \text{ and } \eta \leqslant \alpha\}$. If $B_\alpha = \pi_{\xi\alpha}^{-1}(B_\xi) \cap \pi_{\eta\alpha}^{-1}(B_\eta)$, then $\{B_\alpha\}_{\alpha \in M}$ is a subsystem of $\mathbf{X}$ with inverse limit the subspace

$$\pi_\xi^{-1}(B_\xi) \cap \pi_\eta^{-1}(B_\eta)$$

of $X$.

**7.5 Proposition.** *Let* $\mathbf{X} = \{X_\alpha, \pi_{\alpha\beta}\}_{\alpha, \beta \in \Omega}$ *be an inverse system of topological spaces with inverse limit $X$ and canonical mappings*

$$\pi_\alpha : X \to X_\alpha$$

*for $\alpha$ in $\Omega$. If for each pair $\alpha, \beta$ in $\Omega$ such that $\alpha \leqslant \beta$, $\pi_{\alpha\beta}$ is a continuous closed surjection such that $\pi_{\alpha\beta}^{-1}(x_\alpha)$ is bicompact if $x_\alpha \in X_\alpha$, then for each $\alpha$ in $\Omega$, $\pi_\alpha$ is a continuous closed surjection such that $\pi_\alpha^{-1}(x_\alpha)$ is bicompact if $x_\alpha \in X_\alpha$.*

*Proof.* If $\alpha \in \Omega$ let $M = \{\beta \in \Omega \mid \alpha \leqslant \beta\}$. If $x_\alpha \in X_\alpha$, then as noted above $\pi_\alpha^{-1}(x_\alpha)$ is the inverse limit of the subsystem $\{\pi_{\alpha\beta}^{-1}(x_\alpha)\}_{\beta \in M}$ over $M$ of $\mathbf{X}$. But by hypothesis $\pi_{\alpha\beta}^{-1}(x_\alpha)$ is a non-empty bicompact space if $\beta \in M$. It follows from Proposition 7.2 that $\pi_\alpha^{-1}(x_\alpha)$ is a non-empty bicompact subset of $X$. To complete the proof we must show that the continuous surjection $\pi_\alpha$ is closed. Let $A$ be a closed set of $X$ and suppose that $x_\alpha \in X_\alpha \backslash \pi_\alpha(A)$. Each point of $\pi_\alpha^{-1}(x_\alpha)$ has a neighbourhood which is disjoint from $A$. Hence since $\pi_\alpha^{-1}(x_\alpha)$ is compact, there exists a finite subset $K$ of $\Omega$ and an open set $V_\beta$ of $X_\beta$ for each $\beta$ in $K$ such that $\pi_\alpha^{-1}(x_\alpha) \subset \bigcup_{\beta \in K} \pi_\beta^{-1}(V_\beta) \subset X \backslash A$. Choose $\gamma$ in $\Omega$ such that $\alpha \leqslant \gamma$ and $\beta \leqslant \gamma$ if $\beta \in K$, and let $W_\gamma = \bigcup_{\beta \in K} \pi_{\beta\gamma}^{-1}(V_\beta)$. Then

$$\pi_\alpha^{-1}(x_\alpha) \subset \pi_\gamma^{-1}(W_\gamma) \subset X \backslash A.$$

If $W_\alpha = X_\alpha \backslash \pi_{\alpha\gamma}(X_\gamma \backslash W_\gamma)$, then $W_\alpha$ is open in $X_\alpha$ since $\pi_{\alpha\gamma}$ is closed, $x_\alpha \in W_\alpha$ and $\pi_{\alpha\gamma}^{-1}(W_\alpha) \subset W_\gamma$. It follows that $W_\alpha \cap \pi_\alpha(A) = \varnothing$. Thus $\pi_\alpha(A)$ is a closed set in $X_\alpha$ as required. $\square$

# 2

## PARACOMPACT AND
## PSEUDO-METRIZABLE SPACES

### 1  Paracompact spaces

**1.1  *Definition.*** *A topological space $X$ is said to be* paracompact *if each open covering of $X$ has a locally finite open refinement.*

Clearly a compact space is paracompact. An infinite discrete space is an example of a paracompact space which is not compact.

**1.2  *Lemma.*** *If a covering $\{U_\lambda\}_{\lambda \in \Lambda}$ of a space $X$ has a locally finite open refinement, then there exists a locally finite open covering $\{V_\lambda\}_{\lambda \in \Lambda}$ of $X$ such that $V_\lambda \subset U_\lambda$ for each $\lambda$ in $\Lambda$.*

*Proof.* Let $\{W_\gamma\}_{\gamma \in \Gamma}$ be a locally finite open refinement of $\{U_\lambda\}_{\lambda \in \Lambda}$. Then there exists a function $\tau : \Gamma \to \Lambda$ such that $W_\gamma \subset U_{\tau(\gamma)}$ if $\gamma \in \Gamma$. Let $V_\lambda = \bigcup_{\tau(\gamma) = \lambda} W_\gamma$. Then $\{V_\lambda\}_{\lambda \in \Lambda}$ is an open covering of $X$ and each $V_\lambda \subset U_\lambda$. And $\{V_\lambda\}_{\lambda \in \Lambda}$ is locally finite. For if $x \in X$, there exists an open neighbourhood $G$ of $x$ such that $\Gamma' = \{\gamma \in \Gamma \mid G \cap W_\gamma \neq \varnothing\}$ is a finite set. But $G \cap V_\lambda \neq \varnothing$ if and only if $\lambda = \tau(\gamma)$ for some $\gamma$ in $\Gamma'$. Thus the set $\{\lambda \in \Lambda \mid G \cap V_\lambda \neq \varnothing\}$ is finite. Hence $\{V_\lambda\}_{\lambda \in \Lambda}$ is locally finite. $\square$

We now have the following characterization of paracompactness: a topological space $X$ is paracompact if and only if for each open covering $\{U_\lambda\}_{\lambda \in \Lambda}$ of $X$ there exists a locally finite open covering $\{V_\lambda\}_{\lambda \in \Lambda}$ of $X$ such that $V_\lambda \subset U_\lambda$ for each $\lambda$.

The next lemma allows a generalization of the result that a bicompact space is a $T_4$-space.

**1.3  *Lemma.*** *Let $X$ be a paracompact space, let $A$ be a subset of $X$ and let $B$ be a closed set of $X$ which is disjoint from $A$. If for every $x$ in $B$ there exist disjoint open sets $U_x$ and $V_x$ such that $A \subset U_x$ and $x \in V_x$, then there exist disjoint open sets $U$ and $V$ such that $A \subset U$ and $B \subset V$.*

*Proof.* The open covering of the paracompact space $X$ which consists of $X \backslash B$ together with the sets $V_x$ for $x$ in $B$ has a locally finite open refinement $\{W_\gamma\}_{\gamma \in \Gamma}$. Let

$$\Gamma_1 = \{\gamma \in \Gamma \mid W_\gamma \subset V_x \quad \text{for some} \quad x \text{ in } B\}.$$

If $\gamma \in \Gamma_1$, then $U_x \cap W_\gamma = \varnothing$ for some $x$ so that $A \cap \overline{W}_\gamma = \varnothing$. Now let

$$U = X \backslash \bigcup_{\gamma \in \Gamma_1} \overline{W}_\gamma \quad \text{and} \quad V = \bigcup_{\gamma \in \Gamma_1} W_\gamma.$$

Then $A \subset U$, $B \subset V$ and $U$ and $V$ are disjoint. Clearly $V$ is open. And $U$ is open, since $\{\overline{W}_\gamma\}_{\gamma \in \Gamma_1}$ is a locally finite family so that $\bigcup_{\gamma \in \Gamma_1} \overline{W}_\gamma$ is a closed set.$\square$

**1.4   Proposition.** *Each paracompact regular space is normal and each paracompact Hausdorff space is a $T_4$-space.*

*Proof.* Let $X$ be a paracompact regular space and let $A$ and $B$ be disjoint closed sets of $X$. Since $A$ is a closed set of the regular space $X$, for every $x$ in $B$ there exist disjoint open sets $U_x$ and $V_x$ such that $A \subset U_x$ and $x \in V_x$. It follows from Lemma 1.3 that there exist disjoint open sets $U$ and $V$ such that $A \subset U$ and $B \subset V$. Thus $X$ is a normal space. If $X$ is a paracompact Hausdorff space, $x_0 \in X$ and $B$ is a closed set such that $x_0 \notin B$, then for every $x$ in $B$ there exist disjoint open sets $U_x$ and $V_x$ such that $x_0 \in U_x$ and $x \in V_x$. It follows from Lemma 1.3 that there exist disjoint open sets $U$ and $V$ such that $x_0 \in U$ and $B \subset V$. Thus $X$ is a $T_3$-space and it follows from the first part of this proposition that $X$ is a $T_4$-space.$\square$

**1.5   Proposition.** *A topological space is paracompact and normal if and only if each open covering has a locally finite closed refinement.*

*Proof.* Let $X$ be a paracompact normal space and let $\mathcal{U} = \{U_\lambda\}_{\lambda \in \Lambda}$ be an open covering of $X$. Since $X$ is paracompact, there exists a locally finite open covering $\{V_\lambda\}_{\lambda \in \Lambda}$ of $X$ such that $V_\lambda \subset U_\lambda$ for each $\lambda$, and since $X$ is normal, by Proposition 1.3.9 there exists a closed covering $\mathcal{F} = \{F_\lambda\}_{\lambda \in \Lambda}$ such that $F_\lambda \subset V_\lambda$ for each $\lambda$. The covering $\mathcal{F}$ is a locally finite closed refinement of $\mathcal{U}$.

Now let $X$ be a space with the property that each open covering has a locally finite closed refinement. It follows from Proposition 1.3.9 that $X$ is a normal space. Let $\mathcal{U}$ be an open covering of $X$ and let $\mathcal{F} = \{F_\lambda\}_{\lambda \in \Lambda}$ be a locally finite closed refinement of $\mathcal{U}$. Since $\mathcal{F}$ is locally finite, each point $x$ of $X$ has a neighbourhood $W_x$ such that $\{\lambda \in \Lambda \mid W_x \cap F_\lambda \neq \varnothing\}$ is finite. If $\{E_\gamma\}_{\gamma \in \Gamma}$ is a locally finite closed refinement of the open covering $\{W_x\}_{x \in X}$ of $X$, then for each $\gamma$ in $\Gamma$ the set

$\{\lambda \in \Lambda \mid E_\gamma \cap F_\lambda \neq \varnothing\}$ is finite. It follows from Proposition 1.2.3 that there exists a locally finite family $\{V_\lambda\}_{\lambda \in \Lambda}$ of open sets such that $F_\lambda \subset V_\lambda$ for each $\lambda$. For each $\lambda$ in $\Lambda$, let $U_\lambda$ be a member of $\mathcal{U}$ such that $F_\lambda \subset U_\lambda$. Then $\{V_\lambda \cap U_\lambda\}_{\lambda \in \Lambda}$ is a locally finite open refinement of $\mathcal{U}$. Thus $X$ is a paracompact space. $\square$

**1.6 Proposition.** *A regular space is paracompact and normal if and only if each open covering has a $\sigma$-locally finite open refinement.*

*Proof.* The condition is evidently necessary so we must prove its sufficiency. Let $X$ be a regular space which satisfies the condition and let $\mathcal{U}$ be an open covering of $X$. Since $X$ is a regular space, there exists an open covering $\mathcal{V}$ of $X$ such that the closure of each member of $\mathcal{V}$ is contained in some member of $\mathcal{U}$. Let $\{V_{i\lambda}\}_{i \in \mathbf{N}, \lambda \in \Lambda}$ be an open refinement of $\mathcal{V}$ such that for each positive integer $i$, the family $\{V_{i\lambda}\}_{\lambda \in \Lambda}$ is locally finite. If $n \in \mathbf{N}$ and $\mu \in \Lambda$ let

$$A_{n\mu} = V_{n\mu} \backslash \bigcup_{i<n} \bigcup_{\lambda \in \Lambda} V_{i\lambda}.$$

It is clear that $\mathcal{A} = \{A_{n\mu}\}_{n \in \mathbf{N}, \mu \in \Lambda}$ is a covering of $X$ and a refinement of $\mathcal{V}$. And $\mathcal{A}$ is locally finite. For if $x \in X$, there exists a positive integer $n$ and an element $\mu$ of $\Lambda$ such that $x \in V_{n\mu}$ and $x \notin V_{i\lambda}$ if $i < n$ and $\lambda \in \Lambda$. Then $V_{n\mu}$ is an open neighbourhood of $x$ which is disjoint from $A_{k\lambda}$ if $k > n$ and $\lambda \in \Lambda$. Since the family $\{V_{i\lambda}\}_{\lambda \in \Lambda}$ is locally finite, for each $i \leqslant n$ there exists an open neighbourhood $G_i$ of $x$ such that $\{\lambda \in \Lambda \mid G_i \cap V_{i\lambda} \neq \varnothing\}$ is finite. If $G = V_{n\mu} \cap \bigcap_{i=1}^n G_i$, then $G$ is an open neighbourhood of $x$ which has non-empty intersection with $A_{i\lambda}$ for only finitely many pairs $(i, \lambda)$. It follows that $\{\bar{A}_{n\lambda}\}_{n \in \mathbf{N}, \lambda \in \Lambda}$ is a locally finite closed refinement of $\mathcal{U}$. Hence $X$ is a paracompact normal space by Proposition 1.5. $\square$

The next result establishes the paracompactness of pseudo-metric spaces.

**1.7 Proposition.** *If $\mathcal{U}$ is an open covering of a pseudo-metric space $X$, then there exists a closed covering $\{E_{n\lambda}\}_{n \in \mathbf{N}, \lambda \in \Lambda}$ of $X$ and an open covering $\mathcal{V} = \{V_{n\lambda}\}_{n \in \mathbf{N}, \lambda \in \Lambda}$ of $X$ such that $E_{n\lambda} \subset V_{n\lambda}$ for each $n, \lambda$, the family $\{V_{n\lambda}\}_{\lambda \in \Lambda}$ is discrete for each $n$, and $\mathcal{V}$ is a refinement of $\mathcal{U}$.*

*Proof.* Let $(X, d)$ be a pseudo-metric space and let $\mathcal{U} = \{U_\lambda\}_{\lambda \in \Lambda}$ be an open covering of $X$, where $\Lambda$ is a well-ordered set. If $n \in \mathbf{N}$ and $\lambda \in \Lambda$, let

$$F_{n\lambda} = \{x \in X \mid d(x, X \backslash U_\lambda) \geqslant 1/n\}.$$

Then $F_{n\lambda}$ is a closed set and $F_{n\lambda} \subset U_\lambda$. Let $E_{n\lambda} = F_{n\lambda} \backslash \bigcup_{\mu < \lambda} U_\mu$. Then $E_{n\lambda}$ is a closed set, and if $\lambda \neq \mu$ then $d(E_{n\lambda}, E_{n\mu}) \geqslant 1/n$. For suppose that $\mu < \lambda$. Then $E_{n\lambda} \subset X \backslash U_\mu$, so that if $y \in E_{n\mu}$ then $d(y, E_{n\lambda}) \geqslant 1/n$. Hence $d(E_{n\lambda}, E_{n\mu}) \geqslant 1/n$. Now let $V_{n\lambda} = B_{1/3n}(E_{n\lambda})$. Then $V_{n\lambda}$ is an open set. And $V_{n\lambda} \subset U_\lambda$, for if this were not the case then there would be $x$ in $E_{n\lambda}$ such that $d(x, X \backslash U_\lambda) < 1/3n$, which is impossible since $E_{n\lambda} \subset F_{n\lambda}$. And $d(V_{n\lambda}, V_{n\mu}) \geqslant 1/3n$ if $\lambda \neq \mu$. For if $x' \in V_{n\lambda}$ and $y' \in V_{n\mu}$, then there exist $x$ in $E_{n\lambda}$ and $y$ in $E_{n\mu}$ such that $d(x, x') < 1/3n$ and $d(y, y') < 1/3n$. Hence

$$d(x, y) \leqslant d(x, x') + d(x', y') + d(y', y) < 2/3n + d(x', y').$$

But $d(x, y) \geqslant 1/n$, so that if $x' \in V_{n\lambda}$ and $y' \in V_{n\mu}$ then $d(x', y') > 1/3n$. Thus $d(V_{n\lambda}, V_{n\mu}) \geqslant 1/3n$. It follows that if $n \in \mathbf{N}$ then the family $\{V_{n\lambda}\}_{\lambda \in \Lambda}$ is discrete. For let $x$ be a point of $X$ and suppose that $B_{1/6n}(x)$ meets $V_{n\lambda}$ and $V_{n\mu}$, where $\lambda \neq \mu$. If $y \in B_{1/6n}(x) \cap V_{n\lambda}$ and

$$z \in B_{1/6n}(x) \cap V_{n\mu},$$

then      $d(y, z) \leqslant d(y, x) + d(x, z) < 1/6n + 1/6n = 1/3n,$

which is absurd since $d(V_{n\lambda}, V_{n\mu}) \geqslant 1/3n$. To complete the proof it remains to show that $\{E_{n\lambda}\}_{n \in \mathbf{N}, \lambda \in \Lambda}$ is a covering of $X$. If $x \in X$, then there exists $\lambda$ in $\Lambda$ such that $x \in U_\lambda$ and $x \notin U_\mu$ if $\mu < \lambda$. Since $x \in U_\lambda$ it follows that $d(x, X \backslash U_\lambda) > 0$ and hence $x \in F_{n\lambda}$ for some $n$. Thus $x \in E_{n\lambda}$ as required. $\square$

**1.8   Corollary.** *A pseudo-metric space is paracompact.*

*Proof.* This follows immediately from Propositions 1.6 and 1.7. $\square$

For paracompact normal spaces we have the following stronger version of Proposition 1.3.14.

**1.9   Proposition.** *Let $\{F_\lambda\}_{\lambda \in \Lambda}$ be a locally finite family of closed sets of a paracompact normal space $X$. Then there exists a locally finite family $\{G_\lambda\}_{\lambda \in \Lambda}$ of open sets of $X$ such that $F_\lambda \subset G_\lambda$ for each $\lambda$ and the families $\{F_\lambda\}_{\lambda \in \Lambda}$ and $\{\bar{G}_\lambda\}_{\lambda \in \Lambda}$ are similar.*

*Proof.* Since the family $\{F_\lambda\}_{\lambda \in \Lambda}$ is locally finite there exists an open covering $\mathcal{U}$ of $X$, each member of which meets only a finite number of sets $F_\lambda$. Since $X$ is a paracompact normal space, $\mathcal{U}$ has a locally finite closed refinement $\mathcal{E}$, by Proposition 1.5. Each member of the locally finite closed covering $\mathcal{E}$ meets only a finite number of sets $F_\lambda$ so that, by Proposition 1.2.3, there exists a locally finite family $\{U_\lambda\}_{\lambda \in \Lambda}$ of open sets such that $F_\lambda \subset U_\lambda$ for each $\lambda$. Hence by Proposition 1.3.14 there exists a locally finite family $\{G_\lambda\}_{\lambda\in\Lambda}$ of open sets such that $F_\lambda \subset G_\lambda \subset \bar{G}_\lambda \subset U_\lambda$ and the families $\{F_\lambda\}_{\lambda\in\Lambda}$ and $\{\bar{G}_\lambda\}_{\lambda\in\Lambda}$ are similar. $\square$

A similar argument enables us to establish the next result.

**1.10 Proposition.** *Let $A$ be a closed subspace of a paracompact normal space $X$ and let $\{V_\lambda\}_{\lambda\in\Lambda}$ be a family, locally finite in $A$, of open sets of $A$. Then there exists a locally finite family $\{U_\lambda\}_{\lambda\in\Lambda}$ of open sets of $X$ such that $U_\lambda \cap A = V_\lambda$ for each $\lambda$.*

*Proof.* Since $A$ is closed in $X$, the family $\{V_\lambda\}_{\lambda\in\Lambda}$ is locally finite in $X$. Thus $\{\bar{V}_\lambda\}_{\lambda\in\Lambda}$ is a locally finite family of closed sets of $X$. Using the paracompactness of $X$ and Proposition 1.2.3 we find, as in the proof of the preceding proposition, a locally finite family $\{G_\lambda\}_{\lambda\in\Lambda}$ of open sets of $X$ such that $\bar{V}_\lambda \subset G_\lambda$ for each $\lambda$, and for each $\lambda$ there exists $W_\lambda$ open in $X$ such that $W_\lambda \cap A = V_\lambda$. If $U_\lambda = G_\lambda \cap W_\lambda$, then $\{U_\lambda\}_{\lambda\in\Lambda}$ is a locally finite family of open sets of $X$ and $U_\lambda \cap A = V_\lambda$ for each $\lambda$. $\square$

Subspaces of a paracompact space need not be paracompact. However we have:

**1.11 Proposition.** *A closed subspace of a paracompact space is paracompact.*

*Proof.* Let $A$ be a closed set in a paracompact space $X$ and let

$$\mathscr{V} = \{V_\lambda\}_{\lambda\in\Lambda}$$

be an open covering of the subspace $A$. Then for each $\lambda$ there exists $U_\lambda$ open in $X$ such that $V_\lambda = A \cap U_\lambda$. The open covering of $X$ which consists of the sets $U_\lambda, \lambda\in\Lambda$, together with $X\backslash A$ has a locally finite open refinement $\{W_\gamma\}_{\gamma\in\Gamma}$. The locally finite open covering $\{A \cap W_\gamma\}_{\gamma\in\Gamma}$ of $A$ is a refinement of $\mathscr{V}$. Thus $A$ is paracompact. $\square$

The product of normal spaces need not be normal and the product of paracompact spaces need not be paracompact. However the product of a paracompact (and normal) space with a compact (and normal) space is a paracompact (and normal) space. To establish these two assertions and for later use, we find locally finite refinements of a simple form for open coverings of the product of a paracompact space and a compact space.

**1.12    Lemma.** *If $\mathscr{U}$ is an open covering of the topological product $X \times Y$ of a paracompact space $X$ and a compact space $Y$, then $\mathscr{U}$ has a refinement of the form*

$$\{V_\lambda \times G_{i\lambda} \mid i = 1, \ldots, n_\lambda, \lambda \in \Lambda\},$$

*where $\{V_\lambda\}_{\lambda \in \Lambda}$ is a locally finite open covering of $X$, and for each $\lambda$, $\{G_{i\lambda} \mid i = 1, \ldots, n_\lambda\}$ is a finite open covering of $Y$.*

*Proof.* Let $x$ be a point of $X$. Since $Y$ is compact, there exist an open neighbourhood $W_x$ of $x$ and a finite open covering $\mathscr{G}_x$ of $Y$ such that $W_x \times G$ is contained in some member of $\mathscr{U}$ if $G \in \mathscr{G}_x$. Let $\{V_\lambda\}_{\lambda \in \Lambda}$ be a locally finite open refinement of the open covering $\{W_x\}_{x \in X}$ of the paracompact space $X$. For each $\lambda$ in $\Lambda$ choose $x$ in $X$ such that $V_\lambda \subset W_x$ and let $\mathscr{G}_x = \{G_{i\lambda} \mid i = 1, \ldots, n_\lambda\}$. Then $\{V_\lambda \times G_{i\lambda}\}$ is the required open refinement of $\mathscr{U}$. □

**1.13    Proposition.** *The topological product of a paracompact (normal) space and a compact (normal) space is a paracompact (normal) space.*

*Proof.* Let $X$ be a paracompact space, let $Y$ be a compact space and let $\mathscr{U}$ be an open covering of the topological product $X \times Y$. Then by Lemma 1.12, $\mathscr{U}$ has an open refinement of the form

$$\{V_\lambda \times G_{i\lambda} \mid i = 1, \ldots, n_\lambda, \lambda \in \Lambda\},$$

where $\{V_\lambda\}_{\lambda \in \Lambda}$ is a locally finite open covering of $X$ and

$$\{G_{i\lambda} \mid i = 1, \ldots, n_\lambda\}$$

is an open covering of $Y$ for each $\lambda$. This refinement is obviously locally finite. It follows that $X \times Y$ is a paracompact space. If $X$ and $Y$ are also normal then there exists a closed covering $\{F_\lambda\}_{\lambda \in \Lambda}$ of $X$ such that $F_\lambda \subset V_\lambda$ if $\lambda \in \Lambda$, and for each $\lambda$ there exists a closed covering $\{E_{i\lambda} \mid i = 1, \ldots, n_\lambda\}$ of $Y$ such that $E_{i\lambda} \subset G_{i\lambda}$ for each $i$. The covering $\{F_\lambda \times E_{i\lambda}\}$ of $X \times Y$ is a locally finite closed refinement of $\mathscr{U}$. Hence $X \times Y$ is a paracompact normal space. □

Next a useful characterization of paracompactness in terms of normality of a product will be established.

**1.14    Lemma.** *If $X$ is a topological space such that $X \times I^\tau$ is normal for some infinite cardinal number $\tau$ and $\mathscr{U} = \{U_\lambda\}_{\lambda \in \Lambda}$ is an open covering of $X$ such that $|\Lambda| = \tau$, then there is a closed refinement $\{F_\gamma\}_{\gamma \in \Gamma}$ of $\mathscr{U}$ such that $|\Gamma| = \tau$.*

*Proof.* The proof is by transfinite induction. Suppose that the result is true for all infinite cardinal numbers less than $\tau$. Let $\Lambda$ be well-ordered so that its order type is the first ordinal of cardinality equal to $\tau$, and let $\Omega = \Lambda \cup \{\xi\}$, where $\lambda < \xi$ if $\lambda \in \Lambda$. Let $\Omega$ be given the interval topology. Since $\Omega$ is a bicompact space of weight equal to $\tau$, $\Omega$ can be embedded in $I^\tau$ as a closed subspace. Hence $X \times \Omega$ is homeomorphic with a closed subspace of the normal space $X \times I^\tau$, so that $X \times \Omega$ is a normal space. If $\lambda \in \Lambda$, let $V_\lambda = \bigcup_{\mu \leqslant \lambda} U_\mu$ and consider the subsets

$$A = (X \times \Omega) \backslash \Big( \bigcup_{\lambda \in \Lambda} V_\lambda \times (\lambda, \xi] \Big) \quad \text{and} \quad B = X \times \{\xi\}$$

of $X \times \Omega$. If $x \in X$, then $x \in V_\lambda$ for some $\lambda$ so that $(x, \xi) \in V_\lambda \times (\lambda, \xi]$ and hence $(x, \xi) \notin A$. Thus $A$ and $B$ are disjoint closed sets of the normal space $X \times \Omega$. Hence there exist disjoint open sets $G$ and $H$ of $X \times \Omega$ such that $A \subset G$ and $B \subset H$. If $\lambda \in \Lambda$, let

$$E_\lambda = \{ x \in X \mid (x, \lambda) \notin G \}.$$

Then $E_\lambda$ is a closed set. And $\{E_\lambda\}_{\lambda \in \Lambda}$ is a closed covering of $X$. For if $x \in X$, then $(x, \xi) \in B \subset H$. Since $H$ is an open set, there exists some $\lambda$ such that $\lambda < \xi$ and $\{x\} \times (\lambda, \xi] \subset H$, and it follows that $x \in E_{\lambda+1}$. Furthermore $E_\lambda \subset V_\lambda$ for each $\lambda$. For suppose that $x \notin V_{\lambda_0}$. Then $x \notin V_\lambda$ if $\lambda \leqslant \lambda_0$ since $V_\lambda \subset V_{\lambda_0}$, and $\lambda_0 \notin (\lambda, \xi]$ if $\lambda > \lambda_0$. Thus

$$(x, \lambda_0) \notin V_\lambda \times (\lambda, \xi]$$

if $\lambda \in \Lambda$. It follows that $(x, \lambda_0) \in A \subset G$, so that $x \notin E_{\lambda_0}$. Thus $E_{\lambda_0} \subset V_{\lambda_0}$.

Now let $\lambda$ be an element of $\Lambda$ and let $\sigma$ be the cardinality of the set $\{\mu \in \Lambda \mid \mu \leqslant \lambda\}$. Then $\sigma < |\Lambda| = \tau$. Since $E_\lambda \times I^\sigma$ is homeomorphic with a closed subspace of $X \times I^\tau$ it follows that $E_\lambda \times I^\sigma$ is a normal space. Thus by the induction hypothesis there exists a closed covering $\{F_\gamma\}_{\gamma \in \Gamma(\lambda)}$ of $E_\lambda$ with $|\Gamma(\lambda)| = \sigma$ which is a refinement of the open covering $\{U_\mu \cap E_\lambda\}_{\mu \leqslant \lambda}$. We can suppose that if $\lambda, \mu \in \Lambda$ and $\lambda \neq \mu$ then $\Gamma(\lambda)$ and $\Gamma(\mu)$ are disjoint. If $\Gamma = \bigcup_{\lambda \in \Lambda} \Gamma(\lambda)$, then $|\Gamma| = |\Lambda| = \tau$ and $\{F_\gamma\}_{\gamma \in \Gamma}$ is the required closed refinement of $\mathscr{U}$.

Since $X$ is clearly normal, the above argument applies to the case in which $\tau$ is the first infinite cardinal $\aleph_0$ without the induction hypothesis. Thus the proof by transfinite induction is completed. $\square$

**1.15**   *Proposition. A topological space $X$ is paracompact and normal if and only if the topological product $X \times I^{w(X)}$ is normal.*

*Proof.* It follows from Proposition 1.13 that the condition is necessary. Let $X$ be a space such that $X \times I^{w(X)}$ is normal. Then, since $X$ is homeomorphic with a closed subspace of $X \times I^{w(X)}$, $X$ is normal. Let us consider the copy of $I^{w(X)}$ indexed by a set $\Lambda$ and for each $\lambda$ in $\Lambda$, let $\pi_\lambda \colon I^{w(X)} \to I$ be the projection. If $\mathscr{U}$ is an open covering of $X$, it follows from Lemma 1.14 that $\mathscr{U}$ has a closed refinement $\{F_\lambda\}_{\lambda \in \Lambda}$. For each $\lambda$, choose a member $U_\lambda$ of $\mathscr{U}$ such that $F_\lambda \subset U_\lambda$. Since $X$ is normal, for each $\lambda$ there exists a continuous function $\phi_\lambda \colon X \to I$ such that $\phi_\lambda(x) = 0$ if $x \in F_\lambda$ and $\phi_\lambda(x) = 1$ if $x \notin U_\lambda$. Let $\phi \colon X \to I^{w(X)}$ be the unique continuous function such that $\pi_\lambda \circ \phi = \phi_\lambda$ for every $\lambda$ in $\Lambda$. If $\lambda \in \Lambda$, let $V_\lambda$ be the open set $\pi_\lambda^{-1}([0, 1))$ of $I^{w(X)}$ and let $H = \bigcup_{\lambda \in \Lambda} V_\lambda$. Then $\phi(X) \subset H$ and $F_\lambda \subset \phi^{-1}(V_\lambda) \subset U_\lambda$ for each $\lambda$. Now let

$$D = \{(x, y) \in X \times I^{w(X)} \mid y = \phi(x)\}.$$

Since $I^{w(X)}$ is a Hausdorff space, it follows from Proposition 1.6.3 that $D$ is closed in $X \times I^{w(X)}$, and it is clear that $D$ is contained in the open set $X \times H$. Thus since $X \times I^{w(X)}$ is normal, there exists a continuous function $\Phi \colon X \times I^{w(X)} \to I$ such that $\Phi(x, y) = 1$ if $(x, y) \notin X \times H$ and $\Phi(x, y) = 0$ if $(x, y) \in D$. Since $I^{w(X)}$ is compact, we can define a pseudo-metric $d$ on $X$ by putting

$$d(x, x') = \sup_{y \in I^{w(X)}} |\Phi(x, y) - \Phi(x', y)|$$

if $x, x' \in X$. It is easily checked that $d$ is a continuous function on $X \times X$ using the fact that a continuous real-valued function on a compact space is bounded and attains its bounds. It follows that the pseudo-metric topology induced on the set $X$ by $d$ is smaller than the given topology of $X$. For each $x$ in $X$ let $W_x = \{x' \in X \mid d(x, x') < \frac{1}{2}\}$. Since a pseudo-metric space is paracompact, it follows that $\{W_x\}_{x \in X}$ is an open covering of the space $X$ which has a refinement $\{G_\gamma\}_{\gamma \in \Gamma}$ which is locally finite and open in the pseudo-metric topology, and thus is locally finite and open in the given topology of $X$. If $x, x' \in X$, then

$$\Phi\big(x, \phi(x')\big) = \Phi\big(x, \phi(x')\big) - \Phi\big(x', \phi(x')\big) \leqslant d(x, x').$$

It follows that if $x' \in W_x$, then $\Phi\big(x, \phi(x')\big) < \frac{1}{2}$ and hence

$$\big(\phi(W_x)\big)^- \subset \{y \in Y \mid \Phi(x, y) \leqslant \tfrac{1}{2}\} \subset H.$$

Now suppose that $\gamma \in \Gamma$. Then there exists $x$ in $X$ such that $G_\gamma \subset W_x$ so that $\big(\phi(G_\gamma)\big)^- \subset \big(\phi(W_x)\big)^- \subset H = \bigcup_{\lambda \in \Lambda} V_\lambda$. But $(\phi(G_\gamma))^-$ is a closed set in the bicompact space $I^{w(X)}$, consequently is bicompact, and hence there exists a finite subset $\Lambda(\gamma)$ of $\Lambda$ such that

$$\phi(G_\gamma) \subset \big(\phi(G_\gamma)^-\big) \subset \bigcup_{\lambda \in \Lambda(\gamma)} V_\lambda.$$

It follows that
$$G_\gamma \subset \bigcup_{\lambda \in \Lambda(\gamma)} U_\lambda.$$

Thus the covering $\{G_\gamma \cap U_\lambda \mid \lambda \in \Lambda(\gamma),\ \gamma \in \Gamma\}$ of $X$ is a locally finite open refinement of $\mathscr{U}$. Hence $X$ is paracompact. $\square$

An application of the preceding characterization is in the proof that a space dominated by paracompact normal subspaces is paracompact and normal.

**1.16  Lemma.** *Let $X$ be a space which is dominated by a covering $\{F_\lambda\}_{\lambda \in \Lambda}$ and let $Y$ be a compact regular space. Then the topological product $X \times Y$ is dominated by the covering $\{F_\lambda \times Y\}_{\lambda \in \Lambda}$.*

*Proof.* If $M \subset \Lambda$ then $\bigcup_{\lambda \in M}(F_\lambda \times Y) = (\bigcup_{\lambda \in M} F_\lambda) \times Y$ so that
$$\bigcup_{\lambda \in M}(F_\lambda \times Y)$$
is closed in $X \times Y$. Let $G$ be a subset of $Z = \bigcup_{\lambda \in M}(F_\lambda \times Y)$ such that $G \cap (F_\lambda \times Y)$ is open in $F_\lambda \times Y$ if $\lambda \in M$. Suppose that $(x_0, y_0) \in G$ and $x_0 \in F_{\lambda_0}$, where $\lambda_0 \in M$. If
$$H = \{y \in Y \mid (x_0, y) \in G\},$$
then $H$ is open in $Y$ since $G \cap (F_{\lambda_0} \times Y)$ is open in $F_{\lambda_0} \times Y$. Since $y_0 \in H$ and $Y$ is a regular space, there exists an open set $W$ of $Y$ such that $y_0 \in W \subset \overline{W} \subset H$. Let
$$V = \{x \in \bigcup_{\lambda \in M} F_\lambda \mid \{x\} \times \overline{W} \subset G\}.$$
If $\lambda \in M$ then
$$V \cap F_\lambda = \{x \in F_\lambda \mid \{x\} \times \overline{W} \subset G \cap (F_\lambda \times Y)\}.$$

Since $G \cap (F_\lambda \times Y)$ is open in $F_\lambda \times F$ and $\overline{W}$ is compact, it follows that $V \cap F_\lambda$ is open in $F_\lambda$. Thus $V$ is open in $\bigcup_{\lambda \in M} F_\lambda$. Thus $V \times W$ is open in $Z$ and $(x_0, y_0) \in V \times W \subset G$. Hence $G$ is open in $Z$. Therefore $Z$ has the weak topology with respect to the covering $\{F_\lambda \times Y\}_{\lambda \in M}$. $\square$

**1.17  Proposition.** *If a topological space $X$ is dominated by a covering $\{F_\lambda\}_{\lambda \in \Lambda}$, where each subspace $F_\lambda$ is paracompact and normal, then $X$ is a paracompact normal space.*

*Proof.* It follows from Lemma 1.16 that the topological product $X \times I^{w(X)}$ is dominated by the covering $\{F_\lambda \times I^{w(X)}\}_{\lambda \in \Lambda}$. But $F_\lambda \times I^{w(X)}$ is a normal space for each $\lambda$, by Proposition 1.13. Thus $X \times I^{w(X)}$ is a normal space by Proposition 1.4.3. It now follows from Proposition 1.15 that $X$ is a paracompact normal space. $\square$

**1.18  Definition.** *A topological space is said to be* countably para-compact *if each countable open covering has a locally finite open refinement.*

It follows from Lemma 1.2 that a space is countably paracompact if and only if for each open covering $\{U_i\}_{i \in \mathbb{N}}$ there exists a locally finite open covering $\{V_i\}_{i \in \mathbb{N}}$ such that $V_i \subset U_i$ for every $i$.

**1.19  Proposition.** *A perfectly normal space is countably para-compact.*

*Proof.* Let $X$ be a perfectly normal space and let $\{U_n\}_{n \in \mathbb{N}}$ be a countable open covering of $X$. For each $n$ there exists a continuous real-valued function $f_n$ such that $f_n(x) \geqslant 0$ if $x \in X$, and $f_n(x) > 0$ if and only if $x \in U_n$. Let

$$G_{ni} = \{x \in X \mid f_n(x) > 1/i\}$$

and

$$F_{ni} = \{x \in X \mid f_n(x) \geqslant 1/i\}.$$

Then $G_{ni}$ is an open set of $X$, $F_{ni}$ is a closed set and

$$G_{ni} \subset F_{ni} \subset G_{n, i+1}.$$

Furthermore $\bigcup_{i \in \mathbb{N}} G_{ni} = \bigcup_{i \in \mathbb{N}} F_{ni} = U_n$. Now let

$$V_n = U_n \setminus \bigcup_{r < n} F_{rn}.$$

Then $V_n$ is open in $X$. And if $x \in X$, there exists $n_0$ such that $x \in U_{n_0}$ and $x \notin U_r$ if $r < n_0$. It follows that $x \notin F_{rn_0}$ if $r < n_0$ so that $x \in V_{n_0}$. Hence $\{V_n\}_{n \in \mathbb{N}}$ is an open covering of $X$ and it is clear that $V_n \subset U_n$. If $x \in X$, then $x \in U_{n_0}$ for some $n_0$ so that $x \in G_{n_0 i_0}$ for some $i_0$. And if $y \in G_{n_0 i_0}$, then $y \in F_{n_0 i}$ for all $i \geqslant i_0$. Thus if $i > \max\{n_0, i_0\}$, then $y \in F_{n_0 i}$ and hence $y \in \bigcup_{r < i} F_{ri}$. But $V_i = U_i \setminus \bigcup_{r < i} F_{ri}$ so that $y \notin V_i$ if $i > \max\{n_0, i_0\}$. Hence $\{V_n\}_{n \in \mathbb{N}}$ is a locally finite covering of $X$. $\square$

We can characterize the spaces which are countably paracompact and normal.

**1.20  Proposition.** *A topological space is countably paracompact and normal if and only if each countable open covering is shrinkable.*

*Proof.* If $X$ is a countably paracompact normal space, then for each countable open covering $\{U_i\}_{i \in \mathbb{N}}$ there exists a locally finite open covering $\{V_i\}_{i \in \mathbb{N}}$ such that $V_i \subset U_i$ for every $i$. And since $X$ is normal, this locally finite open covering is shrinkable by Proposition 1.3.9. Hence $\{U_i\}$ is shrinkable.

Conversely let $X$ be a space with the property that each countable open covering is shrinkable. Then $X$ is normal by Proposition 1.3.9. Let $\{U_i\}_{i\in\mathbf{N}}$ be a countable open covering of $X$. Then there exists an open covering $\{W_i\}_{i\in\mathbf{N}}$ such that $\overline{W}_i \subset U_i$ for every $i$. If $i\in\mathbf{N}$, let

$$V_i = U_i \setminus \bigcup_{j<i} \overline{W}_j.$$

Then $\mathscr{V} = \{V_i\}_{i\in\mathbf{N}}$ is an open covering of $X$. For if $x\in X$, there exists a positive integer $i$ such that $x\in U_i$ and $x\notin U_j$ if $j<i$, so that $x\in V_i$. Furthermore, $\mathscr{V}$ is locally finite, for if $x\in X$, there exists $j$ such that $x\in W_j$, and $W_j \cap V_i = \varnothing$ if $i>j$. Clearly $V_i \subset U_i$ for each $i$. Thus $X$ is countably paracompact.$\square$

Next we shall find a sufficient condition for the paracompactness of a countably paracompact normal space.

**1.21  Lemma.** *If $\{V_\lambda\}_{\lambda\in\Lambda}$ is a $\sigma$-locally finite open covering of a countably paracompact space $X$, then there exists a locally finite open covering $\{W_\lambda\}_{\lambda\in\Lambda}$ of $X$ such that $W_\lambda \subset V_\lambda$ for each $\lambda$.*

*Proof.* Let $\Lambda = \bigcup_{i\in\mathbf{N}}\Lambda_i$, where the sets $\Lambda_i$ are disjoint and $\{V_\lambda\}_{\lambda\in\Lambda_i}$ is a locally finite family for each $i$. Let $V_i = \bigcup_{\lambda\in\Lambda_i} V_\lambda$; then $\{V_i\}_{i\in\mathbf{N}}$ is a countable open covering of $X$, so that there is a locally finite open covering $\{W_i\}_{i\in\mathbf{N}}$ such that $W_i \subset V_i$ for each $i$. If $\lambda\in\Lambda_i$, let $W_\lambda = W_i \cap V_\lambda$; then $\{W_\lambda\}_{\lambda\in\Lambda}$ is an open covering of $X$ and $W_\lambda \subset V_\lambda$ for each $\lambda$. If $x\in X$, there is an open set $G_0$ such that $x\in G_0$ and

$$\{i\in\mathbf{N} \mid G_0 \cap W_i \neq \varnothing\} = \{i_1,\ldots,i_m\}.$$

And for each $r = 1,\ldots,m$ there exists an open set $G_r$ of $X$ such that $x\in G_r$, and $\{\lambda\in\Lambda_{i_r} \mid G_r \cap V_\lambda \neq \varnothing\}$ is a finite subset $K_r$ of $\Lambda_{i_r}$. Let

$$G = G_0 \cap \ldots \cap G_m.$$

Then $G$ is an open set and $x\in G$, and if $i\notin\{i_1,\ldots,i_m\}$, then $G\cap W_\lambda = \varnothing$ if $\lambda\in\Lambda_i$, whilst for $r = 1,\ldots,m$, $G\cap W_\lambda = \varnothing$ if $\lambda\in\Lambda_{i_r}\setminus K_r$. Hence $\{W_\lambda\}_{\lambda\in\Lambda}$ is a locally finite family.$\square$

**1.22  Proposition.** *If $X$ is a countably paracompact normal space such that each open covering has an open refinement of the form*

$$\{V_{i\lambda}\}_{i\in\mathbf{N},\,\lambda\in\Lambda},$$

*where for each $i$ the family $\{V_{i\lambda}\}_{\lambda\in\Lambda}$ is disjoint, then $X$ is paracompact.*

*Proof.* Let $\mathcal{U}$ be an open covering of $X$ and let $\{V_{i\lambda}\}_{i\in\mathbf{N},\,\lambda\in\Lambda}$ be an open refinement of $\mathcal{U}$ such that each family $\{V_{i\lambda}\}_{\lambda\in\Lambda}$ is disjoint. If

$$V_i = \bigcup_{\lambda\in\Lambda} V_{i\lambda},$$

then $\{V_i\}_{i\in\mathbf{N}}$ is a countable open covering of $X$. Since $X$ is countably paracompact and normal, by Proposition 1.20 there exists a locally finite open covering $\{W_i\}_{i\in\mathbf{N}}$ of $X$ such that each $\overline{W_i} \subset V_i$. Let

$$W_{i\lambda} = W_i \cap V_{i\lambda}.$$

Then $\mathcal{W} = \{W_{i\lambda}\}_{i\in\mathbf{N},\,\lambda\in\Lambda}$ is an open covering of $X$ and a refinement of $\mathcal{U}$. For a given positive integer $i$, the sets $V_{i\lambda}$, $\lambda\in\Lambda$, together with the set $X\backslash\overline{W_i}$ form an open covering of $X$, and $V_{i\lambda} \cap W_{i\mu} = \varnothing$ if $\mu \neq \lambda$ whilst $(X\backslash\overline{W_i}) \cap W_{i\mu} = \varnothing$ for all $\mu$. Hence for each $i$, the family $\{W_{i\lambda}\}_{\lambda\in\Lambda}$ is discrete. It follows from Lemma 1.21 that $\mathcal{W}$ has a locally finite open refinement. Hence $\mathcal{U}$ has a locally finite open refinement. Thus $X$ is a paracompact space.$\square$

**1.23 Definition.** *A* cushioned refinement *of an open covering* $\{U_\lambda\}_{\lambda\in\Lambda}$ *of a space* $X$ *is a closed covering* $\{E_\lambda\}_{\lambda\in\Lambda}$ *of* $X$ *such that for each subset* $M$ *of* $\Lambda$

$$\left(\bigcup_{\lambda\in M} E_\lambda\right)^{-} \subset \bigcup_{\lambda\in M} U_\lambda.$$

**1.24 Proposition.** *A topological space is paracompact and normal if and only if each open covering has a cushioned refinement.*

*Proof.* If $\mathcal{U} = \{U_\lambda\}_{\lambda\in\Lambda}$ is an open covering of a paracompact normal space $X$, then there exists a locally finite closed covering $\mathcal{E} = \{E_\lambda\}_{\lambda\in\Lambda}$ of $X$ such that $E_\lambda \subset U_\lambda$ for each $\lambda$. The covering $\mathcal{E}$ is evidently a cushioned refinement of $\mathcal{U}$.

Now let $X$ be a space with the property that each open covering has a cushioned refinement. Then each finite open covering of $X$ has a finite closed refinement so that $X$ is a normal space by Proposition 1.3.9. Next $X$ is a countably paracompact space for if $\{U_i\}_{i\in\mathbf{N}}$ is a countable open covering of $X$, then there exists a closed covering $\{F_i\}_{i\in\mathbf{N}}$ such that $F_i \subset U_i$ for each $i$. Since $X$ is a normal space, for each $i$ there exists an open set $V_i$ such that $F_i \subset V_i \subset \overline{V_i} \subset U_i$. Then $\{V_i\}_{i\in\mathbf{N}}$ is an open covering of $X$ such that $\overline{V_i} \subset U_i$ for each $i$. It follows from Proposition 1.20 that $X$ is a countably paracompact space. The proof will now be completed by appealing to Proposition 1.22. Let $\{U_\lambda\}_{\lambda\in\Lambda}$ be an open covering of $X$ and let us suppose that $\Lambda$ is well-ordered. We shall construct by induction closed coverings

$\{E_{i\lambda}\}_{\lambda\in\Lambda}$ for each positive integer $i$, each of which is a cushioned refinement of $\{U_\lambda\}$, with the property that if $i\in\mathbb{N}$ and $\lambda\in\Lambda$ then:

(i) $(\bigcup_{\mu<\lambda}E_{i\mu})^- \cap E_{i+1,\lambda} = \varnothing$;

(ii) $E_{i\lambda} \cap (\bigcup_{\mu>\lambda}E_{i+1,\mu})^- = \varnothing$.

Let $\{E_{1\lambda}\}_{\lambda\in\Lambda}$ be a closed covering of $X$ which is a cushioned refinement of $\{U_\lambda\}$. Suppose that suitable closed coverings $\{E_{i\lambda}\}_{\lambda\in\Lambda}$ have been constructed for $i \leqslant n$. If $\lambda\in\Lambda$, let

$$U_{n\lambda} = U_\lambda \backslash (\bigcup_{\mu<\lambda} E_{n\mu})^-.$$

Then $U_{n\lambda}$ is open and $U_{n\lambda} \subset U_\lambda$, and $\{U_{n\lambda}\}_{\lambda\in\Lambda}$ is a covering of $X$, for if $x\in X$, then $x\in U_{n\lambda}$ for the first $\lambda$ such that $x\in U_\lambda$ since

$$(\bigcup_{\mu<\lambda} E_{n\mu})^- \subset \bigcup_{\mu<\lambda} U_\mu.$$

There exists a covering $\{E_{n+1,\lambda}\}_{\lambda\in\Lambda}$ which is a cushioned refinement of $\{U_{n\lambda}\}_{\lambda\in\Lambda}$ and so of $\{U_\lambda\}_{\lambda\in\Lambda}$. It follows that $E_{n+1,\lambda} \subset U_{n\lambda}$ so that

$$E_{n+1,\lambda} \cap (\bigcup_{\mu<\lambda} E_{n\mu})^- = \varnothing.$$

And if $\mu > \lambda$, then $E_{n\lambda} \cap U_{n\mu} = \varnothing$ so that

$$E_{n\lambda} \cap (\bigcup_{\mu>\lambda} E_{n+1,\mu})^- = \varnothing$$

since $(\bigcup_{\mu>\lambda}E_{n+1,\mu})^- \subset \bigcup_{\mu>\lambda}U_{n\mu}$. Thus conditions (i) and (ii) are satisfied. By induction we can construct the required sequence of closed coverings. Now for each $i, \lambda$ let

$$V_{i\lambda} = X\backslash (\bigcup_{\mu\neq\lambda} E_{i\mu})^-.$$

Then $V_{i\lambda}$ is open and $V_{i\lambda} \subset E_{i\lambda} \subset U_\lambda$. Since $V_{i\lambda} \subset E_{i\lambda}$ and $V_{i\lambda} \cap E_{i\mu} = \varnothing$ if $\mu \neq \lambda$, the family $\{V_{i\lambda}\}_{\lambda\in\Lambda}$ is disjoint for each $i$. In view of Proposition 1.22, the proof will be complete if we show that the family

$$\{V_{i\lambda}\}_{i\in\mathbb{N},\,\lambda\in\Lambda}$$

is a covering of $X$. If $x\in X$, then for each positive integer $i$, there exists $\lambda$ such that $x\in E_{i\lambda}$. Let $\lambda_i$ be the first element of the set

$$\{\lambda\in\Lambda \mid x\in E_{i\lambda}\}$$

and choose $m$ such that $\lambda_m \leqslant \lambda_i$ if $i\in\mathbb{N}$. Then $x\in E_{m\lambda_m}$ so that

$$x \notin (\bigcup_{\lambda>\lambda_m} E_{m+1,\lambda})^-$$

by condition (ii). And $x \in E_{m+2,\lambda}$ for some $\lambda \geqslant \lambda_m$ so that

$$x \notin (\bigcup_{\lambda < \lambda_m} E_{m+1,\lambda})^-$$

by condition (i). It follows that $x \in V_{m+1,\lambda_m}$. Thus $\{V_{i\lambda}\}_{i \in \mathbf{N}, \lambda \in \Lambda}$ is a covering of $X$. $\square$

As an application of the preceding result we can prove:

**1.25   Proposition.** *If $X$ is a paracompact normal space and $f : X \to Y$ is a continuous closed surjection, then $Y$ is a paracompact normal space.*

*Proof.* Let $\mathscr{U} = \{U_\lambda\}_{\lambda \in \Lambda}$ be an open covering of $Y$. Then

$$f^{-1}(\mathscr{U}) = \{f^{-1}(U_\lambda)\}_{\lambda \in \Lambda}$$

is an open covering of $X$ so that by Proposition 1.24 there exists a closed covering $\{E_\lambda\}_{\lambda \in \Lambda}$ of $X$ which is a cushioned refinement of $f^{-1}(\mathscr{U})$. Since $f$ is a closed surjection, $\{f(E_\lambda)\}_{\lambda \in \Lambda}$ is a closed covering of $Y$. Moreover $(f(A))^- = f(\bar{A})$ for each subset $A$ of $X$. Thus if $M$ is a subset of $\Lambda$, then since $(\bigcup_{\lambda \in M} E_\lambda)^- \subset \bigcup_{\lambda \in M} f^{-1}(U_\lambda) = f^{-1}(\bigcup_{\lambda \in M} U_\lambda)$, it follows that

$$\left( \bigcup_{\lambda \in M} f(E_\lambda) \right)^- = \left( f(\bigcup_{\lambda \in M} E_\lambda) \right)^- = f\left( (\bigcup_{\lambda \in M} E_\lambda)^- \right) \subset \bigcup_{\lambda \in M} U_\lambda.$$

Thus $\mathscr{U}$ has a cushioned refinement and it follows from Proposition 1.24 that $Y$ is a paracompact normal space. $\square$

If $\mathscr{B} = \{B_\gamma\}_{\gamma \in \Gamma}$ is a family of subsets of a space $X$ and $A$ is a subset of $X$, we write $\mathrm{St}\,(A, \mathscr{B})$ to denote the union of those sets $B_\gamma$ such that $A \cap B_\gamma \neq \varnothing$, and if $x$ is a point of $X$ we write $\mathrm{St}\,(x, \mathscr{B})$ to denote the union of those sets $B_\gamma$ such that $x \in B_\gamma$.

**1.26   Definition.** *A covering $\mathscr{V} = \{V_\lambda\}_{\lambda \in \Lambda}$ of a space $X$ is called a star-refinement of a covering $\mathscr{U}$ if the covering $\{\mathrm{St}\,(x, \mathscr{V})\}_{x \in X}$ is a refinement of $\mathscr{U}$. The covering $\mathscr{V}$ is called a strong star-refinement of $\mathscr{U}$ if the covering $\{\mathrm{St}\,(V_\lambda, \mathscr{V})\}_{\lambda \in \Lambda}$ is a refinement of $\mathscr{U}$.*

**1.27   Proposition.** *The following statements about a topological space $X$ are equivalent:*
   (a) *$X$ is paracompact and normal;*
   (b) *each open covering of $X$ has an open star-refinement;*
   (c) *each open covering of $X$ has an open strong star-refinement.*

*Proof.* $(a) \Rightarrow (b)$. Let $X$ be a paracompact normal space and let $\mathscr{U} = \{U_\lambda\}_{\lambda \in \Lambda}$ be an open covering of $X$. There exists a closed covering $\{F_\lambda\}_{\lambda \in \Lambda}$ of $X$ and a locally finite open covering $\{V_\lambda\}_{\lambda \in \Lambda}$ such that $F_\lambda \subset V_\lambda \subset U_\lambda$ for each $\lambda$. For each $x$ in $X$, choose an open neighbourhood $V_x$ of $x$ such that

$$\Lambda_x = \{\lambda \in \Lambda \mid V_x \cap V_\lambda \neq \varnothing\}$$

is finite. Let

$$\Lambda_x' = \{\lambda \in \Lambda_x \mid x \in V_\lambda\} \quad \text{and} \quad \Lambda_x'' = \{\lambda \in \Lambda_x \mid x \notin F_\lambda\}.$$

Clearly $\Lambda_x' \cup \Lambda_x'' = \Lambda_x$. For each $x$ in $X$, let

$$W_x = V_x \cap \left( \bigcap_{\lambda \in \Lambda_x'} V_\lambda \right) \cap \left( \bigcap_{\lambda \in \Lambda_x''} (X \backslash F_\lambda) \right).$$

Then $\mathscr{W} = \{W_x\}_{x \in X}$ is an open covering of $X$. If $x \in X$, there exists $\lambda_0$ in $\Lambda$ such that $x \in F_{\lambda_0}$. If $x \in W_y$, then $W_y \cap F_{\lambda_0} \neq \varnothing$ so that $\lambda_0 \in \Lambda_y$ and $\lambda_0 \notin \Lambda_y''$. Thus $\lambda_0 \in \Lambda_y'$ so that $W_y \subset V_{\lambda_0}$. Thus $\mathrm{St}\,(x, \mathscr{W}) \subset V_{\lambda_0} \subset U_{\lambda_0}$. Hence $\mathscr{W}$ is a star-refinement of $\mathscr{U}$.

$(b) \Rightarrow (c)$. Let $X$ be a space such that each open covering has an open star-refinement. If $\mathscr{U}$ is an open covering of $X$, then there exist open coverings $\mathscr{W}$ and $\mathscr{V} = \{V_\gamma\}_{\gamma \in \Gamma}$ such that $\mathscr{W}$ is a star-refinement of $\mathscr{U}$ and $\mathscr{V}$ is a star-refinement of $\mathscr{W}$. Let $\gamma_0 \in \Gamma$. If $\gamma \in \Gamma$ and

$$V_\gamma \cap V_{\gamma_0} \neq \varnothing,$$

then since $\mathscr{V}$ is a star-refinement of $\mathscr{W}$ there exists some member of $\mathscr{W}$ which contains $V_\gamma \cup V_{\gamma_0}$. Thus if $x \in V_{\gamma_0}$ then

$$\mathrm{St}\,(V_{\gamma_0}, \mathscr{V}) \subset \mathrm{St}\,(x, \mathscr{W}).$$

Since $\mathscr{W}$ is a star-refinement of $\mathscr{U}$ it follows that $\mathscr{V}$ is a strong star-refinement of $\mathscr{U}$.

$(c) \Rightarrow (a)$. Let $X$ be a space such that each open covering has an open strong star-refinement. Let $\mathscr{U} = \{U_\lambda\}_{\lambda \in \Lambda}$ be an open covering of $X$. There exists an open covering $\{V_\gamma\}_{\gamma \in \Gamma}$ of $X$ which is a strong star-refinement of $\mathscr{U}$. For each $\gamma$ in $\Gamma$, choose $\tau(\gamma)$ in $\Lambda$ such that

$$\mathrm{St}\,(V_\gamma, \mathscr{V}) \subset U_{\tau(\gamma)}$$

and if $\lambda \in \Lambda$ let

$$E_\lambda = \left( \bigcup_{\tau(\gamma) = \lambda} V_\gamma \right)^-.$$

Then $\{E_\lambda\}_{\lambda \in \Lambda}$ is a closed covering of $X$. Suppose that $M \subset \Lambda$ and that $x \in (\bigcup_{\lambda \in M} E_\lambda)^-$. There exists $\alpha$ in $\Gamma$ such that $x \in V_\alpha$. Thus

$$V_\alpha \cap \left( \bigcup_{\lambda \in M} E_\lambda \right) \neq \varnothing$$

so that

$$V_\alpha \cap ( \bigcup_{\tau(\gamma)=\lambda} V_\lambda)^- \neq \varnothing$$

for some $\lambda$ in $M$, from which it follows that $V_\alpha \cap V_\gamma \neq \varnothing$ for some $\gamma$ such that $\tau(\gamma) = \lambda$. Thus $x \in V_\alpha \subset \operatorname{St}(V_\gamma, \mathscr{V}) \subset U_\lambda$. Thus

$$(\bigcup_{\lambda \in M} E_\lambda)^- \subset \bigcup_{\lambda \in M} U_\lambda.$$

Hence $\{E_\lambda\}_{\lambda \in \Lambda}$ is a cushioned refinement of $\{U_\lambda\}_{\lambda \in \Lambda}$. It follows from Proposition 1.24 that $X$ is a paracompact normal space. $\square$

## 2  Hereditarily paracompact spaces, weakly, strongly and completely paracompact spaces

A subspace of a paracompact space need not be paracompact. By Proposition 1.11, closed subspaces of a paracompact space are paracompact. We can extend this result to normally situated sets in a paracompact regular space.

**2.1  Lemma.** *An $F_\sigma$-set in a paracompact regular space is paracompact.*

*Proof.* Let $A$ be an $F_\sigma$-set in a paracompact regular space $X$ and suppose that $A = \bigcup_{n \in \mathbf{N}} A_n$, where each $A_n$ is a closed set. Let

$$\mathscr{V} = \{V_\lambda\}_{\lambda \in \Lambda}$$

be an open covering of $A$ and let $V_\lambda = U_\lambda \cap A$ for each $\lambda$, where $U_\lambda$ is an open set of $X$. The open covering of $X$ which consists of the sets $U_\lambda$ for $\lambda$ in $\Lambda$, together with $X \backslash A_n$, has a locally finite refinement which consists of open sets $W_{n\lambda}$ for $\lambda$ in $\Lambda$ such that $W_{n\lambda} \subset U_\lambda$, together with an open set $W_n$ such that $W_n \subset X \backslash A_n$. Then $\{W_{n\lambda} \cap A\}_{n \in \mathbf{N}, \lambda \in \Lambda}$ is a $\sigma$-locally finite open covering of $A$, and $W_{n\lambda} \cap A \subset V_\lambda$ for each $n, \lambda$. Since $A$ is a regular space, it follows from Proposition 1.6 that $A$ is paracompact. $\square$

**2.2  Proposition.** *A normally situated set in a paracompact regular space is paracompact.*

*Proof.* Let $M$ be normally situated in the paracompact regular space $X$ and let $\{V_\lambda\}_{\lambda \in \Lambda}$ be an open covering of the subspace $M$. For each $\lambda$, there exists $U_\lambda$ open in $X$ such that $V_\lambda = M \cap U_\lambda$. Thus $\bigcup_{\lambda \in \Lambda} U_\lambda$ is an open neighbourhood of $M$ in $X$. Hence $M \subset G \subset \bigcup_{\lambda \in \Lambda} U_\lambda$, where

$$G = \bigcup_{\gamma \in \Gamma} G_\gamma,$$

each $G_\gamma$ is an open $F_\sigma$-set of $X$ and the family $\{G_\gamma\}_{\gamma\in\Gamma}$ is locally finite in $G$. Since $G$ is normally situated in the paracompact regular and hence normal space $X$, the subspace $G$ is normal by Proposition 1.4.8. Hence there exists a covering $\{F_\gamma\}_{\gamma\in\Gamma}$ of $G$ by closed sets of $G$ such that $F_\gamma \subset G_\gamma$ for each $\gamma$. The subset $F_\gamma$ is closed in $G_\gamma$ so that $F_\gamma$ is an $F_\sigma$-set of $X$. Thus $F_\gamma$ is paracompact by Lemma 2.1. Hence $\{F_\gamma\}_{\gamma\in\Gamma}$ is a locally finite covering of $G$ by closed paracompact normal subsets. Thus the subspace $G$ is paracompact by Proposition 1.17. Now $\{U_\lambda \cap G\}_{\lambda\in\Lambda}$ is an open covering of $G$ so that there exists a locally finite open covering $\{W_\lambda\}_{\lambda\in\Lambda}$ of $G$ such that $W_\lambda \subset U_\lambda \cap G$ for each $\lambda$. Then $\{W_\lambda \cap M\}_{\lambda\in\Lambda}$ is a locally finite open covering of $M$ and each $W_\lambda \cap M \subset V_\lambda$. Thus $M$ is paracompact. $\square$

**2.3  Definition.** *A topological space is said to be* hereditarily paracompact *if every subspace is paracompact.*

The hereditarily paracompact regular spaces have the following characterization:

**2.4  Proposition.** *A paracompact regular space is hereditarily paracompact if and only if it is totally normal.*

*Proof.* In a totally normal space every subset is normally situated. Since a totally normal space is regular, it follows from Proposition 2.2 that a paracompact totally normal space is hereditarily paracompact. For the converse, let $X$ be a hereditarily paracompact regular space and let $G$ be an open set in $X$. Since $X$ is regular, each point of $G$ is contained in an open set whose closure is contained in $G$. Thus since $G$ is paracompact, there exists a family $\{V_\gamma\}_{\gamma\in\Gamma}$ of open sets which is a locally finite covering of $G$ and satisfies $\overline{V}_\gamma \subset G$ for each $\gamma$. Since $G$ is normal, there exists a covering $\{F_\gamma\}_{\gamma\in\Gamma}$ of $G$ by closed sets of $G$ such that each $F_\gamma \subset V_\gamma$, and for each $\gamma$ there exists an open $F_\sigma$-set $G_\gamma$ of $G$ such that $F_\gamma \subset G_\gamma \subset V_\gamma$. Since $G = \bigcup_{\gamma\in\Gamma} F_\gamma$ and $F_\gamma \subset G_\gamma \subset G$, it follows that $G = \bigcup_{\gamma\in\Gamma} G_\gamma$. Since $G_\gamma \subset V_\gamma$ and $\{V_\gamma\}_{\gamma\in\Gamma}$ is locally finite in $G$, it follows that $\{G_\gamma\}_{\gamma\in\Gamma}$ is locally finite in $G$. Also $G_\gamma$ is open in the open set $G$, therefore $G_\gamma$ is open in $X$. Furthermore $G_\gamma$ is an $F_\sigma$-set in $G$ and $G_\gamma \subset \overline{V}_\gamma \subset G$, therefore $G_\gamma$ is an $F_\sigma$-set of the closed set $\overline{V}_\gamma$ of $X$ and hence $G_\gamma$ is an $F_\sigma$-set of $X$. Thus every open set of $X$ is normally situated and it follows that the normal space $X$ is totally normal. $\square$

**2.5  Example.** There exist totally normal spaces which are not perfectly normal.

Let $Z$ be an uncountable set and let $z_0 \in Z$. Let a subset of $Z$ be open if either it does not contain $z_0$ or its complement is finite. Then $Z$ is a hereditarily paracompact space. For suppose that $A \subset Z$. If $z_0 \notin A$, then $A$ is discrete. If $z_0 \in A$ and $\mathscr{U}$ is an open covering of $A$, then there exists a member $U_0$ of $\mathscr{U}$ such that $z_0 \in U_0$ and $U_0$, together with the one-point sets $\{z\}$ for $z$ in $A \backslash U_0$, is a finite open refinement of $\mathscr{U}$. Clearly $Z$ is a Hausdorff space so that $Z$ is a hereditarily paracompact regular space and hence $Z$ is a totally normal space by Proposition 2.4. But $Z$ is not perfectly normal. For an open set containing $z_0$ is of the form $Z \backslash F$, where $F$ is finite, so that a $G_\delta$-set containing $z_0$ is the complement of a countable set. Thus the closed set $\{z_0\}$ of $Z$ is not a $G_\delta$-set.

The concept of paracompactness is modified by both weakening and strengthening the requirement that each open covering has a locally finite open refinement.

**2.6    Definition.** *A topological space is said to be* weakly paracompact *if each open covering has a point-finite open refinement.*

For the definition of strong paracompactness, a new finiteness condition for families of subsets is introduced.

**2.7    Definition.** *A family* $\{A_\lambda\}_{\lambda \in \Lambda}$ *of subsets of some set is said to be* star-finite *if for each* $\lambda_0$ *in* $\Lambda$, *the set* $\{\lambda \in \Lambda \mid A_\lambda \cap A_{\lambda 0} \neq \varnothing\}$ *is finite.*

A star-finite collection of subsets of a topological space is not necessarily locally finite, but a star-finite open covering is locally finite. It follows that a topological space with the property that each open covering has a star-finite open refinement is paracompact.

**2.8    Definition.** *A topological space is said to be* strongly paracompact *if each open covering has a star-finite open refinement.*

Let $\{A_\lambda\}_{\lambda \in \Lambda}$ be a family of non-empty subsets of some set. We can define an equivalence relation $\sim$ in $\Lambda$ as follows: if $\lambda, \lambda' \in \Lambda$, then $\lambda \sim \lambda'$ if there exists a finite sequence $\lambda_1, \ldots, \lambda_m$ of elements of $\Lambda$ such that $\lambda_1 = \lambda$, $\lambda_m = \lambda'$ and $A_{\lambda_i} \cap A_{\lambda_{i+1}} \neq \varnothing$ for $1 \leqslant i < m$. If $\{A_\lambda\}$ is star-finite, then each equivalence class under this relation is countable. For let $M$ be an equivalence class and let $\lambda_0 \in M$. We define inductively a sequence $\{M_i\}_{i\geqslant 0}$ of subsets of $M$. Let $M_0 = \{\lambda_0\}$, and for each $i$ let

$$M_{i+1} = \{\lambda \in M \mid A_\lambda \cap A_{\lambda'} \neq \varnothing \quad \text{for some } \lambda' \text{ in } M_i\},$$

then $M_i \subset M_{i+1} \subset M$ for every $i$. Since $\{A_\lambda\}_{\lambda\in\Lambda}$ is star-finite, $M_{i+1}$ is finite if $M_i$ is finite. Hence every $M_i$ is finite by induction. But

$$M = \bigcup_{i\geqslant 0} M_i$$

so that $M$ is a countable subset of $\Lambda$. It follows that a star-finite collection can be indexed so that it takes the form $\{A_{i\lambda}\}_{i\in\mathbf{N}, \lambda\in\Lambda}$, where $A_{i\lambda} \cap A_{j\mu} = \varnothing$ if $\lambda \neq \mu$. Let $\{V_{i\lambda}\}_{i\in\mathbf{N}, \lambda\in\Lambda}$ be a star-finite open covering of a topological space $X$ such that $V_{i\lambda} \cap V_{j\mu} = \varnothing$ if $\lambda \neq \mu$. If

$$V_\lambda = \bigcup_{i\in\mathbf{N}} V_{i\lambda},$$

then $\{V_\lambda\}_{\lambda\in\Lambda}$ is a disjoint covering of $X$ by open-and-closed sets. Thus if $X$ is a connected space, each star-finite open covering of $X$ is countable. It follows that every open covering of a strongly paracompact connected space contains a countable subcovering. A space which has the property that each open covering contains a countable subcovering is called a *Lindelöf space*. Thus a connected strongly paracompact space is a Lindelöf space. This a partial converse to the following result.

**2.9    Proposition.** *A regular Lindelöf space is strongly paracompact.*

*Proof.* It follows from Proposition 1.6 that a regular Lindelöf space is paracompact and therefore countably paracompact and normal. Let $X$ be a regular Lindelöf space and let $\mathscr{U}$ be an open covering of $X$. Let $\{U_i\}_{i\in\mathbf{N}}$ be a countable family of members of $\mathscr{U}$ which is a covering of $X$. By Proposition 1.20 there exists a closed covering $\{F_i\}_{i\in\mathbf{N}}$ such that $F_i \subset U_i$ for each $i$. For each positive integer $i$, choose an open set $U_{in}$ for each integer $n \geqslant i$ such that

$$F_i \subset U_{in} \subset \overline{U}_{in} \subset U_{i,n+1} \subset U_i.$$

Let us define an open set $G_n$ for every integer $n$ by putting $G_n = \varnothing$ if $n \leqslant 0$, and $G_n = \bigcup_{i\leqslant n} U_{in}$ if $n \in \mathbf{N}$. Then $\{G_n\}$ is an open covering of $X$ since $\bigcup_{i\leqslant n} F_i \subset G_n$ if $n \in \mathbf{N}$. Furthermore if $n \in \mathbf{N}$, then

$$\overline{G}_n = \bigcup_{i\leqslant n} \overline{U}_{in} \subset \bigcup_{i\leqslant n} U_{i,n+1} \subset G_{n+1}.$$

For each positive integer $n$ and each positive integer $i$ such that $i \leqslant n+1$ let

$$V_{in} = U_{i,n+1} \backslash \bar{G}_{n-1}.$$

Then $\mathscr{V} = \{V_{in} \mid i, n \in \mathbf{N} \text{ and } i \leqslant n+1\}$ is an open covering of $X$. For suppose that $x$ is a point of $X$. If $x \in G_1$ then $x \in U_{11} \subset V_{11}$. If $x \notin G_1$, let $n$ be the least integer such that $x \in G_{n+1}$. Then $x \in U_{i,n+1}$ for some $i \leqslant n+1$ and $x \notin \bar{G}_{n-1}$ since $x \notin G_n$. Thus $x \in V_{in}$. Clearly $\mathscr{V}$ is a refinement of $\mathscr{U}$. Finally $\mathscr{V}$ is star-finite. For let $i, j, n, m$ be positive integers such that $i \leqslant n+1$ and $j \leqslant m+1$. Since $V_{jm} \subset G_{m+1}$ and $G_{m+1} \cap V_{in} = \varnothing$ if $m+1 \leqslant n-1$, it follows that $V_{jm} \cap V_{in} = \varnothing$ unless $|n-m| \leqslant 2$. $\square$

**2.10 Proposition.** *The topological product of a strongly paracompact space and a compact space is strongly paracompact.*

*Proof.* By the argument used in the proof of Lemma 1.12 it is easy to see that if $\mathscr{U}$ is an open covering of the topological product $X \times Y$ of a strongly paracompact space $X$ and a compact space $Y$, then $\mathscr{U}$ has a refinement of the form

$$\{V_\lambda \times G_{i\lambda} \mid i = 1, \dots, n_\lambda, \lambda \in \Lambda\},$$

where $\{V_\lambda\}_{\lambda \in \Lambda}$ is a star-finite open covering of $X$ and $\{G_{i\lambda} \mid i = 1, \dots, n_\lambda\}$ is a finite open covering of $Y$ for each $\lambda$. Since a covering of this form is evidently star-finite, it follows that $X \times Y$ is a strongly paracompact space. $\square$

There is a class of spaces intermediate between the classes of strongly paracompact spaces and of paracompact spaces which is significant in dimension theory. Spaces in this class are called completely paracompact. For the definition, the concept of weak refinement is required. A family $\{B_\gamma\}_{\gamma \in \Gamma}$ of subsets of a space $X$ is called a *weak refinement* of a covering $\mathscr{U}$ of $X$ if there exists a subset $\Gamma'$ of $\Gamma$ such that $\{B_\gamma\}_{\gamma \in \Gamma'}$ is a covering of $X$ and a refinement of $\mathscr{U}$. For example a base for the topology of a space $X$ is a weak refinement of every open covering of $X$.

**2.11 Definition.** *A topological space $X$ is said to be completely paracompact if each open covering of $X$ has a weak refinement of the form $\{V_\lambda\}_{\lambda \in \Lambda}$, where $\Lambda = \bigcup_{i \in \mathbf{N}} \Lambda_i$ and $\{V_\lambda\}_{\lambda \in \Lambda_i}$ is a star-finite open covering of $X$ for each $i$.*

**2.12 Proposition.** *A strongly paracompact space is completely paracompact and a completely paracompact regular space is paracompact.*

*Proof.* The first assertion is obviously true. If $X$ is a completely para-compact regular space then each open covering of $X$ has a weak refinement which is $\sigma$-locally finite and hence each open covering of $X$ has an open refinement which is $\sigma$-locally finite. It follows from Proposition 1.6 that $X$ is a paracompact.□

**2.13** *Remark.* In the next section we shall see that there exist metrizable spaces which are not completely paracompact and completely paracompact spaces which are not strongly paracompact.

## 3    Pseudo-metrizable and metrizable spaces

We recall that a topological space $X$ is pseudo-metrizable if there exists a pseudo-metric $d$ on $X$ such that the topology induced by $d$ is the topology of $X$. The main aims of this section are to obtain necessary and sufficient conditions for the pseudo-metrizability of a topological space and to find universal spaces for classes of metrizable spaces of given weight.

If $d$ and $e$ are pseudo-metrics on a set $X$, then $d$ and $e$ are said to be *equivalent* if they induce the same topology on $X$. If $(X, d)$ is a pseudo-metric space and $A$ is a non-empty subset of $X$, we define the *diameter* of $A$, denoted by diam $A$, to be the least upper bound of the set of real numbers $\{d(x, y) \mid x, y \in A\}$ if an upper bound exists, and to equal $\infty$ otherwise.

**3.1    Lemma.** *If $d$ is a pseudo-metric on a set $X$, then there exists an equivalent pseudo-metric $e$ on $X$ such that $(X, e)$ is a pseudo-metric space of diameter at most one.*

*Proof.* If $x, y \in X$, let

$$e(x, y) = \min\{1, d(x, y)\}.$$

Then $e$ is a pseudo-metric on $X$, and clearly $X$ has diameter at most one in $(X, e)$. It remains to show that $d$ and $e$ are equivalent. For a pseudo-metric space the set of all open balls of radius less than one is a base for the topology. But this set of subsets of $X$ is the same for the pseudo-metrics $d$ and $e$. Thus the topologies induced by $d$ and $e$ are identical.□

It is clear that each subspace of a pseudo-metrizable space is pseudo-metrizable. The topological product of a family of pseudo-metrizable spaces is not necessarily pseudo-metrizable. We have however:

**3.2   Proposition.** *The topological product of a countable family of pseudo-metric spaces is pseudo-metrizable.*

*Proof.* Let $\{(X_n, d_n)\}_{n \in \mathbf{N}}$ be a countable family of pseudo-metric spaces. In view of Lemma 3.1 we can suppose that each $X_n$ is of diameter at most one. Let $X$ be the product of the family $\{X_n\}_{n \in \mathbf{N}}$ and let us define $d \colon X \times X \to \mathbf{R}$ by

$$d(x, y) = \sum_{n=1}^{\infty} \frac{1}{2^n} d_n(x_n, y_n)$$

if $x = \{x_n\}$ and $y = \{y_n\}$ are points of $X$. Clearly $d$ is a pseudo-metric and it is not difficult to verify that the topology induced on $X$ by $d$ is the product topology. $\square$

**3.3   Corollary.** *The topological product of a countable family of metrizable spaces is metrizable.* $\square$

Now the *Nagata–Smirnov* theorem which supplies the most important pseudo-metrization criterion can be given.

**3.4   Proposition.** *The following statements about a topological space $X$ are equivalent:*

(a) *$X$ is pseudo-metrizable;*

(b) *$X$ is a regular space and there exists a sequence $\{\mathscr{V}_n\}_{n \in \mathbf{N}}$ of locally finite open coverings of $X$ such that for each point $x$ of $X$ and each open set $U$ such that $x \in U$, there exists an integer $n$ such that* $\mathrm{St}\,(x, \mathscr{V}_n) \subset U$;

(c) *$X$ is a regular space with a $\sigma$-locally finite base for its topology.*

*Proof.* $(a) \Rightarrow (b)$. Let $X$ be a pseudo-metric space. Then $X$ is regular and by Corollary 1.8, $X$ is a paracompact space. Thus for each positive integer $n$, the covering of $X$ by the open balls $B_{1/2^{n+1}}(x)$, $x \in X$, has a locally finite open refinement $\mathscr{V}_n$. Then $\{\mathscr{V}_n\}_{n \in \mathbf{N}}$ is the required sequence of locally finite open coverings of $X$. For let $U$ be an open neighbourhood of a point $x$ in $X$. Then $B_{1/2^n}(x) \subset U$ for some integer $n$. If $V \in \mathscr{V}_n$ and $x \in V$, then $V \subset B_{1/2^{n+1}}(z)$ for some $z$ in $X$, so that if $y \in V$ then

$$d(x, y) \leqslant d(x, z) + d(z, y) < \frac{1}{2^{n+1}} + \frac{1}{2^{n+1}} = \frac{1}{2^n},$$

so that $y \in U$. Thus $\mathrm{St}\,(x, \mathscr{V}_n) \subset U$ as required.

$(b) \Rightarrow (c)$. Obvious.

(c) ⇒ (a). Let $X$ be a regular space and let $\mathscr{B} = \bigcup_{n\in\mathbf{N}}\mathscr{B}_n$ be a base for the topology of $X$, where $\mathscr{B}_n$ is locally finite for each $n$. Then every open covering of $X$ has a refinement which consists of members of $\mathscr{B}$, so that $X$ is a paracompact normal space by Proposition 1.6. For each ordered pair $\mu = (m, n)$ of positive integers, we define a pseudo-metric $d_\mu$ on $X$ as follows. If $U\in\mathscr{B}_m$, let

$$K[U] = \cup\,\{\overline{W}\mid W\in\mathscr{B}_n \quad\text{and}\quad \overline{W}\subset U\}.$$

Since $\mathscr{B}_n$ is locally finite, $K[U]$ is a closed set and it is clear that $K[U]\subset U$. Since $X$ is normal, there exists a continuous function $f_{n,U}X\to I$ such that $f_{n,U}(x) = 0$ if $x\in X\backslash U$ and $f_{n,U}(x) = 1$ if $x\in K[U]$. Each point of $X$ belongs to only finitely many members of $\mathscr{B}_m$. Thus we can define $d_\mu\colon X\times X\to\mathbf{R}$ by putting

$$d_\mu(x,y) = \sum_{U\in\mathscr{B}_m} |f_{n,U}(x)-f_{n,U}(y)| \quad\text{if}\quad x,y\in X.$$

It is clear that $d_\mu$ is a pseudo-metric on $X$. Furthermore for $x_0$ in $X$ the real-valued function $d_\mu(x_0,\ )$ on $X$ is continuous. For if $x_1\in X$, then $x_1$ has an open neighbourhood $N$ which meets only finitely many members of $\mathscr{B}_m$. Let $\{U_1,\dots,U_r\}$ be the finite subset of $\mathscr{B}_m$ consisting of those sets which either contain $x_0$ or have a non-empty intersection with $N$. Then if $x\in N$,

$$d_\mu(x_0,x) = \sum_{i=1}^{r} |f_{n,U_i}(x_0)-f_{n,U_i}(x)|.$$

Thus the restriction of $d_\mu(x_0,\ )$ to the open neighbourhood $N$ of $x_1$ is continuous. It follows that $d_\mu(x_0,\ )$ is continuous. Thus the topology on $X$ induced by the pseudo-metric $d_\mu$ is smaller than the topology of $X$. Thus if $X_\mu$ denotes the set $X$ with the topology induced by $d_\mu$, then the identity function $i_\mu\colon X\to X_\mu$, given by $i_\mu(x) = x$ if $x\in X$, is continuous. Thus we have a countable family $\{X_\mu\}_{\mu\in\mathbf{N}\times\mathbf{N}}$ of pseudo-metric spaces and for each $\mu$ in $\mathbf{N}\times\mathbf{N}$ the identity function $i_\mu\colon X\to X_\mu$ is continuous. We shall appeal to Lemma 1.5.13 to show that $X$ can be embedded in $\tilde{X}$, the topological product of the family $\{X_\mu\}_{\mu\in\mathbf{N}\times\mathbf{N}}$. It is clear that the family $\{i_\mu\}$ separates points of $X$. To show that the family separates points of $X$ from closed sets we must prove that if $x_0\in X$ and $A$ is a closed set of $X$ such that $x_0\notin A$, then there exists $\mu$ in $\mathbf{N}\times\mathbf{N}$ such that $x_0$ does not belong to the closure of $A$ in $X_\mu$. Since $x_0\notin A$, there exists a member $U$ of $\mathscr{B}_m$ for some $m$ such that $x_0\in U$ and $U\cap A = \varnothing$. But $X$ is a regular space so that for some $n$, there exists a member $V$ of $\mathscr{B}_n$ such that $x_0\in V\subset\overline{V}\subset U$. Then

$f_{n,U}(x_0) = 1$ and $f_{n,U}(x) = 0$ if $x \in A$ since $A \subset X \backslash U$. Thus if $\mu = (m,n)$, then $d_\mu(x_0, x) \geqslant |f_{n,U}(x_0) - f_{n,U}(x)| = 1$ if $x \in A$. Thus $d_\mu(x_0, A) \geqslant 1$ and hence $x_0$ does not belong to the closure of $A$ in the space $X_\mu$. Hence $X$ is homeomorphic to a subspace of $\tilde{X}$ and $\tilde{X}$ is pseudo-metrizable by Proposition 3.2. Thus $X$ is a pseudo-metrizable space. $\square$

**3.5 Corollary.** *A topological space is metrizable if and only if it is a $T_3$-space with a $\sigma$-locally finite base for its topology.* $\square$

The next result will be used to establish Bing's metrization theorem and will have another application.

**3.6 Proposition.** *If $X$ is a pseudo-metrizable space, there exists a family $\mathscr{E} = \{E_\gamma\}_{\gamma \in \Gamma}$ of closed sets and a $\sigma$-discrete family $\mathscr{V} = \{V_\gamma\}_{\gamma \in \Gamma}$ of open sets such that $E_\gamma \subset V_\gamma$ for each $\gamma$, and for each point $x$ of $X$ and each open set $U$ such that $x \in U$ there exists $\gamma$ in $\Gamma$ such that*

$$x \in E_\gamma \subset V_\gamma \subset U.$$

*Proof.* Let $X$ be a pseudo-metric space, let $i$ be a positive integer and let $\mathscr{U}_i$ be the covering of $X$ by the open balls $B_{\frac{1}{2}i}(x)$, $x \in X$. By Proposition 1.7 there exists a closed covering $\mathscr{E}_i = \{E_{in\lambda}\}_{n \in \mathbf{N}, \lambda \in \Lambda_i}$ of $X$ and an open covering $\mathscr{V}_i = \{V_{in\lambda}\}_{n \in \mathbf{N}, \lambda \in \Lambda_i}$ of $X$ such that (i) $E_{in\lambda} \subset V_{in\lambda}$ for each $n$, $\lambda$; (ii) the family $\{V_{in\lambda}\}_{\lambda \in \Lambda_i}$ is discrete for each $n$; and (iii) $\mathscr{V}_i$ is a refinement of $\mathscr{U}_i$. If $\mathscr{E} = \{E_{in\lambda}\}_{i, n \in \mathbf{N}, \lambda \in \Lambda_i}$ and

$$\mathscr{V} = \{V_{in\lambda}\}_{i, n \in \mathbf{N}, \lambda \in \Lambda_i},$$

then $\mathscr{E}$ and $\mathscr{V}$ are the required families. $\square$

Proposition 3.7 is due to Bing.

**3.7 Proposition.** *A topological space is pseudo-metrizable if and only if it is a regular space with a $\sigma$-discrete base for its topology.*

*Proof.* It follows from Proposition 3.4 that the condition is sufficient for pseudo-metrizability, and necessity of the condition follows at once from Proposition 3.6. $\square$

We shall require two more pseudo-metrization theorems.

**3.8 Proposition.** *A space $X$ is pseudo-metrizable if and only if there exists a sequence $\{\mathscr{U}_n\}_{n \in \mathbf{N}}$ of open coverings of $X$ such that, for each point $x$ of $X$ and each open set $U$ such that $x \in U$, there exists an open set $V$ and an integer $n$ such that $x \in V$ and $\mathrm{St}\,(V, \mathscr{U}_n) \subset U$.*

*Proof.* If $X$ is a pseudo-metric space and $\mathcal{U}_n$ is the covering of $X$ by the open balls $B_{1/2^n}(x)$, $x \in X$, then the sequence $\{\mathcal{U}_n\}_{n \in \mathbb{N}}$ satisfies the condition, which is therefore necessary for pseudo-metrizability. For sufficiency, let $X$ be a topological space and let $\{\mathcal{U}_n\}_{n \in \mathbb{N}}$ be a sequence of open coverings satisfying the condition. We can suppose that $\mathcal{U}_{n+1}$ is a refinement of $\mathcal{U}_n$ for each $n$. If $U$ is open in $X$ and $x \in U$, there exists an integer $n$ and an open set $V$ such that $x \in V$ and $\operatorname{St}(V, \mathcal{U}_n) \subset U$. If

$$F = X \backslash \bigcup \{W \in \mathcal{U}_n \mid V \cap W = \varnothing\},$$

then $F$ is closed and $V \subset F \subset \operatorname{St}(V, \mathcal{U}_n)$. Thus $x \in V \subset \bar{V} \subset U$ and hence $X$ is a regular space. Furthermore $X$ is paracompact. For let $\mathcal{U}$ be an open covering of $X$. Consider the set $\mathcal{V}$ of open sets of $X$ defined as follows: $V \in \mathcal{V}$ if and only if there exists an integer $n$ such that $V$ is contained in a member of $\mathcal{U}_n$ and $\operatorname{St}(V, \mathcal{U}_n)$ is contained in a member of $\mathcal{U}$, and for each $V$ in $\mathcal{V}$ let $n_V$ be the least integer for which these conditions are satisfied. By hypothesis $\mathcal{V}$ is an open covering of $X$. If $x_0 \in X$, the set of integers $\{n_V \mid V \in \mathcal{V}$ and $x_0 \in V\}$ is non-empty and we put

$$n_0 = \min\{n_V \mid V \in \mathcal{V} \text{ and } x_0 \in V\}.$$

If $V \in \mathcal{V}$ and $x_0 \in V$, then $V$ is contained in some member of $\mathcal{U}_{n_0}$. Thus $\operatorname{St}(x_0, \mathcal{V}) \subset \operatorname{St}(x_0, \mathcal{U}_{n_0})$. Now choose $V_0$ in $\mathcal{V}$ such that $x_0 \in V_0$ and $n_{V_0} = n_0$. Then $\operatorname{St}(x_0, \mathcal{U}_{n_0}) \subset \operatorname{St}(V_0, \mathcal{U}_{n_0})$ and $\operatorname{St}(V_0, \mathcal{U}_{n_0})$ is contained in some member of $\mathcal{U}$. Thus $\operatorname{St}(x_0, \mathcal{V})$ is contained in some member of $\mathcal{U}$ and we see that $\mathcal{V}$ is an open star-refinement of $\mathcal{U}$. It follows from Proposition 1.27 that $X$ is a paracompact space. Now for each $n$ let $\mathcal{B}_n$ be a locally finite open refinement of $\mathcal{U}_n$. It is clear that the sequence $\{\mathcal{B}_n\}$ of locally finite open coverings of $X$ satisfies condition (*b*) of Proposition 3.4. Thus $X$ is a pseudo-metrizable space.□

**3.9 Proposition.** *A space $X$ is pseudo-metrizable if and only if there exists a sequence $\{\mathcal{F}_n\}_{n \in \mathbb{N}}$ of locally finite closed coverings of $X$ such that for each point $x$ of $X$ and each open set $U$ such that $x \in U$, there exists an integer $n$ such that $\operatorname{St}(x, \mathcal{F}_n) \subset U$.*

*Proof.* Since a pseudo-metrizable space is paracompact and normal, necessity of this condition follows from Proposition 3.4. Now let $X$ be a space for which there exists a sequence $\{\mathcal{F}_n\}_{n \in \mathbb{N}}$ of locally finite closed coverings such that, for each open set $U$ and each point $x$ of $U$, there exists an integer $n$ such that $\operatorname{St}(x, \mathcal{F}_n) \subset U$. Without loss of generality we can suppose that $\mathscr{F}_{n+1}$ is a refinement of $\mathscr{F}_n$ for every $n$. If $n \in \mathbf{N}$ and $x \in X$, let

$$V_n(x) = X \backslash \bigcup \{ F \in \mathscr{F}_n \mid x \notin F \}.$$

Then $V_n(x)$ is an open set such that $x \in V_n(x) \subset \operatorname{St}(x, \mathscr{F}_n)$. If

$$\mathscr{V}_n = \{V_n(x)\}_{x \in X},$$

then $\{\mathscr{V}_n\}_{n \in \mathbf{N}}$ is a sequence of open coverings of $X$, satisfying the condition of Proposition 3.8. Let $U$ be open in $X$ and let $x$ be a point of $U$. Then there exist integers $m, n$ such that $m < n$ and

$$V_n(x) \subset \operatorname{St}(x, \mathscr{F}_n) \subset V_m(x) \subset \operatorname{St}(x, \mathscr{F}_m) \subset U.$$

If $V_n(x) \cap V_n(y) \neq \varnothing$, then $V_n(x) \cap \operatorname{St}(y, \mathscr{F}_n) \neq \varnothing$ so that $y \in \operatorname{St}(x, \mathscr{F}_n)$ and hence $y \in V_m(x)$. It follows that $\operatorname{St}(y, \mathscr{F}_m) \subset \operatorname{St}(x, \mathscr{F}_m)$ so that since $V_n(y) \subset \operatorname{St}(y, \mathscr{F}_m)$ we see that $V_n(y) \subset U$. Thus

$$\operatorname{St}(V_n(x), \mathscr{V}_n) \subset U.$$

It follows from Proposition 3.8 that $X$ is a pseudo-metrizable space. $\square$

**3.10   Corollary.** *If $\{F_\gamma\}_{\gamma \in \Gamma}$ is a locally finite closed covering of a space $X$ and each subspace $F_\gamma$ is pseudo-metrizable, then $X$ is pseudo-metrizable.*

*Proof.* If $\gamma \in \Gamma$, let $\{\mathscr{E}_n^\gamma\}_{n \in \mathbf{N}}$ be a sequence of locally finite closed coverings of $F_\gamma$ satisfying the condition of Proposition 3.9, and suppose that $\mathscr{E}_{n+1}^\gamma$ is a refinement of $\mathscr{E}_n^\gamma$ for each $n$. If $\mathscr{E}_n = \bigcup_{\gamma \in \Gamma} \mathscr{E}_n^\gamma$, then $\{\mathscr{E}_n\}_{n \in \mathbf{N}}$ is a sequence of locally finite closed coverings of $X$ satisfying the condition of Proposition 3.9. $\square$

A topological space is said to be a *first-countable space* or to satisfy the first axiom of countability if for each point $x$ of $X$ there exists a countable family $\{V_n\}_{n \in \mathbf{N}}$ of open neighbourhoods of $x$ such that, for each open set $U$ such that $x \in U$, there exists $n$ such that $x \in V_n \subset U$. Clearly a pseudo-metrizable space is first-countable. A first-countable space has the property that its topology is determined by the convergence of sequences. If $X$ is a topological space and $A$ is a subset of $X$, a sequence in $A$ is an indexed family $\{x_n\}_{n \in \mathbf{N}}$ of points of $A$. A sequence $\{x_n\}_{n \in \mathbf{N}}$ in $A$ is said to converge to a point $x$ of $X$ if, for each open set $U$ such that $x \in U$, there exists an integer $N$ such that $x_n \in U$ for every $n$ such that $n \geqslant N$. In a Hausdorff space the point to which a sequence converges is unique if it exists. Clearly, if a sequence in $A$ converges to $x$ then $x \in \bar{A}$. In a first-countable space the converse is true.

**3.11   Proposition.** *If $X$ is a first-countable space, $A$ is a subset of $X$ and $x \in X$, then $x \in \bar{A}$ if and only if there exists a sequence in $A$ which converges to $x$.*

*Proof.* It must be shown that if $x \in \bar{A}$, then there exists a sequence in $A$ which converges to $x$. Let $\{V_n\}_{n \in \mathbf{N}}$ be a countable family of open neighbourhoods of $x$ such that if $U$ is open and $x \in U$, then $x \in V_n \subset U$ for some $n$. Without loss of generality we can suppose that $V_{n+1} \subset V_n$ for all $n$. For each $n$, $V_n \cap A$ is non-empty and we can choose $x_n$ in $V_n \cap A$. Then it is clear that $\{x_n\}_{n \in \mathbf{N}}$ is a sequence in $A$ which converges to $x$. $\square$

The pseudo-metrization theorems make the following definition seem natural.

**3.12   Definition.** *A regular space is said to be* strongly pseudo-metrizable *if it has a base for its topology which is the union of countably many star-finite open coverings.*

Since a star-finite open covering is locally finite, it follows from Proposition 3.4 that a strongly pseudo-metrizable space is pseudo-metrizable.

**3.13   Proposition.** *A pseudo-metrizable space is strongly pseudo-metrizable if and only if it is completely paracompact.*

*Proof.* Let $X$ be a strongly pseudo-metrizable space and let $\mathscr{W}$ be a base for the topology of $X$ which is the union of countably many star-finite coverings. Then $\mathscr{W}$ is a weak refinement of every open covering of $X$ and hence $X$ is completely paracompact. Conversely let $X$ be a completely paracompact pseudo-metric space and for each positive integer $i$ let $\mathscr{W}_i$ be a weak refinement of the covering of $X$ by the open balls $B_{1/2i}(x)$, $x \in X$, such that $\mathscr{W}_i$ is the union of countably many star-finite open coverings. If $\mathscr{W} = \bigcup_{i \in \mathbf{N}} \mathscr{W}_i$, then $\mathscr{W}$ is a base for the topology of $X$ which is the union of countably many star-finite coverings and hence $X$ is strongly pseudo-metrizable. $\square$

The strongly pseudo-metrizable spaces can be characterized as follows:

**3.14   Proposition.** *A space $X$ is strongly pseudo-metrizable if and only if $X$ is regular and there exists a sequence $\{\mathscr{U}_n\}_{n \in \mathbf{N}}$ of star-finite* *open coverings of $X$ such that, for each point $x$ of $X$ and each open set $U$ such that $x \in U$, there exists an integer $n$ such that $\mathrm{St}\,(x, \mathscr{U}_n) \subset U$.*

*Proof.* The condition is clearly sufficient and it remains to prove necessity. Let $X$ be a strongly pseudo-metrizable space. Then $X$ is a regular space and has a base $\mathscr{V} = \bigcup_{n \in \mathbf{N}} \mathscr{V}_n$ for its topology, where for each $n$, $\mathscr{V}_n = \{V_{n\lambda}\}_{\lambda \in \Lambda}$ is a star-finite open covering of $X$. The space $X$ is pseudo-metrizable and we suppose that some pseudo-metric has been chosen which induces the topology of $X$. Let $k$ be a given positive integer. For each positive integer $n$, let $M(n, k)$ be the subset of $\Lambda$ consisting of those indices $\lambda$ for which diam $V_{n\lambda} < 1/k$, and let

$$W_{nk} = \bigcup_{\lambda \in M(n,\,k)} V_{n\lambda}.$$

Then $\mathscr{W}_k = \{W_{nk}\}_{n \in \mathbf{N}}$ is a countable open covering of $X$. It follows from Proposition 1.20 that there exists a closed covering

$$\mathscr{F}_k = \{F_{nk}\}_{n \in \mathbf{N}}$$

such that $F_{nk} \subset W_{nk}$ for each $n$. Now let $\mathscr{U}_{nk}$ be the family of open sets $\{U_{nk\lambda}\}_{\lambda \in \Lambda}$, where

$$U_{nk\lambda} = \begin{cases} V_{n\lambda} & \text{if} \quad \lambda \in M(n, k), \\ V_{n\lambda} \backslash F_{nk} & \text{if} \quad \lambda \in \Lambda \backslash M(n, k). \end{cases}$$

Evidently $\mathscr{U}_{nk}$ is a star-finite open covering of $X$, so that

$$\{\mathscr{U}_{nk}\}_{n \in \mathbf{N},\, k \in \mathbf{N}}$$

is a countable family of star-finite open coverings of $X$. Let $U$ be open in $X$ and let $x \in U$. Then there exists a positive integer $k$ such that $B_{1/k}(x) \subset U$. Since $\mathscr{F}_k$ is a covering of $X$, there exists a positive integer $n$ such that $x \in F_{nk}$. Now if $x \in U_{nk\lambda}$, then $\lambda \in M(n, k)$ so that $U_{nk\lambda} = V_{n\lambda}$ where diam $V_{n\lambda} < 1/k$. It follows that $\mathrm{St}\,(x, \mathscr{U}_{nk}) \subset U$. $\square$

Clearly a discrete space is strongly metrizable. Next we consider some other examples.

**3.15**   *Example.* Let $\Omega$ be a set and let $B$ be the topological product of countably many copies of $\Omega$, where $\Omega$ has the discrete topology. Up to homeomorphism, the space $B$ depends only on the cardinal number $\tau$ of $\Omega$ and we put $B = B(\tau)$. If we wish to be more explicit we say that $B$ is the copy of $B(\tau)$ based on $\Omega$. It follows from Proposition 1.2.11 that $w(B(\tau)) = \tau$ provided that $\tau$ is infinite, and we then call $B(\tau)$ the *Baire space of weight* $\tau$. The space $B(\tau)$ is a countable product of metrizable spaces and is therefore metrizable. We can readily define a metric which induces the topology. Elements of $B(\tau)$ are sequences $\alpha = (\alpha_1, \alpha_2, \ldots)$ of elements of $\Omega$. If $\alpha, \beta \in B(\tau)$, we define $d(\alpha, \beta) = 0$ if $\alpha = \beta$ and otherwise

$$d(\alpha, \beta) = 1/k \quad \text{if} \quad \alpha_i = \beta_i \quad \text{for} \quad i < k \quad \text{and} \quad \alpha_k \neq \beta_k.$$

It is easy to see that $d$ is a metric on $B(\tau)$ which induces the topology of $B(\tau)$. Each open covering of $X$ has a refinement which consists of sets of the form

$$V_k(\alpha) = B_{1/k}(\alpha) = \{\beta \in B(\tau) \mid \beta_i = \alpha_i \quad \text{if} \quad i \leqslant k\},$$

where $\alpha \in B(\tau)$ and $k \in \mathbf{N}$. But if $\alpha, \beta \in B(\tau)$, $k, m \in \mathbf{N}$ and $k \leqslant m$, then either $V_m(\beta) \subset V_k(\alpha)$ or $V_m(\beta)$ and $V_k(\alpha)$ are disjoint. It follows that each open covering of $X$ has a disjoint open refinement. Thus $B(\tau)$ is a strongly paracompact metrizable space. It follows from Proposition 2.10 that $B(\tau) \times I$ is a strongly paracompact metrizable space. Since a subspace of a strongly metrizable space is clearly strongly metrizable, it follows that $B(\tau) \times J$ is a strongly metrizable space, where $J$ is the open unit interval $(0, 1)$. But $B(\tau) \times J$ is not strongly paracompact if $\tau > \aleph_0$. For consider the open covering of $B(\tau) \times J$ which consists of $B(\tau) \times (\frac{1}{2}, 1)$ together with all sets of the form $V_k(\alpha) \times (1/2^{k+1}, 3/2^{k+1})$, where $k$ is a positive integer and $\alpha \in B(\tau)$. Suppose that $\mathscr{U}$ is a star-finite open refinement of this covering. Let $(\alpha, t) \in B(\tau) \times J$. Since $\mathscr{U}$ is star-finite, there exists an open-and-closed set $V$ of $B(\tau) \times J$ such that $(\alpha, t) \in V$, the set $V$ contains only countably many members of $\mathscr{U}$ and each member of $\mathscr{U}$ which is not contained in $V$ is disjoint from $V$. Since $V$ is open and $(\alpha, t) \in V$, there exists a positive integer $k$ such that $(\beta, t) \in V$ if $\beta \in V_k(\alpha)$. Thus if $\beta \in V_k(\alpha)$, then $(\{\beta\} \times J) \cap V$ is a non-empty open-and-closed subset of $\{\beta\} \times J$. Since $J$ is connected it follows that $V_k(\alpha) \times J \subset V$. If $\omega \in \Omega$ let

$$\beta(\omega) = (\beta_1(\omega), \beta_2(\omega), \ldots)$$

be an element of $B(\tau)$ such that $\beta_i(\omega) = \alpha_i$ if $i \leqslant k$ and $\beta_{k+1}(\omega) = \omega$. Let $U_\omega$ be a member of $\mathscr{U}$ which contains $(\beta(\omega), 1/2^{k+1})$. Then

$$U_\omega \subset V_{k+1}(\beta(\omega)) \times (1/2^{k+2}, 3/2^{k+2}),$$

and it follows that if $\omega$ and $\omega'$ are distinct members of $\Omega$, then $U_\omega$ and $U_{\omega'}$ are disjoint. But $U_\omega \subset V_k(\alpha) \times J \subset V$ if $\omega \in \Omega$. If $\Omega$ is uncountable this is absurd since $V$ contains only countably many members of $\mathscr{U}$. Thus if $\tau > \aleph_0$ then $B(\tau) \times J$ is a strongly metrizable space which is not strongly paracompact.

Next we have an example of a metrizable space which is not strongly metrizable.

**3.16** *Example.* Let $\tau$ be a cardinal number and let $\Lambda$ be a set such that $|\Lambda| = \tau$. Let us define an equivalence relation $\sim$ in $I \times \Lambda$ as follows: $(s, \lambda) \sim (t, \mu)$ if $(s, \lambda) = (t, \mu)$ or $s = t = 0$. Let $[s, \lambda]$ denote the equivalence class containing $(s, \lambda)$. We define a metric $d$ on the set of equivalence classes as follows:

$$d([s, \lambda], [t, \mu]) = \begin{cases} |s-t| & \text{if} \quad \lambda = \mu, \\ s+t & \text{if} \quad \lambda \neq \mu. \end{cases}$$

The metric space obtained in this way is independent, up to homeomorphism, of the set $\Lambda$ of cardinality $\tau$ and we denote such a metric space by $J(\tau)$. If we wish to be more explicit we describe the above space as the copy of $J(\tau)$ based on $\Lambda$. For an obvious reason, $J(\tau)$ is sometimes called a 'hedgehog with $\tau$ spines'. If $\tau \geqslant \aleph_0$, then

$$w(J(\tau)) = \tau.$$

For let $A$ be the subset of $J(\tau)$ consisting of those points $[t, \lambda]$ for which $t$ is rational. The set $A$ has cardinality $\tau$ and the open balls of rational radius with centres at points of $A$ form a base for the topology of $J(\tau)$, and this base has cardinality $\tau$. Let $B$ denote the set of points of the form $[1, \lambda]$ for $\lambda$ in $\Lambda$. The subspace $B$ of $J(\tau)$ is discrete and has cardinality $\tau$. Thus $w(J(\tau)) = \tau$. If $\tau > \aleph_0$, then $J(\tau)$ is not strongly metrizable. For $J(\tau)$ is connected so that a star-finite open covering of $J(\tau)$ must be countable. Thus if $J(\tau)$ were strongly metrizable, it would have a countable base for its topology.

Next we consider a countability condition which is equivalent to the property of possessing a countable base for a pseudo-metrizable space.

**3.17** *Definition.* *A topological space is said to be* separable *if it has a countable dense subset.*

**3.18** *Proposition.* *The following statements about a pseudo-metrizable space $X$ are equivalent:*

  (a)  *$X$ has a countable base for its topology;*
  (b)  *$X$ is a Lindelöf space;*
  (c)  *$X$ is a separable space.*

*Proof.* $(a) \Rightarrow (b)$. Let $\mathcal{B}$ be a countable base for the topology of a space $X$ and let $\mathcal{U}$ be an open covering of $X$. There exists a subset $\mathcal{C}$ of $\mathcal{B}$

which is a refinement of $\mathscr{U}$. Choosing a member of $\mathscr{U}$ to contain each member of $\mathscr{C}$, we obtain a countable subset $\mathscr{V}$ of $\mathscr{U}$ which is a covering of $X$.

$(b) \Rightarrow (c)$. Suppose that a pseudo-metric space $X$ is a Lindelöf space, and consider the open covering $\{B_{1/2^n}(x)\}_{x \in X}$, where $n \in \mathbf{N}$. This contains a countable subcovering so that there exists a countable subset $A_n$ of $X$ such that $\{B_{1/2^n}(x)\}_{x \in A_n}$ is a covering of $X$. Let

$$A = \bigcup_{n \in \mathbf{N}} A_n.$$

Then $A$ is a countable subset of $X$ and evidently $\bar{A} = X$.

$(c) \Rightarrow (a)$. Let $A$ be a countable dense subset of the pseudo-metric space $X$, and let $\mathscr{B}$ consist of the subsets of $X$ of the form $B_{1/2^n}(x)$, where $x \in A$ and $n \in \mathbf{N}$. Then $\mathscr{B}$ is clearly countable. If $U$ is open in $X$ and $x_0 \in U$, then there exists $n$ such that $B_{1/2^n}(x_0) \subset U$. Since $\bar{A} = X$ there exists $x$ in $A$ such that $d(x_0, x) < 1/2^{n+1}$. Thus if $y \in B_{1/2^{n+1}}(x)$, then $d(x_0, y) \leqslant d(x_0, x) + d(x, y) < 1/2^n$, so that $y \in U$. Thus there exist an integer $n$ and $x$ in $A$ such that $x_0 \in B_{1/2^{n+1}}(x) \subset U$. Hence $\mathscr{B}$ is a countable base for the topology of $X$. $\square$

The *Urysohn pseudo-metrization theorem* now follows.

**3.19** *Proposition.* A topological space is separable and pseudo-metrizable if and only if it is a regular space with a countable base for its topology.

*Proof.* If $X$ is a separable pseudo-metric space, then $X$ is regular and by Proposition 3.18, $X$ has a countable base for its topology. If $X$ is a regular space with a countable base for its topology, then $X$ is pseudo-metrizable by Proposition 3.4 and $X$ is a separable space by Proposition 3.18. $\square$

**3.20** *Proposition.* A separable pseudo-metrizable space is strongly paracompact and hence strongly pseudo-metrizable.

*Proof.* By Proposition 3.18, a separable pseudo-metrizable space is a regular Lindelöf space and hence is strongly paracompact by Proposition 2.9. $\square$

Next we look at two important examples of separable metric spaces.

**3.21** *Example.* Let $H$ be the set of sequences $\mathbf{x} = \{x_n\}_{n \in \mathbf{N}}$ of real numbers such that $\sum_{n=1}^{\infty} x_n^2$ is convergent. We can define a metric $d$

on $H$ as follows: if $\mathbf{x} = \{x_n\}_{n \in \mathbf{N}}$, $\mathbf{y} = \{y_n\}_{n \in \mathbf{N}}$ are elements of $H$, put

$$d(\mathbf{x}, \mathbf{y}) = \left( \sum_{n=1}^{\infty} (x_n - y_n)^2 \right)^{\frac{1}{2}}.$$

It is easily verified, using Minkowski's inequality, that $d$ is a metric on $H$. This metric space is called *Hilbert space*. Hilbert space is separable. For it is easily seen that the countable subset of $H$ which consists of those elements $\mathbf{x} = \{x_n\}$ such that $x_n$ is rational for all $n$ and the set $\{n \in \mathbf{N} \mid x_n \neq 0\}$ is finite, is dense in $H$. The subspace

$$K = \{\mathbf{x} \in H \mid |x_n| \leqslant 1/n \quad \text{for all } n\}$$

is called the *Hilbert cube*. It is easy to establish that the Hilbert cube $K$ is homeomorphic with the cube $I^{\aleph_0}$.

Since a compact metric space has the Lindelöf property, it follows from Proposition 3.18 that a compact metric space is separable. The following covering property of compact metric spaces is often useful.

**3.22  Proposition.** *Let $(X, d)$ be a compact metric space and let $\{U_\lambda\}_{\lambda \in \Lambda}$ be an open covering of $X$. Then there exists a positive number $\epsilon$ with the property that $x$ and $y$ belong to some $U_\lambda$ if $x, y \in X$ and $d(x, y) < \epsilon$.*

*Proof.* For each point $x$ of $X$, let $\delta(x)$ be a positive real number such that $W(x) = B_{2\delta(x)}(x)$ is contained in some $U_\lambda$. For each $x$ let

$$V(x) = B_{\delta(x)}(x).$$

The open covering $\{V(x)\}_{x \in X}$ of $X$ contains a finite subcovering $\{V(x_1), \ldots, V(x_n)\}$. Let

$$\epsilon = \min \{\delta(x_1), \ldots, \delta(x_n)\}.$$

If $x \in X$, then $x \in V(x_i)$ for some $i$. And if $y \in X$ and $d(x, y) < \epsilon$, then

$$d(x_i, y) \leqslant d(x_i, x) + d(x, y) < \delta(x_i) + \epsilon \leqslant 2\delta(x_i).$$

Thus $x, y \in W(x_i)$ so that $x, y \in U_\lambda$ for some $\lambda$. $\square$

It follows from Proposition 1.5.16 that the Hilbert cube $K$ is a universal space for separable metrizable spaces. The rest of this section is devoted to finding universal spaces for metrizable and strongly metrizable spaces of given weight.

**3.23  Proposition.** *If $\tau$ is an infinite cardinal number, the countable topological product of hedgehogs $J(\tau)$ is a universal space for metrizable spaces of weight $\tau$.*

*Proof.* The countable topological product of hedgehogs $J(\tau)$ is a metrizable space of weight $\tau$. Let $X$ be a metrizable space of weight $\tau$. By Propositions 1.2.10 and 3.7, there exists a base $\mathscr{B} = \{U_{i\lambda}\}_{i\in\mathbf{N},\,\lambda\in\Lambda}$, where $|\Lambda| = \tau$ and for each positive integer $i$, the family $\{U_{i\lambda}\}_{\lambda\in\Lambda}$ is discrete. For each ordered pair $\rho = (m,n)$ of positive integers and each $\lambda$ in $\Lambda$ let

$$V_\lambda^\rho = \bigcup\{U_{m\mu} \mid \overline{U}_{m\mu} \subset U_{n\lambda}\}.$$

Then $\overline{V}_\lambda^\rho \subset U_{n\lambda}$, so that there exists a continuous function $f_\lambda^\rho : X \to I$ such that $f_\lambda^\rho(x) = 0$ if $x \notin U_{n\lambda}$ and $f_\lambda^\rho(x) = 1$ if $x \in \overline{V}_\lambda^\rho$. Let $A_n = \bigcup \overline{U}_{n\lambda}$. Then the subspace $A_n$ is the topological sum of the family $\{\overline{U}_{n\lambda}\}_{\lambda\in\Lambda}$ of subspaces. We consider the copy of $J(\tau)$ based on $\Lambda$ and define

$$f^\rho : A_n \to J(\tau)$$

by $\qquad\qquad f^\rho(x) = [f_\lambda^\rho(x), \lambda] \quad \text{if} \quad x \in \overline{U}_{n\lambda}.$

Let $B_n = X \backslash \bigcup_{\lambda\in\Lambda} U_{n\lambda}$. Then $A_n$ and $B_n$ are closed and $A_n \cup B_n = X$. If $x \in A_n \cap B_n$, then $f^\rho(x) = \mathbf{0}$, where $\mathbf{0}$ is the equivalence class containing $(0,\lambda)$ for all $\lambda$. Thus there exists a continuous function $h^\rho : X \to J(\tau)$ such that $h^\rho|A_n = f^\rho$ and $h^\rho(x) = \mathbf{0}$ if $x \in B_n$. The countable family $\{h^\rho\}_{\rho\in\mathbf{N}\times\mathbf{N}}$ of continuous functions separates points of $X$ from closed sets. For let $A$ be a closed set of $X$ and suppose that $x_0 \notin A$. There exist some $n$ and $\lambda$ such that $x_0 \in U_{n\lambda} \subset X \backslash A$, and there exist some $m$ and $\mu$ such that $x_0 \in \overline{U}_{m\mu} \subset U_{n\lambda}$. Thus if $\rho = (m,n)$, then $x_0 \in \overline{V}_\lambda^\rho \subset U_{n\lambda} \subset X \backslash A$. It follows that $h^\rho(x_0) = [1,\lambda]$, and

$$\{[t,\lambda] \in J(\tau) \mid t > 0\}$$

is an open set containing $h^\rho(x_0)$, which is disjoint from $h^\rho(A)$. Since $X$ is a $T_1$-space, the family $\{h^\rho\}$ also separates points of $X$. It follows from Lemma 1.5.13 that $X$ can be embedded in the countable product of copies of $J(\tau)$. $\square$

**3.24** *Remark.* Let us define $w^*(X)$ to be the least cardinal number $\tau$ such that the metrizable space $X$ has a base for its topology of the form $\{U_{i\lambda}\}_{i\in\mathbf{N},\,\lambda\in\Lambda}$, where $|\Lambda| = \tau$ and for each $i$, the family $\{U_{i\lambda}\}_{\lambda\in\Lambda}$ is discrete. If $w(X) > \aleph_0$, then $w^*(X) = w(X)$, whilst if $w(X) = \aleph_0$, then $w^*(X) = 1$. We have proved that if $X$ is a metrizable space with $w^*(X) = \tau$, then $X$ can be embedded in a countable product of copies of $J(\tau)$. Since $J(1) = I$, we can recover from Proposition 3.23 the embedding of a separable metrizable space in the cube $I^{\aleph_0}$.

For the construction of a universal space for strongly metrizable spaces of given weight we introduce a new concept which will be used frequently.

**3.25   Definition.** *If $\mathscr{U}$ is an open covering of a space $X$, then a continuous function $f: X \to Y$ is called a $\mathscr{U}$-mapping if for each $x$ in $X$ there exists an open set $V_x$ of $Y$ such that $f(x) \in V_x$ and $f^{-1}(V_x)$ is contained in some member of $\mathscr{U}$.*

**3.26   Lemma.** *If $\mathscr{U} = \{U_{n\alpha}\}_{n \in \mathbf{N}, \alpha \in \Omega}$ is a star-finite open covering of a normal space $X$ such that $U_{n\alpha} \cap U_{m\beta} = \varnothing$ if $\alpha \neq \beta$, then there exists a $\mathscr{U}$-mapping $f: X \to I^{\aleph_0} \times \Omega$, where $\Omega$ has the discrete topology.*

*Proof.* Since $X$ is a normal space, there exists a closed covering $\mathscr{F} = \{F_{n\alpha}\}_{n \in \mathbf{N}, \alpha \in \Omega}$ such that $F_{n\alpha} \subset U_{n\alpha}$ for each $n, \alpha$, and for each $n$ and $\alpha$ there exists a continuous function $f_{n\alpha}: X \to I$ such that

$$f_{n\alpha}(x) = 0$$

if $x \in X \backslash U_{n\alpha}$ and $f_{n\alpha}(x) = 1$ if $x \in F_{n\alpha}$. Define $f_\alpha: X \to I^{\aleph_0}$ by putting $f_\alpha(x) = \{f_{n\alpha}(x)\}_{n \in \mathbf{N}}$ if $x \in X$. For each $\alpha$ in $\Omega$ let $U_\alpha = \bigcup_{n \in \mathbf{N}} U_{n\alpha}$. If $x \in X$, there exists a unique $\alpha$ in $\Omega$ such that $x \in U_\alpha$. Define

$$f: X \to I^{\aleph_0} \times \Omega$$

by putting

$$f(x) = \big(f_\alpha(x), \alpha\big) \quad \text{if} \quad x \in U_\alpha.$$

Then $f$ is a continuous function. Moreover $f$ is a $\mathscr{U}$-mapping, for let $x$ be a point of $X$ and suppose that $x \in U_\alpha$; then $f(x) = \big(f_\alpha(x), \alpha\big)$. Since $\mathscr{F}$ is a covering of $X$, it follows that $U_\alpha = \bigcup_{n \in \mathbf{N}} F_{n\alpha}$, so there exists an integer $k$ such that $f_{k\alpha}(x) = 1$. If $W = \{\{t_n\} \in I^{\aleph_0} \mid t_k > 0\}$, then $W \times \{\alpha\}$ is open in $I^{\aleph_0} \times \Omega$, $f(x) \in W \times \{\alpha\}$ and $f^{-1}(W \times \{\alpha\}) \subset U_{k\alpha}$. $\square$

**3.27   Proposition.** *If $\tau$ is an infinite cardinal number, then the topological product $K \times B(\tau)$ of the Hilbert cube $K$ with the Baire space $B(\tau)$ is a universal space for strongly metrizable spaces of weight $\tau$.*

*Proof.* By Proposition 2.10, $K \times B(\tau)$ is strongly paracompact and is therefore a strongly metrizable space, and its weight is $\tau$ by Proposition 1.2.11. Let $X$ be a strongly metrizable space of weight $\tau$. By Proposition 3.14 there exists a sequence $\{\mathscr{U}_i\}_{i \in \mathbf{N}}$ of star-finite open coverings of $X$ such that for each point $x$ of $X$ and each open set $U$ such that $x \in U$, there exists some integer $i$ such that $\mathrm{St}\,(x, \mathscr{U}_i) \subset U$. Since the cardinality of a disjoint open covering of $X$ cannot exceed $\tau$, it follows from Lemma 3.26 that for each positive integer $i$ there exists a $\mathscr{U}_i$-mapping $f_i: X \to Y_i$, where $Y_i$ is the topological product $K \times \Omega$ of the Hilbert cube $K$ with a discrete space $\Omega$ of cardinality $\tau$.

The family $\{f_i\}_{i \in \mathbf{N}}$ separates points of $X$ from closed sets. For let $A$ be a closed set of $X$ and let $x$ be a point of $X \backslash A$. There exists an integer $i$ such that $\mathrm{St}\,(x, \mathcal{U}_i) \subset X \backslash A$. But $f_i \colon X \to Y_i$ is a $\mathcal{U}_i$-mapping. Hence there exists an open set $V$ of $Y_i$ such that $f_i(x) \in V$ and $f_i^{-1}(V)$ is contained in some member of $\mathcal{U}_i$. It follows that

$$f_i^{-1}(V) \subset \mathrm{St}\,(x, \mathcal{U}_i) \subset X \backslash A.$$

Thus $f_i(x) \in V$ and $V \cap f_i(A) = \varnothing$, so that $f_i(x) \notin \big(f_i(A)\big)^-$. Since $X$ is a $T_1$-space, it follows from Lemma 1.5.13 that $X$ can be embedded in the topological product $\Pi_{i \in \mathbf{N}} Y_i$, and $\Pi_{i \in \mathbf{N}} Y_i$ is evidently homeomorphic with $K \times B(\tau)$. $\square$

**3.28** *Remark.* Since $K \times B(\tau)$ is a strongly paracompact metrizable space, $K \times B(\tau)$ is a universal space for strongly paracompact metrizable spaces of weight $\tau$.

## 4   Completeness and the spaces $C_n(X)$

This section is concerned with a topic in the theory of metric spaces rather than of metrizable spaces. The concept of completeness arose in analysis.

**4.1** *Definition.* Let $(X, d)$ be a metric space. A sequence $\{x_n\}_{n \in \mathbf{N}}$ in $X$ is called a Cauchy sequence *if for each positive real number $\epsilon$ there exists a positive integer $N$ such that $d(x_n, x_m) < \epsilon$ for all integers $n, m$ such that $n \geqslant N$ and $m \geqslant N$. A metric space $(X, d)$ is said to be* complete *if every Cauchy sequence in $(X, d)$ converges.*

It should be stressed that completeness is a metric property. If $X$ is a metrizable space, then given sequence in $X$ may be a Cauchy sequence with respect to a metric inducing the topology of $X$, and not a Cauchy sequence with respect to an equivalent metric. A space $X$ may be complete with respect to a metric inducing its topology and not complete with respect to an equivalent metric. For a compact metrizable space however this cannot happen.

**4.2** *Proposition. A compact metric space is complete.*

*Proof.* Let $(X, d)$ be a compact metric space. Let $\{x_n\}_{n \in \mathbf{N}}$ be a Cauchy sequence in $(X, d)$ and let

$$A = \{x \in X \mid x = x_n \quad \text{for some } n\}.$$

If $A$ is finite then the sequence is trivially convergent. Now suppose that $A$ is infinite. Then $A$ cannot be a closed set with the discrete topology since $X$ is compact. Hence there exists a point $x_0$ of $X$, every neighbourhood of which contains infinitely many points of $A$. Let $U$ be an open neighbourhood of $x_0$. There exists a positive number $\epsilon$ such that $B_\epsilon(x_0) \subset U$. There exists a positive integer $N$ such that $d(x_n, x_m) < \epsilon/2$ for all integers $m, n$ such that $m \geqslant N$ and $n \geqslant N$. Finally there exists an integer $k$ such that $k \geqslant N$ and $d(x_0, x_k) < \epsilon/2$. Thus if $n$ is an integer such that $n \geqslant N$, then

$$d(x_0, x_n) \leqslant d(x_0, x_k) + d(x_k, x_n) < \epsilon.$$

Hence the sequence $\{x_n\}$ converges to $x_0$. $\square$

We shall be concerned with complete metric spaces which arise in the following way. Let $X$ be a topological space, let $(Y, d)$ be a metric space and let $C(X, Y)$ denote the set of continuous functions $f: X \to Y$ which are bounded in the sense that if $f \in C(X, Y)$, then there exists a positive real number $M$ such that $d\big(f(x), f(x')\big) \leqslant M$ if $x, x' \in X$. We can define a metric on $C(X, Y)$ as follows: if $f, g \in C(X, Y)$, then

$$d(f, g) = \sup_{x \in X} d\big(f(x), g(x)\big).$$

**4.3**  *Lemma. If $Y$ is a complete metric space, then $C(X, Y)$ is a complete metric space.*

*Proof.* Let $\{f_n\}_{n \in \mathbf{N}}$ be a Cauchy sequence in $C(X, Y)$. If $x \in X$, then $\{f_n(x)\}_{n \in \mathbf{N}}$ is a Cauchy sequence in $Y$ and hence converges to a point, $f(x)$ say, of $Y$. It is not difficult to establish that the function $f: X \to Y$ thus defined is a member of $C(X, Y)$ and that $\{f_n\}_{n \in \mathbf{N}}$ converges to $f$. $\square$

In Chapter 10 we shall use the spaces $C(X, \mathbf{R}^n)$, where for $n > 0$, $\mathbf{R}^n$ is Euclidean $n$-dimensional space and $\mathbf{R}^0$ is the one-point subspace $\{0\}$ of $\mathbf{R}$. For $n \geqslant 0$ let us write $C_n(X) = C(X, \mathbf{R}^n)$. We usually write $C(X)$ instead of $C_1(X)$. For $n > 0$ we shall write

$$\|f - g\| = \sup_{x \in X} \|f(x) - g(x)\|$$

if $f, g \in C_n(X)$.

**4.4**  *Proposition. If $X$ is a topological space and $n$ is a non-negative integer, then $C_n(X)$ is a complete metric space.*

*Proof.* This follows immediately from Lemma 4.3 and the fact that $R^n$ is a complete metric space if $n \geqslant 0$. $\square$

Finally we establish *Baire's theorem* which is the most useful property of complete metric spaces.

**4.5　Proposition.** *The intersection of a countable family of dense open sets in a complete metric space is a dense set.*

*Proof.* Let $\{G_n\}_{n \in \mathbf{N}}$ be a countable family of dense open sets in a complete metric space $(X, d)$. Let $U$ be a non-empty open set in $X$. Then $U \cap G_1 \ne \varnothing$ and we can choose a non-empty open set $V_1$ such that $\overline{V}_1 \subset U \cap G_1$ and $d(x, y) < 1$ if $x, y \in \overline{V}_1$. Proceeding by induction we obtain a sequence of non-empty open sets $\{V_n\}$ such that $\overline{V}_n \subset V_{n-1} \cap G_n$ if $n > 1$ and $d(x, y) < 1/n$ if $x, y \in \overline{V}_n$. Clearly

$$\bigcap_{n \in \mathbf{N}} \overline{V}_n \subset U \cap \bigcap_{n \in \mathbf{N}} G_n.$$

Now if $x_n \in \overline{V}_n$ for each $n$, then it is clear that $\{x_n\}_{n \in \mathbf{N}}$ is a Cauchy sequence in $X$. If $\{x_n\}$ converges to $x_0$, then $x_0 \in \bigcap_{n \in \mathbf{N}} \overline{V}_n$. Thus

$$U \cap \bigcap_{n \in \mathbf{N}} G_n \ne \varnothing. \square$$

**4.6　Corollary.** *If $\{A_n\}_{n \in \mathbf{N}}$ is a countable closed covering of a complete metric space, then at least one set $A_n$ has a non-empty interior.*

*Proof.* Let $\{A_n\}_{n \in \mathbf{N}}$ be a countable closed covering of a complete metric space $X$. Then

$$\bigcap_{n \in \mathbf{N}} (X \backslash A_n) = \varnothing.$$

It follows from Proposition 4.5 that the open set $X \backslash A_n$ is not dense in $X$ for some integer $n$. Since

$$A_n^{\circ} = X \backslash (X \backslash A_n)^{-},$$

it follows that the interior $A_n^{\circ}$ of $A_n$ is non-empty. $\square$

## 5　Perfect mappings

Continuous closed surjections preserve certain topological properties. For example, the closed image of a $T_1$-space is evidently a $T_1$-space. Proposition 1.3.10 states that the closed image of a normal space is normal, and Proposition 1.25 states that the closed image of a paracompact normal space is a paracompact normal space. In this section the more restricted class of perfect mappings is introduced. Perfect mappings preserve many topological properties. First we note the following property of closed mappings.

**5.1   Lemma.** *If* $f:X \to Y$ *is a continuous closed surjection,* $y \in Y$ *and* $U$ *is an open set of* $X$ *such that* $f^{-1}(y) \subset U$, *then there exists an open set* $W$ *of* $Y$ *such that* $y \in W$ *and* $f^{-1}(W) \subset U$.

*Proof.* If $W = Y \backslash f(X \backslash A)$, then $W$ is an open set of $Y$ with the required properties. $\square$

**5.2   Definition.** *A continuous surjection* $f:X \to Y$ *is said to be a* perfect mapping *if* $f$ *is closed, and if* $f^{-1}(y)$ *is a compact subset of* $X$ *if* $y \in Y$.

**5.3   Proposition.** *If* $f:X \to Y$ *is a perfect mapping and* $K$ *is a compact subset of* $Y$, *then* $f^{-1}(K)$ *is a compact subset of* $X$.

*Proof.* Let $\{U_\lambda\}_{\lambda \in \Lambda}$ be a family of open sets of $X$ such that

$$f^{-1}(K) \subset \bigcup_{\lambda \in \Lambda} U_\lambda.$$

If $y \in K$, then there exists a finite subset $M(y)$ of $\Lambda$ such that

$$f^{-1}(y) \subset \bigcup_{\lambda \in M(y)} U_\lambda.$$

Since $f$ is a closed mapping, by Lemma 5.1 there exists an open set $V_y$ of $Y$ such that $y \in V_y$ and $f^{-1}(V_y) \subset \bigcup_{\lambda \in M(y)} U_\lambda$. Since $K$ is compact, there exists a finite subset $B$ of $K$ such that $K \subset \bigcup_{y \in B} V_y$. Hence

$$f^{-1}(K) \subset \bigcup_{y \in B} f^{-1}(V_y) \subset \bigcup_{y \in B} \bigcup_{\lambda \in M(y)} U_\lambda.$$

Thus if $M = \bigcup_{y \in B} M(y)$, then $M$ is a finite subset of $\Lambda$ and

$$f^{-1}(K) \subset \bigcup_{\lambda \in M} U_\lambda.$$

Thus $f^{-1}(K)$ is compact. $\square$

**5.4   Corollary.** *The composite of perfect mappings is a perfect mapping.*

*Proof.* Since the composite of continuous closed surjections is a continuous closed surjection this follows immediately from Proposition 5.3. $\square$

**5.5   Proposition.** *Let* $f:X \to Y$ *be a perfect mapping.*
  (a) *If* $X$ *is a Hausdorff space, then* $Y$ *is a Hausdorff space.*
  (b) *If* $X$ *is a regular space, then* $Y$ *is a regular space.*
  (c) *If the weight of* $X$ *is infinite, then* $w(Y) \leqslant w(X)$.

*Proof.* (a) Let $y$ and $z$ be distinct points of $Y$. Then $f^{-1}(y)$ and $f^{-1}(z)$ are disjoint compact subsets of the Hausdorff space $X$. Hence there exist disjoint open sets $V$ and $W$ of $X$ such that $f^{-1}(y) \subset V$ and $f^{-1}(z) \subset W$. By Lemma 5.1, there exist $G$ and $H$ open in $Y$ such that $y \in G$, $z \in H$, $f^{-1}(G) \subset V$ and $f^{-1}(H) \subset W$, from which it follows that $G$ and $H$ are disjoint. Thus $Y$ is a Hausdorff space.

(b) Let $G$ be open in $Y$ and let $y$ be a point of $G$. Then $f^{-1}(y) \subset f^{-1}(G)$. Since $X$ is regular, for each point $x$ of $f^{-1}(y)$ there exists $V$ open in $X$ such that $x \in V \subset \bar{V} \subset f^{-1}(G)$. Since $f^{-1}(y)$ is compact, there exist $V_1, \ldots, V_k$ open in $X$ such that $f^{-1}(y) \subset V_1 \cup \ldots \cup V_k$ and $\bar{V}_i \subset f^{-1}(G)$ for each $i$. Let $W = V_1 \cup \ldots \cup V_k$. Then $W$ is open in $X$ and

$$f^{-1}(y) \subset W \subset \bar{W} \subset f^{-1}(G).$$

By Lemma 5.1 there exists $H$ open in $Y$ such that $y \in H$ and

$$f^{-1}(H) \subset W,$$

so that $H \subset f(\bar{W}) \subset G$. Since $f(\bar{W})$ is a closed set, it follows that $H$ is an open set of $Y$ such that $y \in H \subset \bar{H} \subset G$. Thus $Y$ is a regular space.

(c) Let $\mathscr{V}$ be a base for the topology of $X$ such that $|\mathscr{V}| = w(X)$ and let $\mathscr{W}$ consist of all sets of the form $\bigcup_{B \in \mathscr{B}} B$, where $\mathscr{B}$ is a finite subset of $\mathscr{V}$. If $W \in \mathscr{W}$, let $H_W = Y \backslash f(X \backslash W)$ and let $\mathscr{H}$ be the set of open sets of $Y$ of the form $H_W$ for some $W$ in $\mathscr{W}$. Since $w(X)$ is infinite, $|\mathscr{W}| = w(X)$, so that the cardinality of the set $\mathscr{H}$ of open sets of $Y$ does not exceed $w(X)$. Furthermore $\mathscr{H}$ is a base for the topology of $Y$. For let $G$ be an open set of $Y$ and let $y$ be a point of $G$. Since

$$f^{-1}(y) \subset f^{-1}(G)$$

and $f^{-1}(y)$ is a compact set, there exists $W$ in $\mathscr{W}$ such that

$$f^{-1}(y) \subset W \subset f^{-1}(G).$$

Then $y \in H_W \subset G$. It follows that $w(Y) \leqslant w(X)$. $\square$

**5.6  Lemma.** *If $f: X \to Y$ is a perfect mapping and $\{A_\lambda\}_{\lambda \in \Lambda}$ is a locally finite family of subsets of $X$, then $\{f(A_\lambda)\}_{\lambda \in \Lambda}$ is a locally finite family of subsets of $Y$.*

*Proof.* Let $y$ be a point of $Y$. If $x \in f^{-1}(y)$, then there exists an open set $V_x$ of $X$ such that $x \in V_x$ and $\Lambda_x = \{\lambda \in \Lambda \mid V_x \cap A_\lambda \neq \varnothing\}$ is finite. Since $f^{-1}(y)$ is compact, there exists a finite subset $B$ of $f^{-1}(y)$ such that $f^{-1}(y) \subset \bigcup_{x \in B} V_x$. By Lemma 5.1, there exists an open set $W$ of $Y$ such that $y \in W$ and $f^{-1}(W) \subset \bigcup_{x \in B} V_x$. If $W \cap f(A_\lambda) \neq \varnothing$, then $f^{-1}(W) \cap A_\lambda \neq \varnothing$, so that $\lambda \in M = \bigcup_{x \in B} \Lambda_x$. Since $M$ is a finite subset of $\Lambda$, it follows that the family $\{f(A_\lambda)\}_{\lambda \in \Lambda}$ is locally finite. $\square$

**5.7 Proposition.** *If $f: X \to Y$ is a perfect mapping and $X$ is a pseudo-metrizable space, then $Y$ is a pseudo-metrizable space.*

*Proof.* Choose a pseudo-metric $d$ on $X$ which induces the topology of $X$. For each positive integer $n$, let $\mathscr{E}_n$ be a locally finite closed refinement of the covering $\{B_{1/2^n}(x)\}_{x \in X}$ of $X$. Then by Lemma 5.6,

$$\mathscr{F}_n = \{f(E) \mid E \in \mathscr{E}_n\}$$

is a locally finite closed covering of $Y$ for each positive integer $n$. We shall complete the proof by showing that the sequence $\{\mathscr{F}_n\}_{n \in \mathbb{N}}$ satisfies the condition of Proposition 3.9. Let $G$ be an open set of $Y$ and let $y$ be a point of $G$. Then $f^{-1}(y)$ is compact and $f^{-1}(y) \subset f^{-1}(G)$. The continuous real-valued function $\phi$ defined on $f^{-1}(y)$, by putting $\phi(x) = d(x, X \backslash f^{-1}(G))$ if $x \in f^{-1}(y)$, is bounded and attains its bounds. Since $\phi(x) > 0$ if $x \in f^{-1}(y)$, it follows that there exists a point $x_0$ of $f^{-1}(y)$ such that $\phi(x) \geqslant \phi(x_0) > 0$ if $x \in f^{-1}(y)$. Thus there exists a positive integer $n$ such that

$$d\big(x, X \backslash f^{-1}(G)\big) \geqslant 1/2^n \quad \text{if} \quad x \in f^{-1}(y).$$

Suppose that $F \in \mathscr{F}_{n+1}$ and $y \in F$. Then $F = f(E)$, where $E \in \mathscr{E}_{n+1}$ and $E \cap f^{-1}(y) \neq \varnothing$. Let $x$ be a point of $E \cap f^{-1}(y)$ and suppose there exists $x''$ in $X$ such that $x'' \in E \cap \big(X \backslash f^{-1}(G)\big)$. Since there exists $x'$ in $X$ such that $E \subset B_{1/2^{n+1}}(x')$, it follows that

$$d(x, x'') \leqslant d(x, x') + d(x', x'') < 1/2^n,$$

which is absurd. Hence $E \subset f^{-1}(G)$ so that $F \subset G$. Thus

$$\text{St}\,(y, \mathscr{F}_{n+1}) \subset G. \,\square$$

**5.8 Corollary.** *If $f: X \to Y$ is a perfect mapping and $X$ is a metrizable space, then $Y$ is a metrizable space.* $\square$

Next we show that the domain of a perfect mapping must have certain 'compactness' properties if the range has that property.

**5.9 Proposition.** *Let $f: X \to Y$ be a perfect mapping.*
    (a) *If $Y$ is a compact space, then $X$ is a compact space.*
    (b) *If $Y$ is a paracompact space, then $X$ is a paracompact space.*
    (c) *If $Y$ is a completely paracompact space, then $X$ is a completely paracompact space.*

*Proof.* Let $\mathscr{U} = \{U_\lambda\}_{\lambda \in \Lambda}$ be an open covering of $X$. If $y \in Y$, there exists a finite subset $\Lambda(y)$ of $\Lambda$ such that $f^{-1}(y) \subset \bigcup_{\lambda \in \Lambda(y)} U_\lambda$. Since $f$

is a closed mapping, there exists an open set $W_y$ of $Y$ such that $y \in W_y$ and $f^{-1}(W_y) \subset \bigcup_{\lambda \in \Lambda(y)} U_\lambda$, and $\mathscr{W} = \{W_y\}_{y \in Y}$ is an open covering of $Y$.

In case $(a)$, in which $Y$ is compact, we can choose a finite subset $B$ of $Y$ such that $\{W_y\}_{y \in B}$ is a covering of $Y$. Then $\{f^{-1}(W_y)\}_{y \in B}$ is a covering of $X$. Hence $\{U_\lambda \mid \lambda \in \Lambda(y), y \in B\}$ is a finite subcovering of $\mathscr{U}$. Thus if $Y$ is compact, then $X$ is compact.

In case $(b)$, in which $Y$ is paracompact, there exists a locally finite open covering $\{V_y\}_{y \in Y}$ of $Y$ such that $V_y \subset W_y$ if $y \in Y$. Then

$$\{f^{-1}(V_y)\}_{y \in Y}$$

is a locally finite open covering of $X$. If $y \in Y$ and $\lambda \in \Lambda(y)$, let

$$U_{y,\lambda} = f^{-1}(V_y) \cap U_\lambda.$$

Then the covering $\{U_{y,\lambda} \mid \lambda \in \Lambda(y), y \in Y\}$ of $X$ is a locally finite open refinement of $\mathscr{U}$. Thus if $Y$ is paracompact, then $X$ is paracompact.

In case $(c)$, in which $Y$ is completely paracompact, there exists a family $\{V_\gamma\}_{\gamma \in \Gamma}$ of open sets of $Y$ such that (i) $\Gamma = \bigcup_{n \in \mathbb{N}} \Gamma(n)$ and the family $\{V_\gamma\}_{\gamma \in \Gamma(n)}$ is a star-finite covering of $Y$ for each $n$, and (ii) there exists a subset $\Gamma'$ of $\Gamma$ such that $\{V_\gamma\}_{\gamma \in \Gamma'}$ is a refinement of $\mathscr{W}$. If $\gamma \in \Gamma'$, choose $y(\gamma)$ in $Y$ such that $V_\gamma \subset W_{y(\gamma)}$. Let $\mathscr{G}$ be the family of open sets of $X$ which consists of all sets $f^{-1}(V_\gamma) \cap U_\lambda$, where $\gamma \in \Gamma'$ and $\lambda \in \Lambda(y(\gamma))$, together with all sets $f^{-1}(V_\gamma)$, where $\gamma \in \Gamma \backslash \Gamma'$. Then $\mathscr{G}$ consists of countably many star-finite open coverings of $X$ and $\mathscr{G}$ is a weak refinement of $\mathscr{U}$ since the subfamily

$$\{f^{-1}(V_\gamma) \cap U_\lambda \mid \lambda \in \Lambda(y(\gamma)), \gamma \in \Gamma'\}$$

is a covering of $X$ and a refinement of $\mathscr{U}$. Thus if $Y$ is completely paracompact, then $X$ is completely paracompact. $\square$

The product of perfect mappings is perfect. For the proof, a lemma is required.

**5.10 Lemma.** *Let $X$ be the topological product of a family $\{X_\lambda\}_{\lambda \in \Lambda}$ of spaces with projections $\pi_\lambda: X \to X_\lambda$ for $\lambda$ in $\Lambda$. If $K = \Pi_{\lambda \in \Lambda} K_\lambda$, where $K_\lambda$ is a compact subset of $X_\lambda$ for each $\lambda$, and $U$ is an open set of $X$ such that $K \subset U$, then there exist a finite subset $M$ of $\Lambda$ and an open set $V_\lambda$ of $X_\lambda$ for each $\lambda$ in $M$ such that*

$$K \subset \bigcap_{\lambda \in M} \pi_\lambda^{-1}(V_\lambda) \subset U.$$

*Proof.* Let $K_1$ and $K_2$ be compact subsets of spaces $X_1$ and $X_2$ respectively, and let $U$ be an open set of $X_1 \times X_2$ such that $K_1 \times K_2 \subset U$.

If $x \in K_1$, then for each $y$ in $K_2$ there exist $G_y$, $H_y$ open in $X_1, X_2$ respectively such that $(x, y) \in G_y \times H_y \subset U$. Since $K_2 \subset \bigcup_{y \in K_2} H_y$, there exists a finite subset $B(x)$ of $K_2$ such that $K_2 \subset \bigcup_{y \in B(x)} H_y$. Let

$$V_1(x) = \bigcap_{y \in B(x)} G_y, \quad V_2(x) = \bigcup_{y \in B(x)} H_y.$$

Then $V_1(x)$, $V_2(x)$ are open in $X_1, X_2$ respectively, $x \in V_1(x)$, $K_2 \subset V_2(x)$ and $V_1(x) \times V_2(x) \subset U$. Since $K_1 \subset \bigcup_{x \in K_1} V_1(x)$, there exists a finite subset $B$ of $K$ such that $K_1 \subset \bigcup_{x \in B} V(x)$. Let

$$V_1 = \bigcup_{x \in B} V_1(x), \quad V_2 = \bigcap_{x \in B} V_2(x).$$

Then $V_1, V_2$ are open in $X_1, X_2$ respectively, $K_1 \subset V_1$, $K_2 \subset V_2$ and $V_1 \times V_2 \subset U$. An easy induction now shows that Lemma 5.10 is true if $\Lambda$ is a finite set.

We now proceed to the proof in the general case. Since $K$ is compact and $U$ is open, $K \subset \bigcup_{i=1}^{n} W_i \subset U$, where $W_i = \bigcap_{\lambda \in M(i)} \pi_\lambda^{-1}(W_{i\lambda})$, $M(i)$ is a finite subset of $\Lambda$, and $W_{i\lambda}$ is open in $X_\lambda$ if $\lambda \in M(i)$. Let $M = \bigcup_{i=1}^{n} M(i)$, let $Y$ be the topological product of the finite family $\{X_\lambda\}_{\lambda \in M}$ with projections $\rho_\lambda : Y \to X_\lambda$ if $\lambda \in M$ and let $\pi : X \to Y$ be the unique continuous function such that $\rho_\lambda \circ \pi = \pi_\lambda$ if $\lambda \in M$. For each $i$, let $G_i = \bigcap_{\lambda \in M(i)} \rho_\lambda^{-1}(W_{i\lambda})$ and let $G = \bigcup_{i=1}^{n} G_i$. Then $G$ is open in $Y$ and since $\pi^{-1}(G_i) = W_i$, it follows that

$$K \subset \pi^{-1}(G) \subset U.$$

Since $\pi(K) = \Pi_{\lambda \in M} K_\lambda \subset G$ and $M$ is finite, it follows from the first part of the proof that there exists $V_\lambda$ open in $X_\lambda$ if $\lambda \in M$, such that $\pi(K) \subset \bigcap_{\lambda \in M} \rho_\lambda^{-1}(V_\lambda) \subset G$. Thus $K \subset \bigcap_{\lambda \in M} \pi_\lambda^{-1}(V_\lambda) \subset \pi^{-1}(G) \subset U$. $\square$

**5.11  Proposition.** If $\{f_\lambda\}_{\lambda \in \Lambda}$ is a family of perfect mappings, then $\Pi_{\lambda \in \Lambda} f_\lambda$ is a perfect mapping.

*Proof.* Let the domain and range of $f_\lambda$ be $X_\lambda$ and $Y_\lambda$ respectively and let $X, Y$ be the topological products of the families $\{X_\lambda\}_{\lambda \in \Lambda}, \{Y_\lambda\}_{\lambda \in \Lambda}$ with projections $\pi_\lambda : X \to X_\lambda$ and $\rho_\lambda : Y \to Y_\lambda$ if $\lambda \in \Lambda$. Let $f = \Pi_{\lambda \in \Lambda} f_\lambda$. Evidently $f : X \to Y$ is a continuous surjection. If $y = \{y_\lambda\} \in Y$, then

$$f^{-1}(y) = \prod_{\lambda \in \Lambda} f_\lambda^{-1}(y_\lambda),$$

which is compact. Finally $f$ is closed. For suppose that $A$ is closed in $X$ and that $y \notin f(A)$, where $y = \{y_\lambda\}$. Then

$$f^{-1}(y) = \prod_{\lambda \in \Lambda} f_\lambda^{-1}(y_\lambda) \subset X \backslash A.$$

It follows from Lemma 5.10 that there exists a finite subset $M$ of $\Lambda$ and an open set $V_\lambda$ of $X_\lambda$ for each $\lambda$ in $M$ such that

$$f^{-1}(y) \subset \bigcap_{\lambda \in M} \pi_\lambda^{-1}(V_\lambda) \subset X \backslash A.$$

If $\lambda \in M$, then $f_\lambda^{-1}(y_\lambda) \subset V_\lambda$ so that, since $f_\lambda$ is a closed mapping, there exists $W_\lambda$ open in $Y_\lambda$ such that $y_\lambda \in W_\lambda$ and $f_\lambda^{-1}(W_\lambda) \subset V_\lambda$. Thus if

$$W = \bigcap_{\lambda \in M} \rho_\lambda^{-1}(W_\lambda),$$

then $W$ is open in $Y$ and $y \in W$. Moreover

$$f^{-1}(W) = \bigcap_{\lambda \in M} f^{-1}\rho_\lambda^{-1}(W_\lambda) = \bigcap_{\lambda \in M} \pi_\lambda^{-1} f_\lambda^{-1}(W_\lambda)$$

$$\subset \bigcap_{\lambda \in M} \pi_\lambda^{-1}(V_\lambda) \subset X \backslash A.$$

Thus $W \cap f(A) = \varnothing$. It follows that $f(A)$ is closed in $Y$.$\quad\square$

There is a useful characterization of the perfect mappings which have Tihonov spaces as domain and range.

**5.12** **Lemma.** *Let $f \colon X \to Y$ be a perfect mapping, where $X$ is a Hausdorff space, and let $S$ be a dense subset of $X$. The continuous surjection of $S$ onto $f(S)$ given by restriction of $f$ is a perfect mapping if and only if $S = f^{-1}f(S)$.*

*Proof.* Let $g \colon S \to f(S)$ be the continuous surjection given by restriction of $f$. Suppose first that $S = f^{-1}f(S)$. If $y \in f(S)$, then $g^{-1}(y) = f^{-1}(y)$, so that $g^{-1}(y)$ is a compact set. If $F$ is a closed set of $S$, then $F = E \cap S$, where $E$ is closed in $X$. If $y \in f(E) \cap f(S)$, then $y = f(x)$, where $x \in E$, and since $f(x) \in f(S)$ it follows that $x \in S$ so that $y \in f(E \cap S)$. Thus $g(F) = f(E) \cap f(S)$ so that $g(F)$ is closed in $f(S)$. Thus $g$ is a perfect mapping. Now suppose that $g$ is a perfect mapping and that there exists $x$ in $X \backslash S$ such that $f(x) = y \in f(S)$. Then $f^{-1}(y) \cap S = g^{-1}(y)$ is a compact set. Since $x \notin f^{-1}(y) \cap S$ and $X$ is a Hausdorff space, it follows that there exists an open set $U$ of $X$ such that $f^{-1}(y) \cap S \subset U$ and $x \notin \overline{U}$. Since $S \backslash U$ is closed in $S$, $f(S \backslash U)$ is a closed set of $f(S)$ and $y \notin f(S \backslash U)$ since $f^{-1}(y) \cap S \subset U$. Let $W$ be an open set of $Y$ such that $W \cap f(S) = f(S) \backslash f(S \backslash U)$. Then $f^{-1}(W) \backslash \overline{U}$ is an open neighbourhood of $x$ in $X$. Since $S$ is dense in $X$, there exists $x'$ in $X$ such that

$$x' \in S \cap (f^{-1}(W) \backslash \overline{U}).$$

Then $x' \in S \backslash U$ and $f(x') \in W$ which is absurd. It follows that

$$f^{-1}f(S) = S.\quad\square$$

**5.13   Proposition.** *If $X$ and $Y$ are Tihonov spaces and $f: X \to Y$ is a continuous surjection, then $f$ is a perfect mapping if and only if*

$$\beta f: \beta X \to \beta Y$$

*satisfies*

$$(\beta f)(\beta X \backslash X) = \beta Y \backslash Y.$$

*Proof.* Let $\bar{f} = \beta f$. Since $f$ is a continuous surjection, $\bar{f}$ is a continuous surjection and hence $\bar{f}$ is a perfect mapping since $\beta X$ and $\beta Y$ are bicompact spaces. Since $\bar{f}(X) = Y$, it follows from Lemma 5.12 that $f$ is perfect if and only if $\bar{f}(\beta X \backslash X) = \beta Y \backslash Y$. $\square$

**5.14   Corollary.** *Let $X$, $Y$ and $Z$ be Tihonov spaces and let $g: X \to Y$ and $h: Y \to Z$ be continuous surjections. If $h \circ g$ is a perfect mapping, then $g$ and $h$ are perfect mappings.*

*Proof.* Let $f = h \circ g$. If $E$ is closed in $Y$, then $h(E) = f(g^{-1}(E))$ so that $h(E)$ is closed in $Z$. And if $z \in Z$, then $h^{-1}(z) = g(f^{-1}(z))$ so that $h^{-1}(z)$ is compact. Thus $h$ is a perfect mapping. If $\bar{f} = \beta f$, $\bar{g} = \beta g$ and $\bar{h} = \beta h$, then $\bar{f} = \bar{h} \circ \bar{g}$. By Proposition 5.13, $\bar{f}(\beta X \backslash X) = \beta Z \backslash Z$ so that $\bar{h} \bar{g}(\beta X \backslash X) = \bar{f}(\beta X \backslash X) = \beta Z \backslash Z$. Since $\bar{h}(Y) = Z$, it follows that

$$\bar{g}(\beta X \backslash X) \subset \beta Y \backslash Y$$

and hence $\bar{g}(\beta X \backslash X) = \beta Y \backslash Y$. Thus $g$ is a perfect mapping by Proposition 5.13. $\square$

Two applications of Corollary 5.14 give results which will be useful later.

**5.15   Proposition.** *Let $X$, $Y_1$ and $Y_2$ be Tihonov spaces and let*

$$f_1: X \to Y_1$$

*and $f_2: X \to Y_2$ be perfect mappings. Let $Y$ be the subspace of $Y_1 \times Y_2$ consisting of those pairs $(y_1, y_2)$ such that for some $x$ in $X$, $y_1 = f_1(x)$ and $y_2 = f_2(x)$. If $\pi_1: Y \to Y_1$ and $\pi_2: Y \to Y_2$ are given by restriction of the projections and $f$ is the unique continuous function such that*

$$\pi_i \circ f = f_i$$

*for $i = 1, 2$, then $f$, $\pi_1$ and $\pi_2$ are perfect mappings.*

*Proof.* Since $f_i$ is a perfect mapping and $f$ and $\pi_i$ are continuous surjections such that $\pi_i \circ f = f_i$, it follows from Corollary 5.14 that $\pi_i$ and $f$ are perfect mappings. $\square$

Before making the second application of 5.14, we note that it follows from Proposition 1.7.5. that for an inverse system of Hausdorff spaces with perfect connecting mappings, the canonical mappings are perfect.

**5.16 Proposition.** *Let* $\mathbf{Y} = \{Y_\alpha, \pi_{\alpha\beta}\}_{\alpha,\beta \in \Omega}$ *be an inverse system of Tihonov spaces, let $Y$ be the inverse limit of $\mathbf{Y}$ and let $\pi_\alpha : Y \to Y_\alpha$ be the canonical mapping if $\alpha \in \Omega$. Suppose that $X$ is a Tihonov space and that for each $\alpha$ in $\Omega$ there exists a perfect mapping $f_\alpha : X \to Y_\alpha$ such that $\pi_{\alpha\beta} \circ f_\beta = f_\alpha$ if $\alpha, \beta \in \Omega$ and $\alpha \leqslant \beta$. Then the unique continuous function $f : X \to Y$, such that $\pi_\alpha \circ f = f_\alpha$ if $\alpha \in \Omega$, is a perfect mapping.*

*Proof.* It follows from the relation $\pi_{\alpha\beta} \circ f_\beta = f_\alpha$, that if $\alpha, \beta \in \Omega$ and $\alpha \leqslant \beta$, then $\pi_{\alpha\beta}$ is a perfect mapping. Thus $\pi_\alpha$ is a perfect mapping if $\alpha \in \Omega$. We show next that $f$ is a surjection. Let $y^\circ = \{y_\alpha^\circ\}_{\alpha \in \Omega}$ be an element of $Y$. The family $\{f_\alpha^{-1}(y_\alpha^\circ)\}_{\alpha \in \Omega}$ of bicompact subsets of $X$ has the finite intersection property. For suppose that $B$ is a finite subset of $\Omega$ and choose $\beta$ in $\Omega$ such that $\alpha \leqslant \beta$ if $\alpha \in B$. There exists $x$ in $X$ such that $f_\beta(x) = y_\beta^\circ$. If $\alpha \in B$, then

$$f_\alpha(x) = \pi_{\alpha\beta} f_\beta(x) = \pi_{\alpha\beta}(y_\beta^\circ) = y_\alpha^\circ.$$

Thus $x \in \bigcap_{\alpha \in B} f_\alpha^{-1}(y_\alpha^\circ)$. It follows that $\bigcap_{\alpha \in \Omega} f_\alpha^{-1}(y_\alpha^\circ)$ is non-empty. If $x^\circ \in \bigcap_{\alpha \in \Omega} f_\alpha^{-1}(y_\alpha^\circ)$, then $f(x^\circ) = y^\circ$. If $\alpha \in \Omega$, then $\pi_\alpha \circ f = f_\alpha$, where $\pi_\alpha$ and $f$ are continuous surjections and $f_\alpha$ is a perfect mapping. It follows from Corollary 5.14 that $f$ is a perfect mapping.$\square$

**5.17 Definition.** *A Hausdorff space $X$ is called a* paracompact $M$-space *if there exists a perfect mapping $f : X \to Z$, where $Z$ is a metrizable space.*

It follows from Proposition 5.9 that a paracompact $M$-space is indeed paracompact. It is evident that every metrizable space is a paracompact $M$-space and that every bicompact space is a paracompact $M$-space. The name 'almost metrizable' has been proposed for paracompact $M$-spaces since they have some properties similar to those of metric spaces. It follows at once, from Corollary 3.3 and Proposition 5.11, that the topological product of a countable family of paracompact $M$-spaces is a paracompact $M$-space. This section will conclude with a characterization of the paracompact $M$-spaces as the inverse limits of inverse systems of metric spaces with perfect connecting mappings.

Let $\Psi = \{\mathcal{U}_i\}_{i \in \mathbf{N}}$ be a sequence of open coverings of a space $X$, where $\mathcal{U}_{i+1}$ is a strong star-refinement of $\mathcal{U}_i$ for each $i$. We define a non-negative function $D$ on $X \times X$ as follows: $D(x,y) = 1$ if $y \notin \mathrm{St}\,(x, \mathcal{U}_1)$; $D(x,y) = 1/2^i$ if $y \in \mathrm{St}\,(x, \mathcal{U}_i)$ and $y \notin \mathrm{St}\,(x, \mathcal{U}_{i+1})$; $D(x,y) = 0$ if $y \in \mathrm{St}\,(x, \mathcal{U}_i)$ for every $i$. Then $D(x,x) = 0$ if $x \in X$ and $D(x,y) = D(y,x)$ if $x, y \in X$. If $x \in X$ and $i \in \mathbf{N}$, then

$$\{y \in X \mid D(x,y) \leqslant 1/2^i\} = \mathrm{St}\,(x, \mathcal{U}_i).$$

The function $D$ does not satisfy the 'triangle inequality', but it is clear that if $x, y, z \in X$, then

$$D(x,z) \leqslant 2 \max\{D(x,y), D(y,z)\}.$$

A pseudo-metric on $X$ can be constructed from $D$. The following property is needed.

**5.18  Lemma.** *If $x_0, \ldots, x_{m+1}$ are points of $X$, then*

$$D(x_0, x_{m+1}) < 4 \sum_{i=0}^{m} D(x_i, x_{i+1}),$$

*provided that $\Sigma_{i=0}^{m} D(x_i, x_{i+1}) > 0$.*

*Proof.* The proof is by induction over $m$. If $m = 1$, then

$$D(x_0, x_2) \leqslant 2 \max\{D(x_0, x_1), D(x_1, x_2)\}$$
$$\leqslant 2\big(D(x_0, x_1) + D(x_1, x_2)\big).$$

Now suppose that the result is true for $m \leqslant k$ and let $x_0, \ldots, x_{k+2}$ be points of $X$. Let $s$ be the largest integer such that $\Sigma_{i=0}^{k+1} D(x_1, x_{i+1}) \leqslant 1/2^{s+1}$. There are two cases. If there exists $j$ such that $D(x_j, x_{j+1}) = 1/2^{s+1}$, then $D(x_0, x_j) = 0$ and $D(x_{j+1}, x_{k+2}) = 0$. If there exists no such $j$, let $p$ be the largest integer such that $\Sigma_{i=0}^{p} D(x_i, x_{i+1}) \leqslant 1/2^{s+2}$. Then $p \leqslant k$ and $\Sigma_{i=p+2}^{k+2} D(x_i, x_{i+1}) \leqslant 1/2^{s+2}$. By the induction hypothesis $D(x_0, x_p) < 1/2^s$, so that $D(x_0, x_p) \leqslant 1/2^{s+1}$ and similarly $D(x_{p+2}, x_{k+2}) \leqslant 1/2^{s+1}$. Also

$$D(x_j, x_{j+2}) \leqslant 1/2^{s+1}.$$

Thus in either case there exist $y, z$ in $X$ such that $D(x_0, y) \leqslant 1/2^{s+1}$, $D(y, z) \leqslant 1/2^{s+1}$ and $D(z, x_{k+2}) \leqslant 1/2^{s+1}$. Thus there exist $U_1, U_2, U_3$ in $\mathcal{U}_{s+1}$ such that $x_0 \in U$, $y \in U_1 \cap U_2$, $z \in U_2 \cap U_3$ and $x_{k+1} \in U_3$. Since $\mathcal{U}_{s+1}$ is a strong star-refinement of $\mathcal{U}_s$, $\mathrm{St}\,(U_2, \mathcal{U}_{s+1})$ is contained in some member $U$ of $\mathcal{U}_s$. Since $x_0 \in U$ and $x_{k+2} \in U$ it follows that

$$D(x_0, x_{k+2}) \leqslant 1/2^s < 4 \sum_{i=0}^{k+1} D(x_i, x_{i+1}). \quad \square$$

We define a non-negative function $d$ on $X \times X$ by taking $d(x,y)$ to be the greatest lower bound of the set of real numbers $\Sigma_{i=0}^{m} D(x_i, x_{i+1})$, where $m$ is a non-negative integer and $\{x_0, \ldots, x_{m+1}\}$ is a sequence of points of $X$ such that $x_0 = x$ and $x_{m+1} = y$. It is evident that $d$ is a pseudo-metric on $X$ and that $d(x,y) \leqslant D(x,y)$ if $x, y \in X$. It follows from Lemma 5.18 that

$$\tfrac{1}{4} D(x,y) \leqslant d(x,y)$$

if $x, y \in X$. Let us define an equivalence relation $\sim$ on $X$ by putting $x \sim y$ if $d(x,y) = 0$. Let $W = W(\Psi)$ be the set of equivalence classes with respect to $\sim$, and let $g \colon X \to W$ be the function defined by taking $g(x)$ to be the equivalence class containing $x$ if $x \in X$. We can define a metric $\rho$ on $W$ by putting $\rho(w, w') = d(x, x')$ if $x \in g^{-1}(w)$ and

$$x' \in g^{-1}(w').$$

Evidently the value of $\rho(w, w')$ is independent of the choice of $x$ and $x'$. It is easily seen that $\rho$ is a metric on $W$. If $X$ has the topology induced by the pseudo-metric $d$ and $W$ has the topology induced by the metric $\rho$, then $g \colon X \to W$ is an identification mapping. Since the pseudo-metric topology on $X$ is smaller than the given topology of $X$, it follows that if $W$ has the metric topology then $g \colon X \to W$ is continuous.

**5.19**    *Lemma. If $W(\Psi)$ has the topology induced by the metric $\rho$, then the continuous surjection $g \colon X \to W(\Psi)$ is a $\mathcal{U}_i$-mapping for every $i$.*

*Proof.* Let $i$ be a positive integer and let $w$ be a point of $W$. Let $H$ be the open ball in $W$ with centre $w$ and radius $1/2^{i+2}$. If $g(x) = w$, then

$$g^{-1}(H) = \{y \in X \mid d(x,y) < 1/2^{i+2}\}.$$

If $x, y \in X$ and $d(x,y) < 1/2^{i+2}$, then $D(x,y) < 1/2^i$ so that $D(x,y) \leqslant 1/2^{i+1}$. It follows that $g^{-1}(H) \subset \mathrm{St}(x, \mathcal{U}_{i+1})$. Hence $g^{-1}(H)$ is contained in some member of $\mathcal{U}_i$. It follows that $g$ is a $\mathcal{U}_i$-mapping. $\square$

**5.20**    *Proposition. Let $X$ be a paracompact $M$-space and let $\mathcal{U}$ be an open covering of $X$. Then there exists a metric space $W$ and a perfect $\mathcal{U}$-mapping $g \colon X \to W$.*

*Proof.* Let $f \colon X \to Z$ be a perfect mapping, where $Z$ is a metric space. For each positive integer $i$ let $\mathcal{G}_i$ be the covering of $Z$ by the open balls of radius $1/2i$, and let $\mathcal{V}_i = \{f^{-1}(G) \mid G \in \mathcal{G}_i\}$. We define inductively a sequence $\Psi = \{\mathcal{U}_i\}_{i \in \mathbb{N}}$ of open coverings of $X$ as follows. Let $\mathcal{U}_1$ be a common refinement of $\mathcal{U}$ and $\mathcal{V}_1$. Suppose we have chosen the open coverings $\mathscr{U}_1, \dots, \mathscr{U}_k$. By Proposition 1.27 we can choose $\mathscr{U}_{k+1}$ to be a strong star-refinement of $\mathscr{U}_k$ and a refinement of $\mathscr{V}_{k+1}$. By induction we obtain a sequence $\Psi = \{\mathscr{U}_i\}_{i \in \mathbb{N}}$ of open coverings of $X$ such that $\mathscr{U}_1$ is a refinement of $\mathscr{U}$ and for each $i$, $\mathscr{U}_i$ is a refinement of $\mathscr{V}_i$ and $\mathscr{U}_{i+1}$ is a strong star-refinement of $\mathscr{U}_i$. Now let $W = W(\Psi)$ and let $g \colon X \to W$ be the continuous surjection defined above. Then $g$ is a $\mathscr{U}$-mapping by Lemma 5.19. If $x, y \in X$ and $f(x) \neq f(y)$, then there exists $i$ such that $f(y) \notin \mathrm{St}\big(f(x), \mathscr{G}_i\big)$. Thus $y \notin \mathrm{St}\,(x, \mathscr{V}_i)$, so that

$$y \notin \mathrm{St}\,(x, \mathscr{U}_i).$$

It follows that $D(x, y) > 0$ so that $d(x, y) > 0$ and hence $g(x) \neq g(y)$. Thus if $x, y \in X$ and $g(x) = g(y)$, then $f(x) = f(y)$. Moreover, if $X$ has the topology induced by the pseudo-metric $d$, then $f \colon X \to Z$ is continuous. For let $H$ be open in $Z$ and let $x \in f^{-1}(H)$. There exists $i$ such that $\mathrm{St}\big(f(x), \mathscr{G}_i\big) \subset H$. If $y \in X$ and $d(x, y) < 1/2^{i+1}$, then $D(x, y) \leqslant 1/2^i$ so that $y \in \mathrm{St}\,(x, \mathscr{V}_i)$. It follows that $f(y) \in H$. Thus $f^{-1}(H)$ is open in the pseudo-metric topology of $X$. It follows that the unique function $h \colon W \to Z$, such that $h \circ g = f$, is continuous. Since $f \colon X \to Z$ is a perfect mapping and $h$ and $g$ are continuous surjections, it follows from Corollary 5.14 that $g$ is a perfect mapping. $\square$

**5.21.** *Proposition.* A topological space $X$ is a paracompact $M$-space if and only if $X$ is homeomorphic with the inverse limit of an inverse system $\{Y_\xi, \pi_{\xi\eta}\}_{\xi, \eta \in \Omega}$, where each space $Y_\xi$ is metrizable and $\pi_{\xi\eta}$ is a perfect mapping if $\xi, \eta \in \Omega$ and $\xi \leqslant \eta$.

*Proof.* Sufficiency of the condition is clear since the canonical mappings are perfect for an inverse system of metrizable spaces with perfect connecting mappings. Now let $X$ be a paracompact $M$-space and let $\Lambda$ be the set of open coverings of $X$ (here we regard an open covering as a subset of the set of open sets). Let $\Omega_0 = \Lambda$, for each integer $n \geqslant 0$ let $\Omega_{n+1} = \Omega_n \times \Omega_n$ and let $\Omega = \bigcup_{n \geqslant 0} \Omega_n$. Let $\Omega_0$ have the trivial partial order so that $\lambda \leqslant \mu$ if and only if $\lambda = \mu$, and if the partial order in $\bigcup_{k \leqslant n} \Omega_k$ has been defined, let us extend it to $\bigcup_{k \leqslant n+1} \Omega_k$ by saying that if $\gamma, \xi \in \bigcup_{k \leqslant n+1} \Omega_k$ and $\xi = (\alpha, \beta) \in \Omega_{n+1}$, then $\gamma \leqslant \xi$ if and only if either $\gamma = \xi$ or $\gamma \in \bigcup_{k \leqslant n} \Omega_k$ and either $\gamma \leqslant \alpha$ or $\gamma \leqslant \beta$. Then $\Omega$, with the partial order defined in this way, is a directed set. For if $\alpha \in \Omega_n$ and $\beta \in \Omega_m$, then there exists $\xi$ in $\Omega_{k+1}$, where

$$k = \max\{n, m\}$$

such that $\alpha \leqslant \xi$ and $\beta \leqslant \xi$. If $\lambda \in \Lambda = \Omega_0$, then by Proposition 5.20 there exist a metric space $Y_\lambda$ and a perfect $\lambda$-mapping $f_\lambda \colon X \to Y_\lambda$. Now suppose that $n \geqslant 0$ and that for each $\alpha$ in $\bigcup_{k \leqslant n} \Omega_k$ we have found a metrizable space $Y_\alpha$ and a perfect mapping $f_\alpha \colon X \to Y_\alpha$. Suppose furthermore that for each pair $\alpha, \beta$ in $\bigcup_{k \leqslant n} \Omega_k$ such that $\alpha \leqslant \beta$ we have found a perfect mapping $\pi_{\alpha\beta} \colon Y_\beta \to Y_\alpha$ such that

$$\pi_{\alpha\beta} \circ f_\beta = f_\alpha,$$

$\pi_{\alpha\alpha}$ is the identity, and if $\alpha, \beta, \gamma \in \bigcup_{k \leqslant n} \Omega_k$ and $\alpha \leqslant \beta \leqslant \gamma$, then

$$\pi_{\alpha\beta} \circ \pi_{\beta\gamma} = \pi_{\alpha\gamma}.$$

Let $\xi = (\alpha, \beta) \in \Omega_{n+1}$. By Proposition 5.15 there exist a metrizable space $Y_\xi$ and perfect mappings $f_\xi \colon X \to Y_\xi$, $\pi_{\alpha\xi} \colon Y_\xi \to Y_\alpha$ and $\pi_{\beta\xi} \colon Y_\xi \to Y_\beta$ such that $\pi_{\alpha\xi} \circ f_\xi = f_\alpha$ and $\pi_{\beta\xi} \circ f_\xi = f_\beta$. Let $\pi_{\xi\xi}$ be the identity mapping on $Y_\xi$. If $\gamma \in \bigcup_{k \leqslant n} \Omega_k$ and $\gamma \leqslant \xi$ then $\gamma \leqslant \alpha$, say, and we can define $\pi_{\gamma\xi} = \pi_{\gamma\alpha} \circ \pi_{\alpha\xi}$. Then it follows that $\pi_{\gamma\xi} \circ f_\xi = f_\gamma$, and since $f_\xi$ is surjective, $\pi_{\gamma\xi}$ is uniquely determined. It follows by induction that if $\xi \in \Omega$, then there exist a metrizable space $Y_\xi$ and a perfect mapping $f_\xi \colon X \to Y_\xi$, and if $\xi, \eta \in \Omega$ and $\xi \leqslant \eta$, then there exists a perfect mapping $\pi_{\xi\eta} \colon Y_\eta \to Y_\xi$ such that $\pi_{\xi\eta} \circ f_\eta = f_\xi$, $\pi_{\xi\xi}$ is the identity mapping, and if $\xi, \eta, \zeta \in \Omega$ and $\xi \leqslant \eta \leqslant \zeta$, then $\pi_{\xi\eta} \circ \pi_{\eta\zeta} = \pi_{\xi\zeta}$. Thus $\{Y_\xi, \pi_{\xi\eta}\}_{\xi, \eta \in \Omega}$ is an inverse system of metrizable spaces over $\Omega$ and the connecting mappings are perfect. If $Y$ is the inverse limit with canonical mappings $\pi_\xi \colon Y \to Y_\xi$, then there exists a unique continuous function $f \colon X \to Y$ such that $\pi_\xi \circ f = f_\xi$ for each $\xi$, and it follows from Proposition 5.16 that $f$ is a perfect mapping. But $f$ is injective. For let $x_1$ and $x_2$ be distinct points of $X$, and let $\lambda = \{X \setminus \{x_1\}, X \setminus \{x_2\}\}$ so that $\lambda \in \Lambda = \Omega_0$. Since $f_\lambda$ is a $\lambda$-mapping it follows that $f_\lambda(x_1) \neq f_\lambda(x_2)$ so that $f(x_1) \neq f(x_2)$. Thus $f \colon X \to Y$ is a continuous closed bijection and hence is a homeomorphism. $\square$

## 6 Simplicial complexes, polyhedra and the nerves of coverings

**6.1 Definition.** *A* simplicial complex *over a set $V$ is a set $K$ of finite subsets of $V$ with the property that if $s \in K$ and $s' \subset s$, then $s' \in K$.*

If $s \in K$, $s$ is called a *simplex* of $K$; if $s \in K$ and $v \in s \subset V$, then $v$ is called a *vertex* of the simplex $s$; a simplex consisting of exactly $q + 1$ vertices is called a *q-simplex*; if $s \in K$ and $s' \subset s$, $s'$ is called a *face* of $s$. If $K$ is a simplicial complex, its dimension $\dim K$ is defined to be equal to $-1$ if $K$ is empty, to equal $n$ if $K$ contains an $n$-simplex but no $(n+1)$-simplex, and to equal $\infty$ if $K$ contains an $n$-simplex for every $n$.

If $K$ is a simplicial complex over $V$, a simplicial complex $K'$ over a subset $V'$ of $V$ is a *subcomplex* of $K$ if each simplex of $K'$ is a simplex of $K$. If $s$ is a simplex of a simplicial complex $K$, then the set $\bar{s}$ of all faces of $s$ is a subcomplex of $K$. If $K$ is a simplicial complex, the set $K_n$ of all $q$-simplexes of $K$ such that $q \leqslant n$ is a subcomplex of $K$ called the *n-skeleton* of $K$.

Let $K$ be a non-empty simplicial complex over a set $V$, and let $|K|$ be the set of all functions $\alpha : V \to I$ such that $\{v \in V \mid \alpha(v) \neq 0\}$ is a simplex of $K$ and $\Sigma_{v \in V} \alpha(v) = 1$. We note that since

$$\{v \in V \mid \alpha(v) \neq 0\}$$

is a simplex of $K$, it follows in particular that if $\alpha \in |K|$ then $\alpha(v) \neq 0$ for only finitely many $v$ in $V$. If $s \in K$, let

$$|s| = \{\alpha \in |K| \mid \alpha(v) \neq 0 \quad \text{only if} \quad v \in s\}.$$

We can define a metric $d$ on $|s|$ as follows: if $\alpha, \beta \in |s|$,

$$d(\alpha, \beta) = \left(\Sigma_{v \in V}(\alpha(v) - \beta(v))^2\right)^{\frac{1}{2}}.$$

If $s$ is a $q$-simplex, then with the topology induced by $d$, $|s|$ is homeomorphic to the compact subset of $\mathbf{R}^{q+1}$

$$\left\{x \in \mathbf{R}^{q+1} \;\middle|\; 0 \leqslant x_i \leqslant 1, \; \sum_{i=0}^{q} x_i = 1\right\}.$$

If $s_1, s_2 \in K$, then clearly either $s_1 \cap s_2 = \varnothing$ in which case $|s_1| \cap |s_2| = \varnothing$ or $s_1 \cap s_2$ is a non-empty face of $s_1$ and $s_2$, in which case

$$|s_1| \cap |s_2| = |s_1 \cap s_2|.$$

Thus the metric topologies for $|s_1|$ and $|s_2|$ induce the same topology on $|s_1| \cap |s_2|$ and $|s_1| \cap |s_2|$ is closed in $|s_1|$ and in $|s_2|$.

**6.2 Definition.** *The geometric realization of a simplicial complex $K$ is the topological space obtained by giving the set $|K|$ the weak topology with respect to the covering $\{|s|\}_{s \in K}$.*

Since $|K|$ has the weak topology with respect to $\{|s|\}_{s \in K}$, for each simplex $s$ of $K$, the topology induced on the subset $|s|$ of $|K|$ is the metric topology, and the topology of $|K|$ is the finest topology with this property. If $Y$ is a topological space, a function $f : |K| \to Y$ is continuous if and only if the restriction $f \mid |s| : |s| \to Y$ is continuous for each $s$ in $K$.

If $L$ is a subcomplex of a simplicial complex $K$ and $s \in K$, then

$|s| \cap |L| = \bigcup_{i=1}^{n} |s_i|$, where $s_1, \ldots, s_n$ are the distinct simplexes of $L$ which are faces of $s$. Thus $|L|$ is closed in $|K|$. We note that if $s \in K$, then $|s| = |\bar{s}|$. It is obvious that if $\{L_\lambda\}$ is a family of subcomplexes of a simplicial complex $K$, then $\bigcup L_\lambda$ is a subcomplex of $K$, and it is also obvious that $|\bigcup L_\lambda| = \bigcup |L_\lambda|$.

**6.3 Proposition.** *If $K$ is a simplicial complex, its geometric realization $|K|$ is dominated by the closed covering $\{|s|\}_{s \in K}$.*

*Proof.* Let $\{s_\lambda\}_{\lambda \in \Lambda}$ be a family of simplexes of $K$. Then

$$\bigcup_{\lambda \in \Lambda} |s_\lambda| = \bigcup_{\lambda \in \Lambda} |\bar{s}_\lambda| = |\bigcup_{\lambda \in \Lambda} \bar{s}_\lambda|.$$

Since $\bigcup_{\lambda \in \Lambda} \bar{s}_\lambda$ is a subcomplex of $K$, $\bigcup_{\lambda \in \Lambda} |s_\lambda|$ is a closed set in $|K|$. Suppose that $F \subset \bigcup_{\lambda \in \Lambda} |s_\lambda|$ and that $F \cap |s_\lambda|$ is closed in $|s_\lambda|$ for each $\lambda$. If $s \in K$, then $|s| \cap \bigcup_{\lambda \in \Lambda} |s_\lambda| = \bigcup_{i=1}^{n} |s_i|$, where $s_1, \ldots, s_n$ are the distinct faces of $s$ which are also faces of some $s_\lambda$, where $\lambda \in \Lambda$. Then $|s| \cap F = \bigcup_{i=1}^{n} (|s_i| \cap F)$ which is closed. Thus $F$ is closed in $|K|$, and hence closed in $\bigcup_{\lambda \in \Lambda} |s_\lambda|$. Therefore $\bigcup_{\lambda \in \Lambda} |s_\lambda|$ has the weak topology with respect to $\{|s_\lambda|\}_{\lambda \in \Lambda}$. $\square$

**6.4 Proposition.** *If $K$ is a simplicial complex, its geometric realization $|K|$ is a paracompact perfectly normal Hausdorff space.*

*Proof.* Since there is an obvious metric topology on the set $|K|$ which is smaller than the topology of $|K|$, it follows that $|K|$ is a Hausdorff space. For each simplex $s$ of $K$, the subspace $|s|$ is compact and metrizable. Since $|K|$ is dominated by the covering $\{|s|\}_{s \in K}$, it follows from Propositions 1.4.20 and 1.17 that $|K|$ is a paracompact perfectly normal space. $\square$

Let $K$ be a simplicial complex over a set $V$. If $s \in K$ the open simplex $\langle s \rangle$ is the subset of $|K|$ defined by

$$\langle s \rangle = \{\alpha \in |K| \mid \alpha(v) \neq 0 \quad \text{if and only if} \quad v \in s\}.$$

Each point $\alpha$ of $|K|$ belongs to a unique open simplex $\langle s \rangle$, where $s = \{v \in V \mid \alpha(v) \neq 0\}$. Thus the open simplexes form a partition of $|K|$. We note that an 'open simplex' is not generally an open set of $|K|$. The 'vertex stars' are however open. If $v \in V$, the *star* of $v$ is defined to be the set

$$\mathrm{St}\,(v) = \{\alpha \in |K| \mid \alpha(v) \neq 0\}.$$

By considering its restriction to each of the subspaces $|s|$, $s \in K$, we see that the function $\theta_v : |K| \to I$ given by $\theta_v(\alpha) = \alpha(v)$ if $\alpha \in |K|$, is continuous. Since $\mathrm{St}\,(v) = \theta_v^{-1}\big((0, 1]\big)$, it follows that $\mathrm{St}\,(v)$ is an open set of $|K|$. Clearly $\{\mathrm{St}\,(v)\}_{v \in V}$ is an open covering of $|K|$. It is clear that $\alpha \in \mathrm{St}\,(v)$ if and only if $\alpha \in \langle s \rangle$ for some simplex $s$ such that $v$ is a vertex of $s$. Therefore

$$\mathrm{St}\,(v) = \bigcup_{v \in s} \langle s \rangle.$$

We can characterize the simplicial complexes with compact geometric realization. A simplicial complex is said to be finite if it consists of finitely many simplexes.

**6.5  Proposition.** *The geometric realization $|K|$ of a simplicial complex $K$ is compact if and only if $K$ is finite.*

*Proof.* Since $|K| = \bigcup_{s \in K} |s|$ and each subspace $|s|$ is compact, the sufficiency of the condition is obvious. Now suppose that $|K|$ is compact. Let $A$ be a subset consisting of one point from each open simplex $\langle s \rangle$ of $K$. If $B$ is any subset of $A$ and $s \in K$, then $B \cap |s|$ is a finite set. Hence $B$ is closed. Thus $A$ is a closed set of $|K|$ and the subspace $A$ has the discrete topology. Thus $A$ is a compact space with the discrete topology and it follows that $A$ is finite. Thus $K$ is finite. □

**6.6  Definition.** *A triangulation of a space $X$ consists of a simplicial complex $K$ and a homeomorphism $h : |K| \to X$. If $X$ has a triangulation, then $X$ is called a* polyhedron.

It follows from Proposition 6.4 that a polyhedron is a paracompact perfectly normal Hausdorff space. Applications of simplicial complexes and polyhedra in general topology result from associating a simplicial complex with each covering of a topological space as follows.

**6.7  Definition.** *Let $\mathcal{U} = \{U_\lambda\}_{\lambda \in \Lambda}$ be a covering of a space $X$. The* nerve $N(\mathcal{U})$ *of $\mathcal{U}$ is the simplicial complex over $\Lambda$ whose simplexes are the finite subsets $M$ of $\Lambda$ such that $\bigcap_{\lambda \in M} U_\lambda$ is non-empty. A continuous function $\phi : X \to |N(\mathcal{U})|$ is called a* canonical mapping *if*

$$\phi^{-1}\big(\mathrm{St}\,(\lambda)\big) \subset U_\lambda$$

*for each $\lambda$.*

We note that a canonical mapping $\phi : X \to |N(\mathcal{U})|$ is a $\mathcal{U}$-mapping. Normality can be characterized by the existence of canonical mappings into the realizations of the nerves of open coverings.

**6.8** *Proposition.* *The following statements about a topological space X are equivalent:*

(a) *X is a normal space;*

(b) *for each locally finite open covering $\mathscr{V}$ of X there exists a canonical mapping $\phi: X \to |N(\mathscr{V})|$;*

(c) *for each finite open covering $\mathscr{V}$ of X there exists a $\mathscr{V}$-mapping $\phi: X \to P$, where $P$ is a polyhedron.*

*Proof.* $(a) \Rightarrow (b)$. Let $\mathscr{V} = \{V_\lambda\}_{\lambda \in \Lambda}$ be a locally finite open covering of a normal space $X$. By Proposition 1.3.9 and 1.3.12 there exists a family $\{\phi_\lambda\}_{\lambda \in \Lambda}$ of continuous mappings of $X$ into $I$ such that the closure $F_\lambda$ of $\{x \in X \mid \phi_\lambda(x) > 0\}$ is contained in $V_\lambda$ for each $\lambda$, and $\Sigma_{\lambda \in \Lambda} \phi_\lambda(x) = 1$. We observe that a point of $|N(\mathscr{V})|$ is a function $\alpha: \Lambda \to I$ such that $\{\lambda \in \Lambda \mid \alpha(\lambda) \neq 0\}$ is a simplex of $N(\mathscr{V})$ and

$$\sum_{\lambda \in \Lambda} \alpha(\lambda) = 1.$$

If $x \in X$, let $\qquad M_x = \{\lambda \in \Lambda \mid x \in F_\lambda\}.$

The subset $M_x$ of $\Lambda$ is finite and since $F_\lambda \subset V_\lambda$ it follows that $x \in V_\lambda$ if $\lambda \in M_x$. Thus $M_x$ is a simplex of $N(\mathscr{V})$. If $\phi_\lambda(x) > 0$ then $\lambda \in M_x$ so that $\{\lambda \in \Lambda \mid \phi_\lambda(x) > 0\} \subset M_x$. Hence $\{\lambda \in \Lambda \mid \phi_\lambda(x) > 0\}$ is a simplex of $N(\mathscr{V})$. It follows that we obtain a point $\phi(x)$ of $|N(\mathscr{V})|$ by putting $\phi(x)(\lambda) = \phi_\lambda(x)$ if $\lambda \in \Lambda$. Thus we obtain a function $\phi: X \to |N(\mathscr{V})|$ which we shall show to be a canonical mapping. We prove first that $\phi$ is continuous. If $x \in X$, there exists an open neighbourhood $W_x$ of $x$ such that $\{\lambda \in \Lambda \mid W_x \cap F_\lambda \neq \varnothing\} = M_x$. Thus if $\lambda \notin M_x$, then $\phi_\lambda(y) = 0$ if $y \in W_x$. Hence $\phi|W_x$ maps $W_x$ into the subspace $|M_x|$ of $|N(\mathscr{V})|$. Since $|M_x|$ has the metric topology, it is easily seen that $\phi|W_x$ is continuous. Since the restriction of $\phi$ to each member of an open covering of $X$ is continuous, it follows that $\phi$ is continuous. Now if $\lambda \in \Lambda$, then $\qquad \phi^{-1}(\operatorname{St}(\lambda)) = \{x \in X \mid \phi_\lambda(x) > 0\} \subset V_\lambda.$

Thus $\phi$ is a canonical mapping.

$(b) \Rightarrow (c)$. Obvious.

$(c) \Rightarrow (a)$. Let $\mathscr{V}$ be a finite open covering of a space $X$, and let $\phi: X \to P$ be a $\mathscr{V}$-mapping, where $P$ is a polyhedron. Since $\phi$ is a $\mathscr{V}$-mapping, there exists an open covering $\mathscr{W}$ of $\phi(X)$ such that

$$\{\phi^{-1}(W) \mid W \in \mathscr{W}\}$$

is a refinement of $\mathscr{V}$. Since $P$ is a paracompact perfectly normal space, $\phi(X)$ is paracompact and perfectly normal. Hence there exists a locally finite closed covering $\mathscr{F}$ of $\phi(X)$ which is a refinement of $\mathscr{W}$.

The locally finite closed covering $\{\phi^{-1}(F) \mid F \in \mathscr{F}\}$ of $X$ is a refinement of $\mathscr{V}$. Thus by Proposition 1.3.9, $X$ is a normal space. $\square$

**6.9** *Remark.* It is easy to see that if the requirements of local finiteness and finiteness respectively are omitted from (*b*) and (*c*), then characterizations of the class of paracompact normal spaces are obtained.

## Notes

The characterizations of paracompactness contained in Propositions 1.5 and 1.6 were obtained by Michael [1953]. Proposition 1.15 is due to Morita [1962] who also applied the characterization to obtain Proposition 1.17. Proposition 1.24 is another characterization of paracompactness obtained by Michael [1959]. Michael [1957] proved that paracompactness is preserved by closed mappings. Tukey [1940] proved that every open covering of a metric space has an open strong star-refinement. A. H. Stone [1948] proved Proposition 1.27 and thus gave the first proof of the paracompactness of metric spaces.

Proposition 2.4 was proved by Dowker [1953]. Morita [1948] proved Proposition 2.9. The class of completely paracompact spaces was introduced by Zarelua [1963].

The metrization condition of Proposition 3.4 was obtained by Nagata [1950] and Smirnov [1951b], and the similar condition of Proposition 3.7 was obtained by Bing [1951]. Proposition 3.8 was obtained by Stone [1960] and Arhangel'skiĭ [1961]. It is closely related to a metrization theorem due to Moore [1935]. Morita [1955] proved Proposition 3.9. The characterization of strong metrizability contained in Proposition 3.13 is due to Zarelua [1963] and Proposition 3.14 is due to Pasynkov [1967]. Kowalsky [1957] found the universal space for metrizable spaces described in Proposition 3.23. Proposition 3.27 seems to be an unpublished result of Morita. The argument here follows that used by Šedivá [1959] in finding the same universal space for the restricted class of strongly paracompact metrizable spaces (Remark 3.28).

Proposition 5.7 is the work of Hanai [1954]. The characterization of perfect mappings given in Proposition 5.13 is due to Henriksen and Isbell [1958]. $M$-spaces were defined by Morita [1964]. The definition taken here of paracompact $M$-spaces comes from Morita's characterization of such spaces. Proposition 5.21 is the work of Kljušin [1964].

# 3

## COVERING DIMENSION

### 1 The definition of dim

The covering dimension of a topological space is defined in terms of the order of open refinements of finite open coverings of the space. The *order* of a family $\{A_\lambda\}_{\lambda \in \Lambda}$ of subsets, not all empty, of some set is the largest integer $n$ for which there exists a subset of $M$ of $\Lambda$ with $n+1$ elements such that $\bigcap_{\lambda \in M} A_\lambda$ is non-empty, or is $\infty$ if there is no such largest integer. A family of empty subsets has order $-1$.

**1.1 Definition.** *The* covering dimension dim $X$ *of a topological space $X$ is the least integer $n$ such that every finite open covering of $X$ has an open refinement of order not exceeding $n$ or is $\infty$ if there is no such integer.*

Thus dim $X = -1$ if and only if $X$ is empty, and dim $X \leqslant n$ if each finite open covering of $X$ has an open refinement of order not exceeding $n$. We have dim $X = n$ if it is true that dim $X \leqslant n$ but it is not true that dim $X \leqslant n-1$. Finally dim $X = \infty$ if for every integer $n$ it is false that dim $X \leqslant n$.

Each of the following two characterizations of covering dimension is sometimes taken as a definition.

**1.2 Proposition.** *If $X$ is a topological space, the following statements about $X$ are equivalent:*

    *(a)* dim $X \leqslant n$;

    *(b) for every finite open covering $\{U_1, \ldots, U_k\}$ of $X$ there is an open covering $\{V_1, \ldots, V_k\}$ of order not exceeding $n$ such that $V_i \subset U_i$ for $i = 1, \ldots, k$;*

    *(c) if $\{U_1, U_2, \ldots, U_{n+2}\}$ is an open covering of $X$, there is an open covering $\{V_1, V_2, \ldots, V_{n+2}\}$ such that each $V_i \subset U_i$ and $\bigcap_{i=1}^{n+2} V_i = \varnothing$.*

*Proof.* $(a) \Rightarrow (b)$. Suppose that dim $X \leqslant n$. The covering $\{U_1, \ldots, U_k\}$ has an open refinement $\mathscr{W}$ of order not exceeding $n$. If $W \in \mathscr{W}$, then $W \subset U_i$ for some $i$. Let each $W$ in $\mathscr{W}$ be associated with one of the sets $U_i$ containing it and let $V_i$ be the union of those members of $\mathcal{W}$ thus associated with $U_i$. Then $V_i$ is open, $V_i \subset U_i$ and each point of $X$ is in some member of $\mathcal{W}$ and hence in some $V_i$. Each point $x$ of $X$ is in at most $n+1$ members of $\mathcal{W}$, each of which is associated with a unique $U_j$, and hence $x$ is in at most $n+1$ members of $\{V_i\}$. Thus $\{V_i\}$ is an open covering of $X$ of order not exceeding $n$.

(b) $\Rightarrow$ (c) is obvious.

(c) $\Rightarrow$ (b). Let $X$ be a space satisfying (c) and let $\{U_1, \ldots, U_k\}$ be a finite open covering of $X$. We can assume that $k > n+1$. Let $G_i = U_i$ if $i \leqslant n+1$, and let $G_{n+2} = \bigcup_{i=n+2}^{k} U_i$. Then $\{G_1, \ldots, G_{n+2}\}$ is an open covering of $X$ and so by hypothesis there is an open covering

$$\{H_1, \ldots, H_{n+2}\}$$

such that each $H_i \subset G_i$ and $\bigcap_{i=1}^{n+2} H_i = \varnothing$. Let $W_i = U_i$ if $i \leqslant n+1$ and let $W_i = U_i \cap H_{n+2}$ if $i > n+1$. Then $\mathcal{W} = \{W_1, \ldots, W_k\}$ is an open covering of $X$, each $W_i \subset U_i$ and $\bigcap_{i=1}^{n+2} W_i = \varnothing$. If there exists a subset $B$ of $\{1, \ldots, k\}$ with $n+2$ elements such that $\bigcap_{i \in B} W_i \neq \varnothing$, let the members of $\mathcal{W}$ be renumbered to give a family $\mathcal{P} = \{P_1, \ldots, P_k\}$ such that $\bigcap_{i=1}^{n+2} P_i \neq \varnothing$. By applying the above construction to $\mathcal{P}$, we obtain an open covering $\mathcal{W}' = \{W_1', \ldots, W_k'\}$ such that each $W_i' \subset P_i$ and $\bigcap_{i=1}^{n+2} W_i' = \varnothing$. Clearly if $C$ is a subset of $\{1, \ldots, k\}$ with $n+2$ elements such that $\bigcap_{i \in C} P_i = \varnothing$, then $\bigcap_{i \in C} W_i' = \varnothing$. Thus by a finite number of repetitions of this process we obtain an open covering $\{V_1, \ldots, V_k\}$ of $X$, of order not exceeding $n$, such that each $V_i \subset U_i$.

(b) $\Rightarrow$ (a) is obvious. $\square$

We notice that if $\dim X = 0$ then $X$ is a normal space. For if $F_1$ and $F_2$ are disjoint closed sets of $X$, then $\{X \backslash F_2, X \backslash F_1\}$ is an open covering of $X$. Hence there exist open sets $V_1$ and $V_2$ such that $V_1 \cup V_2 = X$, $V_1 \subset X \backslash F_2$, $V_2 \subset X \backslash F_1$ and $V_1 \cap V_2 = \varnothing$. Thus $V_1$ and $V_2$ are disjoint open sets such that $F_1 \subset V_1$ and $F_2 \subset V_2$. Hence $X$ is a normal space. Furthermore if there exist open sets $U_1$, $U_2$ such that $U_1 \cup U_2 = X$ and $U_1 \neq X, U_2 \neq X$, then $X$ is not connected. For there exist open sets $V_1, V_2$ such that $V_1 \cup V_2 = X$, $V_1 \subset U_1$, $V_2 \subset U_2$ and $V_1 \cap V_2 = \varnothing$. The nonempty proper subsets $V_1, V_2$ of $X$ are open-and-closed sets. Thus $X$ is not connected. In particular if $X$ is a $T_1$-space such that $\dim X = 0$, then $X$ is not connected. In fact we can make a stronger statement about the disconnectedness of zero-dimensional spaces. A topological space is said to be *totally disconnected* if it contains no connected subspace consisting of more than one point. Thus in a totally disconnected space the component of each point $x$ is the one-point subset $\{x\}$.

**1.3**   *Proposition. Consider the following statements about a topological space $X$:*

   (*a*) $\dim X = 0$;

   (*b*) $X$ *has a base for its topology which consists of open-and-closed sets*;

   (*c*) $X$ *is totally disconnected.*

*For a $T_1$-space $X$, (a) implies (b) and (b) implies (c). For a non-empty bicompact space $X$, (a), (b) and (c) are equivalent.*

*Proof.* Let $X$ be a $T_1$-space such that $\dim X = 0$. If $U$ is an open set of $X$ and $x \in U$, then $\{U, X\backslash\{x\}\}$ is an open covering of $X$. Hence there exist open sets $V$ and $W$ of $X$ such that $V \cup W = X$, $V \subset U$, $W \subset X\backslash\{x\}$ and $V \cap W = \varnothing$. Thus $V$ is an open-and-closed set of $X$ such that $x \in V \subset U$. Hence (*a*) implies (*b*). Now let $X$ be a $T_1$-space which has a base for its topology which consists of open-and-closed sets. Let $A$ be a subset of $X$ containing distinct points $x_1$ and $x_2$. Then $X\backslash\{x_2\}$ is an open neighbourhood of $x_1$, so that there exists an open-and-closed set $V$ such that $x_1 \in V \subset X\backslash\{x_2\}$. The non-empty proper subset $V \cap A$ of $A$ is an open-and-closed set so that $A$ is not connected. Thus (*b*) implies (*c*). Finally suppose that $X$ is a non-empty totally disconnected bicompact space. If $x \in X$, then by Proposition 1.5.5, $\{x\}$ is the intersection of the open-and-closed sets of $X$ containing $x$. Thus by Remark 1.5.2 (3), if $U$ is an open neighbourhood of $x$, then there exists an open-and-closed set $H$ of $X$ such that $x \in H \subset U$. It follows that if $\mathscr{U}$ is an open covering of $X$, then there exists a finite refinement $\{H_1, \ldots, H_k\}$ of $\mathscr{U}$ such that each $H_i$ is an open-and-closed set. For $j = 1, \ldots, k$ let

$$G_j = H_j \backslash \bigcup_{i<j} H_i.$$

Then $\{G_1, \ldots, G_k\}$ is a disjoint open covering of $X$ which is a refinement of $\mathscr{U}$. Hence $\dim X = 0$. $\square$

**1.4**   *Examples.* (i) Let $X = \{a, b, c, d\}$ and let a base for the topology of $X$ consist of the sets $\{a\}$, $\{d\}$, $\{b, d\}$ and $\{c, d\}$. Then

$$\{\{a\}, \{b, d\}, \{c, d\}\}$$

is an open refinement of every open covering of $X$ so that $\dim X \leqslant 1$. But $X$ is non-empty and not normal so that $\dim X > 0$. Hence $\dim X = 1$.

   (ii) Let $I$ be the unit interval and let $\mathscr{U}$ be an open covering of $I$. Then $\mathscr{U}$ has a finite refinement $\mathscr{V}$ which is a covering of $I$ by open intervals. There exists an irreducible subcovering $\mathscr{W}$ of $\mathscr{V}$. Then $\mathscr{W}$ consists, say, of the sets $[0, b_0), (a_1, b_1), \ldots, (a_k, b_k), (a_{k+1}, 1]$, where

$0 \leqslant a_1, b_k \leqslant 1$ and $a_{i+1} < b_i$ for $i = 0, \ldots, k$. Since $\mathscr{W}$ is of order 1, $\dim I \leqslant 1$. Since $I$ is a connected $T_1$-space, $\dim I > 0$ and hence $\dim I = 1$.

(iii) Example 4.1.8 is a totally disconnected separable metric space which does not satisfy condition (b) of Proposition 1.3. In Example 4.3.1, one finds a Tihonov space which satisfies condition (b) of Proposition 1.3 but is not zero-dimensional.

It is not generally true that if $X$ is a topological space and $A$ is a subspace of $X$ then $\dim A \leqslant \dim X$. This question will be examined in more detail in §6. Covering dimension is however monotonic on closed subspaces.

**1.5   Proposition.** *If $A$ is a closed subset of a space $X$ then*

$$\dim A \leqslant \dim X.$$

*Proof.* It is sufficient to show that if $\dim X \leqslant n$, then $\dim A \leqslant n$. Suppose that $\dim X \leqslant n$ and let $\{U_1, \ldots, U_k\}$ be an open covering of $A$. Then for each $i$, $U_i = A \cap V_i$, where $V_i$ is open in $X$. The finite open covering $\{V_1, \ldots, V_k, X \backslash A\}$ of $X$ has an open refinement $\mathscr{W}$ of order not exceeding $n$. Let $\mathscr{V} = \{W \cap A \mid W \in \mathscr{W}\}$. Then $\mathscr{V}$ is an open refinement of $\{U_1, \ldots, U_k\}$ of order not exceeding $n$. $\square$

For normal spaces, covering dimension can be defined in terms of the order of finite closed refinements of finite open coverings.

**1.6   Proposition.** *If $X$ is a normal space, the following statements about $X$ are equivalent:*

(a) $\dim X \leqslant n$;

(b) *for every finite open covering $\{U_1, \ldots, U_k\}$ of $X$ there exists an open covering $\{V_1, \ldots, V_k\}$ such that each $\overline{V}_i \subset U_i$ and the order of $\{\overline{V}_1, \ldots, \overline{V}_k\}$ does not exceed $n$;*

(c) *for every finite open covering $\{U_1, \ldots, U_k\}$ of $X$ there exists a closed covering $\{F_1, \ldots, F_k\}$ such that each $F_i \subset U_i$ and the order of $\{F_1, \ldots, F_k\}$ does not exceed $n$;*

(d) *every finite open covering of $X$ has a finite closed refinement of order not exceeding $n$;*

(e) *if $\{U_1, \ldots, U_{n+2}\}$ is an open covering of $X$ there exists a closed covering $\{F_1, \ldots, F_{n+2}\}$ such that each $F_i \subset U_i$ and $\bigcap_{i=1}^{n+2} F_i = \varnothing$.*

*Proof.* (a) $\Rightarrow$ (b). Suppose that $\dim X \leqslant n$ and let $\{U_1, \ldots, U_k\}$ be an open covering of $X$. Then by Proposition 1.2 there exists an open covering $\{W_1, \ldots, W_k\}$ of order not exceeding $n$ such that each $W_i \subset U_i$. But since $X$ is normal, by Proposition 1.3.9 there exists an open covering $\{V_1, \ldots, V_k\}$ such that $\overline{V}_i \subset W_i$ for each $i$. Then $\{V_1, \ldots, V_k\}$ is an open covering with the required properties.

(b) $\Rightarrow$ (c) and (c) $\Rightarrow$ (d) are obvious.

(d) $\Rightarrow$ (e). Let $X$ be a space satisfying (d) and let $\mathscr{U} = \{U_1, \ldots, U_{n+2}\}$ be an open covering of $X$. The covering $\mathscr{U}$ has a finite closed refinement $\mathscr{E}$ of order not exceeding $n$. If $E \in \mathscr{E}$, then $E \subset U_i$ for some $i$. Let each $E$ in $\mathscr{E}$ be associated with one of the sets $U_i$ containing it and let $F_i$ be the union of those members of $\mathscr{E}$ thus associated with $U_i$. Then $F_i$ is closed, $F_i \subset U_i$ and $\{F_1, \ldots, F_{n+2}\}$ is a covering of $X$ such that $\bigcap_{i=1}^{n+2} F_i = \varnothing$.

(e) $\Rightarrow$ (a). Let $X$ be a space satisfying (e) and let $\{U_1, \ldots, U_{n+2}\}$ be an open covering of $X$. By hypothesis there exists a closed covering $\{F_1, \ldots, F_{n+2}\}$ such that each $F_i \subset U_i$ and $\bigcap_{i=1}^{n+2} F_i = \varnothing$. By Proposition 1.3.14 there exist open sets $V_1, \ldots, V_{n+2}$ such that $F_i \subset V_i \subset U_i$ for each $i$, and $\{V_i\}$ is similar to $\{F_i\}$. Thus $\{V_1, \ldots, V_{n+2}\}$ is an open covering of $X$, each $V_i \subset U_i$ and $\bigcap_{i=1}^{n+2} V_i = \varnothing$. By Proposition 1.2, $\dim X \leqslant n$.$\square$

The following result will have many applications.

**1.7 Proposition.** *If $X$ is a normal space, the following statements about $X$ are equivalent*:

(a) $\dim X \leqslant n$;

(b) *for each family of closed sets $\{F_1, \ldots, F_{n+1}\}$ and each family of open sets $\{U_1, \ldots, U_{n+1}\}$ such that each $F_i \subset U_i$, there exists a family $\{V_1, \ldots, V_{n+1}\}$ of open sets such that $F_i \subset V_i \subset \overline{V}_i \subset U_i$ for each $i$, and $\bigcap_{i=1}^{n+1} \mathrm{bd}\,(V_i) = \varnothing$*;

(c) *for each family of closed sets $\{F_1, \ldots, F_k\}$ and each family of open sets $\{U_1, \ldots, U_k\}$ such that each $F_i \subset U_i$, there exist families $\{V_1, \ldots, V_k\}$ and $\{W_1, \ldots, W_k\}$ of open sets such that*

$$F_i \subset V_i \subset \overline{V}_i \subset W_i \subset U_i$$

*for each $i$, and the order of the family $\{\overline{W}_1 \backslash V_1, \ldots, \overline{W}_k \backslash V_k\}$ does not exceed $n-1$*;

(d) *for each family of closed sets $\{F_1, \ldots, F_k\}$ and each family of open sets $\{U_1, \ldots, U_k\}$ such that each $F_i \subset U_i$, there exists a family $\{V_1, \ldots, V_k\}$ of open sets such that $F_i \subset V_i \subset \overline{V}_i \subset U_i$, and the order of the family*

$$\{\mathrm{bd}\,(V_1), \ldots, \mathrm{bd}\,(V_k)\}$$

*does not exceed $n-1$.*

*Proof.* $(a) \Rightarrow (b)$. Suppose that $\dim X \leqslant n$, let $F_1, \ldots, F_{n+1}$ be closed sets and let $U_1, \ldots, U_{n+1}$ be open sets such that each $F_i \subset U_i$. Since $\dim X \leqslant n$, the open covering of $X$ consisting of sets of the form $\bigcap_{i=1}^{n+1} H_i$, where $H_i = U_i$ or $H_i = X \backslash F_i$ for each $i$, has a finite open refinement $\{W_1, \ldots, W_q\}$ of order not exceeding $n$. Since $X$ is normal, there is a closed covering $\{K_1, \ldots, K_q\}$ such that each $K_r \subset W_r$. For $r = 1; \ldots, q$, let $N_r$ denote the set of integers $i$ such that $F_i \cap W_r \neq \varnothing$. For $r = 1, \ldots, q$, we can find open sets $V_{ir}$ for $i$ in $N_r$ such that

$$K_r \subset V_{ir} \subset \overline{V}_{ir} \subset W_r$$

and $\overline{V}_{ir} \subset V_{jr}$ if $i < j$. Now for each $i = 1, \ldots, n+1$, let

$$V_i = \bigcup_r \{V_{ir} \mid i \in N_r\}.$$

Then $V_i$ is open, and $F_i \subset V_i$, for if $x \in F_i$ and $x \in K_r$, then $i \in N_r$ so that $x \in V_{ir} \subset V_i$. Furthermore if $i \in N_r$ so that $F_i \cap W_r \neq \varnothing$, then $W_r$ is not contained in $X \backslash F_i$ so that $W_r \subset U_i$. Thus if $i \in N_r$, then $V_{ir} \subset U_i$ so that, since $\overline{V}_i = \bigcup_r \{\overline{V}_{ir} \mid i \in N_r\}$, it follows that $\overline{V}_i \subset U_i$. Finally suppose that $x \in \bigcap_{i=1}^{n+1} \mathrm{bd}\,(V_i)$. Since $\mathrm{bd}\,(V_i) \subset \bigcup_r \{\mathrm{bd}\,(V_{ir}) \mid i \in N_r\}$, it follows that for each $i$ there exists $r_i$ such that $x \in \mathrm{bd}\,(V_{ir_i})$. And if $i \neq j$, then $r_i \neq r_j$ for if $r_i = r_j = r$ then $x \in \overline{V}_{ir}$ and $x \in \overline{V}_{jr}$ but $x \notin V_{ir}$ and $x \notin V_{jr}$ which is absurd, since either $\overline{V}_{ir} \subset V_{jr}$ or $\overline{V}_{jr} \subset V_{ir}$. For each $i$, $x \notin V_{ir_i}$ so that $x \notin K_{r_i}$. But $\{K_r\}$ is a covering of $X$ and so there exists $r_0$ different from each of the $r_i$ such that $x \in K_{r_0} \subset W_{r_0}$. Since $x \in \overline{V}_{ir_i}$, it follows that $x \in W_{r_i}$ for $i = 1, \ldots, n+1$ so that $x \in \bigcap_{i=0}^{n+1} W_{r_i}$. Since the order of $\{W_r\}$ does not exceed $n$, this is absurd. Hence $\bigcap_{i=1}^{n+1} \mathrm{bd}\,(V_i) = \varnothing$.

$(b) \Rightarrow (c)$. Let $(b)$ hold, let $F_1, \ldots, F_k$ be closed sets and let $U_1, \ldots, U_k$ be open sets such that $F_i \subset U_i$. We can assume that $k > n+1$ for otherwise there is nothing to prove. Let the subsets of $\{1, \ldots, k\}$ containing $n+1$ elements be enumerated as $C_1, C_2, \ldots, C_q$, where $q = \binom{k}{n+1}$. By use of $(b)$ we can find open sets $V_{i,1}$ for $i$ in $C_1$ such that

$$F_i \subset V_{i,1} \subset \overline{V}_{i,1} \subset U_i \quad \text{and} \quad \bigcap_{i \in C_1} \mathrm{bd}\,(V_{i,1}) = \varnothing.$$

We have a finite family $\{\mathrm{bd}\,(V_{i,1}) \mid i \in C_1\}$ of closed sets of the normal space $X$ and $\mathrm{bd}\,(V_{i,1}) \subset U_i$ for each $i$ in $C_1$. Thus by Proposition 1.3.14, for each $i$ in $C_1$ there exists an open set $G_i$ such that

$$\mathrm{bd}\,(V_{i,1}) \subset G_i \subset \overline{G}_i \subset U_i,$$

and $\{\overline{G}_i\}_{i \in C_1}$ is similar to $\{\mathrm{bd}\,(V_{i,1})\}_{i \in C_1}$, so that in particular $\bigcap_{i \in C_1} \overline{G}_i = \varnothing$. Let $W_{i,1} = V_{i,1} \cup G_i$ if $i \in C_1$. Then $\overline{V}_{i,1} \subset W_{i,1} \subset \overline{W}_{i,1} \subset U_i$ and since

$$\overline{W}_{i,1} \backslash V_{i,1} \subset \overline{G}_i$$

we have $\bigcap_{i \in C_1} (\overline{W}_{i,1} \backslash V_{i,1}) = \varnothing$. If $i \notin C_1$, let $V_{i,1}$ be an open set such that $F_i \subset V_{i,1} \subset \overline{V}_{i,1} \subset U_i$ and let $W_{i,1} = U_i$. Then for $i = 1, \ldots, k$ we have open sets $V_{i,1}$ and $W_{i,1}$ such that

$$F_i \subset V_{i,1} \subset \overline{V}_{i,1} \subset W_{i,1} \subset U_i \quad \text{and} \quad \bigcap_{i \in C_1} (\overline{W}_{i,1} \backslash V_{i,1}) = \varnothing.$$

Suppose that $1 < m \leqslant q$ and for $i = 1, \ldots, k$ we have found open sets $V_{i,m-1}$ and $W_{i,m-1}$ such that

$$F_i \subset V_{i,m-1} \subset \overline{V}_{i,m-1} \subset W_{i,m-1} \subset U_i$$

and $\qquad \bigcap_{i \in C_j} (\overline{W}_{i,m-1} \backslash V_{i,m-1}) = \varnothing \quad \text{if} \quad 1 \leqslant j \leqslant m-1.$

By the above argument we can find open sets $V_{i,m}$ and $W_{i,m}$ such that

$$\overline{V}_{i,m-1} \subset V_{i,m} \subset \overline{V}_{i,m} \subset W_{i,m} \subset W_{i,m-1}$$

and $\qquad \bigcap_{i \in C_m} (\overline{W}_{i,m} \backslash V_{i,m}) = \varnothing.$

Since $\overline{W}_{i,m} \backslash V_{i,m} \subset \overline{W}_{i,m-1} \backslash V_{i,m-1}$ we have

$$\bigcap_{i \in C_j} (\overline{W}_{i,m} \backslash V_{i,m}) = \varnothing \quad \text{if} \quad j \leqslant m.$$

Thus by induction for $i = 1, \ldots, k$, we can find open sets $V_i, W_i$ ($= V_{i,q}, W_{i,q}$ respectively) such that

$$F_i \subset V_i \subset \overline{V}_i \subset W_i \subset U_i \quad \text{and} \quad \bigcap_{i \in C_j} (\overline{W}_i \backslash V_i) = \varnothing$$

for $j = 1, \ldots, q$. Thus the order of the family $\{\overline{W}_1 \backslash V_1, \ldots, \overline{W}_k \backslash V_k\}$ does not exceed $n-1$.

(c) $\Rightarrow$ (d) is obvious.

(d) $\Rightarrow$ (a). Let (d) hold and let $\{U_1, \ldots, U_{n+2}\}$ be an open covering of $X$. Since $X$ is normal there exists a closed covering $\{F_1, \ldots, F_{n+2}\}$ of $X$ such that $F_i \subset U_i$ for each $i$. By hypothesis there exists a family of open sets $\{V_1, \ldots, V_{n+2}\}$ such that

$$F_i \subset V_i \subset \overline{V}_i \subset U_i$$

for each $i$, and the family $\{\mathrm{bd}\,(V_1), \ldots, \mathrm{bd}\,(V_{n+2})\}$ has order not exceeding $n-1$. Let $L_j = \overline{V}_j \backslash \bigcup_{i<j} V_i$ for $j = 1, \ldots, n+2$. For each $j$, $L_j$ is closed, and $\{L_1, \ldots, L_{n+2}\}$ is a closed covering of $X$, for if $x \in X$, there exists $j$ such that $x \in V_j$ and $x \notin V_i$ for $i < j$ so that $x \in L_j$. Now

$$L_j = \overline{V}_j \cap \bigcap_{i<j} (X \backslash V_i)$$

so that

$$\bigcap_{j=1}^{n+2} L_j = \bigcap_{j=1}^{n+2} \overline{V}_j \cap \bigcap_{i=1}^{n+1} (X \backslash V_i) \subset \bigcap_{j=1}^{n+1} \mathrm{bd}\,(V_j) = \varnothing.$$

Thus $\{L_1, \ldots, L_{n+2}\}$ is a closed covering of $X$, $L_j \subset \overline{V}_j \subset U_j$ and

$$\bigcap_{j=1}^{n+2} L_j = \varnothing.$$

Hence by Proposition 1.6, $\dim X \leqslant n$.□

We can extend the result of Proposition 1.7 to countable families.

**1.8** *Proposition.* Let $X$ be a normal space such that $\dim X \leqslant n$. For each family $\{F_i\}_{i \in \mathbf{N}}$ of closed sets and each family $\{U_i\}_{i \in \mathbf{N}}$ of open sets such that $F_i \subset U_i$ for each $i$, there exists a family $\{V_i\}_{i \in \mathbf{N}}$ of open sets such that $F_i \subset V_i \subset U_i$ for each $i$ and the family $\{\mathrm{bd}\,(V_i)\}_{i \in \mathbf{N}}$ is of order not exceeding $n - 1$.

*Proof.* For each positive integer $i$ there exists an open set $V_{i,0}$ such that

$$F_i \subset V_{i,0} \subset \overline{V}_{i,0} \subset U_i.$$

Let $W_{i,0} = U_i$ for each $i$. The non-empty finite subsets of $\mathbf{N}$ form a countable set and we enumerate them as $\Gamma_j$, $j \in \mathbf{N}$. We shall prove that there exists families $\{V_{i,j}\}_{i,j \in \mathbf{N}}$ and $\{W_{i,j}\}_{i,j \in \mathbf{N}}$ of open sets of $X$ such that for each positive integer $k$ the following condition $(\alpha_k)$ is satisfied: for each $i$

$$\overline{V}_{i,k-1} \subset V_{i,k} \subset \overline{V}_{i,k} \subset W_{i,k} \subset W_{i,k-1}$$

and the finite family $\{\overline{W}_{i,k} \backslash V_{i,k}\}_{i \in \Gamma_k}$ is of order not exceeding $n - 1$. The proof is by induction. Suppose we have constructed families of open sets $\{V_{i,j} \mid i \in \mathbf{N}, j = 1, \ldots, k\}$ and $\{W_{i,j} \mid i \in \mathbf{N}, j = 1, \ldots, k\}$ such that the condition $(\alpha_j)$ is satisfied for $j = 1, \ldots, k$. We apply Proposition 1.7 to the finite family $\{\overline{V}_{i,k}\}_{i \in \Gamma_{k+1}}$ of closed sets and the family $\{W_{i,k}\}_{i \in \Gamma_{k+1}}$ of open sets such that $\overline{V}_{i,k} \subset W_{i,k}$ for each $i$ in $\Gamma_{k+1}$. We obtain open sets $V_{i,k+1}$ and $W_{i,k+1}$ for $i$ in $\Gamma_{k+1}$ such that

$$\overline{V}_{i,k} \subset V_{i,k+1} \subset \overline{V}_{i,k+1} \subset W_{i,k+1} \subset W_{i,k}$$

and the family $\{\overline{W}_{i,k+1} \backslash V_{i,k+1}\}_{i \in \Gamma_{k+1}}$ is of order not exceeding $n - 1$. If $i \in \mathbf{N} \backslash \Gamma_{k+1}$ we choose open sets $V_{i,k+1}$ and $W_{i,k+1}$ such that

$$\overline{V}_{i,k} \subset V_{i,k+1} \subset \overline{V}_{i,k+1} \subset W_{i,k+1} \subset W_{i,k}.$$

We then have families $\{V_{i,k+1}\}_{i \in \mathbb{N}}, \{W_{i,k+1}\}_{i \in \mathbb{N}}$ such that condition $(\alpha_{k+1})$ is satisfied. The method of construction of the families

$$\{V_{i,1}\}_{i \in \mathbb{N}}, \{W_{i,1}\}_{i \in \mathbb{N}}$$

is obvious, and the existence of the required families

$$\{V_{i,j}\}_{i,j \in \mathbb{N}}, \{W_{i,j}\}_{i,j \in \mathbb{N}}$$

now follows by induction.

For each positive integer $i$ we define $V_i = \bigcup_{j \in \mathbb{N}} V_{i,j}$. Then $V_i$ is open and it is clear that $F_i \subset V_i \subset U_i$ for each $i$. Suppose that the order of $\{\mathrm{bd}\,(V_i)\}_{i \in \mathbb{N}}$ is not less than $n$. Then there exists a subset $\Gamma_k$ of $\mathbb{N}$ with $n+1$ elements such that $\bigcap_{i \in \Gamma_k} \mathrm{bd}\,(V_i) \neq \varnothing$. But

$$V_{i,k} \subset V_i \subset W_{i,k}$$

for each $i$, so that $\overline{V_i} \subset \overline{W_{i,k}}$ and $\mathrm{bd}\,(V_i) \subset \overline{W_{i,k}} \backslash V_{i,k}$ and we have a contradiction of $(\alpha_k)$. Thus the family $\{\mathrm{bd}\,(V_i)\}_{i \in \mathbb{N}}$ is of order not exceeding $n-1$. $\square$

If $E, F$ are disjoint closed sets in a space $X$, a closed set $C$ is said to *separate* $E$ and $F$ if $X \backslash C = G \cup H$, where $G$ and $H$ are disjoint open sets and $E \subset G, F \subset H$. From Proposition 1.7 we obtain two more useful characterizations of $n$-dimensionality.

**1.9 Proposition.** *If $X$ is a normal space, the following statements about $X$ are equivalent:*

(a) $\dim X \leqslant n$;

(b) *for each family of $n+1$ pairs of closed sets*

$$\{(E_1, F_1), \ldots, (E_{n+1}, F_{n+1})\},$$

*where $E_i \cap F_i = \varnothing$ for each $i$, there exist $n+1$ continuous functions $f_i \colon X \to [-1, 1]$, $i = 1, \ldots, n+1$, such that for each $i$, $f_i(x) = 1$ if $x \in E_i$, $f_i(x) = -1$ if $x \in F_i$ and $\bigcap_{i=1}^{n+1} f_i^{-1}(0) = \varnothing$;*

(c) *for each family of $n+1$ pairs of closed sets*

$$\{(E_1, F_1), \ldots, (E_{n+1}, F_{n+1})\},$$

*where $E_i \cap F_i = \varnothing$ for each $i$, there exists a family $\{C_1, \ldots, C_{n+1}\}$ of closed sets such that each $C_i$ separates $E_i$ and $F_i$ in $X$ and $\bigcap_{i=1}^{n+1} C_i = \varnothing$.*

*Proof.* (a) $\Rightarrow$ (b). Let $X$ be a normal space such that $\dim X \leqslant n$ and let $\{(E_1, F_1), \ldots, (E_{n+1}, F_{n+1})\}$ be a family of pairs of disjoint closed sets.

By Proposition 1.7 there exist open sets $V_1, \ldots, V_{n+1}$ and $W_1, \ldots, W_{n+1}$ such that

$$E_i \subset V_i \subset \overline{V}_i \subset W_i \subset X \backslash F_i \quad \text{and} \quad \bigcap_{i=1}^{n+1} (\overline{W}_i \backslash V_i) = \varnothing.$$

By Urysohn's lemma, for each $i$ there exists a continuous function $f_i : X \to [-1, 1]$ such that $f_i(x) = 1$ if $x \in \overline{V}_i$, and $f_i(x) = -1$ if $x \notin W_i$. We note that $f_i^{-1}(0) \subset W_i \backslash \overline{V}_i \subset \overline{W}_i \backslash V_i$. Thus we have $n + 1$ continuous functions $f_i : X \to [-1, 1]$ such that $f_i(x) = 1$ if $x \in E_i$, $f_i(x) = -1$ if $x \in F_i$ and $\bigcap_{i=1}^{n+1} f_i^{-1}(0) = \varnothing$.

$(b) \Rightarrow (c)$. Let $(b)$ hold and let $\{(E_1, F_1), \ldots, (E_{n+1}, F_{n+1})\}$ be a family of pairs of disjoint closed sets. Then there exist continuous functions $f_i : X \to [-1, 1]$ such that $f_i(x) = 1$ if $x \in E_i$, $f_i(x) = -1$ if $x \in F_i$ and $\bigcap_{i=1}^{n+1} f_i^{-1}(0) = \varnothing$. Let

$$C_i = \{x \in X \mid f_i(x) = 0\},$$

$$G_i = \{x \in X \mid f_i(x) > 0\},$$

$$H_i = \{x \in X \mid f_i(x) < 0\}.$$

Then $C_i$ is a closed set, $G_i$ and $H_i$ are disjoint open sets such that $X \backslash C_i = G_i \cup H_i$ and $E_i \subset G_i$, $F_i \subset H_i$. Thus $C_i$ separates $E_i$ and $F_i$ in $X$ for each $i$, and $\bigcap_{i=1}^{n+1} C_i = \varnothing$.

$(c) \Rightarrow (a)$. Let $(c)$ hold, let $F_1, \ldots, F_{n+1}$ be closed sets and let

$$U_1, \ldots, U_{n+1}$$

be open sets such that each $F_i \subset U_i$. Then there exist closed sets $C_1, \ldots, C_{n+1}$ such that for each $i$, $C_i$ separates the disjoint closed sets $F_i$ and $X \backslash U_i$ in $X$ and $\bigcap_{i=1}^{n+1} C_i = \varnothing$. Since $C_i$ separates $F_i$ and $X \backslash U_i$, there exists an open set $V_i$ such that $F_i \subset V_i$ and $V_i \cup C_i$ is a closed set contained in $U_i$. Thus there exist open sets $V_1, \ldots, V_{n+1}$ such that $F_i \subset V_i \subset \overline{V}_i \subset U_i$ and $\mathrm{bd}\,(V_i) \subset C_i$ so that $\bigcap_{i=1}^{n+1} \mathrm{bd}\,(V_i) = \varnothing$. It follows from Proposition 1.7 that $\dim X \leqslant n$. $\square$

We conclude this section by showing the equivalence of $\dim X \leqslant n$ for a normal space $X$ with an apparently stronger condition.

**1.10   Proposition.** *A normal space $X$ satisfies $\dim X \leqslant n$ if and only if each finite open covering of $X$ has a star-refinement which is a finite open covering of $X$ of order not exceeding $n$.*

*Proof.* The condition is obviously sufficient and we prove its necessity. Let $\mathscr{U} = \{U_1, \ldots, U_k\}$ be an open covering of a normal space $X$ such that $\dim X \leqslant n$. Since $X$ is normal, there exists a closed covering

$\{F_1, \ldots, F_k\}$ of $X$ such that each $F_i \subset U_i$. Let $\Gamma$ be the set of non-empty subsets of $\{1, \ldots, k\}$. For each $\gamma$ in $\Gamma$ let

$$V_\gamma = \left( \bigcap_{i \in \gamma} U_i \right) \cap \left( \bigcap_{i \notin \gamma} (X \backslash F_i) \right).$$

If $x \in X$, then $x \in V_\gamma$, where $\gamma = \{i \mid x \in U_i\}$. Thus $\mathscr{V} = \{V_\gamma\}_{\gamma \in \Gamma}$ is a finite open covering of $X$. Furthermore, $\mathscr{V}$ is a star-refinement of $\mathscr{U}$, for if $x \in X$, then $x \in F_j$ for some $j$, and if $x \in V_\gamma$, then $j \in \gamma$ so that $V_\gamma \subset U_j$. Hence $\mathrm{St}\,(x, \mathscr{V}) \subset U_j$. Now since $\dim X \leqslant n$, there exists a finite open refinement $\mathscr{W}$ of $\mathscr{V}$ such that the order of $\mathscr{W}$ does not exceed $n$. Since $\mathscr{V}$ is a star-refinement of $\mathscr{U}$, it follows that $\mathscr{W}$ is a star-refinement of $\mathscr{U}$.☐

## 2   The dimension of Euclidean space

In this section we shall prove that the Euclidean space $\mathbf{R}^n$ is $n$-dimensional in the sense of covering dimension. The main step in the proof will be a characterization of the covering dimension of a normal space in terms of extension of mappings into spheres. For this result there are some preliminaries.

Let us say that two continuous functions $f_0, f_1 : X \to Y$ are *homotopic* if there exists a continuous function $h : X \times I \to Y$ such that

$$h(x, 0) = f_0(x), \quad h(x, 1) = f_1(x) \quad \text{if} \quad x \in X.$$

The continuous function $h$ is called a homotopy between $f_0$ and $f_1$. Let us say that two continuous functions $f_0, f_1 : X \to S^n$ are *uniformly homotopic* if they are homotopic and there exists a homotopy

$$h : X \times I \to S^n$$

between them which has the property that for each positive real number $\epsilon$ there exists a positive real number $\delta$ such that

$$\| h(x, s) - h(x, t) \| < \epsilon \quad \text{if} \quad |s - t| < \delta,$$

for all $x$ in $X$. The homotopy $h$ is then called a uniform homotopy between $f_0$ and $f_1$. Since $S^n$ is compact, the concept of uniform homotopy of mappings into $S^n$ does not depend on the particular metric on $S^n$, in the sense that if $\rho$ is an equivalent metric on $S^n$, then the continuous functions $f_0, f_1 : X \to S^n$ are uniformly homotopic if and only if there exists a homotopy $h$ between them which satisfies the condition: for each positive real number $\epsilon$ there exists a positive real number $\delta$ such that $\rho(h(x, s), h(x, t)) < \epsilon$ if $|s - t| < \delta$ for all $x$ in $X$.

**2.1 Lemma.** *Let $X$ be a normal space, let $A$ be a closed subspace of $X$ and let the continuous functions $f_0, f_1 \colon A \to S^n$ be uniformly homotopic. If $f_0$ has an extension $g_0 \colon X \to S^n$, then $f_1$ has an extension $g_1 \colon X \to S^n$.*

*Proof.* Let $h$ be a uniform homotopy between $f_0$ and $f_1$. Then there exists $\delta > 0$ such that for all $x$ in $A$

$$\|h(y,s) - h(y,t)\| < 1 \quad \text{if} \quad |s-t| < \delta.$$

Choose a positive integer $m$ such that $1/m < \delta$, and for $r = 0, \ldots, m$ let $e_r \colon A \to S^n$ be defined by

$$e_r(x) = h(x, r/m) \quad \text{if} \quad x \in A.$$

Then $e_0, \ldots, e_m$ are continuous functions such that $e_0 = f_0$, $e_m = f_1$ and $e_{r-1}$ and $e_r$ are uniformly homotopic for $r = 1, \ldots, m$. Furthermore if $x \in A$, then $\|e_{r-1}(x) - e_r(x)\| < 1$. Therefore it is sufficient to prove the result under the assumption that

$$\|f_0(x) - f_1(x)\| < 1 \quad \text{for all} \quad x \text{ in } A.$$

Since $X$ is a normal space, by the Tietze–Urysohn extension theorem $f$ has an extension $g \colon X \to E^{n+1}$. Let

$$V = \{x \in X \mid \|g_0(x) - g(x)\| < 1\}.$$

Then $V$ is an open set and since we assume that $\|f_0(x) - f_1(x)\| < 1$ if $x \in A$, it follows that $A \subset V$. By Urysohn's lemma there exists a continuous function $\phi \colon X \to I$ such that $\phi(x) = 0$ if $x \notin V$ and $\phi(x) = 1$ if $x \in A$. Let $h' \colon X \times I \to E^{n+1}$ be given by

$$h'(x,t) = (1-t) g_0(x) + t g(x) \quad \text{if} \quad (x,t) \in X \times I.$$

If $x \in V$, then $h'(x,t) \neq \mathbf{0}$. For

$$\|h'(x,t)\| \geqslant \|g_0(x)\| - \|g_0(x) - h'(x,t)\|$$
$$\geqslant \|g_0(x)\| - \|g_0(x) - g(x)\|,$$

and $g_0(x) \in S^n$ so that $\|g_0(x)\| = 1$, whilst if $x \in V$, then

$$\|g_0(x) - g(x)\| < 1.$$

Thus if $x \in V$, then $\|h'(x,t)\| > 0$. It follows that $h'(x, \phi(x)) \neq \mathbf{0}$ if $x \in X$, for if $x \notin V$ then $\phi(x) = 0$ and $h'(x, 0) = g_0(x) \in S^n$. Thus we can define $g_1 \colon X \to S^n$ by

$$g_1(x) = h'(x, \phi(x)) / \|h'(x, \phi(x))\| \quad \text{if} \quad x \in X.$$

The continuous function $g_1$ is the required extension of $f$. For if $x \in A$,

$$h'(x, \phi(x)) = h'(x, 1) = g(x) = f_1(x). \quad \square$$

**2.2 Theorem.** *If $X$ is a normal space, then $\dim X \leqslant n$ if and only if for each closed set $A$ of $X$, each continuous function $f: A \to S^n$ has an extension $g: X \to S^n$.*

*Proof.* Let $X$ be a normal space such that $\dim X \leqslant n$, let $A$ be a closed subspace of $X$ and let $f: A \to S^n$ be a given continuous function. We regard $S^n$ as the boundary of the cube $Q^{n+1}$ in $\mathbf{R}^{n+1}$, where

$$Q^{n+1} = \{t \in \mathbf{R}^{n+1} \mid \|t_i\| \leqslant 1 \quad \text{for} \quad i = 1, \ldots, n+1\}.$$

If $x \in A$ let

$$f(x) = \big(f_1(x), \ldots, f_{n+1}(x)\big),$$

and for $i = 1, \ldots, n+1$ let

$$E_i = \{x \in A \mid f_i(x) = 1\}, \quad F_i = \{x \in A \mid f_i(x) = -1\}.$$

Then $E_i$ and $F_i$ are disjoint sets, closed in $A$ and hence in $X$, and $A = \bigcup_{i=1}^{n+1}(E_i \cup F_i)$. By Proposition 1.9, there exist continuous functions $\phi_i: X \to [-1, 1]$, $i = 1, \ldots, n+1$, such that $\phi_i(x) = 1$ if $x \in E_i$, $\phi_i(x) = -1$ if $x \in F_i$ and $\bigcap_{i=1}^{n+1}\phi_i^{-1}(0) = \varnothing$. Let $\phi: X \to Q^{n+1}$ be given by

$$\phi(x) = \big(\phi_1(x), \ldots, \phi_{n+1}(x)\big) \quad \text{if} \quad x \in X.$$

If $x \in A$, then either $x \in E_i$ for some $i$ so that $\phi_i(x) = f_i(x) = 1$, or $x \in F_j$ for some $j$ so that $\phi_j(x) = f_j(x) = -1$. It follows that we can define continuous functions $\psi: A \to S^n$ and $h: A \times I \to S^n$ by putting $\psi(x) = \phi(x)$ if $x \in A$ and

$$h(x, t) = (1-t)\phi(x) + tf(x) \quad \text{if} \quad (x, t) \in A \times I.$$

If $x \in A$ and $s, t \in I$, then

$$\|h(x, s) - h(x, t)\| = |s - t| \, \|\phi(x) - f(x)\|.$$

Since $\|\phi(x) - f(x)\| \leqslant 2\sqrt{(n+1)}$ if $x \in A$, it follows that $h$ is a uniform homotopy between $\psi$ and $f$. Since $\bigcap_{i=1}^{n+1}\phi_i^{-1}(0) = \varnothing$, it follows that $\phi(X) \subset Q^{n+1}\backslash\{0\}$ so that $\psi$ has an extension to $X$ since $S^n$ is a retract of $Q^{n+1}\backslash\{0\}$. It now follows from Lemma 2.1 that $f$ has an extension $g: X \to S^n$.

Conversely let $X$ be a normal space which satisfies the condition and let $(E_1, F_1), \ldots, (E_{n+1}, F_{n+1})$ be $n+1$ pairs of disjoint closed sets. By Urysohn's lemma, for $i = 1, \ldots, n+1$ there exists a continuous function $f_i: X \to [-1, 1]$ such that $f_i(x) = 1$ if $x \in E_i$, and $f_i(x) = -1$ if $x \in F_i$. Let us regard $S^n$ as the boundary of the cube $Q^{n+1}$ in $\mathbf{R}^{n+1}$. If

$A = \bigcup_{i=1}^{n+1}(E_i \cup F_i)$, then $A$ is a closed set of $X$ and we can define a continuous function $f: A \rightarrow S^n$ by

$$f(x) = \big(f_1(x), \ldots, f_{n+1}(x)\big) \quad \text{if} \quad x \in A.$$

By hypothesis $f$ has an extension $g: X \rightarrow S^n$. For $x$ in $X$ let

$$g(x) = \big(g_1(x), \ldots, g_{n+1}(x)\big).$$

Then for $i = 1, \ldots, n+1$ we have a continuous function

$$g_i: X \rightarrow [-1, 1]$$

such that $g_i(x) = 1$ if $x \in E_i$, $g_i(x) = -1$ if $x \in F_i$ and

$$\bigcap_{i=1}^{n+1} g_i^{-1}(0) = \{x \in X \mid g(x) = 0\} = \varnothing.$$

Thus $\dim X \leqslant n$ by Proposition 1.9. $\square$

We have the following extension of Theorem 2.2.

**2.3   Proposition.** *Let $A$ be a closed set of a normal space $X$ such that $\dim F \leqslant n$ for each closed $F$ of $X$ which is disjoint from $A$. Then each continuous function $f: A \rightarrow S^n$ has an extension $g: X \rightarrow S^n$.*

*Proof.* Since $X$ is normal, by Proposition 1.3.8 there exists an open set $U$ such that $A \subset U$ and a mapping $\phi: U \rightarrow S^n$ which extends $f$, and there exists an open set $V$ such that

$$A \subset V \subset \overline{V} \subset U.$$

The set $\overline{V} \backslash V$ is closed in $X \backslash V$ and $\dim (X \backslash V) \leqslant n$ since $X \backslash V$ is a closed set of $X$ disjoint from $A$. Hence by Theorem 2.2 there exists a continuous function $\psi: X \backslash V \rightarrow S^n$ such that $\psi | \overline{V} \backslash V = \phi | \overline{V} \backslash V$. Let $g: X \rightarrow S^n$ be defined as follows:

$$g(x) = \begin{cases} \phi(x) & \text{if} \quad x \in \overline{V}, \\ \psi(x) & \text{if} \quad x \in X \backslash V. \end{cases}$$

The definition is meaningful and the continuous function $g$ is the required extension of $f$. $\square$

Next we shall establish a 'sum theorem'. The proof will be by induction using Theorem 2.2 and the following lemma.

**2.4   Lemma.** *Let $X$ be a normal space and let $A$ be a closed set of $X$ such that $\dim A \leqslant n$. If $B$ is a closed set of $X$ and $\phi: B \rightarrow S^n$ is continuous,* *then there exist an open set $V$ such that $B \subset V$, and a continuous function $\psi: A \cup \overline{V} \to S^n$ such that $\psi|B = \phi$.*

*Proof.* By Proposition 1.3.8 there exist an open set $U$ such that $B \subset U$, and a continuous function $g: U \to S^n$ such that $g|B = \phi$. There exists an open set $V$ such that $B \subset V \subset \overline{V} \subset U$. If $\overline{V}$ does not meet $A$, then let $\psi: A \cup \overline{V} \to S^n$ be a mapping such that $\psi|A$ is constant and $\psi|\overline{V} = g|\overline{V}$. If $V$ meets $A$, then $g|A \cap \overline{V}$ has an extension

$$h: A \to S^n$$

since $\dim A \leqslant n$. Let $\psi: A \cup \overline{V} \to S^n$ be the unique mapping such that $\psi|A = h$ and $\psi|\overline{V} = g|\overline{V}$. In both cases $\psi$ is the required extension. $\square$

**2.5 Theorem.** *(The countable sum theorem.) If $X$ is a normal space and $X = \bigcup_{i \in \mathbf{N}} A_i$, where each $A_i$ is closed in $X$ and $\dim A_i \leqslant n$, then $\dim X \leqslant n$.*

*Proof.* Let $A$ be a closed set of $X$ and let $f: A \to S^n$ be a continuous function. For each positive integer $j$, let $B_j = \bigcup_{i \leqslant j} A_i$. We shall construct by induction open sets $V_j$ for $j$ in $\mathbf{N}$ such that $A \subset V_1$ and $B_{j-1} \cup \overline{V}_{j-1} \subset V_j$ for $j > 1$, and continuous functions $f_j: B_j \cup \overline{V}_j \to S^n$ such that $f_1|A = f$ and $f_j|B_{j-1} \cup \overline{V}_{j-1} = f_{j-1}$ for $j > 1$.

By Lemma 2.4 there exist an open set $V_1$ such that $A \subset V_1$, and a continuous function $f_1: A \cup \overline{V}_1 \to S^n$ such that $f_1|A = f$. Suppose that there exist open sets $V_j$ and continuous functions $f_j$ with the required properties if $j \leqslant k$. By Lemma 2.4 there exist an open set $V_{k+1}$ such that $B_k \cup \overline{V}_k \subset V_{k+1}$, and a continuous function

$$f_{k+1}: A_{k+1} \cup \overline{V}_{k+1} \to S^n$$

such that $f_{k+1}|B_k \cup \overline{V}_k = f_k$. Since $B_{k+1} = A_{k+1} \cup B_k$ and $B_k \subset V_{k+1}$, we have $A_{k+1} \cup \overline{V}_{k+1} = B_{k+1} \cup \overline{V}_{k+1}$. It follows by induction that there exist open sets $V_j$ and continuous functions $f_j$ with the stated properties for $j \in \mathbf{N}$.

Let $g_j: V_j \to S^n$ be given by $g_j = f_j|V_j$. Then $A \subset V_1$ and $g_1|A = f$. Furthermore $V_j \subset V_{j+1}$ for each $j$ and $g_{j+1}|V_j = g_j$. The sets $V_j$ are open and since $B_j \subset V_{j+1}$ for each $j$, it follows that $\bigcup_{j \in \mathbf{N}} V_j = X$. Thus we can define a continuous function $g: X \to S^n$ by $g(x) = g_j(x)$ if $x \in V_j$. Then $g|A = f$ so that $\dim X \leqslant n$ by Theorem 2.2. $\square$

Sum theorems will be discussed in more detail in §5. Next we have a 'product theorem' needed for the determination of the dimension of Euclidean space. Chapter 9 will be devoted to a full discussion of product theorems.

**2.6 Proposition.** *If $X$ and $Y$ are bicompact spaces at least one of which is non-empty, then*

$$\dim X \times Y \leqslant \dim X + \dim Y.$$

*Proof.* It is enough to prove that if $\dim X \leqslant m$ and $\dim Y \leqslant n$, then $\dim X \times Y \leqslant m+n$. Thus we suppose that $\dim X \leqslant m$, $\dim Y \leqslant n$ and let $\mathscr{U}$ be a finite open covering of $X \times Y$. Then $\mathscr{U}$ has an open refinement

$$\{V_i \times W_{ij} \,|\, j = 1, \ldots, n_i, i = 1, \ldots, k\},$$

where $\{V_1, \ldots, V_k\}$ is an open covering of $X$ and $\{W_{i1}, \ldots, W_{in_i}\}$ is an open covering of $Y$ for each $i$. The space $Y$ is normal, so that for each $i$ there exists a closed covering $\{K_{i1}, \ldots, K_{in_i}\}$ of $Y$ such that $K_{ij} \subset W_{ij}$. But $\dim Y \leqslant n$ so that by Proposition 1.7 there exist open sets $G_{ij}$ for $j = 1, \ldots, n_i$ and $i = 1, \ldots, k$ such that

$$K_{ij} \subset G_{ij} \subset \bar{G}_{ij} \subset W_{ij},$$

and the order of the family

$$\{\mathrm{bd}(G_{ij}) \,|\, j = 1, \ldots, n_i, i = 1, \ldots, k\}$$

does not exceed $n-1$. For $j = 1, \ldots, n_i$ and $i = 1, \ldots, k$ let

$$L_{ij} = \bar{G}_{ij} \backslash \bigcup_{r<j} G_{ir}.$$

Then $L_{ij} \subset W_{ij}$ and $\{L_{ij} \,|\, j = 1, \ldots, n_i\}$ is a closed covering of $Y$ for each $i$. Since $\dim X \leqslant m$, by Proposition 1.6 there exists a closed covering $\{F_1, \ldots, F_k\}$ of $X$, of order not exceeding $m$, such that $F_i \subset V_i$ for each $i$. If

$$\mathscr{F} = \{F_i \times L_{ij} \,|\, j = 1, \ldots, n_i, i = 1, \ldots, k\},$$

then $\mathscr{F}$ is a finite closed covering of $X \times Y$ and a refinement of $\mathscr{U}$. The space $X \times Y$ is bicompact and so normal, so that by Proposition 1.6 the proof will be complete if it is shown that the order of $\mathscr{F}$ does not exceed $m+n$. Suppose that $\bigcap_{i=1}^{r} \bigcap_{\nu=1}^{s(i)} (F_{\mu_i} \times L_{\mu_i j_{i\nu}})$ is non-empty, where $\mu_1 < \ldots < \mu_r$ and $j_{i1} < \ldots < j_{is(i)}$ for each $i$. Then $r \leqslant m+1$ and since

$$\bigcap_{\nu=1}^{s(i)} L_{\mu_i j_{i\nu}} = \bigcap_{\nu=1}^{s(i)} \bar{G}_{\mu_i j_{i\nu}} \cap \bigcap_{j<j_{u(i)}} (X \backslash G_{\mu_i j})$$

$$\subset \bigcap_{\nu=1}^{s(i)-1} \mathrm{bd}\,(G_{\mu_i j_{i\nu}}),$$

it follows that $\Sigma_{i=1}^{r}\big(s(i)-1\big) \leqslant n$. Thus

$$\sum_{i=1}^{r} s(i) \leqslant n+r \leqslant m+n+1.$$

Hence the order of $\mathscr{F}$ does not exceed $m+n$. □

**2.7    Theorem.** $\dim \mathbf{R}^n = n$.

*Proof.* Let

$$I^n = \{t \in \mathbf{R}^n \,|\, 0 \leqslant t_i \leqslant 1, \quad i = 1,\ldots,n\}.$$

We saw in Example 1.4 that $\dim I = 1$. Since $I^n$ is homeomorphic with $I^{n-1} \times I$ for $n > 1$, Proposition 2.6 allows an easy inductive proof that $\dim I^n \leqslant n$. For each $n$-tuple $(k_1,\ldots,k_n)$ of integers let

$$J_{(k_1,\ldots k_n)} = \{t \in \mathbf{R}^n \,|\, k_i \leqslant t_i \leqslant k_i+1 \text{ for } i = 1,\ldots,n\}.$$

Then $J_{(k_1,\ldots,k_n)}$ is a subset of $\mathbf{R}^n$ homeomorphic with $I^n$ and the collection of such subsets is a countable closed covering of $\mathbf{R}^n$. It follows from Theorem 2.5 that $\dim \mathbf{R}^n \leqslant n$. Since $\dim I = 1$ it follows that $\dim \mathbf{R} = 1$. Now suppose that for some $n > 1$, $\dim \mathbf{R}^n \leqslant n-1$ so that $\dim E^n \leqslant n-1$. It follows from Theorem 2.2 that the identity mapping $S^{n-1} \to S^{n-1}$ has an extension to $E^n$. Thus $S^{n-1}$ is a retract of $E^n$ which, by Brouwer's theorem, is absurd. It follows that $\dim \mathbf{R}^n = n$ for all positive integers $n$. □

## 3    Inessential mappings and unstable values

In this section there are two more characterizations of covering dimension of normal spaces which are closely related to Theorem 2.2. In different presentations of the theory, each of them might occupy the fundamental position of that theorem.

Let

$$Q^{n+1} = \{t \in \mathbf{R}^{n+1} \,|\, |t_i| \leqslant 1 \quad \text{for} \quad i = 1,\ldots,n+1\},$$

and let the boundary of $Q^{n+1}$ be denoted by $S^n$. Clearly $S^n$ is homeomorphic with the unit sphere in $\mathbf{R}^{n+1}$. If $X$ is a topological space, a continuous function $f \colon X \to Q^{n+1}$ is said to be *inessential* if the mapping $f^{-1}(S^n) \to S^n$ given by restriction of $f$ has an extension $g \colon X \to S^n$.

**3.1    Proposition.** *If $X$ is a normal space, then $\dim X \leqslant n$ if and only if every continuous mapping of $X$ to $Q^{n+1}$ is inessential.*

*Proof.* Let $X$ be a normal space such that $\dim X \leqslant n$ and let

$$f \colon X \to Q^{n+1}$$

be a continuous function. Since $f^{-1}(S^n)$ is a closed set of $X$, it follows from Theorem 2.2 that there exists a continuous function $g: X \to S^n$ such that $g(x) = f(x)$ if $x \in f^{-1}(S^n)$. Thus $f$ is inessential.

Conversely let $X$ be a normal space such that every continuous mapping of $X$ into $Q^{n+1}$ is inessential. Let $A$ be a closed set of $X$ and let $f: A \to S^n$ be a continuous function. By the Tietze–Urysohn extension theorem there exists $\phi: X \to Q^{n+1}$ such that $\phi(x) = f(x)$ if $x \in A$. By hypothesis $\phi$ is inessential so that there exists a continuous function $g: X \to S^n$ such that $g(x) = \phi(x)$ if $x \in \phi^{-1}(S^n)$. Since

$$A \subset \phi^{-1}(S^n),$$

it follows that $g \mid A = f$. Thus $\dim X \leqslant n$ by Theorem 2.2. $\square$

Let $f: X \to Q^{n+1}$ be a continuous function. A point $y_0$ of $f(X)$ is called an *unstable value* of $f$ if for each positive real number $\epsilon$ there exists a continuous function $g: X \to Q^{n+1}$ such that $y_0 \notin g(X)$ and

$$\|f(x) - g(x)\| < \epsilon$$

for every $x$ in $X$.

**3.2   Proposition.** *The following statements about a normal space $X$ are equivalent:*

(a) $\dim X \leqslant n$;

(b) *some interior point $y_0$ of $Q^{n+1}$ is an unstable value of every continuous mapping $f: X \to Q^{n+1}$ such that $y_0 \in f(X)$;*

(c) *all values of every continuous mapping of $X$ into $Q^{n+1}$ are unstable.*

*Proof.* (a) $\Rightarrow$ (b). Suppose that $\dim X \leqslant n$ and let $f: X \to Q^{n+1}$ be a continuous function such that the origin $\mathbf{0}$ is a point of $f(X)$. Let $\epsilon > 0$ be given and let

$$E = \{x \in X \mid \|f(x)\| \leqslant \tfrac{1}{2}\epsilon\}, \quad F = \{x \in X \mid \|f(x)\| \geqslant \tfrac{1}{2}\epsilon\}.$$

If $A = E \cap F$ and

$$S = \{t \in \mathbf{R}^{n+1} \mid \|t\| = \tfrac{1}{2}\epsilon\},$$

we can define a continuous function $\phi: A \to S$ by $\phi(x) = f(x)$ for $x$ in $A$. Since $A$ is a closed set of $X$ and $S$ is homeomorphic with $S^n$, by Theorem 2.2 there exists a continuous function $\psi: X \to S$ such that $\psi \mid A = \phi$. Let $g: X \to Q^{n+1}$ be defined by

$$g(x) = \begin{cases} \psi(x) & \text{if } x \in E, \\ f(x) & \text{if } x \in F. \end{cases}$$

This definition is meaningful and since the restrictions $g|E$ and $g|F$ are continuous, $g$ is a continuous function. Clearly $\|g(x)-f(x)\| < \epsilon$ for all $x$ and $\mathbf{0} \notin g(X)$. Thus $\mathbf{0}$ is an unstable value of $f$.

$(b) \Rightarrow (c)$. If $y_0$ and $y_1$ are interior points of $Q^{n+1}$, then there exists a uniform homeomorphism $h$ of $Q^{n+1}$ with itself such that $h(y_0) = y_1$. Thus if one interior point of $Q^{n+1}$ is an unstable value whenever it is attained, then every interior point of $Q^{n+1}$ has this property. To complete the proof we must show that if $f:X \to Q^{n+1}$ is a continuous function, then every boundary point of $Q^{n+1}$ is an unstable value of $f$. Given $\epsilon > 0$, let $g:X \to Q^{n+1}$ be the continuous function given by

$$g(x) = (1-\epsilon/n+1)f(x)$$

for $x$ in $X$. If $x \in X$, then $\|g(x)-f(x)\| < \epsilon$ and $g(X)$ contains no boundary point of $Q^{n+1}$.

$(c) \Rightarrow (a)$. Let $\{(A_1, B_1), \ldots, (A_{n+1}, B_{n+1})\}$ be a family of pairs of disjoint closed sets of a normal space $X$ which has the property that all values of every continuous mapping of $X$ into $Q^{n+1}$ are unstable. By Urysohn's lemma, for $i = 1, \ldots, n+1$, there exists a continuous function $f_i:X \to [-1, 1]$ such that $f_i(x) = 1$ if $x \in A_i$, and $f_i(x) = -1$ if $x \in B_i$. Let us define $f:X \to Q^{n+1}$ by putting

$$f(x) = \big(f_1(x), \ldots, f_{n+1}(x)\big) \quad \text{if} \quad x \in X.$$

Since $\mathbf{0}$ is an unstable value of $f$ if it is attained, it follows that there exists a continuous function $g:X \to Q^{n+1}$ such that $\|f(x)-g(x)\| < \tfrac{1}{2}$ for every $x$ and $\mathbf{0} \notin g(X)$. Let

$$g(x) = \big(g_1(x), \ldots, g_{n+1}(x)\big) \quad \text{if} \quad x \in X,$$

and for $i = 1, \ldots, n+1$ let

$$U_i = \{x \in X \mid g_i(x) < 0\},$$

$$V_i = \{x \in X \mid g_i(x) > 0\},$$

$$C_i = \{x \in X \mid g_i(x) = 0\}.$$

Then for each $i$, $C_i$ is a closed set and since $A_i \subset U_i$ and $B_i \subset V_i$ we see that $C_i$ separates $A_i$ and $B_i$. And since $\mathbf{0} \notin g(X)$, we have

$$\bigcap_{i=1}^{n+1} C_i = \varnothing.$$

Thus $\dim X \leqslant n$ by Proposition 1.9.$\square$

## 4   Locally finite coverings

The concept of dimension would be different if based on arbitrary open coverings instead of finite coverings. There follows an example of a zero-dimensional normal space which would have infinite dimension if dimension were defined in terms of the order of refinements of arbitrary open coverings.

**4.1**   *Example.* The space $T$ of ordinals less than the first uncountable ordinal satisfies dim $T = 0$, but there is an open covering of $T$ which has no open refinement of finite order.

Let $\mathscr{U} = \{U_1, \ldots, U_k\}$ be a finite open covering of $T$. Then by Lemma 1.4.13, there is some $\alpha_0$ in $T$ and some $j$ such that $(\alpha_0, \omega_1) \subset U_j$, where $(\alpha_0, \omega_1) = \{\alpha \in T \mid \alpha > \alpha_0\}$. We construct by induction a decreasing sequence $\{\alpha_r\}_{r \in \mathbf{N}}$ of ordinals. We choose $i_0$ such that $\alpha_0 \in U_{i_0}$. If

$$[0, \alpha_0] \subset U_{i_0}$$

we put $\alpha_r = 0$ for all $r$ in $\mathbf{N}$. Otherwise we choose $\alpha_1$ to be the least ordinal such that $(\alpha_1, \alpha_0] \subset U_{i_0}$. If $\alpha_1, \ldots, \alpha_m$ have been constructed, we choose $i_m$ such that $\alpha_m \in U_{i_m}$. If $[0, \alpha_m] \subset U_{i_m}$, we put $\alpha_r = 0$ for $r > m$. Otherwise we choose $\alpha_{m+1}$ to be the least ordinal such that $(\alpha_{m+1}, \alpha_m] \subset U_{i_m}$. Since $T$ can contain no strictly decreasing infinite sequence, there exist ordinals $\alpha_1, \ldots, \alpha_m$ such that $\alpha_m < \ldots < \alpha_1 < \alpha_0$ and integers $i_0, \ldots, i_m$ such that $(\alpha_{r+1}, \alpha_r] \subset U_{i_r}$ for $r = 0, \ldots, m-1$, and $[0, \alpha_m] \subset U_{i_m}$. Then $[0, \alpha_m]$, $(\alpha_0, \omega_1)$ and the sets $(\alpha_{r+1}, \alpha_r]$ for $r = 0, \ldots, m-1$ form a disjoint open covering of $T$ which is a refinement of $\mathscr{U}$. Thus dim $T = 0$. But by Lemma 1.4.13, if $\mathscr{W}$ is an open covering of $T$ which has a point-countable open refinement, then some member of $\mathscr{W}$ must contain $(\alpha, \omega_1)$ for some $\alpha$ in $T$. Thus the open covering $\{[0, \alpha)\}_{\alpha \in T \setminus \{0\}}$ has no point-countable open refinement and in particular no open refinement of finite order.

We shall show that for normal spaces the concept of dimension can be based on locally finite open coverings. First we need the following result which is of independent interest.

**4.2**   **Proposition.** *Let $\{F_\gamma\}_{\gamma \in \mathbf{N}}$ be a locally finite closed covering of a space $X$ such that dim $F_\gamma \leqslant n$ for each $\gamma$, and let $\{G_\lambda\}_{\lambda \in \Lambda}$ be a locally finite open covering of $X$ such that no $F_\gamma$ meets infinitely many $G_\lambda$.*

*Then there exists an open covering $\{H_\lambda\}_{\lambda\in\Lambda}$ of $X$ of order not exceeding $n$ such that $H_\lambda \subset G_\lambda$ if $\lambda \in \Lambda$.*

*Proof.* Let $\Gamma$ be well-ordered and let $\alpha$ be an element of $\Gamma$. Suppose that we have found a closed set $A_{\gamma\lambda}$ and an open set $G_{\gamma\lambda}$ for each $\gamma$ in $\Gamma$ such that $\gamma < \alpha$ and each $\lambda$ in $\Lambda$ to satisfy:

(a) $A_{\gamma\lambda} \subset F_\gamma$,

(b) $G_{\gamma\lambda} = G_\lambda\backslash\bigcup_{\beta<\gamma}A_{\beta\lambda}$,

(c) $\bigcup_{\lambda\in\Lambda}(G_{\gamma\lambda}\backslash A_{\gamma\lambda}) = X$,

(d) the order of $\{(F_\gamma \cap G_{\gamma\lambda})\backslash A_{\gamma\lambda}\}_{\lambda\in\Lambda}$ does not exceed $n$. We shall define $G_{\alpha\lambda}$ and $A_{\alpha\lambda}$ for each $\lambda$ in $\Lambda$. Let

$$G_{\alpha\lambda} = G_\lambda\backslash\bigcup_{\gamma<\alpha} A_{\gamma\lambda}.$$

Since $A_{\gamma\lambda} \subset F_\gamma$, it follows that $\{A_{\gamma\lambda}\}_{\gamma<\alpha}$ is a locally finite family of closed sets so that $\bigcup_{\gamma<\alpha}A_{\gamma\lambda}$ is closed. Hence $G_{\alpha\lambda}$ is an open set for each $\lambda$.

Moreover $\bigcup_{\lambda\in\Lambda}G_{\alpha\lambda} = X$. For suppose that $x\in X\backslash\bigcup_{\lambda\in\Lambda}G_{\alpha\lambda}$ and let $M_x = \{\lambda\in\Lambda\,|\,x\in G_\lambda\}$. For each $\lambda$ in $M_x$ there exists $\gamma(\lambda)$ in $\Gamma$ such that $\gamma(\lambda) < \alpha$ and $x\in A_{\gamma(\lambda)\lambda}$. Since $M_x$ is finite, there exists $\delta$ in $\Gamma$ such that $\delta < \alpha$ and $\gamma(\lambda) \leqslant \delta$ for each $\lambda$ in $M_x$. If $\lambda\in M_x$ and $\gamma(\lambda) = \delta$, then $x\in A_{\delta\lambda}$ so that $x\notin G_{\delta\lambda}\backslash A_{\delta\lambda}$. If $\lambda\in M_x$ and $\gamma(\lambda) < \delta$, then $x\in\bigcup_{\beta<\delta}A_{\beta\lambda}$ so that $x\notin G_{\delta\lambda}$ and hence $x\notin G_{\delta\lambda}\backslash A_{\delta\lambda}$. It is clear that $x\notin G_{\delta\lambda}\backslash A_{\delta\lambda}$ if $\lambda\notin M_x$. Thus we have a contradiction of (c). It follows that $\{G_{\alpha\lambda}\}_{\lambda\in\Lambda}$ is an open covering of $X$.

The set $F_\alpha$ meets only finitely many sets $G_\lambda$ so that $\{F_\alpha \cap G_{\alpha\lambda}\}_{\lambda\in\Lambda}$ is an open covering of $F_\alpha$, only finitely many members of which are non-empty. But $\dim F_\alpha \leqslant n$ so that there exists an open covering $\{U_\lambda\}_{\lambda\in\Lambda}$ of $F_\alpha$ such that $U_\lambda \subset F_\alpha \cap G_{\alpha\lambda}$ for each $\lambda$, and the order of $\{U_\lambda\}_{\lambda\in\Lambda}$ does not exceed $n$. Let

$$A_{\alpha\lambda} = ((F_\alpha \cap G_{\alpha\lambda})\backslash U_\lambda)^-.$$

Then $A_{\alpha\lambda}$ is a closed set of $X$ and $A_{\alpha\lambda} \subset F_\alpha$ for each $\lambda$. Since

$$(F_\alpha \cap G_{\alpha\lambda})\backslash A_{\alpha\lambda} = U_\lambda,$$

it follows that the order of $\{(F_\alpha \cap G_{\alpha\lambda})\backslash A_{\alpha\lambda}\}_{\lambda\in\Lambda}$ does not exceed $n$. Furthermore since $\{U_\lambda\}_{\lambda\in\Lambda}$ is a covering of $F_\alpha$, it follows that

$$F_\alpha \subset \bigcup_{\lambda\in\Lambda} (G_{\alpha\lambda}\backslash A_{\alpha\lambda}).$$

Since $\{G_{\alpha\lambda}\}_{\lambda\in\Lambda}$ is a covering of $X$ and $A_{\alpha\lambda} \subset F_\alpha$ for each $\lambda$, it follows that $X\backslash F_\alpha \subset \bigcup_{\lambda\in\Lambda}(G_{\alpha\lambda}\backslash A_{\alpha\lambda})$. Hence

$$\bigcup_{\lambda\in\Lambda} (G_{\alpha\lambda}\backslash A_{\alpha\lambda}) = X.$$

It follows by transfinite induction that we can construct $G_{\gamma\lambda}$ and $A_{\gamma\lambda}$ for $\gamma$ in $\Gamma$, $\lambda$ in $\Lambda$ to satisfy $(a)$, $(b)$, $(c)$ and $(d)$.

Now let

$$H_\lambda = G_\lambda \backslash \bigcup_{\gamma\in\Gamma} A_{\gamma\lambda}.$$

Since $A_{\gamma\lambda} \subset F_\gamma$, it follows that $\{A_{\gamma\lambda}\}_{\gamma\in\Gamma}$ is a locally finite family of closed sets so that $\bigcup_{\gamma\in\Gamma} A_{\gamma\lambda}$ is closed. Hence $H_\lambda$ is an open set for each $\lambda$. The assumption that $X \backslash \bigcup_{\lambda\in\Lambda} H_\lambda$ is non-empty leads to a contradiction of $(c)$ by an argument used above. Thus $\{H_\lambda\}_{\lambda\in\Lambda}$ is an open covering of $X$ such that $H_\lambda \subset G_\lambda$ if $\lambda \in \Lambda$. For each $\gamma$ in $\Gamma$ the order of the family $\{(F_\gamma \cap G_{\gamma\lambda}) \backslash A_{\gamma\lambda}\}_{\lambda\in\Lambda}$ does not exceed $n$. But

$$F_\gamma \cap H_\lambda = (F_\gamma \cap G_\lambda) \backslash \bigcup_{\gamma\in\Gamma} A_{\gamma\lambda}$$

and

$$(F_\gamma \cap G_{\gamma\lambda}) \backslash A_{\gamma\lambda} = (F_\gamma \cap G_\lambda) \backslash \bigcup_{\beta\leqslant\gamma} A_{\beta\lambda}.$$

Thus $F_\gamma \cap H_\lambda \subset (F_\gamma \cap G_{\gamma\lambda}) \backslash A_{\gamma\lambda}$ and it follows that the order of

$$\{F_\gamma \cap H_\lambda\}_{\lambda\in\Lambda}$$

does not exceed $n$. Since $\{F_\gamma\}_{\gamma\in\Gamma}$ is a covering of $X$, it follows that the order of $\{H_\lambda\}_{\lambda\in\Lambda}$ does not exceed $n$. $\square$

**4.3** **Theorem.** *If $X$ is a normal space, then $\dim X \leqslant n$ if and only if each locally finite open covering has an open refinement of order not exceeding $n$.*

*Proof.* The condition is clearly sufficient. Thus let $X$ be a normal space such that $\dim X \leqslant n$, and let $\{G_\lambda\}_{\lambda\in\Lambda}$ be a locally finite open covering of $X$. By Proposition 1.3.13 there exist a locally finite open refinement $\{H_\gamma\}_{\gamma\in\Gamma}$ of $\{G_\lambda\}_{\lambda\in\Lambda}$ and a locally finite open covering $\{U_\gamma\}_{\gamma\in\Gamma}$ of $X$ such that no $U_\gamma$ meets infinitely many $H_\beta$, and there exists a locally finite closed covering $\{F_\gamma\}_{\gamma\in\Gamma}$ of $X$ such that each $F_\gamma \subset U_\gamma$. Since $F_\gamma$ is closed and $\dim X \leqslant n$, it follows that $\dim F_\gamma \leqslant n$ for each $\gamma$, and no $F_\gamma$ meets infinitely many $H_\beta$. By Proposition 3.2 there exists an open covering $\{W_\gamma\}_{\gamma\in\Gamma}$ of $X$ such that $W_\gamma \subset H_\gamma$ for $\gamma$ in $\Gamma$ and the order of $\{W_\gamma\}_{\gamma\in\Gamma}$ does not exceed $n$. Clearly $\{W_\gamma\}_{\gamma\in\Gamma}$ is a refinement of $\{G_\lambda\}_{\lambda\in\Lambda}$. $\square$

**4.4** **Corollary.** *If $X$ is a normal space such that $\dim X \leqslant n$ and $\{G_\lambda\}_{\lambda\in\Lambda}$ is a locally finite open covering of $X$, then there exists an open covering $\{H_\lambda\}_{\lambda\in\Lambda}$ of $X$ such that each $H_\lambda \subset G_\lambda$ and the order of $\{H_\lambda\}_{\lambda\in\Lambda}$ does not exceed $n$.*

*Proof.* There exists an open refinement $\{W_\gamma\}_{\gamma \in \Gamma}$ of $\{G_\lambda\}_{\lambda \in \Lambda}$ which has order not exceeding $n$. We can define a function $\tau : \Gamma \to \Lambda$ such that $W_\gamma \subset G_{\tau(\gamma)}$ for each $\gamma$. Let $H_\lambda = \bigcup_{\tau(\gamma)=\lambda} W_\gamma$. Then $\{H_\lambda\}_{\lambda \in \Lambda}$ is an open covering of $X$, each $H_\lambda \subset G_\lambda$ and the order of $\{H_\lambda\}_{\lambda \in \Lambda}$ does not exceed $n$. $\square$

**4.5** *Remark.* It is now clear that if $X$ is a paracompact normal space, then $\dim X \leqslant n$ if and only if each open covering of $X$ has an open refinement of order not exceeding $n$.

The following result, which complements Proposition 1.7 will be useful in the discussion of the dimension of pseudo-metrizable spaces.

**4.6** *Proposition.* Let $\{G_\lambda\}_{\lambda \in \Lambda}$ be a locally finite family of open sets of a normal space $X$ such that $\dim X \leqslant n$ and let $\{F_\lambda\}_{\lambda \in \Lambda}$ be a family of closed sets such that each $F_\lambda \subset G_\lambda$. Then there exists a family $\{V_\lambda\}_{\lambda \in \Lambda}$ of open sets such that

$$F_\lambda \subset V_\lambda \subset \overline{V}_\lambda \subset G_\lambda,$$

and the order of the family $\{\mathrm{bd}\,(V_\lambda)\}_{\lambda \in \Lambda}$ does not exceed $n-1$.

*Proof.* Let $\Lambda$ be well-ordered and let $\Gamma$ denote the set of finite subsets of $\Lambda$. For each $\gamma$ in $\Gamma$ let

$$W_\gamma = \bigcap_{\lambda \in \gamma} G_\lambda \cap \bigcap_{\lambda \notin \gamma} (X \backslash F_\lambda).$$

Since $\bigcup_{\lambda \notin \gamma} F_\lambda$ is a closed set, it follows that $\bigcap_{\lambda \notin \gamma} (X \backslash F_\lambda)$ is open and hence $W_\gamma$ is an open set. It is clear that $\{W_\gamma\}_{\gamma \in \Gamma}$ is a covering of $X$. And if $x \in X$, there exists an open neighbourhood $N$ of $x$ and $\gamma_0$ in $\Gamma$ such that $N \cap G_\lambda = \varnothing$ if $\lambda \notin \gamma_0$. It follows that $N \cap W_\gamma = \varnothing$ if $\gamma$ is not a subset of $\gamma_0$. Hence the set $\{\gamma \in \Gamma \mid N \cap W_\gamma \neq \varnothing\}$ is finite. Thus $\{W_\gamma\}_{\gamma \in \Gamma}$ is a locally finite open covering of $X$. Since $X$ is normal and $\dim X \leqslant n$, it follows from Corollary 4.4 that there is a locally finite open covering $\{H_\gamma\}_{\gamma \in \Gamma}$ of order not exceeding $n$ such that each $H_\gamma \subset W_\gamma$, and there exists a closed covering $\{K_\gamma\}_{\gamma \in \Gamma}$ of $X$ such that each $K_\gamma \subset H_\gamma$.

For $\gamma$ in $\Gamma$ and each $\lambda$ in $\gamma$ we can find an open set $V_{\lambda\gamma}$ such that

$$K_\gamma \subset V_{\lambda\gamma} \subset \overline{V}_{\lambda\gamma} \subset H_\gamma,$$

and if $\lambda, \mu \in \gamma$ and $\lambda < \mu$ then $\overline{V}_{\lambda\gamma} \subset V_{\mu\gamma}$. For $\lambda$ in $\Lambda$ let

$$\Gamma_\lambda = \{\gamma \in \Gamma \mid \lambda \in \gamma\}$$

and let

$$V_\lambda = \bigcup_{\gamma \in \Gamma_\lambda} V_{\lambda\gamma}.$$

Suppose that $x \in F_\lambda$. Since $\{K_\gamma\}$ is a covering of $X$, it follows that $x \in K_\gamma$ for some $\gamma$ and $\gamma \in \Gamma_\lambda$ since $K_\gamma \cap F_\lambda = \varnothing$ if $\lambda \notin \gamma$. But if $\gamma \in \Gamma_\lambda$, then $K_\gamma \subset V_{\lambda\gamma}$. Thus $x \in V_\lambda$ so that $F_\lambda \subset V_\lambda$. If $\gamma \in \Gamma_\lambda$, then $\overline{V}_{\lambda\gamma} \subset H_\gamma \subset G_\lambda$ and $\{\overline{V}_{\lambda\gamma}\}_{\gamma \in \Gamma_\lambda}$ is locally finite so that $\overline{V}_\lambda \subset G_\lambda$. Hence for each $\lambda$ in $\Lambda$ we have found an open set $V_\lambda$ such that $F_\lambda \subset V_\lambda \subset \overline{V}_\lambda \subset G_\lambda$.

It remains to show that the order of $\{\mathrm{bd}\,(V_\lambda)\}_{\lambda \in \Lambda}$ does not exceed $n-1$. Suppose that $x \in \bigcap_{i=1}^{n+1} \mathrm{bd}\,(V_{\lambda_i})$, where $\lambda_1, \ldots, \lambda_{n+1}$ are all different. Since

$$\mathrm{bd}\,(V_\lambda) \subset \bigcup_{\gamma \in \Gamma_\lambda} \mathrm{bd}\,(V_{\lambda\gamma}),$$

it follows that for each $i$ there exists $\gamma_i$ such that $\lambda_i \in \gamma_i$ and $x \in \mathrm{bd}\,(V_{\lambda_i \gamma_i})$. If $i \neq j$, then $\gamma_i \neq \gamma_j$, for if $\gamma_i = \gamma_j = \gamma$, then $\lambda_i, \lambda_j \in \gamma$ and

$$x \in \overline{V}_{\lambda_i \gamma}, x \in \overline{V}_{\lambda_j \gamma},$$

but $x \notin V_{\lambda_i \gamma}, x \notin V_{\lambda_j \gamma}$, which is absurd since either $\overline{V}_{\lambda_i \gamma} \subset V_{\lambda_j \gamma}$ or $\overline{V}_{\lambda_j \gamma} \subset V_{\lambda_i \gamma}$. Since $x \notin V_{\lambda_i \gamma_i}$, it follows that $x \notin K_{\gamma_i}$. But $\{K_\gamma\}_{\gamma \in \Gamma}$ is a covering of $X$, so that there exists $\gamma$ in $\Gamma$, different from each of the $\gamma_i$, such that $x \in K_\gamma \subset H_\gamma$. Since $x \in \overline{V}_{\lambda_i \gamma_i}$ it follows that $x \in H_{\gamma_i}$ for each $i$ so that $x \in H_\gamma \cap \bigcap_{i=1}^{n+1} H_{\gamma_i}$. Since the order of $\{H_\gamma\}_{\gamma \in \Gamma}$ does not exceed $n$, this is absurd. Hence $\bigcap_{i=1}^{n+1} \mathrm{bd}\,(V_{\lambda_i}) = \varnothing$. $\square$

**4.7   Corollary.** *Let $\{G_\lambda\}_{\lambda \in \Lambda}$ be a locally finite family of open sets of a normal space $X$ such that $\dim X \leqslant n$, and let $\{F_\lambda\}_{\lambda \in \Lambda}$ be a family of closed sets such that each $F_\lambda \subset G_\lambda$. Then there exist open sets $V_\lambda, W_\lambda$ for $\lambda$ in $\Lambda$ such that*

$$F_\lambda \subset V_\lambda \subset \overline{V}_\lambda \subset W_\lambda \subset G_\lambda,$$

*and the order of the family $\{\overline{W}_\lambda \backslash V_\lambda\}_{\lambda \in \Lambda}$ does not exceed $n-1$.*

*Proof.* By Proposition 4.6 there exist open sets $V_\lambda$ such that

$$F_\lambda \subset V_\lambda \subset \overline{V}_\lambda \subset G_\lambda$$

and the order of $\{\overline{V}_\lambda \backslash V_\lambda\}_{\lambda \in \Lambda}$ does not exceed $n-1$. Since $X$ is normal and $\{G_\lambda\}_{\lambda \in \Lambda}$ is locally finite, by Proposition 1.3.14 there exist open sets $H_\lambda$ such that $\overline{V}_\lambda \backslash V_\lambda \subset H_\lambda \subset G_\lambda$ for each $\lambda$, and the families

$$\{\overline{V}_\lambda \backslash V_\lambda\}_{\lambda \in \Lambda}$$

and $\{\overline{H}_\lambda\}_{\lambda \in \Lambda}$ are similar. For each $\lambda$ in $\Lambda$, let $W_\lambda = V_\lambda \cup H_\lambda$. Since $\overline{V}_\lambda \backslash V_\lambda \subset \overline{W}_\lambda \backslash V_\lambda \subset \overline{H}_\lambda$ for each $\lambda$, the families $\{\overline{V}_\lambda \backslash V_\lambda\}_{\lambda \in \Lambda}$ and $\{\overline{W}_\lambda \backslash V_\lambda\}_{\lambda \in \Lambda}$ are similar. Thus $F_\lambda \subset V_\lambda \subset \overline{V}_\lambda \subset W_\lambda \subset G_\lambda$ and the family $\{\overline{W}_\lambda \backslash V_\lambda\}_{\lambda \in \Lambda}$ is of order not exceeding $n-1$. $\square$

## 5  Sum theorems

This section is devoted to 'sum theorems'. We seek determinations of the dimension of a space in terms of the dimension of the members of some covering of the space. We already have the countable sum theorem (Theorem 2.5) which states that if $\{A_i\}_{i\in\mathbf{N}}$ is a countable closed covering of a normal space $X$, then $\dim X = \sup \dim A_i$.

**5.1  Proposition.** *If $\{A_\lambda\}_{\lambda\in\Lambda}$ is a locally finite closed covering of a space $X$ and $\dim A_\lambda \leqslant n$ for each $\lambda$ in $\Lambda$, then $\dim X \leqslant n$.*

*Proof.* By Proposition 4.2, each finite open covering of $X$ has an open refinement of order not exceeding $n$. Hence $\dim X \leqslant n$.□

For normal spaces we have:

**5.2  Proposition.** *If $\{A_\lambda\}_{\lambda\in\Lambda}$ is a locally finite covering of a normal space $X$, where each $A_\lambda$ is an $F_\sigma$-set and $\dim A_\lambda \leqslant n$, then $\dim X \leqslant n$.*

*Proof.* For each $\lambda, A_\lambda = \bigcup_{i\in\mathbf{N}}A_{i\lambda}$, where $A_{i\lambda}$ is a closed set. Since $A_{i\lambda}$ is a closed set of $A_\lambda$, it follows that $\dim A_{i\lambda} \leqslant n$. Since $\{A_{i\lambda}\}_{\lambda\in\Lambda}$ is a locally finite family of closed sets, it follows that $B_i = \bigcup_{\lambda\in\Lambda}A_{i\lambda}$ is a closed set of $X$ for each $i$ and $\dim B_i \leqslant n$ by Proposition 5.1. Since $\{B_i\}_{i\in\mathbf{N}}$ is a countable closed covering of the normal space $X$, by the countable sum theorem $\dim X \leqslant n$.□

The countable sum theorem has a similar generalization.

**5.3  Proposition.** *If $X$ is a normal space and $X = \bigcup_{k\in\mathbf{N}}A_k$, where each $A_k$ is an $F_\sigma$-set and $\dim A_k \leqslant n$, then $\dim X \leqslant n$.*

*Proof.* For each $k, A_k = \bigcup_{i\in\mathbf{N}}A_{ik}$, where $A_{ik}$ is closed in $X$ and hence closed in $A_k$. It follows that $\dim A_{ik} \leqslant n$, and $\{A_{ik}\}$ is a covering of $X$. Hence by the countable sum theorem $\dim X \leqslant n$.□

For paracompact spaces, Propositions 5.2 and 5.3 have a common generalization. A family $\{A_\lambda\}_{\lambda\in\Lambda}$ of subsets of a space $X$ is said to be locally countable if for each $x$ in $X$ there exists an open neighbourhood $N$ of $x$ such that the set $\{\lambda\in\Lambda \mid N\cap A_\lambda \neq \varnothing\}$ is countable.

**5.4  Proposition.** *If $\{A_\lambda\}_{\lambda\in\Lambda}$ is a locally countable covering of a paracompact normal space $X$, where each $A_\lambda$ is an $F_\sigma$-set and $\dim A_\lambda \leqslant n$, then $\dim X \leqslant n$.*

*Proof.* Since $\{A_\lambda\}_{\lambda\in\Lambda}$ is locally countable, there exists an open covering $\{U_\gamma\}_{\gamma\in\Gamma}$ of $X$ such that for each $\gamma$ in $\Gamma$, the set $\{\lambda\in\Lambda\mid U_\gamma\cap A_\lambda\neq\varnothing\}$ is countable. Since $X$ is a paracompact normal space there exists a locally finite closed covering $\{K_\gamma\}_{\gamma\in\Gamma}$ of $X$ such that $K_\gamma\subset U_\gamma$ for each $\gamma$. The set $K_\gamma\cap A_\lambda$ is closed in $A_\lambda$, so that $\dim K_\gamma\cap A_\lambda\leqslant n$. The family $\{K_\gamma\cap A_\lambda\}_{\lambda\in\Lambda}$ is a covering of the normal space $K_\gamma$ by $F_\sigma$-sets, and only countably many of these sets are non-empty. Hence $\dim K_\gamma\leqslant n$ by Proposition 5.3. It now follows from Proposition 5.1 that $\dim X\leqslant n$.□

We recall that if $A$ is an $F_\sigma$-set in a normal space $X$, then the subspace $A$ is normal. It follows from the countable sum theorem that

$$\dim A\leqslant\dim X.$$

For suppose that $\dim X\leqslant n$ and that $A=\bigcup_{\lambda\in\mathbf{N}}A_i$, where each $A_i$ is a closed set. Then $\{A_i\}_{i\in\mathbf{N}}$ is a countable closed covering of the normal space $A$ and $\dim A_i\leqslant n$ for each $i$. Hence $\dim A\leqslant n$.

**5.5  Proposition.** *Let $X$ be a perfectly normal space and let $\{H_i\}_{i\in\mathbf{N}}$ be a covering of $X$ such that $\bigcup_{i\leqslant k}H_i$ is an open set and $\dim H_k\leqslant n$ for each $k$. Then $\dim X\leqslant n$.*

*Proof.* For each $k$, let

$$F_k=H_k\backslash\bigcup_{i<k}H_i.$$

Since $X$ is a perfectly normal space, $\bigcup_{i\leqslant k}H_i$ is an $F_\sigma$-set of $X$ so that

$$F_k=(\bigcup_{i\leqslant k}H_i)\cap(X\backslash\bigcup_{i<k}H_i)$$

is a closed subset of an $F_\sigma$-set and hence is an $F_\sigma$-set of $X$. Since $F_k\subset H_k$, it follows that $F_k$ is an $F_\sigma$-set of the normal space $H_k$ so that $\dim F_k\leqslant\dim H_k\leqslant n$. Since $\{F_k\}_{k\in\mathbf{N}}$ is a countable covering of $X$ by $F_\sigma$-sets it follows from Proposition 5.3 that $\dim X\leqslant n$.□

The next result will have many applications in the rest of the work.

**5.6  Proposition.** *Let $A$ be a closed set of a normal space $X$. If $\dim A\leqslant n$ and if $\dim F\leqslant n$ for each closed $F$ of $X$ which does not meet $A$, then $\dim X\leqslant n$.*

*Proof.* Let $B$ be a closed set of $X$ and let $f:B\to S^n$ be a continuous function. It follows from Lemma 2.4 that $f$ has an extension

$$g:A\cup B\to S^n.$$

By hypothesis if $F$ is a closed set of $X$ disjoint from $A \cup B$, then $\dim F \leqslant n$ so that by Proposition 2.3, $g$ has an extension $h \colon X \to S^n$. Then $h$ is an extension of $f$. Thus $\dim X \leqslant n$ by Theorem 2.2. $\square$

**5.7 Proposition.** *If $X$ is a normal space and $X = A \cup B$, where $A$ is a closed set, $\dim A \leqslant n$ and $\dim B \leqslant n$, then $\dim X \leqslant n$.*

*Proof.* If $F$ is a closed set of $X$ which is disjoint from $A$, then $F$ is a closed set of $B$ so that $\dim F \leqslant n$. It follows from Proposition 5.6 that $\dim X \leqslant n$. $\square$

**5.8 Corollary.** *If $A$ is a closed set of a normal space $X$, then*

$$\dim X \leqslant \max\{\dim A, \dim(X \backslash A)\}. \square$$

Next we shall prove that if $\{U_\lambda\}_{\lambda \in \Lambda}$ is a point-finite open covering of a normal space $X$ such that $\dim U_\lambda \leqslant n$ for each $\lambda$, then $\dim X \leqslant n$. The corresponding statement for closed coverings is false since any $T_1$-space has a point-finite covering by zero-dimensional closed sets, namely its points. For the proof we establish the following property of point-finite open coverings which will also be used later.

**5.9 Lemma.** *If $\mathcal{U} = \{U_\lambda\}_{\lambda \in \Lambda}$ is a point-finite open covering of a topological space $X$, and for each non-negative integer $k$, the set of points $x$ of $X$ such that $\{\lambda \in \Lambda \mid x \in U_\lambda\}$ has at most $k$ elements is denoted by $E_k$, then $\{E_k\}_{k \geqslant 0}$ is a closed covering of $X$. Furthermore if $k > 0$ and $F$ is a closed set of $E_k$ disjoint from $E_{k-1}$, then there is a discrete closed covering of $F$, each member of which is contained in some member of $\mathcal{U}$.*

*Proof.* If $x \notin E_k$, then there exists a subset of $\gamma$ of $\Lambda$ with $k+1$ members such that $x \in U_\lambda$ for each $\lambda$ in $\gamma$. Thus $\bigcap_{\lambda \in \gamma} U_\lambda$ is an open neighbourhood of $x$ disjoint from $E_k$. Thus $E_k$ is a closed set. Since $\mathcal{U}$ is a point-finite covering, it is clear that $\{E_k\}_{k \geqslant 0}$ is a covering of $X$. Now let $k$ be positive and let $F$ be a closed set of $E_k$ disjoint from $E_{k-1}$. Let $\Gamma$ be the set of subsets of $\Lambda$ which have $k$ elements and for each $\gamma$ in $\Gamma$ let $V_\gamma = \bigcap_{\lambda \in \gamma} U_\lambda$. The family $\{V_\gamma \cap F\}_{\gamma \in \Gamma}$ is a disjoint open covering of $F$ and thus is a discrete closed covering of $F$, and $V_\gamma \cap F \subset U_\lambda$ for each $\lambda$ in $\gamma$. $\square$

**5.10 Proposition.** *If $\{U_\lambda\}_{\lambda \in \Lambda}$ is a point-finite open covering of a normal space $X$ such that $\dim U_\lambda \leqslant n$ for each $\lambda$, then $\dim X \leqslant n$.*

*Proof.* Let $\{E_k\}_{k \geqslant 0}$ be the closed covering of $X$ given by Lemma 5.9. We shall prove by induction that $\dim E_k \leqslant n$ for all $k$. The set $E_0$ is empty. Let us suppose that $k \geqslant 1$ and that $\dim E_{k-1} \leqslant n$. If $F$ is a closed set of $E_k$ disjoint from $E_{k-1}$ then, by Lemma 5.9, $F = \bigcup_{\gamma \in \Gamma} F_\gamma$, where $\{F_\gamma\}_{\gamma \in \Gamma}$ is a discrete family of closed sets of $F$ such that for each $\gamma$ in $\Gamma$ there exists some $\lambda$ in $\Lambda$ such that $F_\gamma \subset U_\lambda$. Since $F_\gamma$ is closed in $F$ which is closed in $X$, it follows that $F_\gamma$ is closed in $U_\lambda$ so that $\dim F_\gamma \leqslant n$ for each $\gamma$. Thus $\dim F \leqslant n$ by Proposition 5.1 and it follows from Proposition 5.7 that $\dim E_k \leqslant n$. The sets $E_k$ form a countable closed covering of the normal space $X$. Hence $\dim X \leqslant n$.□

Finally we obtain a 'Urysohn inequality' for covering dimension.

**5.11  Proposition.** *If $X$ is a normal space and $X = A \cup B$, then*

$$\dim X \leqslant \dim A + \dim B + 1.$$

*Proof.* It will be enough to prove that if $\dim A \leqslant m$ and $\dim B \leqslant n$, then $\dim X \leqslant m+n+1$. Since $X = \bar{A} \cup B$, in view of Proposition 5.7 it is enough to prove that $\dim \bar{A} \leqslant m+n+1$. Let $\{G_1, \ldots, G_k\}$ be a finite covering of $\bar{A}$ by open sets of $\bar{A}$. Since $\bar{A}$ is normal, there exists a covering $\{H_1, \ldots, H_k\}$ of $\bar{A}$ by open sets of $\bar{A}$ such that each $\bar{H}_i \subset G_i$. Since $\dim A \leqslant m$, there exists a covering $\{W_1, \ldots, W_k\}$ of $A$ by open sets of $A$ such that each $W_i \subset H_i$ and the order of $\{W_1, \ldots, W_k\}$ does not exceed $m$. For each $i$, let $V_i$ be an open set in $\bar{A}$ such that $A \cap V_i = W_i$. Then since $A$ is dense in $\bar{A}$, the order of $\{V_1, \ldots, V_k\}$ does not exceed $m$. Furthermore since $\bar{V}_i = \bar{W}_i$, it follows that $V_i \subset \bar{W}_i \subset \bar{H}_i \subset G_i$. Let

$$D = \bar{A} \setminus \bigcup_{i=1}^{k} V_i.$$

Then $D$ is a closed set of $X$ so that $D$ is normal. Since $D \subset B$, it follows that $\dim D \leqslant n$. Hence there exists a covering $\{F_1, \ldots, F_k\}$ of $D$ by closed sets of $D$ such that each $F_i \subset G_i$ and the order of $\{F_1, \ldots, F_k\}$ does not exceed $n$. Since $F_1, \ldots, F_k$ are closed in $\bar{A}$, by Proposition 1.3.14 there exist open sets $U_1, \ldots, U_k$ of $\bar{A}$ such that $F_i \subset U_i \subset G_i$ and $\{U_1, \ldots, U_k\}$ is similar to $\{F_1, \ldots, F_k\}$. Thus the order of $\{U_1, \ldots, U_k\}$ does not exceed $n$. Hence $\{U_1, \ldots, U_k, V_1, \ldots, V_k\}$ is a covering of $\bar{A}$ by open sets of $\bar{A}$ which is a refinement of $\{G_1, \ldots, G_k\}$ and is of order not exceeding $m+n+1$. Hence $\dim \bar{A} \leqslant m+n+1$.□

## 6   The monotonicity of covering dimension

We return now to the question of monotonicity of dimension. We wish to know under what conditions we can assert that if $M$ is a subspace of a space $X$, then $\dim M \leqslant \dim X$. In §1 we saw that if $M$ is closed in $X$, then $\dim M \leqslant \dim X$, and in the preceding section we saw that if $M$ is an $F_\sigma$-set in a normal space $X$, then $\dim M \leqslant \dim X$. We shall prove that if $M$ is normally situated in a normal space $X$, then $\dim M \leqslant \dim X$. First we have an example of failure of monotonocity.

**6.1** *Example.* There exists a completely normal zero-dimensional space containing an open subspace of dimension equal to 1.

Let $I$ be the unit interval and let $J$ be a space consisting of the points of $I$ and a special point $j_0$. The open sets of $J$ are $J$ and the open sets of $I$. The space $J$ is completely normal. For if $Y$ is a subspace of $J$, then either $j_0 \in Y$, in which case $Y$ is trivially normal since it contains no disjoint pairs of non-empty closed sets, or $j_0 \notin Y$, in which case $Y$ is a subspace of $I$ and so is normal. Since $J$ is non-empty and $\{J\}$ is a refinement of every open covering of $J$, we see that $\dim J = 0$. But $I$ is a subspace of $J$ and $\dim I = 1$.

Of course $J$ is neither Hausdorff nor regular. In the next chapter there will be an example of a $T_4$-space with a non-normal subspace of higher dimension, and in Chapter 5 there will be an example of a $T_4$-space with a normal subspace of higher dimension. It is an open question whether the subset theorem, that $\dim M \leqslant \dim X$ if $M \subset X$, holds for all completely normal regular spaces. Here the subset theorem is proved for the class of totally normal spaces.

**6.2** *Proposition.* If $M$ *is normally situated in a normal space* $X$, *then* $\dim M \leqslant \dim X$.

*Proof.* Let us suppose that $\dim X \leqslant n$ and let $\{V_1, \ldots, V_k\}$ be a covering of $M$ by open sets of $M$. For each $i = 1, \ldots, k$, there exists $U_i$ open in $X$ such that $V_i = M \cap U_i$. If $U = \bigcup_{i=1}^k U_i$, then $U$ is an open set of $X$ and $M \subset U$. Hence there exists an open set $G$ such that $M \subset G \subset U$ and $G = \bigcup_{\lambda \in \Lambda} G_\lambda$, where $\{G_\lambda\}_{\lambda \in \Lambda}$ is a family, locally finite in $G$, of open $F_\sigma$-sets of $X$. The subspace $G$ is normal and $\{G_\lambda\}_{\lambda \in \Lambda}$ is a locally finite covering of $G$ by $F_\sigma$-sets. For each $\lambda$, $\dim G_\lambda \leqslant n$ since $G_\lambda$ is an $F_\sigma$-set of $X$. Hence $\dim G \leqslant n$ by Proposition 5.2. Now

$$\{U_1 \cap G, \ldots, U_k \cap G\}$$

is an open covering of $G$. Hence there exists an open covering $\{P_1, \ldots, P_k\}$ of $G$, of order not exceeding $n$, such that each $P_i \subset U_i \cap G$. If

$$W_i = P_i \cap M,$$

then each $W_i$ is open in $M$, $W_i \subset V_i$ and $\{W_1, \ldots, W_k\}$ is a covering of $M$ of order not exceeding $n$. Hence $\dim M \leqslant n$. $\square$

**6.3   Corollary.** *If M is a generalized $F_\sigma$-set in a normal space X, then* dim $M \leqslant$ dim $X$.

*Proof.* This follows from the preceding result since by Proposition 1.4.5 a generalized $F_\sigma$-set in a normal space $X$ is normally situated in $X$.□

**6.4   Theorem.** (*The subset theorem.*) *If M is a subspace of a totally normal space X, then* dim $M \leqslant$ dim $X$.

*Proof.* A totally normal space is a normal space in which every subset is normally situated (Definition 1.4.10). Hence this result follows from Proposition 6.2.□

The following result is slightly more general.

**6.5   Proposition.** *Let X be a completely normal space in which each open set is the union of a point-finite family of open $F_\sigma$-sets of X. If $M \subset X$, then* dim $M \leqslant$ dim $X$.

*Proof.* It follows from Proposition 5.10 that if $U$ is an open set in a space $X$ with this property, then dim $U \leqslant$ dim $X$. It is clear from the proof of Proposition 6.2 that this implies the validity of the subset theorem for $X$.□

For the last subset theorem we require a lemma which also provides an alternative proof that dimension is monotonic on generalized $F_\sigma$-sets in a normal space.

**6.6   Lemma.** *Let X be a normal space and let M be a subspace of X. Then* dim $M \leqslant$ dim $X$ *if for each open set U of X such that $M \subset U$ there exists a disjoint covering $\{G_\lambda\}_{\lambda \in \Lambda}$ of M by open sets of M, and a family $\{F_\lambda\}_{\lambda \in \Lambda}$ of $F_\sigma$-sets of X, such that $G_\lambda \subset F_\lambda \subset U$ for each $\lambda$.*

*Proof.* Let us suppose that dim $X \leqslant n$ and let $\{V_1, \ldots, V_k\}$ be an open covering of $M$. There exist $U_1, \ldots, U_k$ open in $X$ such that $M \cap U_i = V_i$ for $i = 1, \ldots, k$. Let $U = \bigcup_{i=1}^k U_i$. Then $M \subset U$. Thus $M = \bigcup_{\lambda \in \Lambda} G_\lambda$, where the $G_\lambda$ are disjoint and open in $M$ and $G_\lambda \subset F_\lambda \subset U$, where $F_\lambda$ is an $F_\sigma$-set of $X$. It follows that dim $F_\lambda \leqslant n$. Since $F_\lambda \subset \bigcup_{i=1}^k U_i$, there exist open set $W_{i\lambda}$ for $i = 1, \ldots, k$ such that $F_\lambda \subset \bigcup_{i=1}^k W_{i\lambda}$, each $W_{i\lambda} \subset U_i$, and the order of $\{F_\lambda \cap W_{1\lambda}, \ldots, F_\lambda \cap W_{k\lambda}\}$ does not exceed $n$. For $i = 1, \ldots k$ let

$$W_i = \bigcup_{\lambda \in \Lambda} (G_\lambda \cap W_{i\lambda}).$$

Then $\{W_1, \ldots, W_k\}$ is an open covering of $M$, of order not exceeding $n$, and each $W_i \subset V_i$. Thus dim $M \leqslant n$. $\square$

**6.7** **Proposition.** *If $X$ is a normal regular space and $M$ is a strongly paracompact subspace, then* dim $M \leqslant$ dim $X$.

*Proof.* Let $U$ be an open set such that $M \subset U$. Since $X$ is a regular space, each point $x$ of $M$ has an open neighbourhood $V_x$ of $x$ in $X$ such that $\bar{V}_x \subset U$. Since $M$ is strongly paracompact, the open covering $\{V_x \cap M\}_{x \in M}$ of $M$ has a refinement of the form $\{G_{i\lambda}\}_{i \in \mathbf{N}, \lambda \in \Lambda}$, where $G_{i\lambda}$ is open in $M$ and $G_{i\lambda}, G_{j\mu}$ are disjoint if $\lambda \neq \mu$. For $\lambda$ in $\Lambda$ let

$$G_\lambda = \bigcup_{i \in \mathbf{N}} G_{i\lambda}$$

and let $F_\lambda = \bigcup_{i \in \mathbf{N}} \bar{G}_{i\lambda}$, where $\bar{G}_{i\lambda}$ is the closure of $G_{i\lambda}$ in $X$. Then $\{G_\lambda\}_{\lambda \in \Lambda}$ is a disjoint open covering of $M$, and for each $\lambda$, $G_\lambda \subset F_\lambda$ and $F_\lambda$ is an $F_\sigma$-set of $X$. Finally each $G_{i\lambda} \subset V_x$ for some $x$, so that $\bar{G}_{i\lambda} \subset U$ and hence $F_\lambda \subset U$. It follows from Lemma 6.6 that dim $M \leqslant$ dim $X$. $\square$

**6.8** *Remark.* The relation dim $M \leqslant$ dim $X$ is not true in general for a paracompact subspace $M$ of a normal regular space $X$. In fact in the Notes on Chapter 7, we shall find a bicompact space with a metrizable subspace of higher dimension.

We can make a slight improvement in the preceding subset theorem. A space $Y$ is said to be locally strongly paracompact if each point $y$ of $Y$ has an open neighbourhood $V$ whose closure $\bar{V}$ is strongly paracompact.

**6.9** **Proposition.** *Let $X$ be a normal regular space and let $M$ be a subspace of $X$ which is weakly paracompact, normal and locally strongly paracompact. Then* dim $M \leqslant$ dim $X$.

*Proof.* Let us suppose that dim $X \leqslant n$. Since $M$ is locally strongly paracompact, each point of $M$ has a neighbourhood whose closure in $M$ is strongly paracompact. Since $M$ is weakly paracompact, there is a point-finite covering $\{U_\lambda\}_{\lambda \in \Lambda}$ of $M$, where each $U_\lambda$ is open in $M$ and the closure $C_\lambda$ of $U_\lambda$ in $M$ is strongly paracompact. A point-finite open covering of a normal space is shrinkable so that there is a closed covering $\{F_\lambda\}_{\lambda \in \Lambda}$ of $M$ such that $F_\lambda \subset U_\lambda$ for each $\lambda$. For each $\lambda$, there exists an open $F_\sigma$-set $H_\lambda$ of $M$ such that $F_\lambda \subset H_\lambda \subset U_\lambda$. Since $C_\lambda$ is strongly paracompact, dim $C_\lambda \leqslant n$ by Proposition 6.7. But $H_\lambda$ is an

$F_\sigma$-set of the normal space $C_\lambda$, so that $\dim H_\lambda \leqslant n$. Hence the normal space $M$ has a point-finite open covering $\{H_\lambda\}_{\lambda \in \Lambda}$ such that $\dim H_\lambda \leqslant n$ for each $\lambda$. It follows from Proposition 5.10 that $\dim M \leqslant n$.□

## 7  The dimension of polyhedra

In this section we shall prove that if $K$ is a simplicial complex, then the dimension of its geometric realization $\dim |K|$ is equal to its combinatorial dimension $\dim K$. For this purpose we obtain a sum theorem for normal spaces which is more general than Proposition 5.1.

**7.1  Proposition.** *If a topological space $X$ is dominated by a covering $\{A_\lambda\}_{\lambda \in \Lambda}$, where the subspace $A_\lambda$ is normal and $\dim A_\lambda \leqslant n$, then $X$ is a normal space and $\dim X \leqslant n$.*

*Proof.* It follows from Proposition 1.4.3 that $X$ is a normal space. Let $B$ be a closed set of $X$ and let $f: B \to S^n$ be a continuous function. Let $\mathscr{G}$ be the set of continuous functions $g$ which are extensions of $f$ and have domains of the form $B \cup \bigcup_{\lambda \in \Lambda'} A_\lambda$ for some subset $\Lambda'$ of $\Lambda$. It follows from Lemma 2.4 that $\mathscr{G}$ is a non-empty set. Let $\mathscr{G}$ be partially ordered by the relation: if $g, h \in \mathscr{G}$, then $g \leqslant h$ if $h$ is an extension of $g$. It follows from Proposition 1.2.5 that every chain in $\mathscr{G}$ has an upper bound in $\mathscr{G}$. Thus by Zorn's lemma, $\mathscr{G}$ contains a maximal element, $g$ say. Let $D$ be the domain of $g$ and suppose that $D \neq X$. Choose $\lambda$ such that $A_\lambda$ is not contained in $D$. Since $\dim A_\lambda \leqslant n$, it follows from Lemma 2.4 that $g$ has an extension to $D \cup A_\lambda$. But this is a contradiction of the maximality of $g$ in $\mathscr{G}$. Thus $D = X$ and $g: X \to S^n$ is an extension of $f$. Thus $\dim X \leqslant n$ by Theorem 2.2.□

**7.2  Proposition.** *If $K$ is a simplicial complex, then $\dim |K| = \dim K$.*

*Proof.* If $s$ is an $n$-simplex of $K$, then $|s|$ is a closed set of $|K|$ which is homeomorphic with the closed $n$-ball $E^n$, so that $\dim |s| = n$. Hence $\dim |K| \geqslant n$ if $K$ contains an $n$-simplex, and it follows that

$$\dim |K| \geqslant \dim K.$$

It only remains to show that if $\dim K$ is finite then $\dim |K| \leqslant \dim K$. But $|K|$ is dominated by the covering $\{|s|\}_{s \in K}$ and $\dim |s| \leqslant \dim K$ for each $s$ in $K$. It follows from Proposition 7.1 that $\dim |K| \leqslant \dim K$. Hence $\dim |K| = \dim K$.□

Dimension can now be characterized in terms of mappings into polyhedra. A lemma is needed.

**7.3 Lemma.** *If $X$ is a space with the property that for each finite open covering $\mathcal{W}$ of $X$ there exists a paracompact normal space $Y$ such that dim $Y \leqslant n$ and a $\mathcal{W}$-mapping of $X$ onto $Y$, then dim $X \leqslant n$.*

*Proof.* Let $\mathcal{W}$ be a finite open covering of $X$. By hypothesis there exists a paracompact normal space $Y$ such that dim $Y \leqslant n$ and a surjective $\mathcal{W}$-mapping $\phi: X \to Y$. Thus there exists an open covering $\mathcal{G}$ of $Y$ such that $\phi^{-1}(G)$ is contained in some member of $\mathcal{W}$ for each $G$ in $\mathcal{G}$, and since dim $Y \leqslant n$, there exists an open refinement $\mathcal{H}$ of $\mathcal{G}$ of order not exceeding $n$. Then $\{\phi^{-1}(H) \mid H \in \mathcal{H}\}$ is an open covering of $X$ of order not exceeding $n$ which is a refinement of $\mathcal{W}$. Hence dim $X \leqslant n$. $\square$

**7.4 Proposition.** *The following statements about a topological space $X$ are equivalent:*

*(a) $X$ is a normal space such that dim $X \leqslant n$;*

*(b) for each locally finite open covering $\{W_\lambda\}_{\lambda \in \Lambda}$ of $X$ there exists a continuous function $\phi: X \to P$, where $P$ is a polyhedron such that dim $P \leqslant n$, and an open covering $\{G_\lambda\}_{\lambda \in \Lambda}$ of $P$ such that $\phi^{-1}(G_\lambda) \subset W_\lambda$ for each $\lambda$;*

*(c) for each finite open covering $\mathcal{W}$ of $X$ there exists a $\mathcal{W}$-mapping of $X$ into a polyhedron $P$ such that dim $P \leqslant n$.*

*Proof.* $(a) \Rightarrow (b)$. By Corollary 4.4 there exists a locally finite open covering $\mathcal{V} = \{V_\lambda\}_{\lambda \in \Lambda}$ of $X$ of order not exceeding $n$ such that $V_\lambda \subset W_\lambda$ for each $\lambda$. By Proposition 2.6.8, there exists a canonical mapping $\phi: X \to |N(\mathcal{V})|$. The nerve $N(\mathcal{V})$ is a simplicial complex of combinatorial dimension not exceeding $n$, so that $|N(\mathcal{V})|$ is a polyhedron of covering dimension not exceeding $n$. If $G_\lambda = \mathrm{St}\,(\lambda)$ for each $\lambda$, then $\{G_\lambda\}$ is an open covering of $|N(\mathcal{V})|$ and $\phi^{-1}(G_\lambda) \subset V_\lambda \subset W_\lambda$ for each $\lambda$.

$(b) \Rightarrow (c)$ is obvious.

$(c) \Rightarrow (a)$. By Proposition 2.6.8, if $(c)$ holds then $X$ is a normal space. For each finite open covering $\mathcal{W}$ of $X$ there exist a polyhedron $P$ such that dim $P \leqslant n$, and a $\mathcal{W}$-mapping of $X$ onto a subspace $Y$ of $P$. By Proposition 2.6.4, $P$ is a paracompact perfectly normal space, so that by Proposition 2.2.4, $Y$ is a paracompact normal space, and dim $Y \leqslant n$ by Theorem 6.4. Hence dim $X \leqslant n$ by Lemma 7.3. $\square$

**7.5 Remark.** We can also make the following characterization of covering dimension: if $X$ is a normal space, then dim $X \leqslant n$ if and only if for every finite open covering $\mathcal{W}$ of $X$ there is a $\mathcal{W}$-mapping of $X$ into a compact polyhedron $P$ such that dim $P \leqslant n$. To see this one needs to make only trivial modifications to the above proof. The mapping 'into' can in fact be replaced by a mapping 'onto': a normal space $X$ satisfies dim $X \leqslant n$ if and only if for each finite open covering $\mathscr{W}$ there exists a $\mathscr{W}$-mapping of $X$ onto a compact polyhedron $P$ such that dim $P \leqslant n$. We need only prove necessity of the condition. By an argument similar to that in the proof of Proposition 7.4, if $X$ is a normal space such that dim $X \leqslant n$ and $\mathscr{W}$ is a finite open covering of $X$, then there exists a finite simplicial complex $K$ such that dim $K \leqslant n$ and a continuous function $\phi: X \to |K|$ such that $\phi^{-1} \mathrm{St}(v)$ is contained in some member of $\mathscr{W}$ for each vertex $v$ of $K$. Consider the set $\mathscr{S}$ of subcomplexes $M$ of $K$ which have the property that there exists a continuous function $\theta: X \to |M|$ such that

$$\theta^{-1} \mathrm{St}_M(v)$$

is contained in some member of $\mathscr{W}$ for each vertex $v$ of $M$, where $\mathrm{St}_M(v)$ is the star of the vertex $v$ in $|M|$. Since the set of subcomplexes of $K$ is finite, there exists a subcomplex $Q$ of $K$ which is a minimal member of $\mathscr{S}$ in the sense that $Q \in \mathscr{S}$ and if $L$ is a proper subcomplex of $Q$, then $L \notin \mathscr{S}$. There exists a continuous function $\psi: X \to |Q|$ such that $\psi^{-1} \mathrm{St}_Q(v)$ is contained in some member of $\mathscr{W}$ for each vertex $v$ of $Q$. Suppose that $\psi$ is not surjective so that there exists a point $\alpha_0$ of $|Q|$ such that $\alpha_0 \notin \psi(X)$. If $L$ is the subcomplex of $Q$ consisting of those simplexes $s$ such that $\alpha_0 \notin |s|$, then there is a retraction of $|Q| \backslash \{\alpha_0\}$ onto $|L|$ whose restriction to $\mathrm{St}_Q(v)$ is a retraction of $\mathrm{St}_Q(v)$ onto $\mathrm{St}_L(v)$ for each vertex $v$ of $L$. If $\psi': X \to |L|$ is the composite of $\psi$ and the retraction, we see that $(\psi')^{-1} \mathrm{St}_L(v)$ is contained in some member of $\mathscr{W}$ for each vertex $v$ of $L$ so that $L \in \mathscr{S}$, which contradicts the minimality of $Q$. Thus the $\mathscr{W}$-mapping $\psi: X \to |Q|$ is a surjection. Since $|Q|$ is a compact polyhedron and dim $|Q| \leqslant n$, this completes the proof.

## 8 Rank and Dimension

A family $\{V_\lambda\}_{\lambda \in \Lambda}$ of subsets of a set $X$ is said to be *independent* if for each pair $\lambda, \mu$ of distinct elements of $M$ neither $V_\lambda \subset V_\mu$ nor $V_\mu \subset V_\lambda$ is true. Let $\mathscr{V} = \{V_\lambda\}_{\lambda \in \Lambda}$ be a family of subsets of a non-empty set $X$. If $x \in X$, let

$$\Lambda_x = \{\lambda \in \Lambda \mid x \in V_\lambda\}.$$

We recall that the order of $\mathscr{V}$ does not exceed $n$ if $\Lambda_x$ has at most

$n + 1$ members for each $x$ in $X$. We say that the rank of $\mathscr{V}$ does not exceed $n$ if for each $x$ in $X$, each subset $M$ of $\Lambda_x$ such that $\{V_\lambda\}_{\lambda \in M}$ is independent, has at most $n + 1$ members. The *rank* of $\mathscr{V}$, denoted by $\mathrm{r}\mathscr{V}$, is the least integer $n$ for which the statement that the rank of $\mathscr{V}$ does not exceed $n$ is true. If there exists no such integer then we put $\mathrm{r}\mathscr{V} = \infty$. It is evident that the rank of a family does not exceed its order.

Let $X$ be a space such that $\dim X \leqslant n$. Then since rank does not exceed order, each finite open covering of $X$ has an open refinement of rank not exceeding $n$. In this section we shall prove that this condition is also sufficient for $\dim X \leqslant n$.

Let $\mathscr{V} = \{V_\lambda\}_{\lambda \in \Lambda}$ be a family of subsets of a space $X$ and let $\Gamma$ be the set of subsets $\gamma$ of $\Lambda$ with the property that $\{V_\lambda\}_{\lambda \in \gamma}$ is a chain with respect to inclusion so that if $\lambda, \mu \in \gamma$, then either $V_\lambda \subset V_\mu$ or $V_\mu \subset V_\lambda$. For each $\gamma$ in $\Gamma$, let $W_\gamma = \bigcup_{\lambda \in \gamma} V_\lambda$ and let $\tilde{\mathscr{V}} = \{W_\gamma\}_{\gamma \in \Gamma}$. Since each singleton subset of $\Lambda$ is a member of $\Gamma$ we can regard $\mathscr{V}$ as a subfamily of $\tilde{\mathscr{V}}$.

**8.1** **Lemma.** *If $\mathscr{V}$ is a family of subsets of a set $X$, then $\tilde{\mathscr{V}}$ has the same rank as $\mathscr{V}$. Furthermore if $\mathscr{V}$ is a refinement of some covering of $X$, then so is $\tilde{\mathscr{V}}$.*

*Proof.* Since $\mathscr{V}$ is a subfamily of $\tilde{\mathscr{V}}$ it follows that $\mathrm{r}\mathscr{V} \leqslant \mathrm{r}\tilde{\mathscr{V}}$, so that the proof of the first assertion will be complete if we show that $\mathrm{r}\tilde{\mathscr{V}} \leqslant \mathrm{r}\mathscr{V}$. Let $\mathscr{V} = \{V_\lambda\}_{\lambda \in \Lambda}$. Then $\tilde{\mathscr{V}} = \{W_\gamma\}_{\gamma \in \Gamma}$, where $\Gamma$ is the set of subsets $\gamma$ of $\Lambda$ such that $\{V_\lambda\}_{\lambda \in \gamma}$ is a chain, and $W_\gamma = \bigcup_{\lambda \in \gamma} V_\lambda$ for each $\gamma$ in $\Gamma$. The required inequality follows at once if we prove that the following statement $(\alpha_k)$ is true for all non-negative integers $k$:

$(\alpha_k)$ If $\gamma_0, \dots, \gamma_k \in \Gamma$ and $W_{\gamma_0}, \dots, W_{\gamma_k}$ are independent sets containing the point $x$ of $X$, then there exists $\lambda_i$ in $\gamma_i$ for $i = 0, \dots, k$ such that $V_{\lambda_0}, \dots, V_{\lambda_k}$ are independent sets containing $x$.

The proof is by induction. The statement $(\alpha_0)$ is obviously true. We suppose that $(\alpha_{k-1})$ is true, where $k > 0$, and prove $(\alpha_k)$. For each $i = 0, \dots, k-1$, $W_{\gamma_i} \cap (X \backslash W_{\gamma_k}) \neq \varnothing$ so that we can choose an element $\mu_i$ of $\gamma_i$ such that $V_{\mu_i} \cap (X \backslash W_{\gamma_k}) \neq \varnothing$. For $i = 0, \dots, k-1$, let

$$\beta_i = \{\lambda \in \gamma \mid V_\lambda \supset V_{\mu_i}\}.$$

Since $W_{\beta_i} = W_{\gamma_i}$, the sets $W_{\beta_0}, \dots, W_{\beta_{k-1}}$ are independent and contain $x$.

Hence there exists $\lambda_i$ in $\beta_i$ for $i = 0, \ldots, k-1$ such that $V_{\lambda_0}, \ldots, V_{\lambda_{k-1}}$ are independent and contain $x$. Since $\{V_\lambda\}_{\lambda \in \gamma_k}$ is a chain and

$$W_{\gamma_k} \cap (X \backslash W_{\gamma_i}) \neq \varnothing$$

for $i = 0, \ldots, k-1$, we can choose an element $\lambda_k$ of $\gamma_k$ such that $x \in V_{\lambda_k}$ and $V_{\lambda_k} \cap (X \backslash W_{\gamma_i}) \neq \varnothing$ for $i = 0, \ldots, k-1$. Then

$$\{V_{\lambda_0}, \ldots, V_{\lambda_{k-1}}, V_{\lambda_k}\}$$

is an independent family each member of which contains $x$. It follows by induction that $(\alpha_k)$ is true for all non-negative integers $k$.

Now let $\mathcal{U} = \{U_1, \ldots, U_k\}$ be a covering of $X$ and suppose that $\mathcal{V}$ is a refinement of $\mathcal{U}$. If $\gamma \in \Gamma$ and $W_\gamma$ is not contained in any member of $\mathcal{U}$, then for each $i = 1, \ldots, k$ there exists $\lambda_i$ in $\gamma$ such that

$$V_{\lambda_i} \cap (X \backslash U_i) \neq \varnothing.$$

If $V_\mu$ is the largest of the sets $V_{\lambda_1}, \ldots, V_{\lambda_k}$, then $V_\mu$ is not contained in any member of $\mathcal{U}$. This is a contradiction of the fact that $\mathcal{V}$ is a refinement of $\mathcal{U}$. Thus if $\mathcal{V}$ is a refinement of $\mathcal{U}$, then $\tilde{\mathcal{V}}$ is a refinement of $\mathcal{U}$. $\square$

If $\mathcal{V} = \{V_\lambda\}_{\lambda \in \Lambda}$ is a family of subsets of a set $X$, let us denote by $\mathcal{V}^*$ the subset of the set $\mathcal{P}(X)$ of subsets of $X$ given by

$$\mathcal{V}^* = \{V \in \mathcal{P}(X) \mid V = V_\lambda \quad \text{for some } \lambda \text{ in } \Lambda\}.$$

It is obvious that if $\mathcal{V}^*$ is regarded as a family indexed by itself, then $\mathrm{r}\mathcal{V}^* = \mathrm{r}\mathcal{V}$.

**8.2   Lemma.** *If the family $\mathcal{V}$ is a covering of a set $X$ such that*

$$\mathcal{V}^* = \tilde{\mathcal{V}}^*,$$

*then there exists an independent subfamily of $\mathcal{V}$ which is a covering of $X$.*

*Proof.* Let $\mathcal{V} = \{V_\lambda\}_{\lambda \in \Lambda}$ and let $M$ be the subset of $\Lambda$ consisting of those $\mu$ such that $V_\mu$ is maximal in the sense that there exists no $\lambda$ in $\Lambda$ distinct from $\mu$ such that $V_\lambda \supset V_\mu$. Clearly $\{V_\mu\}_{\mu \in M}$ is an independent subfamily of $\mathcal{V}$ and it remains to show that it is a covering of $X$. Let $x_0$ be a point of $X$ and choose $\lambda_0$ in $\Lambda$ such that $x_0 \in V_{\lambda_0}$. By the Kuratowski lemma there exists a subset $\gamma$ of $\Lambda$ such that $\lambda_0 \in \gamma$ and $\{V_\lambda\}_{\lambda \in \gamma}$ is a maximal chain. Then $\bigcup_{\lambda \in \gamma} V_\lambda \in \tilde{\mathcal{V}}^*$, so that since $\mathcal{V}^* = \tilde{\mathcal{V}}^*$ there exists an element of $\mu_0$ of $\Lambda$ such that $V_{\mu_0} = \bigcup_{\lambda \in \gamma} V_\lambda$. Evidently $\mu_0 \in M$ and $x_0 \in V_{\mu_0}$. Thus $\{V_\mu\}_{\mu \in M}$ is a covering of $X$. $\square$

**8.3 Proposition.** *If $X$ is a topological space, then* $\dim X \leqslant n$ *if and only if each finite open covering of $X$ has an open refinement of rank not exceeding $n$.*

*Proof.* The condition is trivially necessary for if $\dim X \leqslant n$, then each finite open covering has an open refinement of order, and hence of rank, not exceeding $n$. Conversely let $X$ be a space such that each finite open covering has an open refinement of rank not exceeding $n$. Let $\mathscr{U}$ be a finite open covering of $X$ and let $\mathscr{V}$ be an open refinement of $\mathscr{U}$ such that $r\mathscr{V} = m \leqslant n$. We shall construct a transfinite sequence $\{\mathscr{V}_\alpha\}$ of open refinements of $\mathscr{U}$, each of rank equal to $m$. Let $\mathscr{V}_0 = \mathscr{V}$. Suppose that $\mathscr{V}_\alpha$ has been defined for $\alpha < \beta$ and let

$$\mathscr{W}_\beta = \bigcup_{\alpha < \beta} \mathscr{V}_\alpha^*.$$

We regard $\mathscr{W}_\beta$ as a family indexed by itself and define $\mathscr{V}_\beta = \widetilde{\mathscr{W}}_\beta$. In this way we obtain a transfinite sequence of open coverings $\mathscr{V}_\alpha$ of $X$, and $\mathscr{V}_\alpha$ is a refinement of $\mathscr{U}$ for each ordinal $\alpha$ by Lemma 8.1. We shall prove by transfinite induction that $r\mathscr{V}_\beta = m$ for all ordinals $\beta$. Suppose that this has been established for all $\alpha$ such that $\alpha < \beta$. By Lemma 8.1, $r\mathscr{V}_\beta = r\mathscr{W}_\beta$. If $\beta = \gamma + 1$ then $\mathscr{W}_\beta = \mathscr{V}_\gamma^*$ so that $r\mathscr{W}_\beta = r\mathscr{V}_\gamma = m$. Now suppose that $\beta$ is a limit ordinal. If $\{W_0, \ldots, W_k\}$ is an independent subfamily of $\mathscr{W}_\beta$, then there exists $\gamma$ such that $\gamma < \beta$ and $\{W_0, \ldots, W_k\}$ is a subset of $\mathscr{V}_\gamma^*$. Since $r\mathscr{V}_\gamma = m$, it follows that $k \leqslant m$. Thus $r\mathscr{W}_\beta \leqslant m$. Hence $r\mathscr{V}_\beta \leqslant m$ for all ordinals $\beta$. Since it is clear that $r\mathscr{V} \leqslant r\mathscr{V}_\beta$ this completes the proof that $r\mathscr{V}_\beta = m$ for all ordinals $\beta$. Now $\{\mathscr{V}_\alpha^*\}$ is an increasing transfinite sequence of subsets of the set $\mathscr{P}(X)$ of all subsets of $X$ and so is ultimately constant. Thus there exists a first ordinal $\xi$ such that $\mathscr{V}_\xi^* = \mathscr{V}_{\xi+1}^*$. Since $\mathscr{V}_{\xi+1} = (\mathscr{V}_\xi^*)^\sim$ it follows from Lemma 8.2 that there exists an independent subfamily $\mathscr{W}$ of $\mathscr{V}_\xi^*$ which is a covering of $X$. Clearly $r\mathscr{W} \leqslant m \leqslant n$. But since $\mathscr{W}$ is independent its rank is equal to its order. Thus $\mathscr{W}$ is an open refinement of $\mathscr{U}$ which has order not exceeding $n$. Hence $\dim X \leqslant n$. $\square$

## Notes

The theory of dimension grew from attempts to establish the topological invariance of the dimension of Euclidean spaces. The first proof that $\mathbf{R}^n$ and $\mathbf{R}^m$ are not homeomorphic if $n \neq m$ is due to Brouwer

[1911]. Brouwer's proof did not explicitly reveal a topological property of the Euclidean space $R^n$ which might serve as a definition of $n$-dimensionality. Subsequently he introduced [1913] a topologically invariant dimension function which takes the value $n$ on Euclidean $n$-space $R^n$. This dimension function coincides with small inductive dimension (see Chapter 4) for the class of locally connected separable metric spaces. Lebesgue [1911] suggested another approach to the question which gave rise to the definition of covering dimension. He observed that the $n$-dimensional cube $I^n$ can be covered by arbitrarily small closed sets so that not more than $n + 1$ of these sets meet. The usual pattern of brickwork shows the way this can be achieved in the two-dimensional case. He formulated the theorem which asserts that every covering of the cube $I^n$ by closed sets of sufficiently small diameter has order not less than $n$. The proof of this theorem was given by Brouwer [1913]. Lebesgue's theorem allows a topologically invariant definition of the dimension of Euclidean spaces. When the definition is cast in a form applicable to an arbitrary topological space, the concept of covering dimension is obtained. This definition was made by Čech [1933]. The first theory of dimension was, however, that of small inductive dimension to be discussed in the next chapter. Many of the theorems which occur in Chapter 3 were first proved for the small inductive dimension of compact metric or separable metric spaces. The rest of this book will show that a preference for covering dimension, or its modification to be discussed in Chapter 10, is now natural.

The development of the theory of covering dimension for normal spaces is due to Aleksandrov [1947], Dowker [1947], Hemmingsen [1946] and Morita [1950]. They obtained the important characterization of dimension in terms of extension of mappings into spheres (Theorem 2.2). The first result of this type was obtained by Aleksandrov [1932] and it was extended to separable metric spaces by Hurewicz [1935]. The extension of Theorem 2.2 which is formulated in Proposition 2.3 is due to Aleksandrov. Lemma 2.1 is an extension due to Dowker of a result of Borsuk [1937]. Propositions 1.7, 1.8 and 1.9 were obtained by Morita. Hemmingsen also proved Proposition 1.9. The first result of this type for separable metric spaces was found by Eilenberg and Otto [1938]. The countable sum theorem (Theorem 2.5) was first proved for normal spaces by Čech [1933]. Morita and Hemmingsen gave new proofs of this theorem. The proof given here is essentially due to Wallace [1945]. The product theorem for bicompact spaces (Proposition 2.6) was obtained by Hemmingsen.

Dowker [1947] showed that covering dimension for normal spaces could be based on locally finite open coverings. The proof given here is due to Katětov [1952]. There is also a proof of this result by Morita [1950a] and he obtained Proposition 4.6. Ostrand [1971] has shown that covering dimension can be based on locally finite open coverings for all topological spaces and obtained other interesting results for the covering dimension of general spaces. Morita [1950a] proved Proposition 5.1 for normal spaces. Proposition 5.6 was obtained by Dowker [1955]. The sum theorem for point-finite open coverings is due to Zarelua [1963]. The 'Urysohn inequality', so called because Urysohn [1925] gave the first theorem of this type for the small inductive dimension of compact metric spaces, was established by Zarelua [1963a].

The class of totally normal spaces was introduced by Dowker [1953]. In [1955] he proved Theorem 6.4. Čech [1933] proved the subset theorem for perfectly normal spaces. Smirnov [1951a] proved that dimension is monotonic on generalized $F_\sigma$-sets in a normal space. Morita [1953] proved that dimension is monotonic on strongly paracompact subspaces. The extension of this result given in Proposition 6.9 is due essentially to Pupko [1961].

The concept of the nerve of a covering was introduced by Aleksandrov [1928, 1928a]. In [1947] he obtained the characterization of covering dimension given in Remark 7.5. Proposition 7.4 is based on an extension due to Dowker [1948] of Aleksandrov's mapping theorem. The fact that a space dominated by $n$-dimensional subspaces is $n$-dimensional was established by Morita [1953a].

Proposition 8.3 was proved by Arhangel'skiĭ [1963]. The concept of rank was introduced by Nagata [1961] who proved that a metrizable space $X$ satisfies dim $X \leqslant n$ if and only if $X$ has a base for its topology of rank not exceeding $n$. Arhangel'skiĭ also proved that a bicompact space which has a base of rank equal to 1 is metrizable.

# 4

## INDUCTIVE DIMENSION

### 1 Small inductive dimension

The *small inductive dimension* of a space $X$, ind $X$, is defined inductively as follows. A space $X$ satisfies ind $X = -1$ if and only if $X$ is empty. If $n$ is a non-negative integer, then ind $X \leqslant n$ means that for each point $x$ of $X$ and each open set $G$ such that $x \in G$ there exists an open set $U$ such that $x \in U \subset G$ and ind bd $(U) \leqslant n-1$. We put ind $X = n$ if it is true that ind $X \leqslant n$, but it is not true that ind $X \leqslant n-1$. If there exists no integer $n$ for which ind $X \leqslant n$ then we put ind $X = \infty$.

**1.1 Proposition.** *A space $X$ satisfies* ind $X = 0$ *if and only if it is non-empty and has a base for its topology which consists of open-and-closed sets.*

*Proof.* This follows immediately from the definition since an open set $U$ is an open-and-closed set if and only if bd $(U)$ is empty. $\square$

It follows that a space $X$ which satisfies ind $X = 0$ is regular, and if ind $X = 0$ and $X$ does not have the trivial topology, then $X$ is not connected. Moreover it follows from Proposition 3.1.3 that if $X$ is a $T_1$-space such that ind $X = 0$, then $X$ is totally disconnected; if $X$ is a $T_1$-space such that dim $X = 0$, then ind $X = 0$, and if $X$ is a bicompact space then the statements ind $X = 0$, dim $X = 0$ and $X$ is totally disconnected, are equivalent.

**1.2 Example.** If $B$ is a non-empty subspace of the space $\mathbf{R}$ of real numbers and the interior of $B$ in $\mathbf{R}$ is empty, then ind $B = 0$. For let $x$ be a point of $B$ and let $G$ be an open set of $B$ such that $x \in G$. There exists an open interval $V$ such that $x \in V \cap B \subset G$. Since the interior of $B$ is empty there exist $y, z$ in $V$ such that $y < x < z$ and $y \notin B$, $z \notin B$. If
$$U = \{t \in B \mid y < t < z\},$$
then $U$ is an open-and-closed set of $B$ and $x \in U \subset G$. Hence ind $B = 0$.

We recall that the *Cantor set* is the subspace $C$ of the unit interval $I$ consisting of those real numbers $t$ which have a ternary expansion

$\Sigma_{n=1}^{\infty} t_n/3^n$ such that no $t_n$ is equal to 1. There is a useful geometrical description of the Cantor set. If the open middle third $(\frac{1}{3}, \frac{2}{3})$ is deleted from $I$, then the open middle thirds $(\frac{1}{9}, \frac{2}{9})$ and $(\frac{7}{9}, \frac{8}{9})$ are deleted from each of the remaining intervals and this process is continued so that infinitely many open intervals are deleted from $I$, then the resulting subspace of $I$ is the Cantor set $C$. It follows from Example 1.2 that $\mathrm{ind}\, C = 0$. It is a special case of Proposition 1.3 that $C$ is a universal space for the class of separable metrizable spaces $X$ such that $\mathrm{ind}\, X = 0$.

Let $\tau$ be a cardinal number and let $\Lambda$ be a set such that $|\Lambda| = \tau$. For each $\lambda$ in $\Lambda$ let $D_\lambda$ be the set $\{0, 1\}$ with the discrete topology. Up to homeomorphism, the topological product $\Pi_{\lambda \in \Lambda} D_\lambda$ is independent of the set $\Lambda$ such that $|\Lambda| = \tau$ and we denote this space by $D^\tau$. It is easily seen that the Cantor set $C$ is homeomorphic with $D^{\aleph_0}$.

**1.3 Proposition.** *If $\tau$ is an infinite cardinal number, then $D^\tau$ is a universal space for the class of $T_1$-spaces $X$ such that $w(X) = \tau$ and $\mathrm{ind}\, X = 0$.*

*Proof.* Since $D^\tau$ has a base for its topology which consists of open-and-closed sets, it follows that $\mathrm{ind}\, D^\tau = 0$ and it follows from Proposition 1.2.11 that $w(D^\tau) = \tau$. Now let $X$ be a $T_1$-space such that $w(X) = \tau$ and $\mathrm{ind}\, X = 0$. Let $\mathscr{U}$ be a base for the topology of $X$ which consists of open-and-closed sets. By Proposition 1.2.10 there exists a subset $\mathscr{W}$ of $\mathscr{U}$ such that $|\mathscr{W}| = \tau$ and $\mathscr{W}$ is a base for the topology of $X$. If $W \in \mathscr{W}$ let $f_W \colon X \to \{0, 1\}$ be the function defined by $f_W(x) = 0$ if $x \in W$ and $f_W(x) = 1$ if $x \notin W$. If $\{0, 1\}$ has the discrete topology then $f_W$ is continuous. Since $X$ is a $T_1$-space, it follows from Lemma 1.5.13 that to complete the proof it is enough to show that the family $\{f_W\}_{W \in \mathscr{W}}$ separates points of $X$ from closed sets. Let $A$ be a closed set of $X$ and let $x$ be a point of $X \backslash A$. There exists a member $W$ of $\mathscr{W}$ such that $x \in W \subset X \backslash A$. Then $f_W(x) = 0$ and $f_W(A) = \{1\}$. $\square$

Small inductive dimension is monotonic.

**1.4 Proposition.** *If $A$ is a subspace of a space $X$, then $\mathrm{ind}\, A \leqslant \mathrm{ind} X$.*

*Proof.* We shall prove that if $\mathrm{ind}\, X \leqslant n$ and $A \subset X$, then $\mathrm{ind}\, A \leqslant n$. The proof is by induction. The statement is obviously true if $n = -1$. We assume it is true for spaces of small inductive dimension not exceeding $n - 1$. Let $X$ be a space such that $\mathrm{ind}\, X \leqslant n$ and let $A$ be a subspace of $X$. Let $x$ be a point of $A$ and let $G$ be an open set of $A$ such that $x \in G$. There exists $H$ open in $X$ such that $G = H \cap A$.

Since $\operatorname{ind} X \leqslant n$, there exists $V$ open in $X$ such that $x \in V \subset H$ and $\operatorname{ind} \operatorname{bd}(V) \leqslant n-1$. Let $U = V \cap A$. Then $U$ is open in $A$, $x \in U \subset G$ and $\operatorname{bd}_A(U) \subset \operatorname{bd}(V) \cap A$. By the induction hypothesis,

$$\operatorname{ind} \operatorname{bd}_A(U) \leqslant n-1.$$

Therefore $\operatorname{ind} A \leqslant n.\square$

**1.5** *Example.* If $B$ is a subspace of the space $\mathbf{R}$ of real numbers, then $\operatorname{ind} B \leqslant 1$ and $\operatorname{ind} B = 1$ if and only if the interior of $B$ in $\mathbf{R}$ is non-empty. We show first that $\operatorname{ind} \mathbf{R} = 1$. If $x \in \mathbf{R}$ and $G$ is an open set such that $x \in G$, then there exists an open interval $U$ such that $x \in U \subset G$. Since $\operatorname{bd}(U)$ is a pair of points with the discrete topology, it follows that $\operatorname{ind} \operatorname{bd}(U) = 0$. Hence $\operatorname{ind} \mathbf{R} \leqslant 1$. But $\mathbf{R}$ is connected so that $\operatorname{ind} \mathbf{R} = 1$. Now it follows from Proposition 1.4 that if $B \subset R$ then $\operatorname{ind} B \leqslant 1$. If the interior of $B$ is non-empty then $B$ contains an open interval. Thus $B$ has a subspace which is homeomorphic with $\mathbf{R}$ and so $\operatorname{ind} B \geqslant 1$ by Proposition 1.4. Thus $\operatorname{ind} B = 1$. If the interior of $B$ is empty, then $\operatorname{ind} B \leqslant 0$ as was established in Example 1.2. Thus $\operatorname{ind} B = 1$ if and only if the interior of $B$ is non-empty.

**1.6** *Remarks.* (1) It will be proved later in this chapter that if $X$ is a separable metrizable space, then $\dim X = \operatorname{ind} X$. It follows that $\operatorname{ind} \mathbf{R}^n = n$ for each positive integer $n$.

(2) By an argument similar to that used to obtain the small inductive dimension of $\mathbf{R}$, it can be shown that if $X$ is a linearly ordered space, then $\operatorname{ind} X \leqslant 1$ and $\operatorname{ind} X = 1$ if $X$ is connected.

We recall from §3 of Chapter 2 that Hilbert space $H$ consists of the sequences $\mathbf{x} = \{x_i\}_{i \in \mathbf{N}}$ of real numbers such that $\Sigma_{i=1}^{\infty} x_i^2$ is convergent, with topology induced by the metric $d$ on $H$ given by

$$d(\mathbf{x}, \mathbf{y}) = \sqrt{\left( \sum_{i=1}^{\infty} (x_i - y_i)^2 \right)} \quad \text{if} \quad \mathbf{x}, \mathbf{y} \in H.$$

The Hilbert cube $K$ is the subspace of $H$ defined by

$$K = \{\mathbf{x} \in H \mid |x_i| \leqslant 1/i \quad \text{for all} \quad i\}.$$

For separable metrizable spaces, small inductive dimension coincides with covering dimension. Since $K$ is homeomorphic with $I^{\aleph_0}$, it follows that $K$ has subspaces homeomorphic with $I^n$ for each positive integer $n$. Thus

$$\operatorname{ind} K = \operatorname{ind} H = \infty.$$

We shall determine the small inductive dimension of some subspaces of $K$ and $H$.

**1.7** *Example.* Let $P$ be the subspace of $K$ consisting of the points $\mathbf{x} = \{x_i\}_{i \in \mathbb{N}}$ such that $x_i$ is rational for each $i$. Then

$$\operatorname{ind} P = 0.$$

For let $\mathbf{a} = \{a_i\}_{i \in \mathbb{N}}$ be a point of $P$ and let $G$ be an open neighbourhood of $\mathbf{a}$ in $P$. There exists $\epsilon > 0$ such that

$$\{\mathbf{x} \in P \mid d(\mathbf{x}, \mathbf{a}) < \epsilon\} \subset G.$$

Choose $n$ sufficiently large that $\Sigma_{i=n+1}^{\infty} 1/i^2 < \epsilon^2/8$ and choose $\delta > 0$ such that $n\delta^2 < \tfrac{1}{2}\epsilon^2$. For each positive integer $i$ such that $i \leqslant n$, choose irrational numbers $p_i$ and $q_i$ such that $p_i < a_i < q_i$ and $q_i - p_i < \delta$ and furthermore either $|p_i| < 1/i$ for every $i$, or $|q_i| < 1/i$ for every $i$. Let

$$V = \{\mathbf{x} \in P \mid p_i < x_i < q_i \quad \text{for} \quad i \leqslant n\}.$$

The set $V$ is open in $P$, $\mathbf{a} \in V$ and if $\mathbf{x} \in V$ then

$$d(\mathbf{x}, \mathbf{a})^2 = \sum_{i=1}^{\infty} (x_i - a_i)^2 < n\delta^2 + \sum_{i=n+1}^{\infty} \left(\frac{2}{i}\right)^2 < \frac{\epsilon^2}{2} + \frac{\epsilon^2}{2} = \epsilon^2,$$

so that $V \subset G$. Finally $V$ is the intersection with $P$ of the set

$$\{\mathbf{x} \in K \mid p_i < x_i < q_i \quad \text{for} \quad i \leqslant n\},$$

and every point $\mathbf{y}$ of the boundary of this set in $K$ has $y_i$ irrational for at least one $i$ such that $i \leqslant n$. Hence $V$ has an empty boundary in $P$. Thus $\operatorname{ind} P = 0$.

In contrast with the preceding example, the set $Q$ of points of the Hilbert space $H$ with all coordinates rational is not zero-dimensional.

**1.8** *Example.* Let $Q$ be the subspace of $H$ consisting of the points $\mathbf{x} = \{x_i\}$ such that $x_i$ is rational for each $i$. Then

$$\operatorname{ind} Q = 1.$$

Let $M$ be the subspace of $H$ defined by

$$M = \{\mathbf{x} \in H \mid 1/x_i \text{ is a positive integer for each } i\}.$$

We shall show that $\operatorname{ind} M > 0$. Let $\mathbf{a} = \{a_i\}$ be a point of $M$ and suppose that $1/a_i = n_i > 1$ for each $i$. Let

$$G = \{\mathbf{x} \in M \mid d(\mathbf{a}, \mathbf{x}) < \tfrac{1}{2}\}$$

and let $U$ be any open set such that $\mathbf{a} \in U \subset G$. For each $k$-tuple $(m_1, \ldots, m_k)$ of positive integers let $\mathbf{a}(m_1, \ldots, m_k)$ be the point $\mathbf{x} = \{x_i\}$

of $M$ such that $x_i = 1/m_i$ if $i \leqslant k$ and $x_i = a_i$ if $i > k$. We shall define by induction a sequence $\{m_i\}_{i \in \mathbb{N}}$ of positive integers. Let $m_1$ be the least positive integer such that $\mathbf{a}(m_1) \in U$. Suppose that we have defined $m_i$ if $i < k$ such that $\mathbf{a}(m_1, \ldots, m_{k-1}) \in U$ and choose $m_k$ to be the least positive integer such that $\mathbf{a}(m_1, \ldots, m_k) \in U$. For each $k$, let $\mathbf{a}_k = \mathbf{a}(m_1, \ldots, m_k)$. If $\mu = \frac{1}{2} + \sqrt{(\Sigma_{i=1}^\infty a_i^2)}$, then for each point $\mathbf{x} = \{x_i\}$ of $U$ we have $\Sigma_{i=1}^\infty x_i^2 < \mu^2$. Since $\mathbf{a}_k \in U$ we have $\Sigma_{i=1}^k 1/m_i^2 < \mu^2$ for each positive integer $k$, and so $\Sigma_{i=1}^\infty 1/m_i^2$ converges. Thus we can define a point $\mathbf{b} = \{b_i\}$ of $M$ by putting $b_i = 1/m_i$ for each $i$. Since each neighbourhood of $\mathbf{b}$ in $M$ contains $\mathbf{a}_k$ for all sufficiently large $k$ it follows that $\mathbf{b} \in \overline{U}$, where $\overline{U}$ denotes the closure of $U$ in $M$. Since $\mathbf{a}_k \in U$ we have $d(\mathbf{a}, \mathbf{a}_k) < \frac{1}{2}$ for each $k$. Thus since

$$d(\mathbf{a}, \mathbf{a}_k) \geqslant \left| \frac{1}{n_k} - \frac{1}{m_k} \right|$$

and $n_k > 1$, it follows that $m_k > 1$. Thus for each positive integer $k$ we can define a point $\mathbf{c}_k$ of $M \setminus U$ by putting $\mathbf{c}_k = \mathbf{a}(m_1, \ldots, m_{k-1}, m_k - 1)$. Now

$$d(\mathbf{a}_k, \mathbf{c}_k) = 1/m_k(m_k - 1),$$

so that since $1/m_k \to 0$ as $k \to \infty$ we see that each neighbourhood of $\mathbf{b}$ in $M$ contains $\mathbf{c}_k$ for all sufficiently large $k$. Thus $\mathbf{b} \in M \setminus U$ and hence $\mathbf{b} \in \mathrm{bd}(U)$. Thus $\mathrm{bd}(U)$ is non-empty and it follows that $\mathrm{ind}\, M > 0$.

Next we show that $\mathrm{ind}\, Q \leqslant 1$. Let

$$S = \left\{ \mathbf{x} \in Q \mid \sum_{i=1}^\infty x_i^2 = 1 \right\}.$$

Since each open neighbourhood of a point in $Q$ contains an open neighbourhood of the point with boundary homeomorphic with $S$, it is sufficient to prove that $\mathrm{ind}\, S = 0$. Let $\mathbf{a} = \{a_i\}$ be a point of $S$ and let $G$ be an open set of $S$ such that $\mathbf{a} \in G$. Then there exists $\epsilon > 0$ such that

$$\{\mathbf{x} \in S \mid d(\mathbf{a}, \mathbf{x}) < \epsilon\} \subset G.$$

Choose an irrational number $\delta$ such that $0 < \delta < \min\{1, \epsilon^4/16\}$. And choose $n$ such that $\Sigma_{i=n+1}^\infty a_i^2 < \delta$. Let

$$V = \left\{ \mathbf{x} \in H \mid \sum_{i=1}^n a_i x_i > 1 - \delta \right\}.$$

Clearly $V$ is an open set in $H$. If $\mathbf{x} \in \mathrm{bd}(V)$ then

$$\sum_{i=1}^n a_i x_i = 1 - \delta,$$

so that $x_1, \ldots, x_n$ are not all rational. Thus bd $(V)$ contains no points of $Q$. If $U = V \cap S$, then $U$ is an open set of $S$ and its boundary in $S$ is empty. Since $\mathbf{a} \in S$, it follows that $\Sigma_{i=1}^{\infty} a_i^2 = 1$. But $\Sigma_{i=n+1}^{\infty} a_i^2 < \delta$ so that $\Sigma_{i=1}^{n} a_i^2 > 1 - \delta$ and hence $\mathbf{a} \in U$. Finally if $\mathbf{x} \in U$, then

$$\sum_{i=1}^{\infty} (a_i - x_i)^2 = \sum_{i=1}^{\infty} a_i^2 + \sum_{i=1}^{\infty} x_i^2 - 2 \sum_{i=1}^{n} a_i x_i - 2 \sum_{i=n+1}^{\infty} a_i x_i$$

$$< 1 + 1 - 2(1 - \delta) + 2 \left| \sum_{i=n+1}^{\infty} a_i x_i \right|.$$

But by Cauchy's inequality

$$\left| \sum_{i=n+1}^{\infty} a_i x_i \right| \leqslant \left( \sum_{i=n+1}^{\infty} a_i^2 \right)^{\frac{1}{2}} \left( \sum_{i=n+1}^{\infty} x_i^2 \right)^{\frac{1}{2}} < \delta^{\frac{1}{2}}.$$

Hence

$$\sum_{i=1}^{\infty} (a_i - x_i)^2 < 2\delta + 2\delta^{\frac{1}{2}} < 4\delta^{\frac{1}{2}} < \epsilon^2,$$

so that $d(\mathbf{a}, \mathbf{x}) < \epsilon$. Thus $\mathbf{a} \in U \subset G$. Since the boundary of $U$ in $S$ is empty it follows that ind $S = 0$.

Thus we have ind $Q \leqslant 1$ and ind $M \geqslant 1$. Since $M \subset Q$ it follows from Proposition 1.4 that

$$\text{ind } M = \text{ind } Q = 1.$$

**1.9** *Remark.* The space $Q$ studied in Example 1.8 is totally disconnected. For let $\mathbf{a}$ and $\mathbf{b}$ be two distinct points of $Q$. There exists some $i$ such that $a_i < b_i$, say. If $c$ is an irrational number such that $a_i < c < b_i$, then

$$U = \{ \mathbf{x} \in Q \,|\, x_i < c \}$$

is an open-and-closed set of $Q$ such that $\mathbf{a} \in U$ and $\mathbf{b} \notin U$. It follows that no connected subset of $Q$ contains more than one point. Thus $Q$ is an example of a totally disconnected space of positive dimension.

## 2   Large inductive dimension

The *large inductive dimension* of a space $X$, Ind $X$, is defined inductively as follows. A space $X$ satisfies Ind $X = -1$ if and only if $X$ is empty. If $n$ is a non-negative integer, then Ind $X \leqslant n$ means that for each closed set $E$ and each open set $G$ of $X$ such that $E \subset G$ there exists an open set $U$ such that $E \subset U \subset G$ and Ind bd $(U) \leqslant n - 1$. We put Ind $X = n$ if it is true that Ind $X \leqslant n$ but it is not true that Ind $X \leqslant n - 1$. If there exists no integer $n$ for which Ind $X \leqslant n$, then we put Ind $X = \infty$.

If $X$ is a space with the property that each open set is the union of closed sets, then $\operatorname{ind} X \leqslant \operatorname{Ind} X$. To prove this, it is sufficient to show that if $\operatorname{Ind} X \leqslant n$ then $\operatorname{ind} X \leqslant n$, and this is established by an easy induction. The result is clearly true if $n = -1$ and we suppose that for some non-negative integer $n$ the result is true for such spaces of large inductive dimension not exceeding $n-1$. Let $X$ be such a space such that $\operatorname{Ind} X \leqslant n$, let $x$ be a point of $X$ and let $G$ be an open set such that $x \in G$. Then there exists a closed set $E$ such that $x \in E \subset G$. There exists an open set $U$ such that $E \subset U \subset G$ and $\operatorname{Ind} \operatorname{bd}(U) \leqslant n-1$. By the induction hypothesis $\operatorname{ind} \operatorname{bd}(U) \leqslant n-1$ and we see that $\operatorname{ind} X \leqslant n$. Thus in particular the inequality

$$\operatorname{ind} X \leqslant \operatorname{Ind} X$$

holds if $X$ is a $T_1$-space or a regular space. The inequality does not hold in general as the following example shows. Let $X = \{a, b\}$ and let the topology for $X$ be $\{\varnothing, \{a\}, X\}$. Then $\operatorname{Ind} X = 0$ but $\operatorname{ind} X = 1$.

**2.1  Proposition.** *If $X$ is a compact space and $\operatorname{ind} X = 0$, then $\operatorname{Ind} X = 0$.*

*Proof.* Let $X$ be a compact space such that $\operatorname{ind} X = 0$, let $E$ be a closed set of $X$ and let $G$ be an open set such that $E \subset G$. Since $\operatorname{ind} X = 0$, for each point $x$ of $E$ there exists an open-and-closed set $U_x$ such that $x \in U_x \subset G$. Hence $E \subset \bigcup_{x \in E} U_x \subset G$. Since $E$ is closed in the compact space $X$, $E$ is compact. Hence there exist points $x_1, \ldots, x_k$ of $E$ such that

$$E \subset \bigcup_{i=1}^{k} U_{x_i} \subset G.$$

Since $\bigcup_{i=1}^{k} U_{x_i}$ is an open-and-closed set of $X$, it follows that

$$\operatorname{Ind} X = 0. \square$$

**2.2  Corollary.** *If $X$ is a bicompact space, then $\operatorname{Ind} X = 0$ if and only if $\operatorname{ind} X = 0$.*

*Proof.* If $X$ is a bicompact space then $\operatorname{ind} X \leqslant \operatorname{Ind} X$ so that if $\operatorname{Ind} X = 0$ then $\operatorname{ind} X = 0$. But by Proposition 2.1, if $\operatorname{ind} X = 0$ then $\operatorname{Ind} X = 0. \square$

It is clear that if $X$ is a space such that $\operatorname{Ind} X = 0$, then $X$ is normal. Moreover we have:

**2.3  Proposition.** *If $X$ is a topological space, then* $\operatorname{Ind} X = 0$ *if and only if* $\dim X = 0$.

*Proof.* Let $X$ be a space such that $\dim X = 0$. Let $E$ be a closed set and let $G$ be an open set such that $E \subset G$. Then $\{G, X \backslash E\}$ is an open covering of $X$ so that there exist disjoint open sets $U$ and $V$ such that $U \subset G$, $V \subset X \backslash E$, and $U \cup V = X$. Since $V \subset X \backslash E$, it follows that $E \subset U$. Thus $E \subset U \subset G$, where $U$ is an open-and-closed set of $X$. Thus $\operatorname{Ind} X = 0$.

Conversely let $X$ be a space such that $\operatorname{Ind} X = 0$ and let $\{U_1, U_2\}$ be an open covering of $X$. If $A_1 = X \backslash U_2$, then $A_1$ is a closed set of $X$ and $A_1 \subset U_1$. Since $\operatorname{Ind} X = 0$ there exists an open-and-closed set $V_1$ such that $A_1 \subset V_1 \subset U_1$. If $V_2 = X \backslash V_1$, then $\{V_1, V_2\}$ is an open covering of $X$, $V_1 \cap V_2 = \varnothing$ and $V_1 \subset U_1$, whilst $V_2 \subset X \backslash A_1 = U_2$. It follows from Proposition 3.1.2 that $\dim X \leqslant 0$ and $\dim X = 0$ since $X$ is non-empty.□

Next we show that for a bicompact space $X$, $\operatorname{Ind} X = 1$ if and only if $\operatorname{ind} X = 1$. In dimensions greater than 1, small and large inductive dimensions do not in general coincide for bicompact spaces. In a later chapter there is an example of a bicompact space $X$ for which $\operatorname{ind} X = 2$ and $\operatorname{Ind} X = 3$.

**2.4  Proposition.** *If $X$ is a bicompact space, then* $\operatorname{Ind} X = 1$ *if and only if* $\operatorname{ind} X = 1$.

*Proof.* If $X$ is a bicompact space such that $\operatorname{Ind} X = 1$, then $\operatorname{ind} X \leqslant 1$. But by Proposition 2.2, if $\operatorname{ind} X = 0$ then $\operatorname{Ind} X = 0$. Hence $\operatorname{ind} X = 1$.

Conversely let $X$ be a bicompact space such that $\operatorname{ind} X = 1$. Let $E$ be a closed set of $X$ and let $G$ be an open set such that $E \subset G$. Since $E$ is compact and $\operatorname{ind} X = 1$ there exist open sets $V_1, \ldots, V_k$ such that $E \subset \bigcup_{i=1}^{k} V_i \subset G$ and $\operatorname{ind} \operatorname{bd}(V_i) \leqslant 0$ for $i = 1, \ldots, k$. Since $\operatorname{bd}(V_i)$ is bicompact it follows from Propositions 2.1 and 2.3 that

$$\operatorname{ind} \operatorname{bd}(V_i) = \operatorname{Ind} \operatorname{bd}(V_i) = \dim \operatorname{bd}(V_i).$$

It follows that $\dim \bigcup_{i=1}^{k} \operatorname{bd}(V_i) \leqslant 0$, so that if $V = \bigcup_{i=1}^{k} V_i$ then $\dim \operatorname{bd}(V) \leqslant 0$ since $\operatorname{bd}(V)$ is a closed subset of $\bigcup_{i=1}^{k} \operatorname{bd}(V_i)$. Thus we have an open set $V$ such that $E \subset V \subset G$ and $\operatorname{Ind} \operatorname{bd}(V) \leqslant 0$ by Proposition 2.3. Hence $\operatorname{Ind} X \leqslant 1$. But $\operatorname{Ind} X \geqslant \operatorname{ind} X$, so that $\operatorname{Ind} X = \operatorname{ind} X = 1$.□

The subset theorem, that $A \subset X$ implies that $\operatorname{Ind} A \leqslant \operatorname{Ind} X$, does not hold in general. The completely normal space $J$ considered in Example 3.6.1 provides a counterexample. Recall that $J$ contains an open subspace homeomorphic with the unit interval and satisfies $\dim J = 0$. Since $\dim J = 0$ it follows that $\operatorname{Ind} J = 0$. But $\operatorname{ind} I = 1$ by Example 1.5 so that $\operatorname{Ind} I = 1$ by Proposition 2.4. In the next section there is an example of a bicompact space for which the subset theorem holds neither for covering dimension nor for large inductive dimension. Large inductive dimension is however monotonic on closed sets.

**2.5  Proposition.** *If $A$ is a closed subspace of a space $X$, then*

$$\operatorname{Ind} A \leqslant \operatorname{Ind} X.$$

*Proof.* It is sufficient to show that if $\operatorname{Ind} X \leqslant n$ and $A$ is closed in $X$, then $\operatorname{Ind} A \leqslant n$. This is trivially true in dimension $-1$ and we assume it is true for dimension $n-1$. Let $X$ be a space such that $\operatorname{Ind} X \leqslant n$, let $A$ be closed in $X$ and suppose that $E \subset G \subset A$, where $E$ is closed in $A$ and $G$ is open in $A$. Then $E$ is closed in $X$ and there exists $H$ open in $X$ such that $H \cap A = G$. Since $E \subset H$, there exists $V$ open in $X$ such that $E \subset V \subset H$ and $\operatorname{Ind} \operatorname{bd}(V) \leqslant n-1$. Let $U = V \cap A$. Then $U$ is open in $A$ and $E \subset U \subset G$, and the boundary of $U$ in $A$, $\operatorname{bd}_A(U)$, is closed in $A$, hence closed in $X$, and is contained in $\operatorname{bd}(V)$. Thus $\operatorname{bd}_A(U)$ is a closed subset of $\operatorname{bd}(V)$ so that $\operatorname{Ind} \operatorname{bd}_A(U) \leqslant n-1$ by the induction hypothesis. Therefore $\operatorname{Ind} A \leqslant n$. The proof is completed by induction.□

In §4 we shall examine the subset theorem for Ind in more detail and show that if the subset theorem is satisfied for a space, then so is a countable sum theorem. Next there are some characterizations of large inductive dimension.

**2.6  Proposition.** *The following statements about a space $X$ are equivalent:*
  (a) $\operatorname{Ind} X \leqslant n$;
  (b) *if $E \subset G \subset X$, where $E$ is closed and $G$ is open, then $X$ is the union of three disjoint sets $U$, $V$ and $C$ such that $U$ and $V$ are open,*

$$E \subset U \subset G \quad \text{and} \quad \operatorname{Ind} C \leqslant n-1.$$

*Proof* (a) $\Rightarrow$ (b). If $\operatorname{Ind} X \leqslant n$ and $E \subset G$, where $E$ is closed and $G$ is open in $X$, then there exists an open set $U$ such that $E \subset U \subset G$

and $\operatorname{Ind} \operatorname{bd}(U) \leqslant n-1$. Let us put $C = \operatorname{bd}(U) = \overline{U}\backslash U$ and $V = X\backslash \overline{U}$. Then $U$, $V$ and $C$ are disjoint, $X = U \cup V \cup C$, $U$ and $V$ are open and $\operatorname{Ind} C \leqslant n-1$.

(b) $\Rightarrow$ (a). Suppose that $E \subset G \subset X$, where $E$ is closed and $G$ is open. If (b) holds then $X$ is the union of disjoint sets $U$, $V$ and $C$ such that $U$ and $V$ are open, $E \subset U \subset G$ and $\operatorname{Ind} C \leqslant n-1$. Since $U$ is contained in the closed set $U \cup C$, it follows that $\overline{U} \subset U \cup C$ and hence $\overline{U}\backslash U = \operatorname{bd}(U) \subset C$. Since $\operatorname{bd}(U)$ is closed, $\operatorname{Ind} \operatorname{bd}(U) \leqslant n-1$ by Proposition 2.5. Hence $\operatorname{Ind} X \leqslant n$.□

**2.7 Proposition.** *The following statements about a space $X$ are equivalent*:

(a) *$X$ is normal and $\operatorname{Ind} X \leqslant n$*;

(b) *if $E$ and $F$ are disjoint closed sets of $X$, then $E$ and $F$ can be separated by a closed set $C$ such that $\operatorname{Ind} C \leqslant n-1$.*

*Proof.* (a) $\Rightarrow$ (b). Let $X$ be a normal space such that $\operatorname{Ind} X \leqslant n$ and let $E$ and $F$ be disjoint closed sets of $X$. Since $X$ is normal, there exists an open set $G$ such that

$$E \subset G \subset \overline{G} \subset X\backslash F.$$

Since $\operatorname{Ind} X \leqslant n$, there exists an open set $U$ such that $E \subset U \subset G$ and $\operatorname{Ind} \operatorname{bd}(U) \leqslant n-1$. Let $C = \operatorname{bd}(U)$ and $V = X\backslash \overline{U}$. Then $U$, $V$ and $C$ are disjoint sets, $X = U \cup V \cup C$ and $\operatorname{Ind} C \leqslant n-1$. Furthermore $E \subset U$ and since $U \subset G$, it follows that $\overline{U} \subset \overline{G} \subset X\backslash F$ so that $F \subset X\backslash \overline{U} = V$.

(b) $\Rightarrow$ (a). Let (b) be satisfied and suppose that $E \subset G$, where $E$ is closed and $G$ is open. Let $F = X\backslash G$, so that $E$ and $F$ are disjoint closed sets. Hence $X$ is the union of disjoint sets $U$, $V$ and $C$ such that $U$ and $V$ are open, $E \subset U$, $F \subset V$ and $\operatorname{Ind} C \leqslant n-1$. Since

$$X\backslash G = F \subset V \subset X\backslash U,$$

it follows that $U \subset G$. Hence $\operatorname{Ind} X \leqslant n$ by Proposition 2.6. Normality of $X$ is obvious.□

The following general result holds.

**2.8 Proposition.** *If $X$ is the topological sum of spaces $X_\lambda$ such that $\operatorname{Ind} X_\lambda \leqslant n$ for each $\lambda$, then $\operatorname{Ind} X \leqslant n$.*

*Proof.* This is trivially true for dimension $-1$ and we assume it true for dimension $n-1$, where $n$ is a non-negative integer. Suppose that

$E \subset G \subset X$, where $E$ is closed and $G$ is open in $X$. Then for each $\lambda$, $E \cap X_\lambda \subset G \cap X_\lambda$, where $E \cap X_\lambda$ is closed in $X_\lambda$ and $G \cap X_\lambda$ is open in $X_\lambda$. Since $\operatorname{Ind} X_\lambda \leqslant n$, it follows from Proposition 2.6 that $X_\lambda$ is the union of disjoint sets $U_\lambda$, $V_\lambda$ and $C_\lambda$ such that $U_\lambda$ and $V_\lambda$ are open in $X_\lambda$, $E \cap X_\lambda \subset U_\lambda \subset G \cap X_\lambda$ and $\operatorname{Ind} C_\lambda \leqslant n-1$. Let $U = \bigcup U_\lambda$, $V = \bigcup V_\lambda$ and $C = \bigcup C_\lambda$. Then $U$, $V$ and $C$ are disjoint, their union is $X$, and $U$ and $V$ are open sets. Also

$$E = \bigcup (E \cap X_\lambda) \subset \bigcup U_\lambda = U \subset \bigcup (G \cap X_\lambda) = G.$$

Finally the subspace $C$ is the topological sum of the subspaces $C_\lambda$ and $\operatorname{Ind} C_\lambda \leqslant n-1$ for each $\lambda$. Hence $\operatorname{Ind} C \leqslant n-1$ by the induction hypothesis. Thus $\operatorname{Ind} X \leqslant n$. The proof is now completed by induction. $\square$

We conclude this section by obtaining an important relation between dim and Ind: if $X$ is a normal space, then $\dim X \leqslant \operatorname{Ind} X$. This is an immediate consequence of Lemma 2.9 which will have numerous other applications. There are normal spaces for which dim and Ind differ. In the next section there is an example of a bicompact space $S$ such that $\dim S = 1$ and $\operatorname{Ind} S = 2$. In §5 of this chapter it is proved that if $X$ is a metrizable space then $\dim X = \operatorname{Ind} X$. In particular $\operatorname{Ind} \mathbf{R}^n = n$ for each positive integer $n$.

**2.9   Lemma.** *If $X$ is a normal space, then $\dim X \leqslant n$ if $X$ has the property that for each closed set $E$ and each open set $G$ of $X$ such that $E \subset G$ there exists an open set $U$ such that $E \subset U \subset G$ and*

$$\dim \operatorname{bd}(U) \leqslant n-1.$$

*Proof.* Let $\{G_1, \ldots, G_{n+2}\}$ be an open covering of $X$. Then since $X$ is normal there exists a closed covering $\{E_1, \ldots, E_{n+2}\}$ such that each $E_i \subset G_i$. By hypothesis, for each $i$ there exists an open set $U_i$ such that $E_i \subset U_i \subset G_i$ and $\dim \operatorname{bd}(U_i) \leqslant n-1$. Let $B = \bigcup_{i=1}^{n+2} \operatorname{bd}(U_i)$. Then $B$ is a normal space since $B$ is a closed set of $X$. It follows from the sum theorem that $\dim B \leqslant n-1$. We have an open covering $\{G_1 \cap B, \ldots, G_{n+2} \cap B\}$ of $B$ so that there exists a closed covering $\{F_1, \ldots, F_{n+2}\}$ of $B$, of order not exceeding $n-1$, such that each $F_i \subset G_i \cap B$. Since $F_i$ is a closed set of $X$, it follows from Corollary 1.3.15 that there exist open sets $P_1, \ldots, P_{n+2}$ of $X$ such that

$$F_i \subset P_i \subset G_i$$

for each $i$ and $\{P_1, \ldots, P_{n+2}\}$ is similar to $\{F_1, \ldots, F_{n+2}\}$. Next let

$$Q_i = U_i \backslash \bigcup_{j<i} \overline{U}_j$$

for $i = 1, \ldots, n+2$. Then $\{Q_1, \ldots, Q_{n+2}\}$ is a family of disjoint open sets of $X$ and each $Q_i \subset G_i$. Furthermore $X \backslash B \subset \bigcup_{i=1}^{n+2} Q_i$. For let $x$ be a point of $X \backslash B$ and let $i$ be the first integer such that $x \in U_i$. If $j < i$, then $x \notin \mathrm{bd}\,(U_j)$ and $x \notin U_j$ so that $x \notin \overline{U}_j$. Hence $x \in Q_i$. Thus if

$$H_i = P_i \cup Q_i,$$

then $\{H_1, \ldots, H_{n+2}\}$ is an open covering of $X$ and $H_i \subset G_i$ for each $i$. Since the order of the family $\{P_1, \ldots, P_{n+2}\}$ does not exceed $n-1$ and $\{Q_1, \ldots, Q_{n+2}\}$ is a disjoint family, it follows that $\bigcap_{i=1}^{n+2} H_i = \varnothing$. Thus $\dim X \leqslant n$ by Proposition 3.1.2. $\square$

**2.10  Proposition.** *If $X$ is a normal space, then* $\dim X \leqslant \mathrm{Ind}\, X$.

*Proof.* It is enough to show that if $\mathrm{Ind}\, X \leqslant n$ then $\dim X \leqslant n$. The proof is by induction, the case $n = -1$ being trivial. Let us suppose that the result is true in dimension $n-1$, where $n$ is a non-negative integer, and let $X$ be a normal space such that $\mathrm{Ind}\, X \leqslant n$. If $E$ is closed in $X$, $G$ is open in $X$ and $E \subset G$, then there exists an open set $U$ of $X$ such that $E \subset U \subset G$ and $\mathrm{Ind}\, \mathrm{bd}\,(U) \leqslant n-1$. Since $\mathrm{bd}\,(U)$ is closed, the subspace $\mathrm{bd}\,(U)$ of $X$ is normal. Thus $\dim \mathrm{bd}\,(U) \leqslant n-1$ by the induction hypothesis. It follows from Lemma 2.9 that $\dim X \leqslant n$. The proof is completed by induction. $\square$

# 3  Two examples

**3.1**  *Example* There is a zero-dimensional bicompact space for which the subset theorem fails both for covering dimension and for large inductive dimension.

Let $X$ be the topological product $P \times N$ of the linearly ordered space $P$ consisting of the ordinals not exceeding the first uncountable ordinal $\omega_1$ and the linearly ordered space $N$ consisting of the ordinals not exceeding the first infinite ordinal $\omega_0$. It follows from Proposition 1.5.7 that $P$ and $N$ are bicompact spaces. Thus $X$ is a bicompact space which is called the *Tihonov plank*.

Since $X$ is bicompact, $X$ is a normal space. But $X$ is not completely normal. For let $Y = X \backslash \{(\omega_1, \omega_0)\}$ and let

$$A = \{\omega_1\} \times [0, \omega_0), \quad U = P \times [0, \omega_0).$$

Then $A$ is closed in $Y$, $U$ is open in $Y$ and $A \subset U$. Let $V$ be an open set in $Y$ such that $A \subset V$. If $k \in [0, \omega_0)$, there exists $\beta_k$ in $P \backslash \{\omega_1\}$ such that $[\beta_k, \omega_1] \times \{k\} \subset V$. If $\beta$ is the least upper bound of the countable family $\{\beta_k\}$ of elements of $P$, then $\beta < \omega_1$ and $\{\beta\} \times [0, \omega_0) \subset V$. Hence $(\beta, \omega_0) \in \overline{V}$ so that $\overline{V} \not\subset U$. Thus there exists no open set $V$ such that $A \subset V \subset \overline{V} \subset U$. It follows that $Y$ is not normal.

If $x = (\alpha, k) \in X$ and $G$ is an open set of $X$ such that $x \in G$, then there exist $\beta$ in $P$ and a non-negative integer $m$ such that $\beta < \alpha$, $m \leqslant k$ and

$$U = (\beta, \alpha] \times [m, k] \subset G.$$

Since $U$ is an open-and-closed set such that $x \in U \subset G$, it follows that ind $X = 0$. It now follows from Proposition 2.1 that Ind $X = 0$, and from Proposition 2.3 that dim $X = 0$. Since $Y$ is not normal, it follows that dim $Y > 0$ and Ind $Y > 0$. Thus the subset theorems are not generally true for bicompact spaces. It is not difficult to establish that dim $Y = 1$ and Ind $Y = 1$.

The construction of the second example of this section requires some preliminary discussion. Let $C$ denote the Cantor set. When we refer to the ternary expansion $\Sigma_{n=1}^{\infty} t_n/3^n$ of a point of $C$ we shall understand the unique ternary expansion such that $t_n \neq 1$ if $n \in \mathbb{N}$. We can define a surjection $\phi : C \to I$ as follows: if $t \in C$ and $t = \Sigma_{n=1}^{\infty} t_n/3^n$ then

$$\phi(t) = \sum_{n=1}^{\infty} t_n/2^{n+1}.$$

If $t, t' \in C$ and $|t - t'| < 3^{-n}$, then $|\phi(t) - \phi(t')| < 2^{-n}$. It follows that $\phi$ is continuous. Furthermore if $s, t \in C$ and $s < t$, then $\phi(s) \leqslant \phi(t)$ and $\phi(s) < \phi(t)$ except for countably many pairs $s, t$. Let $D$ be the subset of $C$ consisting of those points $t$ such that $0 < t < 1$ which have ternary expansion $\Sigma_{n=1}^{\infty} t_n/3^n$, where either $t_n = 0$ for all but finitely many values of $n$, or $t_n = 2$ for all but finitely many values of $n$. The countable set $D$ is dense in $C$ and $\phi(D)$ is the countable dense set of dyadic rationals in the open unit interval $(0, 1)$. If $s, t \in D$ and

$$s = \sum_{n=1}^{\infty} s_n/3^n, \quad t = \sum_{n=1}^{\infty} t_n/3^n,$$

let us say that $s$ and $t$ are related if there exists a positive integer $m$ such that $s_n = t_n$ if $n < m$, $s_m = 0$, $t_m = 2$, and $s_n = 2$, $t_n = 0$ if $n > m$. Evidently $D$ is partitioned into related pairs. And if $s, t \in C$ then $\phi(s) = \phi(t)$ if and only if $s$ and $t$ are related points of $D$. If one regards the Cantor set as obtained from the unit interval by deleting open intervals, then $D$ consists of the end points of deleted intervals, a pair of points being related if and only if they are the end points of some deleted interval. The Cantor set is bicompact so that $\phi : C \to I$

is a continuous closed surjection and hence an identification mapping. Thus the space obtained from the Cantor set by identifying pairwise the end points of the deleted intervals is the unit interval.

Let $L$ consist of the first uncountable ordinal $\omega_1$ together with all pairs $(\alpha, t)$, where $\alpha$ is an ordinal less than $\omega_1$ and $t$ is a real number such that $0 \leqslant t < 1$. Let us define a linear order on $L$ as follows: if $\xi \in L \backslash \{\omega_1\}$, then $\xi < \omega_1$, and if $\xi_1 = (\alpha_1, t_1)$ and $\xi_2 = (\alpha_2, t_2)$, where $\alpha_1$ and $\alpha_2$ are ordinals less than $\omega_1$, then $\xi_1 < \xi_2$ if either (i) $\alpha_1 < \alpha_2$ or (ii) $\alpha_1 = \alpha_2$ and $t_1 < t_2$. If $L$ is given the interval topology, then $L$ is a bicompact space by Proposition 1.5.7. The linearly ordered space $L$ is usually called the *long line*. If $\alpha$ is an ordinal less than $\omega_1$, $t$ is a real number such that $0 \leqslant t < 1$ and $\xi_0 = (\alpha, t)$, $\xi_1 = (\alpha+1, t)$, then the subspace $[\xi_0, \xi_1]$ of $L$ is homeomorphic with the unit interval $I$.

Now let $R$ be the subset of the cartesian product $L \times I$ given by

$$R = \{(\xi, t) \in L \times I \mid \xi = \omega_1 \quad \text{or} \quad t \in C\}.$$

If $\xi_0 \in L \backslash \{\omega_1\}$ and $u, v$ are real numbers such that $u \leqslant v$, let

$$C_u^v(\xi_0) = \{(\xi, t) \in R \mid \xi = \xi_0 \quad \text{and} \quad u \leqslant t \leqslant v\}.$$

Consider the surjection

$$f : L \times C \to R$$

of the topological product $L \times C$ onto the set $R$ given by $f(\xi, t) = (\xi, t)$ if $\xi < \omega_1$ and $f(\omega_1, t) = (\omega_1, \phi(t))$, and let $R$ have the identification topology with respect to $f$. Since $R$ is the continuous image of a bicompact space, $R$ is compact and it is easy to see that $R$ is a Hausdorff space. Since $f$ is a closed mapping, the subspace $\{\omega_1\} \times I$ of $R$ has the identification topology with respect to the surjection

$$\{\omega_1\} \times C \to \{\omega_1\} \times I$$

given by restriction of $f$. Thus the copy $\{\omega_1\} \times I$ of the unit interval has its usual topology.

**3.2 Proposition.** *The bicompact space $R$ satisfies*

$$\dim R = \operatorname{ind} R = \operatorname{Ind} R = 1.$$

*Proof.* Each point of $R \backslash (\{\omega_1\} \times I)$ has arbitrarily small neighbourhoods of the form

$$V = \{(\xi, t) \in R \mid \xi_1 < \xi < \xi_2, u < t < v\},$$

where $\xi_1 < \xi_2 < \omega_1$ and $u$ and $v$ are real numbers not in $C$ such that $u < v$. Since

$$\mathrm{bd}\,(V) = C_u^v(\xi_1) \cup C_u^v(\xi_2),$$

therefore $\mathrm{ind}\,\mathrm{bd}\,(V) = 0$. If $x = (\omega_1, t_0) \in R$, where $0 < t_0 < 1$ and $G$ is an open neighbourhood of $x$, then there exist dyadic rationals $d, e$ such that $0 < d < t_0 < e < 1$ and $\{\omega_1\} \times [d, e] \subset G$. If

$$d = \phi(u) = \phi(u'),$$

where $u' < u$, and $e = \phi(v) = \phi(v')$, where $v < v'$, then there exists $\xi_0$ in $L \backslash \{\omega_1\}$ such that

$$V_0 = \{(\xi, t) \in R \mid \xi_0 < \xi < \omega_1 \quad \text{and} \quad u \leqslant t \leqslant v\}$$

is contained in $G$. Then $V = V_0 \cup \big(\{\omega_1\} \times (d, e)\big)$ is an open set of $R$ such that $x \in V \subset G$ and since

$$\mathrm{bd}\,(V) = C_u^v(\xi_0) \cup \{(\omega_1, d), (\omega_1, e)\},$$

it follows that $\mathrm{ind}\,\mathrm{bd}\,(V) = 0$. Trivial modifications to the argument are needed to show that $(\omega_1, 0)$ and $(\omega_1, 1)$ have arbitrarily small neighbourhoods with boundaries which are zero-dimensional. Thus $\mathrm{ind}\,R = 1$. Since $R$ is bicompact it follows from Proposition 2.4 that $\mathrm{Ind}\,R = 1$. Hence $\dim R \leqslant 1$ and $\dim R \neq 0$ since $\mathrm{Ind}\,R \neq 0$, so that $\dim R = 1$. $\square$

Let $M$ be a countable dense subset of the open unit interval $(0, 1)$. There exists a homeomorphism $h$ of $I$ with itself such that $h(s) < h(t)$ if $s < t$ and $M$ is the image under $h$ of the set of dyadic rationals in the open unit interval. If $\psi = h \circ \phi$, then we have a continuous surjection $\psi: C \to I$ such that $\psi(s) \leqslant \psi(t)$ if $s, t \in C$ and $s < t$, and $\psi(s) = \psi(t)$ if and only if $s$ and $t$ are related points of $D$. Furthermore $\psi(D) = M$. If we define

$$g: L \times C \to R$$

by $g(\xi, t) = (\xi, t)$ if $\xi < \omega_1$ and $g(\omega_1, t) = \big(\omega_1, \psi(t)\big)$, and let the set $R$ have the identification topology with respect to $g$, we obtain a space $R_M$, which is homeomorphic with the space $R$. The following property of $R_M$ is needed.

**3.3**    *Lemma. Let $U$ be an open set of $R_M$ such that $U \cap (\{\omega_1\} \times I) \neq \varnothing$ and let $c$ be the least upper bound of the set of real numbers $t$ such that $(\omega_1, t) \in U$. If $c \notin M$ and $c \neq 1$ then $\mathrm{ind}\,\mathrm{bd}\,(U) \geqslant 1$.*

*Proof.* Let $c = \psi(a)$, where $a \in C \backslash D$ and $a \neq 1$. Let $\{t_n\}$ be a sequence of points of $C$ with limit $a$ such that $(\omega_1, \psi(t_n)) \in U$ for each $n$. For each positive integer $n$ there exists $\xi_n$ in $L \backslash \{\omega_1\}$ such that

$$[\xi_n, \omega_1) \times \{t_n\} \subset U.$$

If $\xi_0$ is the least upper bound of the sequence $\{\xi_n\}$ of elements of $L$, then $\xi_0 < \omega_1$ and it is clear that $[\xi_0, \omega_1) \times \{a\} \subset \overline{U}$. Thus either (i) there exists $\xi$ such that $\xi_0 \leqslant \xi < \omega_1$ and $[\xi, \omega_1) \times \{a\} \subset \mathrm{bd}\,(U)$, or (ii) there exists a cofinal subset $A$ of $L \backslash \{\omega_1\}$ such that $(\xi, a) \in U$ if $\xi \in A$. In case (i) it is clear that $\mathrm{ind}\,\mathrm{bd}\,(U) \geqslant 1$. If case (ii) arises then for each $\xi$ in $A$ we can choose $b(\xi)$ in $C$ such that $a < b(\xi)$ and $C_a^{b(\xi)}(\xi) \subset U$. Since $A$ is a cofinal set, there exist an element $b$ of $C$ such that $a < b$, and a subset $B$ of $A$ which is cofinal in $L \backslash \{\omega_1\}$ such that $b(\xi) \geqslant b$ if $\xi \in B$. If $\psi(b) = d$, then $c < d$ since $c \notin M$, and it is clear that

$$\{\omega_1\} \times [c, d] \subset \overline{U}.$$

Hence $\{\omega_1\} \times [c, d] \subset \mathrm{bd}\,(U)$ so that $\mathrm{ind}\,\mathrm{bd}\,(U) \geqslant 1.\square$

**3.4**  *Example.* There exists a bicompact space $S$ such that

$$\dim S = 1 \quad \text{and} \quad \mathrm{ind}\,S = \mathrm{Ind}\,S = 2.$$

Furthermore $S = S_1 \cup S_2$, where $S_1$ and $S_2$ are closed sets of $S$ such that

$$\mathrm{ind}\,S_1 = \mathrm{Ind}\,S_1 = \mathrm{ind}\,S_2 = \mathrm{Ind}\,S_2 = 1.$$

We construct $S$ by identifying points in two spaces each of which is homeomorphic with $R$. Let $Y = Y_1 \cup Y_2$, where $Y_1$ and $Y_2$ are linearly ordered sets, each order-isomorphic to the long line, such that $Y_1 \cap Y_2 = \{y_0\}$, where $y_0$ is the last element of $Y_1$ and of $Y_2$. Let

$$S = \{(y, t) \in Y \times I \mid y = y_0 \quad \text{or} \quad t \in C\}.$$

And for $i = 1, 2$ let

$$S_i = \{(y, t) \in S \mid y \in Y_i\}.$$

Let $M_1$ and $M_2$ be countable dense subsets of the open unit interval $(0, 1)$ which are disjoint, and for $i = 1, 2$ let $\psi_i : C \to I$ be a continuous order-preserving surjection such that $\psi_i(D) = M_i$ and $\psi_i(s) = \psi_i(t)$ if and only if $s$ and $t$ are related points in $D$. For $i = 1, 2$ let $Y_i$ have the interval topology, let

$$g_i : Y_i \times C \to S_i$$

be given by $g_i(y, t) = (y, t)$ if $y < y_0$, whilst $g_i(y_0, t) = (y_0, \psi_i(t))$, and let $S_i$ have the identification topology with respect to the surjection $g_i$.

Then $S_1$ and $S_2$ are bicompact spaces, each of which is homeomorphic with $R$. The set $S_1 \cap S_2 = \{y_0\} \times I$ is closed in $S_1$ and in $S_2$ and this copy of $I$ has its usual topology both as a subspace of $S_1$ and as a subspace of $S_2$. Thus we can give $S$ the weak topology with respect to $\{S_1, S_2\}$. Then $S$ is a bicompact space and $S_1$ and $S_2$ are closed subspaces of $S$. Since $S_1$ and $S_2$ are homeomorphic with $R$, it follows from Proposition 3.2 that for $i = 1, 2$,

$$\dim S_i = \operatorname{ind} S_i = \operatorname{Ind} S_i = 1.$$

Since $S = S_1 \cup S_2$, where $S_1$ and $S_2$ are closed in $S$, it follows that $\dim S = 1$. Let $G$ be an open set of $S$ such that

$$G \cap \left( \{y_0\} \times I \right) = \{y_0\} \times (\tfrac{1}{4}, \tfrac{3}{4})$$

and let $U$ be an open neighbourhood of $(y_0, \tfrac{1}{2})$ such that $U \subset G$. Let $c$ be the least upper bound of the set of real numbers $t$ such that $(y_0, t) \in U$. Then $c \neq 1$ and since $M_1$ and $M_2$ are disjoint, $c$ cannot belong to both $M_1$ and $M_2$. Let us suppose that $c \notin M_1$. It follows from Lemma 3.3 that $\operatorname{ind} \operatorname{bd}_{S_1}(U \cap S_1) \geqslant 1$. Hence $\operatorname{ind} \operatorname{bd}(U) \geqslant 1$. It follows that $\operatorname{ind} S \geqslant 2$. It is not difficult to see that $\operatorname{Ind} S \leqslant 2$. Since $\operatorname{ind} S \leqslant \operatorname{Ind} S$, it follows that $\operatorname{ind} S = \operatorname{Ind} S = 2$.

## 4 Subset and sum theorems

The examples of the preceding section show that even for bicompact spaces neither the subset theorem nor the finite sum theorem is generally true for Ind. We shall show that for a completely normal space the countable sum theorem for Ind is satisfied if an open subset theorem is satisfied and that these theorems hold for totally normal spaces. The following lemma will be used frequently.

**4.1 Lemma.** Let $\{X_i\}_{i \in \mathbb{N}}$ be a family of open sets in a completely normal space $X$ such that $X_1 = X$, $X_{i+1} \subset X_i$ and $\operatorname{Ind}(X_i \backslash X_{i+1}) \leqslant n$ for each $i$ and $\bigcap_{i \in \mathbb{N}} X_i = \varnothing$. Then $\operatorname{Ind} X \leqslant n$.

*Proof.* The proof is by induction. The assertion is trivially true for dimension $-1$ and we assume it is true for dimension $n-1$. Let $E$ and $F$ be disjoint closed sets of $X$. Since $X$ is normal there exist open sets $U_0$ and $V_0$ such that $E \subset U_0$, $F \subset V_0$ and $\overline{U}_0 \cap \overline{V}_0 = \varnothing$.

Let $D_i = X_i \backslash X_{i+1}$. We construct disjoint sets $U_i$, $V_i$ and $C_i$ for $i = 1, 2, \ldots$, such that $U_i$ and $V_i$ are open in $X_i$, and hence open in $X$, and $C_i \subset D_i \subset U_i \cup V_i \cup C_i$, $\operatorname{Ind} C_i \leqslant n-1$, $\overline{U}_i \cap \overline{V}_i \cap X_{i+1} = \varnothing$,

$$U_i \supset \overline{U}_{i-1} \cap X_i \quad \text{and} \quad V_i \supset \overline{V}_{i-1} \cap X_i.$$

Assume that we have open sets $U_{i-1}$ and $V_{i-1}$ such that

$$\overline{U}_{i-1} \cap \overline{V}_{i-1} \cap X_i = \varnothing.$$

We construct the sets $U_i$, $V_i$ and $C_i$ as follows. The sets $\overline{U}_{i-1} \cap D_i$ and $\overline{V}_{i-1} \cap D_i$ are disjoint and closed in $D_i$ and $\text{Ind}\, D_i \leqslant n$. Hence $D_i$ is the union of disjoint sets $G_i$, $H_i$ and $C_i$ such that $G_i$ and $H_i$ are open in $D_i$ and $\overline{U}_{i-1} \cap D_i \subset G_i$, $\overline{V}_{i-1} \cap D_i \subset H_i$ and $\text{Ind}\, C_i \leqslant n-1$; and $C_i$ is closed in $D_i$ which is closed in $X_i$. Hence $X_i \backslash C_i$ is open in $X_i$ and hence open in $X$. The sets $G_i$ and $H_i$ are closed in $D_i \backslash C_i$ and hence closed in $X_i \backslash C_i$. Let

$$E_i = (\overline{U}_{i-1} \cup G_i) \cap (X_i \backslash C_i), \quad F_i = (\overline{V}_{i-1} \cup H_i) \cap (X_i \backslash C_i).$$

Then $E_i$ and $F_i$ are closed sets of $X_i \backslash C_i$. Since $\overline{V}_{i-1} \cap D_i \subset H_i$ and $G_i \subset D_i \backslash H_i$, it follows that $\overline{V}_{i-1} \cap G_i = \varnothing$ and similarly $\overline{U}_{i-1} \cap H_i = \varnothing$. Hence, since

$$\overline{U}_{i-1} \cap \overline{V}_{i-1} \cap (X_i \backslash C_i) \subset \overline{U}_{i-1} \cap \overline{V}_{i-1} \cap X_i = \varnothing$$

and $G_i \cap H_i = \varnothing$, we see that $E_i$ and $F_i$ are disjoint. Since $X$ is completely normal, $X_i \backslash C_i$ is normal. Hence there exist sets $U_i$ and $V_i$ open in $X_i \backslash C_i$ and hence open in $X$ such that $E_i \subset U_i$, $F_i \subset V_i$ and $U_i \cap V_i = \varnothing$ and such that moreover $\overline{U}_i \cap \overline{V}_i \cap (X_i \backslash C_i) = \varnothing$. Since $X_{i+1} \subset X_i \backslash C_i$, it follows that $\overline{U}_i \cap \overline{V}_i \cap X_{i+1} = \varnothing$, and since $U_i$ and $V_i$ are disjoint and contained in $X_i \backslash C_i$, the sets $U_i$, $V_i$ and $C_i$ are disjoint. Since

$$D_i = G_i \cup H_i \cup C_i,$$

$G_i \subset E_i \subset U_i$ and $H_i \subset F_i \subset V_i$, it follows that

$$C_i \subset D_i \subset U_i \cup V_i \cup C_i.$$

Also $\overline{U}_{i-1} \cap D_i \subset G_i$, therefore $\overline{U}_{i-1} \cap C_i = \varnothing$ so that

$$\overline{U}_{i-1} \cap X_i = \overline{U}_{i-1} \cap (X_i \backslash C_i) \subset E_i \subset U_i.$$

Similarly $V_i \supset \overline{V}_{i-1} \cap X_i$. Thus the sets $U_i$, $V_i$ and $C_i$ have the required properties. Since $\overline{U}_0 \cap \overline{V}_0 \cap X_1 = \overline{U}_0 \cap \overline{V}_0 = \varnothing$, the above method can be used to construct $U_1$, $V_1$ and $C_1$. It follows by induction that for $i = 1, 2, \ldots$, we can find sets $U_i$, $V_i$ and $C_i$ with the required properties.

Let $U = \bigcup_{i \in \mathbf{N}} U_i$, $V = \bigcup_{i \in \mathbf{N}} V_i$ and $C = \bigcup_{i \in \mathbf{N}} C_i$ and let $Z_i = \bigcup_{j \geqslant i} C_j$ if $i \in \mathbf{N}$. Then $U$ and $V$ are open sets, $E \subset U_0 \subset U_1 \subset U$ and similarly $F \subset V$. Every point of $X$ is in some $D_i$ and hence in $U_i$, $V_i$ or $C_i$ so that $X = U \cup V \cup C$. If $i \leqslant j$, then $U_i \cap X_j \subset U_j$ so that

$$U_i \cap V_j \subset U_j \cap V_j = \varnothing$$

and similarly $U_j \cap V_i = \varnothing$. Hence $U \cap V = \varnothing$. If $i \leqslant j$, then

$$U_i \cap C_j \subset U_j \cap C_j = \varnothing$$

and if $i > j$, then $U_i \cap C_j \subset X_i \cap C_j = \varnothing$. Hence $U \cap C = \varnothing$. Similarly $V \cap C = \varnothing$. Thus the sets $U$, $V$ and $C$ are disjoint. Since $X$ is a completely normal space, $C$ is completely normal. For each $i$,

$$Z_i = C \cap X_i$$

so that $Z_i$ is open in $C$ and clearly $Z_i \supset Z_{i+1}$. Furthermore $Z_1 = C$ and $\bigcap_{i \in \mathbf{N}} Z_i \subset \bigcap_{i \in \mathbf{N}} X_i = \varnothing$. For each $i$, $Z_i = C_i \cup Z_{i+1}$ and

$$C_i \cap Z_{i+1} = \varnothing.$$

Therefore $C_i = Z_i \backslash Z_{i+1}$ and

$$\mathrm{Ind}\,(Z_i \backslash Z_{i+1}) = \mathrm{Ind}\, C_i \leqslant n-1.$$

Therefore $\mathrm{Ind}\, C \leqslant n-1$ by the induction hypothesis. It follows from Proposition 2.7 that $\mathrm{Ind}\, X \leqslant n$. The proof is now completed by induction. $\square$

**4.2   Corollary.** *If $A$ is a closed set of a completely normal space $X$, then*

$$\mathrm{Ind}\, X \leqslant \max\,\{\mathrm{Ind}\, A, \mathrm{Ind}\,(X \backslash A)\}.$$

*Proof.* Let $\max\,\{\mathrm{Ind},\, A\, \mathrm{Ind}\,(X \backslash A)\} = n$. Consider the family $\{X_i\}_{i \in \mathbf{N}}$ of open sets of $X$ defined as follows: $X_1 = X$, $X_2 = X \backslash A$ and $X_i = \varnothing$ if $i \geqslant 3$. Then $X_i \supset X_{i+1}$ for each $i$ and $\bigcap_{i \in \mathbf{N}} X_i = \varnothing$. Furthermore $\mathrm{Ind}\,(X_1 \backslash X_2) = \mathrm{Ind}\, A \leqslant n$, $\mathrm{Ind}\,(X_2 \backslash X_3) = \mathrm{Ind}\,(X \backslash A) \leqslant n$ whilst

$$X_i \backslash X_{i+1} = \varnothing \quad \text{if} \quad i \geqslant 3.$$

Therefore $\mathrm{Ind}\, X \leqslant n$ by Lemma 4.1. $\square$

In order to discuss the relations between the subset theorem, the open subset theorem and the sum theorem, we consider the following conditions which a space $X$ may satisfy.

$(a_n)$ If $B \subset A \subset X$ and $\mathrm{Ind}\, A \leqslant n$ then $\mathrm{Ind}\, B \leqslant n$.

$(b_n)$ If $G \subset A \subset X$, where $G$ is open in $A$ and $\mathrm{Ind}\, A \leqslant n$, then $\mathrm{Ind}\, G \leqslant n$.

$(c_n)$ If $A = B \cup C \subset X$, where $B$ is closed in $A$, $\mathrm{Ind}\, B \leqslant n$ and $\mathrm{Ind}\, C \leqslant n$, then $\mathrm{Ind}\, A \leqslant n$.

$(d_n)$ If $A = \bigcup_{i \in \mathbf{N}} A_i \subset X$, where each $A_i$ is closed in $A$ and $\mathrm{Ind}\, A_i \leqslant n$, then $\mathrm{Ind}\, A \leqslant n$.

If $Y$ is a subspace of $X$ and $X$ satisfies one of these conditions then $Y$ satisfies the same condition. Clearly $(a_n)$ implies $(b_n)$.

**4.3   Lemma.** *If a space satisfies* $(a_{n-1})$ *and* $(b_n)$, *then it satisfies* $(a_n)$.

*Proof.* Let $X$ satisfy conditions $(a_{n-1})$ and $(b_n)$, and suppose that $B \subset A \subset X$, where $\operatorname{Ind} A \leqslant n$. Let $E$ be a closed set of $B$ and let $G$ be an open set of $B$ such that $E \subset G$. There exist $E_1$ closed in $A$ and $G_1$ open in $A$ such that $E_1 \cap B = E$ and $G_1 \cap B = G$. Let

$$H = (A \backslash E_1) \cup G_1$$

so that $B = (B \backslash E) \cup G \subset H$. Since $H$ is open in $A$ and $(b_n)$ is satisfied, it follows that $\operatorname{Ind} H \leqslant n$. The set $E_1 \cap H$ is closed in $H$, $G_1$ is open in $H$ and

$$E_1 \cap H = E_1 \cap \big((A \backslash E_1) \cup G_1\big) = E_1 \cap G_1 \subset G_1.$$

Therefore, since $\operatorname{Ind} H \leqslant n$, there exists $U_1$ open in $H$ such that $E_1 \cap H \subset U_1 \subset G_1$ and $\operatorname{Ind}(\overline{U}_1 \cap H \backslash U_1) \leqslant n-1$. Let $U = U_1 \cap B$ so that $U$ is open in $B$. Furthermore $E = E_1 \cap B = E_1 \cap H \cap B \subset U_1 \cap B = U$ and $U = U_1 \cap B \subset G_1 \cap B = G$. Thus $E \subset U \subset G$. The boundary of $U$ in $B$ is

$$\operatorname{bd}_B(U) = \overline{U} \cap B \backslash U = \overline{U} \cap B \backslash U_1 \cap B \subset (\overline{U}_1 \backslash U_1) \cap B.$$

But since $B \subset H$ it follows that

$$(\overline{U}_1 \backslash U_1) \cap B = (\overline{U}_1 \cap H \backslash U_1) \cap B \subset \overline{U}_1 \cap H \backslash U_1.$$

Thus $\operatorname{Ind} \operatorname{bd}_B(U) \leqslant n-1$ since $(a_{n-1})$ is satisfied. Thus $\operatorname{Ind} B \leqslant n$ and condition $(a_n)$ is satisfied. $\square$

**4.4   Lemma.** *If a completely normal space satisfies* $(b_n)$, *then it satisfies* $(c_n)$ *and* $(d_n)$.

*Proof.* Let $X$ be a completely normal space satisfying $(b_n)$. Let $A = B \cup C \subset X$, where $B$ is closed in $A$, $\operatorname{Ind} B \leqslant n$ and $\operatorname{Ind} C \leqslant n$. Since $A \backslash B$ is open in $A$, it follows that $A \backslash B$ is open in $C$ and hence $\operatorname{Ind}(A \backslash B) \leqslant n$ by $(b_n)$. The subspace $A$ is completely normal since $X$ is completely normal. Furthermore $B$ is closed in $A$, $\operatorname{Ind} B \leqslant n$ and $\operatorname{Ind}(A \backslash B) \leqslant n$. Hence $\operatorname{Ind} A \leqslant n$ by Corollary 4.2. Thus condition $(c_n)$ is satisfied.

Next let $A = \bigcup_{i \in \mathbf{N}} A_i \subset X$, where $A_i$ is closed in $A$ and $\operatorname{Ind} A_i \leqslant n$. Let $D_i = A_i \backslash \bigcup_{j < i} A_j$ and let $Y_i = \bigcup_{j \geqslant i} D_j = A \backslash \bigcup_{j < i} A_j$. Then $Y_i \supset Y_{i+1}$ and $\bigcap_{i \in \mathbf{N}} Y_i = A \backslash \bigcup_{j \in \mathbf{N}} A_j = \varnothing$. Since $\bigcup_{j < i} A_j$ is closed in $A$, it follows that $D_i$ is open in $A_i$ and that $Y_i$ is open in $A$. Thus $\operatorname{Ind} D_i \leqslant n$ since $(b_n)$ is satisfied. The sets $D_i$ are disjoint and $Y_i = D_i \cup Y_{i+1}$ so that $D_i = Y_i \backslash Y_{i+1}$. Hence $\mathrm{Ind}\,(Y_i \backslash Y_{i+1}) \leqslant n$. Since $A$ is completely normal it follows from Lemma 4.1 that $\mathrm{Ind}\,A \leqslant n$. Thus condition $(d_n)$ is satisfied.□

**4.5   Proposition.** *If $X$ is a completely normal space satisfying condition $(b_n)$ for all $n$, then $X$ also satisfies $(a_n)$, $(c_n)$ and $(d_n)$ for all $n$.*

*Proof.* Condition $(a_{-1})$ is trivially satisfied. Thus an inductive argument based on Lemma 4.3 shows that $X$ satisfies $(a_n)$ for every $n$. Lemma 4.4 implies that $(c_n)$ and $(d_n)$ are satisfied for every $n$.□

**4.6   Lemma.** *Let $X$ be a normal space satisfying the condition $(d_{n-1})$. Let $\{F_i\}_{i \in \mathbf{N}}$ be a countable closed covering of $X$ and suppose that for each $i$ there exist an open set $V_i$ and a closed set $C_i$ such that $F_i \subset V_i \subset C_i$ and $\mathrm{Ind}\,C_i \leqslant n$. Then $\mathrm{Ind}\,X \leqslant n$.*

*Proof.* Let $E$ be a closed set of $X$ and let $G$ be an open set such that $E \subset G$. Since $X$ is a normal space, there is a sequence $\{W_i\}$ of open sets such that

$$E \subset \overline{W}_{i+1} \subset W_i \subset G.$$

Let $K = \bigcap_{i \in \mathbf{N}} W_i = \bigcap_{i \in \mathbf{N}} \overline{W}_i$. Then $K$ is closed and

$$E \subset K \subset W_i \subset \overline{W}_i \subset G.$$

Now $F_i \cap K \subset V_i \cap W_i, F_i \cap K$ is closed and $V_i \cap W_i$ is open and contained in $C_i$. Hence, since $\mathrm{Ind}\,C_i \leqslant n$, there exists $U_i$ open in $C_i$ such that $F_i \cap K \subset U_i \subset V_i \cap W_i$ and $\mathrm{Ind}\,(\overline{U}_i \cap C_i \backslash U_i) \leqslant n-1$. Since $U_i$ is open in $C_i$ and $U_i \subset V_i$, it follows that $U_i$ is open in $V_i$ and hence open in $X$, and $\overline{U}_i \cap C_i = \overline{U}_i$ since $U_i \subset C_i$ and $C_i$ is closed. Hence

$$\mathrm{Ind\,bd}\,(U_i) \leqslant n-1.$$

If $U = \bigcup_{i \in \mathbf{N}} U_i$, then $U$ is open in $X$ and $E \subset U \subset G$ for

$$E \subset K = \bigcup (F_i \cap K) \subset \bigcup U_i = U \subset \bigcup W_i \subset G.$$

To complete the proof we show that $\mathrm{Ind\,bd}\,(U) \leqslant n-1$. If $x \notin K$, then $x \notin \overline{W}_j$ for some $j$ and hence, since each $U_i \subset \overline{W}_i$, the neighbourhood $X \backslash \overline{W}_j$ of $x$ has non-empty intersection with only a finite number of the sets $U_i$. Thus if $x \in \overline{U}$ and $x \notin K$, then $x \in \overline{U}_i$ for some $i$. Since $K \subset \bigcup_{i \in \mathbf{N}} \overline{U}_i$ it follows that $\overline{U} \subset \bigcup_{i \in \mathbf{N}} \overline{U}_i$ so that $\overline{U} = \bigcup_{i \in \mathbf{N}} \overline{U}_i$. Thus $\mathrm{bd}\,(U) \subset \bigcup_{i \in \mathbf{N}} \mathrm{bd}\,(U_i)$. Each set $\mathrm{bd}\,(U_i)$ is closed and

$$\mathrm{Ind\,bd}\,(U_i) \leqslant n-1.$$

Hence $\operatorname{Ind} \bigcup_{i\in\mathbb{N}} \operatorname{bd}(U_i) \leqslant n-1$ since condition $(d_{n-1})$ is satisfied. But $\operatorname{bd}(U)$ is closed so that $\operatorname{Ind} \operatorname{bd}(U) \leqslant n-1$. Thus $\operatorname{Ind} X \leqslant n$. $\square$

**4.7 Lemma.** *If a totally normal space satisfies condition $(d_{n-1})$, then it satisfies $(b_n)$.*

*Proof.* Let $X$ be a totally normal space satisfying condition $(d_{n-1})$ and suppose that $G \subset A \subset X$, where $G$ is open in $A$ and $\operatorname{Ind} A \leqslant n$. Since $A$ is a totally normal space, by Lemma 1.4.6, for each positive integer $i$ there is a family $\{W_{i\lambda}\}_{\lambda\in\Lambda}$, locally finite in $G$, of disjoint open sets of $A$, and a family $\{F_{i\lambda}\}_{\lambda\in\Lambda}$ of closed sets of $A$, such that $F_{i\lambda} \subset W_{i\lambda} \subset G$ and $\bigcup_{i\in\mathbb{N}}\bigcup_{\lambda\in\Lambda} F_{i\lambda} = G$. Since $A$ is normal, there exist $V_{i\lambda}$ open in $A$ and $C_{i\lambda}$ closed in $A$ such that $F_{i\lambda} \subset V_{i\lambda} \subset C_{i\lambda} \subset W_{i\lambda}$. Let $F_i = \bigcup_{\lambda\in\Lambda} F_{i\lambda}$ so that $\{F_i\}_{i\in\mathbb{N}}$ is a countable closed covering of $G$. For each $i$, let $V_i = \bigcup_{\lambda\in\Lambda} V_{i\lambda}$ and $C_i = \bigcup_{\lambda\in\Lambda} C_{i\lambda}$. Then $V_i$ is open in $G$, $C_i$ is closed in $G$ and $F_i \subset V_i \subset C_i$. For each $i$, the family $\{C_{i\lambda}\}_{\lambda\in\Lambda}$ is discrete in $G$. Hence the subspace $C_i$ of $G$ is the topological sum of the subspaces $C_{i\lambda}$. But $C_{i\lambda}$ is a closed set of $A$, so that $\operatorname{Ind} C_{i\lambda} \leqslant n$. Hence $\operatorname{Ind} C_i \leqslant n$ by Proposition 2.8. Since $X$ is totally normal, $G$ is normal, and since $X$ satisfies $(d_{n-1})$, so does $G$. Hence $\operatorname{Ind} G \leqslant n$ by Lemma 4.6. Thus $X$ satisfies condition $(b_n)$. $\square$

**4.8 Theorem.** (i) *If $A$ is a subset of a totally normal space $X$, then $\operatorname{Ind} A \leqslant \operatorname{Ind} X$.*

(ii) *If $X$ is a totally normal space and $X = A \cup B$, where $A$ is closed, $\operatorname{Ind} A \leqslant n$ and $\operatorname{Ind} B \leqslant n$, then $\operatorname{Ind} X \leqslant n$.*

(iii) *If $X$ is a totally normal space and $X = \bigcup_{i\in\mathbb{N}} A_i$, where each $A_i$ is closed and $\operatorname{Ind} A_i \leqslant n$, then $\operatorname{Ind} X \leqslant n$.*

*Proof.* Let $X$ be a totally normal space. Then $X$ is completely normal and hence $(b_n)$ implies $(d_n)$, by Lemma 4.4. By Lemma 4.7, $(d_{n-1})$ implies $(b_n)$. Hence since $(b_{-1})$ and $(d_{-1})$ are trivially satisfied, $(b_n)$ and $(d_n)$ hold for all $n$. Hence $(a_n)$ and $(c_n)$ hold for all $n$ by Proposition 4.5. It follows from $(a_n)$, $(c_n)$ and $(d_n)$ respectively that (i), (ii) and (iii) hold. $\square$

**4.9 Proposition.** *If $X$ is a totally normal space which is dominated by a covering $\{A_\lambda\}_{\lambda\in\Lambda}$ such that $\operatorname{Ind} A_\lambda \leqslant n$ if $\lambda \in \Lambda$, then $\operatorname{Ind} X \leqslant n$.*

*Proof.* The assertion is clearly true if $n = -1$. We assume the assertion true in dimension $n-1$ and prove it for dimension $n$. Let $\Lambda$ be well-ordered with last element $\xi$, and if $\mu \in \Lambda$ let $B_\mu = \bigcup_{\lambda\leqslant\mu} A_\lambda$. Let $E$ and $F$ be disjoint closed sets of $X$, let $\mu$ be an element of $\Lambda$ and suppose that for each $\lambda$ such that $\lambda < \mu$ we have found disjoint sets $G_\lambda$, $C_\lambda$ and $H_\lambda$ such that $G_\lambda$ and $H_\lambda$ are open in $B_\lambda$ and (1) $B_\lambda = G_\lambda \cup C_\lambda \cup H_\lambda$; (2) $E \cap B_\lambda \subset G_\lambda$ and $F \cap B_\lambda \subset H_\lambda$; (3) if $\nu < \lambda$ then $G_\nu = G_\lambda \cap B_\nu$, $C_\nu = C_\lambda \cap B_\nu$ and $H_\nu = H_\lambda \cap B_\nu$; (4) Ind $C_\lambda \leqslant n-1$. We shall construct $G_\mu$, $C_\mu$ and $H_\mu$ to satisfy the same conditions.

Let $C = \bigcup_{\lambda < \mu} C_\lambda$. Then $C \subset \bigcup_{\lambda < \mu} B_\lambda = \bigcup_{\lambda < \mu} A_\lambda$ and $C \cap B_\lambda = C_\lambda$ if $\lambda < \mu$, so that $C \cap A_\lambda = C_\lambda \cap A_\lambda$. But $C_\lambda$ is closed in $B_\lambda$ and hence is closed in $X$. Thus $C \cap A_\lambda$ is closed if $\lambda < \mu$, so that $C$ is a closed set. Furthermore $C$ is dominated by the covering $\{C \cap A_\lambda\}_{\lambda < \mu}$, and $C \cap A_\lambda \subset C_\lambda$, so that Ind $C \cap A_\lambda \leqslant n-1$ if $\lambda < \mu$. It follows from the induction hypothesis that Ind $C \leqslant n-1$. Next let

$$G = \bigcup_{\lambda < \mu} G_\lambda, \quad H = \bigcup_{\lambda < \mu} H_\lambda, \quad B = \bigcup_{\lambda < \mu} B_\lambda.$$

Then $G$ and $H$ are open in $B$. Moreover $G$, $H$ and $C$ are disjoint and

$$B = G \cup C \cup H.$$

Furthermore $G$ and $H$ are closed sets of $X \backslash C$. Thus since $E \cap B \subset G$ and $F \cap B \subset H$, it follows that $E \cup G$ and $F \cup H$ are disjoint closed sets of $X \backslash C$. Since $X \backslash C$ is normal, there exist disjoint sets $P$ and $Q$ open in $X \backslash C$ and hence open in $X$ such that

$$E \cup G \subset P, \quad F \cup H \subset Q, \quad \bar{P} \cap \bar{Q} \subset C.$$

The sets $\bar{P} \cap (A_\mu \backslash B)$ and $\bar{Q} \cap (A_\mu \backslash B)$ are disjoint and closed in the normal space $A_\mu \backslash B$. Since $A_\mu$ is a totally normal space and Ind $A_\mu \leqslant n$, it follows from Theorem 4.8 that Ind $(A_\mu \backslash B) \leqslant n$. Hence by Proposition 2.7 there exist disjoint sets $G'$, $C'$ and $H'$ such that $G'$ and $H'$ are open in $A_\mu \backslash B$ and $A_\mu \backslash B = G' \cup C' \cup H'$ and furthermore $\bar{P} \cap (A_\mu \backslash B) \subset G'$, $\bar{Q} \cap (A_\mu \backslash B) \subset H'$ and Ind $C' \leqslant n-1$. Finally let

$$G_\mu = G \cup G', \quad C_\mu = C \cup C', \quad H_\mu = H \cup H'.$$

Then $G_\mu$, $C_\mu$ and $H_\mu$ are disjoint. Since $P \cap B = G$ and $P \cap (A_\mu \backslash B) \subset G'$ it follows that $G_\mu = G' \cup (P \cap B_\mu)$. Since $G'$ is open in $A_\mu \backslash B = B_\mu \backslash B$, it follows that $G'$ is open in $B_\mu$ and hence $G_\mu$ is open in $B_\mu$. Similarly $H_\mu$ is open in $B_\mu$. The conditions (1), (2) and (3) are obviously satisfied. Furthermore $C_\mu = C \cup C'$, where $C$ is closed in $X$ and hence closed in $C_\mu$, and also Ind $C \leqslant n-1$ and Ind $C' \leqslant n-1$. Since $C_\mu$ is totally normal it follows from Theorem 4.8 that Ind $C_\mu \leqslant n-1$ so that condition (4) is satisfied.

By transfinite induction we see that $E$ and $F$ can be separated by a closed set $C_\xi$ such that Ind $C_\xi \leqslant n-1$. It follows from Proposition 2.7 that Ind $X \leqslant n$. The proof is now completed by induction. $\square$

**4.10   Corollary.** *If $X$ is a totally normal space and $\{A_\lambda\}_{\lambda \in \Lambda}$ is a $\sigma$-locally finite closed covering of $X$ such that $\operatorname{Ind} A_\lambda \leqslant n$ for each $\lambda$ then $\operatorname{Ind} X \leqslant n$.*

*Proof.* Let $\Lambda = \bigcup_{i \in \mathbf{N}} \Lambda(i)$, where $\{A_\lambda\}_{\lambda \in \Lambda(i)}$ is a locally finite family for each $i$, and let $B_i = \bigcup_{\lambda \in \Lambda(i)} A_\lambda$. Then $B_i$ is a closed set and $\{A_\lambda\}_{\lambda \in \Lambda(i)}$ is a locally finite closed covering of $B_i$. Since a space is dominated by a locally finite closed covering, it follows from Proposition 4.9 that $\operatorname{Ind} B_i \leqslant n$. Since $X$ is a totally normal space and $X = \bigcup_{i \in \mathbf{N}} B_i$, it follows from Theorem 4.8 that $\operatorname{Ind} X \leqslant n$.$\square$

We can now establish an analogue of Proposition 3.5.6 for large inductive dimension.

**4.11   Proposition.** *Let $A$ be a closed set of a totally normal space $X$. If $\operatorname{Ind} A \leqslant n$ and if $\operatorname{Ind} F \leqslant n$ for each closed set $F$ of $X$ which does not meet $A$, then $\operatorname{Ind} X \leqslant n$.*

*Proof.* The open subset $X \setminus A$ of the totally normal space $X$ is the union of a family $\{G_\lambda\}_{\lambda \in \Lambda}$, locally finite in $X \setminus A$, of open $F_\sigma$-sets of $X$. Thus $G_\lambda = \bigcup_{i \in \mathbf{N}} F_{i\lambda}$, where each set $F_{i\lambda}$ is closed in $X$ and disjoint from $A$. Thus $\operatorname{Ind} F_{i\lambda} \leqslant n$ and since $\{F_{i\lambda}\}_{\lambda \in \mathbf{N}, \lambda \in \Lambda}$ is a $\sigma$-locally finite closed covering of the totally normal space $X \setminus A$, it follows from Corollary 4.10 that $\operatorname{Ind}(X \setminus A) \leqslant n$. It now follows from Corollary 4.2 that $\operatorname{Ind} X \leqslant n$.$\square$

For open coverings we have the following sum theorem.

**4.12   Proposition.** *If $\{U_\lambda\}_{\lambda \in \Lambda}$ is a point-finite open covering of a totally normal space $X$ such that $\operatorname{Ind} U_\lambda \leqslant n$ for each $\lambda$, then $\operatorname{Ind} X \leqslant n$.*

*Proof.* The proof is exactly the same as that of Proposition 3.5.10 using Proposition 4.11 instead of Proposition 3.5.6.$\square$

We recall that Example 3.4 shows that not even the finite sum theorem for Ind is true for all normal spaces. We obtain next some finite sum theorems for normal spaces by imposing conditions on the intersection of the summands.

**4.13   Proposition.** *If $X$ is a normal space and $X = A \cup B$, where $A$ and $B$ are closed, $\operatorname{Ind} A \leqslant n$, $\operatorname{Ind} B \leqslant n$ and $\operatorname{Ind} A \cap B \leqslant 0$, then $\operatorname{Ind} X \leqslant n$.*

*Proof.* Let $E$ be a closed set of $X$ and let $G$ be an open set such that $E \subset G$. Let $C = A \cap B$. Since $\operatorname{Ind} C \leqslant 0$, there exists an open-and-closed set $P$ in $C$ such that

$$E \cap C \subset P \subset G \cap C.$$

Let $F_1 = (A \cap E) \cup P$ and $H_1 = (G \backslash B) \cup P$. Then $F_1 \subset H_1$, since

$$F_1 = (E \backslash B) \cup P.$$

Since $P$ is a closed subset of $C$, it follows that $P$ is closed in $A$. Hence $F_1$ is a closed set of $A$. And

$$A \backslash H_1 = \big(A \backslash (G \backslash B)\big) \cap (A \backslash P) = \big(C \cup (A \backslash G)\big) \cap (A \backslash P)$$
$$= (C \backslash P) \cup (A \backslash G).$$

The set $C \backslash P$ is closed in $C$ and hence in $A$ and $A \backslash G$ is a closed set of $A$. Thus $A \backslash H_1$ is closed in $A$, so that $H_1$ is open in $A$. Since $A$ is normal and $\operatorname{Ind} A \leqslant n$, there exists $V_1$ open in $A$ such that

$$F \subset V_1 \subset \overline{V}_1 \subset H \quad \text{and} \quad \operatorname{Ind}(\overline{V}_1 \backslash V_1) \leqslant n-1,$$

where $\overline{V}_1$ denotes the closure of $V_1$ in $A$ and hence in $X$. We note that $A \cap E \subset V_1 \subset A \cap G$ and $V_1 \cap C = \overline{V}_1 \cap C = P$. By similar arguments there exists a set $V_2$, open in $B$, such that $B \cap E \subset V_2 \subset B \cap G$ and $\operatorname{Ind}(\overline{V}_2 \backslash V_2) \leqslant n-1$ and moreover $V_2 \cap C = \overline{V}_2 \cap C = P$.

Let $V = V_1 \cup V_2$. Clearly $E \subset V \subset G$, and $X \backslash V = (X \backslash V_1) \cap (X \backslash V_2)$. But $X \backslash V_1 = (A \backslash V_1) \cup (B \backslash V_1)$ and since $V_1 \cap B = V_2 \cap A$, it follows that $(A \backslash V_1) \cap (X \backslash V_2) = A \backslash V_1$ and $(B \backslash V_1) \cap (X \backslash V_2) = B \backslash V_2$. Thus

$$X \backslash V = (A \backslash V_1) \cup (B \backslash V_2).$$

The set $A \backslash V_1$ is closed in $A$ and hence in $X$ and similarly $B \backslash V_2$ is a closed set of $X$. Thus $V$ is open in $X$. Finally

$$\operatorname{bd}(V) = (\overline{V}_1 \cup \overline{V}_2) \backslash (V_1 \cup V_2) = (\overline{V}_1 \backslash V_1) \cup (\overline{V}_2 \backslash V_2)$$

since $V_2 \cap A \subset V_1$ and $V_1 \cap B \subset V_2$. But since $V_1 \cap C = \overline{V}_1 \cap C$, it follows that $\overline{V}_1 \backslash V_1 \subset A \backslash C$. Similarly $\overline{V}_2 \backslash V_2 \subset B \backslash C$. Thus $\operatorname{bd}(V)$ is the topological sum of $\overline{V}_1 \backslash V_1$ and $\overline{V}_2 \backslash V_2$. It follows that $\operatorname{Ind} \operatorname{bd}(V) \leqslant n-1$. Hence $\operatorname{Ind} X \leqslant n$. $\square$

We have a generalization of Corollary 4.2.

**4.14 Proposition.** *Let $X$ be a normal space and let $A$ be a closed set of $X$ such that $X \backslash F$ is normal if $F$ is a closed set which is contained in $A$. Then*

$$\operatorname{Ind} X \leqslant \max \{\operatorname{Ind} A, \operatorname{Ind}(X \backslash A)\}.$$

*Proof.* Let $(\alpha_n)$ be the statement: if $X$ is a normal space, then $\text{Ind}\, X \leqslant n$ if $\text{Ind}\, A \leqslant n$ and $\text{Ind}\,(X \backslash A) \leqslant n$, where $A$ is a closed set of $X$ with the property that $X \backslash F$ is normal if $F$ is a closed set such that $F \subset A$. We shall prove by induction that $(\alpha_n)$ holds for all integers $n \geqslant -1$. The statement $(\alpha_{-1})$ is clearly true.

Let $X$ be a normal space, let $A$ be a closed set such that $X \backslash F$ is normal for each closed set $F$ contained in $A$ and suppose that $\text{Ind}\, A \leqslant n$ and $\text{Ind}\,(X \backslash A) \leqslant n$, where $n$ is a non-negative integer. Let $E$ be a closed set of $X$, and $G$ an open set such that $E \subset G$. Since $\text{Ind}\, A \leqslant n$, there exists $P$ open in $A$ such that

$$E \cap A \subset P \subset G \cap A \quad \text{and} \quad \text{Ind}\,\text{bd}_A(P) \leqslant n-1.$$

Since $A$ is a closed set, $\text{bd}_A(P) = \bar{P} \backslash P$.

Let $Y = X \backslash \text{bd}_A(P)$. By hypothesis the subspace $Y$ is normal. Since $P = \bar{P} \cap Y$ and $A \backslash \bar{P} = (A \backslash P) \cap Y$, we see that $P \cup E$ and $(A \backslash \bar{P}) \cup (Y \backslash G)$ are disjoint closed sets of $Y$. Hence there exist open sets $L$, $M$ of $Y$ such that $P \cup E \subset L$, $(A \backslash \bar{P}) \cup (Y \backslash G) \subset M$ and the closures $\tilde{L}$ and $\tilde{M}$ of $L$ and $M$ respectively in $Y$ are disjoint. If

$$N = L \backslash A \quad \text{and} \quad H = M \backslash A,$$

then the closures $\tilde{N}$ and $\tilde{H}$ of $N$ and $H$ respectively in $X \backslash A$ are disjoint. Since $X \backslash A$ is normal and $\text{Ind}\,(X \backslash A) \leqslant n$, there exists $Q$ open in $X \backslash A$ such that $\tilde{N} \subset Q$, $\tilde{Q} \cap \tilde{H} = \varnothing$ and $\text{Ind}\,\text{bd}_{X \backslash A}(Q) \leqslant n-1$, where $\tilde{Q}$ is the closure of $Q$ in $X \backslash A$.

Let $U = P \cup Q$. Then $U$ is open in $X$. For $P = L \cap A$ so that $P \cup Q = (L \cap A) \cup Q = L \cup Q$ since $L \backslash A \subset Q$. The sets $L$ and $Q$ are open in $Y$ and $X \backslash A$ respectively and hence are open in $X$. Thus $U$ is open. Since $E \subset L$, it follows that $E \subset U$ and since $P \subset G$ and $Q \subset Y \backslash M \subset G$ it follows that $U \subset G$. Finally

$$\text{bd}\,(U) = \bar{U} \backslash U = (\bar{P} \cup \bar{Q}) \cap (X \backslash P) \cap (X \backslash Q).$$

Also $\bar{P} \subset A$ and $Q \subset X \backslash A$, therefore

$$\bar{P} \cap (X \backslash P) \cap (X \backslash Q) = \bar{P} \backslash P = \text{bd}_A(P).$$

Since $\bar{Q} = (\bar{Q} \cap A) \cup \tilde{Q}$ we see that

$$\bar{Q} \cap (X \backslash P) \cap (X \backslash Q) = [\bar{Q} \cap A \cap (X \backslash P)] \cup [\tilde{Q} \cap (X \backslash Q)].$$

But $\tilde{Q} \cap (X \backslash Q) = \text{bd}_{X \backslash A}(Q)$, and $Q \subset X \backslash M$ so that $A \cap \bar{Q} \subset A \backslash M \subset \bar{P}$. Thus $\bar{Q} \cap A \cap (X \backslash P) \subset \text{bd}_A(P)$. Therefore

$$\text{bd}\,(U) = \text{bd}_A(P) \cup \text{bd}_{X \backslash A}(Q).$$

Now $\operatorname{bd}(U)$ is a normal space. If $K$ is closed and $K \subset \operatorname{bd}_A(P)$, then $\operatorname{bd}(U)\backslash K = \operatorname{bd}(U) \cap (X\backslash K)$, which is a closed subspace of the normal space $X\backslash K$. Thus $\operatorname{bd}(U)\backslash K$ is normal. If $(\alpha_{n-1})$ is true, then

$$\operatorname{Ind}\operatorname{bd}(U) \leqslant n-1$$

since $\operatorname{Ind}\operatorname{bd}_A(P) \leqslant n-1$ and $\operatorname{Ind}\operatorname{bd}_{X\backslash A}(Q) \leqslant n-1$. Hence $\operatorname{Ind} X \leqslant n$ and $(\alpha_n)$ is true. The proof is completed by induction. $\square$

**4.15  Proposition.** *Let $X$ be a normal space such that $X = A \cup B$, where $A$ and $B$ are closed, $\operatorname{Ind} A \leqslant n$ and $\operatorname{Ind} B \leqslant n$. If $X\backslash F$ is normal, $\operatorname{Ind}(A\backslash F) \leqslant n$ and $\operatorname{Ind}(B\backslash F) \leqslant n$ for each closed set $F$ of $A \cap B$, then $\operatorname{Ind} X \leqslant n$.*

*Proof.* Let $E$ be a closed set of $X$ and let $G$ be an open set such that $E \subset G$. Let $C = A \cap B$. Since $\operatorname{Ind} C \leqslant n$, there exists $P$ open in $C$ such that

$$E \cap C \subset P \subset G \cap C \quad \text{and} \quad \operatorname{Ind}\operatorname{bd}_C(P) \leqslant n-1.$$

Let $Y = X\backslash\operatorname{bd}_C(P)$ and let $Y_1 = A \cap Y$ and $Y_2 = B \cap Y$. Then by hypothesis $Y$ is normal, so that $Y_1$ and $Y_2$ are normal, and also $\operatorname{Ind} Y_1 \leqslant n$ and $\operatorname{Ind} Y_2 \leqslant n$. Let $F_1 = (E \cap A) \cup P$ and $H_1 = (G\backslash B) \cup P$. Then $F_1$ is closed in $Y_1$ since $P$ is closed in $Y_1$, and $H_1$ is open in $Y_1$ since $Y_1\backslash H_1 = (Y_1\backslash G) \cup (C\backslash P)$. Furthermore $F_1 \subset H_1$. Hence there exists $V_1$ open in $Y_1$ such that

$$F_1 \subset V_1 \subset \tilde{V}_1 \subset H_1 \quad \text{and} \quad \operatorname{Ind}\operatorname{bd}_{Y_1}(V_1) \leqslant n-1,$$

where $\tilde{V}_1$ is the closure of $V_1$ in $Y_1$. Similarly there exists $V_2$ open in $Y_2$ such that

$$(E \cap B) \cup P \subset V_2 \subset \tilde{V}_2 \subset (G\backslash A) \cup P \quad \text{and} \quad \operatorname{Ind}\operatorname{bd}_{Y_2}(V_2) \leqslant n-1,$$

where $\tilde{V}_2$ is the closure of $V_2$ in $Y_2$. Let $V = V_1 \cup V_2$. Then

$$Y\backslash V = (Y_1\backslash V_1) \cup (Y_2\backslash V_2),$$

so that $V$ is open in $Y$ and hence open in $X$. Clearly $E \subset V \subset G$. Now $\tilde{V}_1 = \bar{V}_1 \cap Y_1 = \bar{V}_1\backslash\operatorname{bd}_C(P)$. But $\operatorname{bd}_C(P) \subset \bar{V}_1$ so that

$$\bar{V}_1 = \tilde{V}_1 \cup \operatorname{bd}_C(P).$$

Thus $\bar{V} = \tilde{V}_1 \cup \tilde{V}_2 \cup \operatorname{bd}_C(P)$ so that

$$\operatorname{bd}(V) = \operatorname{bd}_{Y_1}(V_1) \cup \operatorname{bd}_{Y_2}(V_2) \cup \operatorname{bd}_C(P).$$

If $K$ is a closed subset of $\text{bd}_C(P)$, then $\text{bd}(V)\backslash K$ is a closed subset of $X\backslash K$ and hence is normal. Also $\text{Ind}\,\text{bd}_C(P) \leqslant n-1$. Since

$$\text{bd}_{Y_1}(V_1) \subset X\backslash B \quad \text{and} \quad \text{bd}_{Y_2}(V_2) \subset X\backslash A,$$

the subspace $\text{bd}(V)\backslash\text{bd}_C(P)$ is the topological sum of $\text{bd}_{Y_1}(V_1)$ and $\text{bd}_{Y_2}(V_2)$ so that $\text{Ind}\big(\text{bd}(V)\backslash\text{bd}_C(P)\big) \leqslant n-1$. It follows from Proposition 4.14 that $\text{Ind}\,\text{bd}(V) \leqslant n-1$. Hence $\text{Ind}\,X \leqslant n$. $\square$

Finally we obtain a 'Urysohn inequality' for large inductive dimension. A lemma is needed.

**4.16  Lemma.** *Let $X$ be a completely normal space and let $A$ be a subspace of $X$ such that $\text{Ind}\,A \leqslant m$. If $E$ is closed in $X$, $G$ is open in $X$ and $E \subset G$, then there exists $U$ open in $X$ such that $E \subset U \subset G$ and $\text{Ind}\big(A \cap \text{bd}(U)\big) \leqslant m-1$.*

*Proof.* Since $X$ is normal, there exists an open set $L$ such that

$$E \subset L \subset \bar{L} \subset G.$$

Since $\bar{L} \cap A$ is closed in $A$, $G \cap A$ is open in $A$ and $\bar{L} \cap A \subset G \cap A$, there exists $H$ open in $A$ such that

$$\bar{L} \cap A \subset H \subset G \cap A \quad \text{and} \quad \text{Ind}\,\text{bd}_A(H) \leqslant m-1.$$

Let $P = E \cup H$ and $Q = A\backslash\bar{H}$. Then $\bar{P} = E \cup \bar{H}$ and it is clear that $\bar{P} \cap Q = \varnothing$. Since $Q \subset A\backslash H$ which is closed in $A$, it follows that $\bar{Q} \cap A \subset A\backslash H$ so that $\bar{Q} \cap H = \varnothing$. Since $Q \subset A\backslash H \subset X\backslash L$ it follows that $\bar{Q} \subset X\backslash L$ and hence $\bar{Q} \cap E = \varnothing$. Thus $P \cap \bar{Q} = \varnothing$. Since $X$ is completely normal there exists $W$ open in $X$ such that

$$E \cup H \subset W \quad \text{and} \quad \bar{W} \cap (A\backslash\bar{H}) = \varnothing.$$

Let $U = G \cap W$. Then $U$ is open in $X$ and $E \subset U \subset G$. Since $U \subset W$. it follows that $\bar{U} \cap (A\backslash\bar{H}) = \varnothing$ so that $\bar{U} \cap A \subset \bar{H} \cap A$. But $H \subset U$ so that $\bar{U} \cap A = \bar{H} \cap A$. Hence

$$\text{bd}(U) \cap A = \bar{U} \cap A\backslash U = \bar{H} \cap A\backslash U \subset \bar{H} \cap A\backslash H = \text{bd}_A(H),$$

and $\text{bd}(U) \cap A$ is closed in $A$ and therefore closed in $\text{bd}_A(H)$. Thus $\text{Ind}\,\text{bd}(U) \cap A \leqslant m-1$. $\square$

**4.17  Proposition.** *If $X$ is a completely normal space and $X = A \cup B$, then*

$$\text{Ind}\,X \leqslant \text{Ind}\,A + \text{Ind}\,B + 1.$$

*Proof.* It is enough to prove that if $\text{Ind}\,A \leqslant m$ and $\text{Ind}\,B \leqslant n$, then $\text{Ind}\,X \leqslant m+n+1$. The proof is by induction over $m$. The statement is true if $m = -1$ and we assume it is true if one of the summands has dimension $m-1$. Let $X$ be a completely normal space and suppose that $X = A \cup B$, where $\operatorname{Ind} A \leqslant m$ and $\operatorname{Ind} B \leqslant n$. If $E$ is closed in $X$, $G$ is open in $X$ and $E \subset G$, then by Lemma 4.16 there exists $U$ open in $X$ such that $E \subset U \subset G$ and $\operatorname{Ind} \operatorname{bd}(U) \cap A \leqslant m-1$. Since

$$\operatorname{bd}(U) \cap B$$

is closed in $B$, it follows that $\operatorname{Ind} \operatorname{bd}(U) \cap B \leqslant n$. Hence

$$\operatorname{Ind} \operatorname{bd}(U) \leqslant m+n$$

by the induction hypothesis. Thus $\operatorname{Ind} X \leqslant m+n+1$. The proof is now completed by induction. $\square$

## 5   Dimension of pseudo-metrizable spaces

In this section it will be shown that if $X$ is a pseudo-metrizable space, then $\dim X = \operatorname{Ind} X$.

**5.1   Lemma.** *If a space $X$ has a $\sigma$-locally finite base for its topology consisting of open-and-closed sets, then $\operatorname{Ind} X \leqslant 0$.*

*Proof.* Let $\mathscr{V}$ be a base consisting of open-and-closed sets and let $\mathscr{V} = \bigcup_{i \in \mathbf{N}} \mathscr{V}_i$, where each $\mathscr{V}_i$ is locally finite. Let $E$ be a closed set of $X$ and let $G$ be an open set such that $E \subset G$. If $x \in X$, then either $x \in G$ or $x \in X \setminus E$. It follows that there exists $V$ in $\mathscr{V}$ such that $x \in V$ and either $V \subset G$ or $V \cap E = \varnothing$. Thus if we define

$$U_{2i-1} = \bigcup \{ V \in \mathscr{V}_i \mid V \subset G \}$$

and

$$U_{2i} = \bigcup \{ V \in \mathscr{V}_i \mid V \cap E = \varnothing \},$$

then $\{U_i\}_{i \in \mathbf{N}}$ is a countable covering of $X$. Furthermore $U_i$ is an open-and-closed set for each $i$ since $\mathscr{V}_i$ is a locally finite family of open-and-closed sets. If $j \in \mathbf{N}$, let

$$W_j = U_j \setminus \bigcup_{i<j} U_i.$$

Then $\{W_j\}_{j \in \mathbf{N}}$ is a countable covering of $X$ consisting of disjoint sets which are open-and-closed. If $i \in \mathbf{N}$, then $W_{2i-1} \subset G$ and $W_{2i} \subset X \setminus E$. Let $W = \bigcup_{i \in \mathbf{N}} W_{2i-1}$. Then $E \subset W \subset G$ and $W$ is an open-and-closed set of $X$. Thus $\operatorname{Ind} X \leqslant 0$. $\square$

**5.2   Lemma.** *If a regular space $X$ has a $\sigma$-locally finite base $\mathscr{V}$ for its topology such that $\operatorname{Ind} \operatorname{bd}(V) \leqslant n-1$ if $V \in \mathscr{V}$, then $\operatorname{Ind} X \leqslant n$.*

*Proof.* It follows from Proposition 2.3.4 that $X$ is a pseudo-metrizable space. Let $A = \bigcup_{V \in \mathscr{V}} \mathrm{bd}\,(V)$. Then $A$ is a totally normal space and $\{\mathrm{bd}\,(V)\}_{V \in \mathscr{V}}$ is a $\sigma$-locally finite closed covering of $A$ such that

$$\mathrm{Ind}\,\mathrm{bd}\,(V) \leqslant n - 1$$

for each $V$ in $\mathscr{V}$. It follows from Corollary 4.10 that $\mathrm{Ind}\,A \leqslant n - 1$. If $B = X \backslash A$, then $\{V \cap B\}_{V \in \mathscr{V}}$ is a $\sigma$-locally finite base for the topology of $B$ and consists of open-and-closed sets of $B$. It follows from Lemma 5.1 that $\mathrm{Ind}\,B \leqslant 0$. Since $X = A \cup B$, where $\mathrm{Ind}\,A \leqslant n - 1$ and $\mathrm{Ind}\,B \leqslant 0$, it follows from Proposition 4.17 that $\mathrm{Ind}\,X \leqslant n$. $\square$

**5.3 Proposition.** *The following statements about a topological space $X$ are equivalent:*

(a) $X$ *is a pseudo-metrizable space such that* $\dim X \leqslant n$;

(b) $X$ *is a regular space with a $\sigma$-locally finite base $\mathscr{V}$ for its topology such that the order of* $\{\mathrm{bd}\,(V) \mid V \in \mathscr{V}\}$ *does not exceed* $n - 1$;

(c) $X$ *is a regular space with a $\sigma$-locally finite base $\mathscr{V}$ for its topology such that* $\mathrm{Ind}\,\mathrm{bd}\,(V) \leqslant n - 1$ *if* $V \in \mathscr{V}$;

(d) $X$ *is a pseudo-metrizable space such that* $\mathrm{Ind}\,X \leqslant n$.

*Proof.* (a) $\Rightarrow$ (b). Let $X$ be a pseudo-metrizable space such that $\dim X \leqslant n$. It follows from Proposition 2.3.4 that there exists a sequence $\{\mathscr{U}_i\}_{i \in \mathbf{N}}$ of locally finite open coverings of $X$ such that for each point $x$ of $X$ and each open set $U$ such that $x \in U$, there exists some $i$ such that $\mathrm{St}\,(x, \mathscr{U}_i) \subset U$. For each $i$, let $\mathscr{U}_i = \{U_\lambda\}_{\lambda \in \Lambda(i)}$, where we suppose that the sets $\Lambda(i)$ and $\Lambda(j)$ are disjoint if $i \neq j$. For each $i$ there exists a closed covering $\{F_\lambda\}_{\lambda \in \Lambda(i)}$ such that $F_\lambda \subset U_\lambda$ for every $\lambda$. For each positive integer $j$ and each $\lambda$ in $\bigcup_{i \leqslant j} \Lambda(i)$ we shall find open sets $V_\lambda^j$ and $W_\lambda^j$ such that

$$F_\lambda \subset V_\lambda^j \subset V_\lambda^{j+1} \subset \overline{V}_\lambda^{j+1} \subset W_\lambda^{j+1} \subset W_\lambda^j \subset U_\lambda,$$

and furthermore the order of the family $\{\overline{W}_\lambda^j \backslash V_\lambda^j \mid \lambda \in \bigcup_{i \leqslant j} \Lambda(i)\}$ does not exceed $n - 1$.

The proof is by induction. By Corollary 3.4.7 there exist families of open sets $\{V_\lambda\}_{\lambda \in \Lambda(1)}$ and $\{W_\lambda\}_{\lambda \in \Lambda(1)}$ such that if $\lambda \in \Lambda(1)$ then

$$F_\lambda \subset V_\lambda^1 \subset \overline{V}_\lambda^1 \subset W_\lambda^1 \subset U_\lambda,$$

and the order of $\{\overline{W}_\lambda^1 \backslash V_\lambda^1\}_{\lambda \in \Lambda(1)}$ does not exceed $n - 1$. Suppose now that $j > 1$ and we have found open sets $V_\lambda^{j-1}$ and $W_\lambda^{j-1}$ for $\lambda$ in $\bigcup_{i < j} \Lambda(i)$ such that if $\lambda \in \bigcup_{i < j} \Lambda(i)$ then

$$F_\lambda \subset V_\lambda^{j-1} \subset \overline{V}_\lambda^{j-1} \subset W_\lambda^{j-1} \subset U_\lambda.$$

The family $\{G_\lambda \mid \lambda \in \bigcup_{i<j} \Lambda(i)\}$ is locally finite, where $G_\lambda = W_\lambda^{j-1}$ if $\lambda \in \bigcup_{i<j} \Lambda(i)$ and $G_\lambda = U_\lambda$ if $\lambda \in \Lambda(j)$. It follows from Corollary 3.4.7 that there exist families $\{V_\lambda^j \mid \lambda \in \bigcup_{i\leqslant j} \Lambda(i)\}$ and $\{W_\lambda^j \mid \lambda \in \bigcup_{i\leqslant j} \Lambda(i)\}$ of open sets such that if $\lambda \in \bigcup_{i<j} \Lambda(i)$ then

$$\bar{V}_\lambda^{j-1} \subset V_\lambda^j \subset \bar{V}_\lambda^j \subset W_\lambda^j \subset W_\lambda^{j-1},$$

if $\lambda \in \Lambda(j)$ then

$$F_\lambda \subset V_\lambda^j \subset \bar{V}_\lambda^j \subset W_\lambda^j \subset U_\lambda,$$

and the order of the family $\{\bar{W}_\lambda^j \backslash V_\lambda^j \mid \lambda \in \bigcup_{i\leqslant j} \Lambda(i)\}$ does not exceed $n-1$. By induction there exist families of open sets with the properties stated above.

If $i \in \mathbf{N}$ and $\lambda \in \Lambda(i)$ let $V_\lambda = \bigcup_{j \geqslant i} V_\lambda^j$. Clearly $F_\lambda \subset V_\lambda \subset U_\lambda$. If $\Lambda = \bigcup_{i \in \mathbf{N}} \Lambda(i)$, then $\mathscr{V} = \{V_\lambda\}_{\lambda \in \Lambda}$ is a $\sigma$-locally finite family of open sets. Furthermore $\mathscr{V}$ is a base for the topology of $X$, for if $x \in X$ and $U$ is an open set such that $x \in U$, there exists some $i$ such that

$$\mathrm{St}\,(x, \mathscr{U}_i) \subset U;$$

and there exists $\lambda$ in $\Lambda(i)$ such that $x \in V_\lambda$ so that

$$x \in V_\lambda \subset U_\lambda \subset \mathrm{St}\,(x, \mathscr{U}_i) \subset U.$$

Thus $\mathscr{V}$ is a $\sigma$-locally finite base for the topology of $X$. Finally we show that the order of the family $\{\mathrm{bd}\,(V_\lambda)\}_{\lambda \in \Lambda}$ does not exceed $n-1$. Let $\lambda(0), \ldots, \lambda(n)$ be distinct elements of $\Lambda$ and suppose that $\lambda(r) \in \Lambda(i_r)$ for $r = 0, \ldots, n$. Let $j = \max\{i_0, \ldots, i_n\}$. Then for $r = 0, \ldots, n$

$$V_{\lambda(r)}^j \subset V_{\lambda(r)} \subset W_{\lambda(r)}^j$$

and hence $\mathrm{bd}\,(V_{\lambda(r)}) \subset \bar{W}_{\lambda(r)}^j \backslash V_{\lambda(r)}^j$. Since the order of

$$\{\bar{W}_\lambda^j \backslash V_\lambda^j \mid \lambda \in \bigcup_{i\leqslant j} \Lambda(i)\}$$

does not exceed $n-1$, it follows that $\bigcap_{r=0}^n \mathrm{bd}\,(V_{\lambda(r)}) = \varnothing$ as required.

$(b) \Rightarrow (c)$. The proof is by induction. The implication holds if $n = 0$, for a family has order $-1$ if and only if it consists of empty sets. Now let $\mathscr{V}$ be a $\sigma$-locally finite base for the topology of a regular space $X$ such that the order of $\{\mathrm{bd}\,(V) \mid V \in \mathscr{V}\}$ does not exceed $n-1$, where $n$ is a positive integer. Let $V_0$ be a member of $\mathscr{V}$ and let $B_0 = \mathrm{bd}\,(V_0)$. If $\mathscr{W} = \{V \cap B_0 \mid V \in \mathscr{V}\}$, then $\mathscr{W}$ is a $\sigma$-locally finite base for the topology of the regular space $B_0$. If $V \in \mathscr{V}$, then

$$\mathrm{bd}_{B_0}(V \cap B_0) \subset \mathrm{bd}\,(V) \cap B_0.$$

Since $V_0 \cap B_0 = \varnothing$ we see that the order of $\{\mathrm{bd}_{B_0}(V \cap B_0) \mid V \in \mathscr{V}\}$ does not exceed $n-2$. From the appropriate induction hypothesis and Lemma 5.2, it follows that $\operatorname{Ind} B_0 \leqslant n-1$. Thus $\mathscr{V}$ is a $\sigma$-locally finite base for the topology of $X$ such that $\operatorname{Ind} \operatorname{bd}(V) \leqslant n-1$ if $V \in \mathscr{V}$.

$(c) \Rightarrow (d)$. Lemma 5.2.

$(d) \Rightarrow (a)$. This is obvious since if $X$ is a normal space then

$$\dim X \leqslant \operatorname{Ind} X$$

by Proposition 2.10. $\square$

**5.4   Theorem.** *If $X$ is a pseudo-metrizable space, then*

$$\dim X = \operatorname{Ind} X.$$

*Proof.* This is an immediate consequence of Proposition 5.3. $\square$

As another consequence of Proposition 5.3, we obtain a product theorem for the dimension of pseudo-metrizable spaces.

**5.5   Proposition.** *If $X$ and $Y$ are pseudo-metrizable spaces, at least one of which is non-empty, then*

$$\operatorname{Ind} X \times Y \leqslant \operatorname{Ind} X + \operatorname{Ind} Y.$$

*Proof.* We shall prove that if $\operatorname{Ind} X \leqslant m$ and $\operatorname{Ind} Y \leqslant n$, then

$$\operatorname{Ind} X \times Y \leqslant m+n.$$

The proof is by induction over $m+n$. If $m+n = -1$ the statement is trivially true. Suppose that the statement is true for the product of two pseudo-metrizable spaces for which the sum of the large inductive dimensions is equal to $k$, and suppose that $\operatorname{Ind} X \leqslant m$ and $\operatorname{Ind} Y \leqslant n$, where $m+n = k+1$. By Proposition 5.3 there exist $\sigma$-locally finite bases $\mathscr{V}$ and $\mathscr{W}$ for the topologies of $X$ and $Y$ respectively such that $\operatorname{Ind} \operatorname{bd}(V) \leqslant m-1$ if $V \in \mathscr{V}$, and $\operatorname{Ind} \operatorname{bd}(W) \leqslant n-1$ if $W \in \mathscr{W}$. Then $\mathscr{U} = \{V \times W \mid V \in \mathscr{V}, W \in \mathscr{W}\}$ is a $\sigma$-locally finite base for the topology of $X \times Y$, and if $V \in \mathscr{V}$ and $W \in \mathscr{W}$ then

$$\operatorname{bd}(V \times W) = (\overline{V} \times \operatorname{bd}(W)) \cup (\operatorname{bd}(V) \times \overline{W}).$$

It follows from Theorem 4.8 that

$$\operatorname{Ind} \operatorname{bd}(V \times W) \leqslant \max\{\operatorname{Ind}(\overline{V} \times \operatorname{bd}(W)), \operatorname{Ind}(\operatorname{bd}(V) \times \overline{W})\}.$$

But $\operatorname{Ind} \overline{V} \leqslant m$ and $\operatorname{Ind} \operatorname{bd}(W) \leqslant n-1$, so that

$$\operatorname{Ind}(\overline{V} \times \operatorname{bd}(W)) \leqslant m+n-1$$

by the induction hypothesis. Similarly $\operatorname{Ind}\big(\operatorname{bd}(V)\times\overline{W}\big)\leqslant m+n-1$. Thus $\mathscr{U}$ is a $\sigma$-locally finite base for the topology of the pseudo-metrizable space $X\times Y$ such that $\operatorname{Ind}\operatorname{bd}(U)\leqslant m+n-1$ if $U\in\mathscr{U}$. It follows from Proposition 5.3 that $\operatorname{Ind} X\times Y\leqslant m+n$.$\square$

In Chapter 7 there is an example of a metrizable space for which small inductive dimension differs from covering dimension and large inductive dimension. Here we shall prove that all three dimensions coincide for strongly pseudo-metrizable spaces. We shall need the following two lemmas both of which will have other applications.

**5.6  Lemma.** *If $\mathscr{U}$ is an open covering of a topological space and $\mathscr{U}$ has a weak refinement which consists of countably many star-finite open coverings, then $\mathscr{U}$ has a $\sigma$-discrete open refinement $\mathscr{W}$ such that the boundary of each member of $\mathscr{W}$ is contained in the boundary of some member of $\mathscr{U}$.*

*Proof.* Let $\mathscr{V}$ be a weak refinement of $\mathscr{U}$, where $\mathscr{V}=\bigcup_{i\in\mathbf{N}}\mathscr{V}_i$ and each $\mathscr{V}_i$ is a star-finite open covering of $X$. For each $i$, let $\mathscr{V}_i=\{V_{m\lambda}\}_{\lambda\in\Lambda(i)}$, where $V_{m\lambda}$ and $V_{n\mu}$ are disjoint if $\lambda,\mu\in\Lambda(i)$ and $\lambda\neq\mu$. We can suppose that $\Lambda(i)$ and $\Lambda(j)$ are disjoint if $i\neq j$. If $\lambda\in\Lambda(i)$, let $V_\lambda=\bigcup_{m\in\mathbf{N}}V_{m\lambda}$. Then $\{V_\lambda\}_{\lambda\in\Lambda(i)}$ is a disjoint covering of $X$ by open-and-closed sets. For each pair of positive integers $m,i$ let $L(m,i)$ be the subset of $\Lambda(i)$ consisting of those elements $\lambda$ such that $V_{m\lambda}$ is contained in some member of $\mathscr{U}$. If $\lambda\in L(m,i)$ choose $U_{m\lambda}$ in $\mathscr{U}$ such that $V_{m\lambda}\subset U_{m\lambda}$ and let $W_{m\lambda}=V_\lambda\cap U_{m\lambda}$. Since $\mathscr{V}$ is a weak refinement of $\mathscr{U}$, it follows that $\mathscr{W}=\{W_{m\lambda}\}_{\lambda\in L(m,i),\,m\in\mathbf{N},\,i\in\mathbf{N}}$ is an open covering of $X$. Clearly $\mathscr{W}$ is a refinement of $\mathscr{U}$. Since $W_{m\lambda}\subset V_\lambda$ if $\lambda\in L(m,i)$, the family $\{W_{m\lambda}\}_{\lambda\in L(m,i)}$ is discrete and hence $\mathscr{W}$ is $\sigma$-discrete covering. Finally since

$$W_{m\lambda}=V_\lambda\cap U_{m\lambda}$$

and $V_\lambda$ is an open-and-closed set, it follows from Proposition 1.2.4 that $\operatorname{bd}(W_{m\lambda})\subset\operatorname{bd}(U_{m\lambda})$.$\square$

**5.7  Lemma.** *If $X$ is a completely paracompact regular space such that $\operatorname{ind} X\leqslant n$, $E$ is a closed set of $X$ and $G$ is an open set such that $E\subset G$, then there exists an open set $V$ such that $E\subset V\subset G$ and $\operatorname{bd}(V)=\bigcup_{k\in\mathbf{N}}F_k$, where $F_k$ is a closed set such that $\operatorname{ind} F_k\leqslant n-1$ for each $k$.*

*Proof.* If $x\in X$, there exists an open neighbourhood $V_x$ of $x$ such that $\operatorname{ind}\operatorname{bd}(V_x)\leqslant n-1$ and $V_x\subset G$ if $x\in E$, whilst $V_x\subset X\backslash E$ if $x\in X\backslash E$. By Lemma 5.6, the open covering $\{V_x\}_{x\in X}$ of $X$ has a $\sigma$-discrete open refinement $\mathscr{A}$ such that the boundary of each member of $\mathscr{A}$ is contained in bd $(V_x)$ for some $x$ in $X$. Thus if $U \in \mathscr{A}$, then ind bd $(U) \leqslant n-1$. Since $X$ is a regular space, each point $x$ of $X$ has an open neighbourhood $W_x$ such that $\overline{W}_x$ is contained in some member of $\mathscr{A}$ and

$$\text{ind bd } (W_x) \leqslant n-1.$$

By applying Lemma 5.6 again, we obtain a $\sigma$-discrete open covering $\mathscr{B}$ of $X$ such that ind bd $(W) \leqslant n-1$ and $\overline{W}$ is contained in some member of $\mathscr{A}$ if $W \in \mathscr{B}$.

Let $\mathscr{A} = \bigcup_{i \in \mathbf{N}} \mathscr{A}_i$, where each $\mathscr{A}_i$ is discrete, and let $\mathscr{B} = \bigcup_{i \in \mathbf{N}} \mathscr{B}_i$, where each $\mathscr{B}_i$ is discrete. Let $\mathscr{W}_{ij}$ be the set of those members of $\mathscr{B}_i$ with closure contained in a member of $\mathscr{A}_j$ and let $\mathscr{U}_{ij} = \mathscr{A}_j$. By re-indexing we have $\mathscr{A} = \bigcup_{k \in \mathbf{N}} \mathscr{U}_k$, where $\mathscr{U}_k$ is discrete for each $k$, and $\mathscr{B} = \bigcup_{k \in \mathbf{N}} \mathscr{W}_k$, where $\mathscr{W}_k$ is discrete for each $k$ and the closure of each member of $\mathscr{W}_k$ is contained in a member of $\mathscr{U}_k$.

Let $U_k$ be the union of those members of $\mathscr{U}_k$ which meet $E$ and let $W_k$ be the union of all members of $\mathscr{W}_k$. Let

$$V_k = U_k \setminus \bigcup_{i<k} \overline{W}_i,$$

and let $V = \bigcup_{k \in \mathbf{N}} V_k$. Then $V$ is an open set. If $U \in \mathscr{U}_k$ and $U \cap E \neq \varnothing$, then $U \subset G$ so that $U_k \subset G$ for each $k$ and hence $V_k \subset G$ for each $k$. If $x \in E$ then there exists $k$ such that $x \in U_k$ and $x \notin U_i$ if $i < k$. Since $E \cap \overline{W}_i \subset U_i$, it follows that $x \notin \overline{W}_i$ if $i < k$ and hence $x \in V_k$. Thus $E \subset V \subset G$. Since $\{W_i\}$ is an open covering of $X$ and $W_i \cap V_k = \varnothing$ if $k > i$, the family $\{V_k\}_{k \in \mathbf{N}}$ is locally finite and therefore

$$\text{bd } (V) \subset \bigcup_{k \in \mathbf{N}} \text{bd } (V_k).$$

Furthermore
$$\text{bd } (V_k) \subset \text{bd } (U_k) \cup \bigcup_{i<k} \text{bd } (W_i).$$
Thus
$$\text{bd } (V) \subset \bigcup_{k \in \mathbf{N}} \text{bd } (U_k) \cup \bigcup_{i \in \mathbf{N}} \text{bd } (W_i).$$

If $W \in \mathscr{W}_i$, then ind bd $(W) \leqslant n-1$. But $\mathscr{W}_i$ is a discrete family so that bd $(W_i)$ is the topological sum of the subspaces bd $(W)$ for $W$ in $\mathscr{W}_i$. Hence ind bd $(W_i) \leqslant n-1$ for each positive integer $i$. Similarly ind bd $(U_k) \leqslant n-1$ for each $k$. $\square$

**5.8 Proposition.** *If $X$ is a completely paracompact regular space, then*

$$\dim X \leqslant \text{ind } X.$$

*If additionally $X$ is totally normal then*

$$\text{ind } X = \text{Ind } X.$$

*Proof.* The first statement is true if $\text{ind}\, X = -1$ and we suppose it established for completely paracompact regular spaces of small inductive dimension less than $n$, where $n$ is a non-negative integer. Let $X$ be a completely paracompact regular space such that $\text{ind}\, X \leqslant n$, let $E$ be a closed set of $X$ and let $G$ be an open set such that $E \subset G$. Then by Lemma 5.7, there exists an open set $V$ such that $E \subset V \subset G$ and $\text{bd}\,(V) = \bigcup_{k \in \mathbb{N}} F_k$, where each $F_k$ is a closed set such that

$$\text{ind}\, F_k \leqslant n-1.$$

By the induction hypothesis, $\dim F_k \leqslant n-1$ for each $k$ so that $\dim \text{bd}\,(V) \leqslant n-1$ by the countable sum theorem (Theorem 3.2.5). It follows from Lemma 2.9 that $\dim X \leqslant n$ and we have an inductive proof of the first assertion.

The same argument employing Theorem 4.8 and the definition of large inductive dimension shows that if $X$ is a completely paracompact totally normal space, then $\text{Ind}\, X \leqslant \text{ind}\, X$. But $\text{ind}\, X \leqslant \text{Ind}\, X$ since $X$ is a regular space and hence $\text{ind}\, X = \text{Ind}\, X$.☐

**5.9  Proposition.** *If $X$ is a strongly pseudo-metrizable space, then*

$$\dim X = \text{ind}\, X = \text{Ind}\, X.$$

*Proof.* Since $X$ is a pseudo-metrizable space, $\dim X = \text{Ind}\, X$ by Theorem 5.4. But by Proposition 2.3.13, a strongly pseudo-metrizable space is completely paracompact so that $\text{ind}\, X = \text{Ind}\, X$ by Proposition 5.8.☐

**5.10  Corollary.** *If $X$ is a separable pseudo-metrizable space, then*

$$\dim X = \text{ind}\, X = \text{Ind}\, X.$$

*Proof.* By Proposition 2.3.20, a separable pseudo-metrizable space is strongly pseudo-metrizable.☐

The last two results of this section extend the classes of spaces for which we can assert that the principal dimension functions coincide. We observe that if $X$ is a normal space and $X = \bigcup_{i \in \mathbb{N}} X_i$, where each $X_i$ is closed and perfectly normal, then $X$ is a perfectly normal space. For if $U$ is an open set of $X$, then $U = \bigcup_{i \in \mathbb{N}} U \cap X_i$. For each $i$, $U \cap X_i$ is an $F_\sigma$-set of $X_i$ and hence is an $F_\sigma$-set of $X$. Thus $U$ is an $F_\sigma$-set of $X$.

**5.11  Proposition.** *If $X$ is a normal space which is the union of countably many closed pseudo-metrizable subspaces, then*

$$\dim X = \operatorname{Ind} X.$$

*Proof.* Let $X = \bigcup_{i \in \mathbb{N}} X_i$, where each $X_i$ is a pseudo-metrizable closed subset. Then by Theorem 5.4, $\operatorname{Ind} X_i = \dim X_i \leqslant \dim X$. Since $X$ is perfectly normal, it follows from Theorem 4.8 that $\operatorname{Ind} X \leqslant \dim X$. Thus $\dim X = \operatorname{Ind} X$. □

**5.12  Proposition.** *If $X$ is a normal space which is the union of countably many closed strongly pseudo-metrizable subspaces, then*

$$\dim X = \operatorname{ind} X = \operatorname{Ind} X.$$

*Proof.* Let $X = \bigcup_{i \in \mathbb{N}} X_i$, where each $X_i$ is closed and strongly pseudo-metrizable. By Proposition 5.8, $\operatorname{Ind} X_i = \operatorname{ind} X_i \leqslant \operatorname{ind} X$, so that since $X$ is perfectly normal, it follows from Theorem 4.8 that

$$\operatorname{Ind} X \leqslant \operatorname{ind} X.$$

Therefore $\operatorname{ind} X = \operatorname{Ind} X$, and $\dim X = \operatorname{Ind} X$ by Proposition 5.11. □

## Notes

An inductive definition of dimension was proposed, in general terms, by Poincare [1912], writing in a philosophical journal. Brouwer [1913] gave a precise inductive definition of dimension. The theory of small inductive dimension for the class of compact metric spaces was formulated and developed by Urysohn [1922, 1925, 1926] and Menger [1923, 1924]. The theory was extended to the class of separable metric spaces by Hurewicz [1927] and Tumarkin [1926, 1928]. The largely complete dimension theory of separable metrizable spaces, containing the first forms of most of the important theorems in this book, was elegantly described by Hurewicz and Wallman [1941].

Vedenisov [1939] proved that $D^{\tau}$ is a universal space for $T_1$-spaces of weight not exceeding $\tau$ and small inductive dimension equal to zero. Erdös [1940] determined the dimension of the rational points in Hilbert space (Example 1.8).

The definition of large inductive dimension is due to Čech [1932]. The fact that $\dim X \leqslant \operatorname{Ind} X$ if $X$ is a normal space was established by Vedenisov [1941]. Aleksandrov [1941] proved that $\dim X \leqslant \operatorname{ind} X$

if $X$ is bicompact. The first example of a bicompact space such that $\dim X < \operatorname{ind} X$ is due to Lunc [1949]. Shortly afterwards Lokucievskiĭ [1949] constructed Example 3.4, which is simpler than Lunc's example. These spaces are not first-countable and their dimensional properties are closely related to the failure of the first axiom of countability. Fedorčuk [1968] constructed a first-countable bicompact space $B$ such that $\dim B = 2$ and $3 \leqslant \operatorname{ind} B \leqslant 4$. And Filippov [1969a] then constructed a first-countable bicompact space $B$ such that $\dim B = 1$ and $\operatorname{ind} B = \operatorname{Ind} B = 2$ and furthermore $B$ is the union of two closed sets which are 1-dimensional in every sense. There is more information about the relation between the three dimension functions for bicompact spaces in Chapter 8. The Tihonov plank (Example 3.1) was constructed by Tihonov [1930] and used by Hurewicz and Wallman [1941] and Vedenisov [1939] to show the failure of the subset theorems for dim and Ind.

Čech established sum and subset theorems for the large inductive dimension of perfectly normal spaces. Theorem 4.8 and the results which lead up to it are due to Dowker [1953]. Nagami [1969] proved Proposition 4.9. The restricted finite sum theorems for the large inductive dimension of normal spaces (Propositions 4.13, 4.14 and 4.15) were obtained by Lifanov [1967]. The 'Urysohn inequality' (Proposition 4.17) was established by Smirnov [1951].

The coincidence of all three dimension functions for separable metric spaces is one of the most important facts of classical dimension theory. The equality of covering dimension and large inductive dimension for metric spaces was established by Katětov [1951, 1952] and Morita [1954]. The method of proof used here is essentially that of Morita. As consequences of characterizations of the dimension of metric spaces, which will be discussed in Chapter 7, Dowker and Hurewicz [1956] and Nagami [1969] found new proofs of the equality. Arhangel'skiĭ [1969] has given a proof in which no use is made of sum theorems for large inductive dimension. The class of completely paracompact spaces was introduced by Zarelua [1963] and he proved Proposition 5.8.

If $X$ is a space such that $\operatorname{ind} X = n$, where $n$ is a positive integer, then $X$ contains a closed subspace $Y$ such that $\operatorname{ind} Y = p$ if $p$ is a positive integer not exceeding $n$. In 1926, Tumarkin asked whether a compact metric space of infinite dimension contains closed subspaces of every positive dimension. Henderson [1967] has answered the question negatively by giving an example of a compact metric space having no positive-dimensional closed subspaces.

The definitions of small and large inductive dimension can be naturally extended by transfinite induction so that the dimensions can be any ordinal number, finite or transfinite. Small transfinite inductive dimension was studied by Toulmin [1954], and Smirnov [1959] studied large transfinite inductive dimension. Further properties of large transfinite inductive dimension were obtained by Levšenko [1965] and Henderson [1968]. Landau [1969] and Pears [1971] obtained results for large transfinite inductive dimension analogous to some of Toulmin's theorems.

Menger [1929] posed the problem of characterizing axiomatically the small inductive dimension function and solved the problem on subspaces of the plane. Menger's axioms are not adequate to determine dimension on the class of all subspaces of Euclidean spaces (Kuz'minov [1968]). Nishiura [1966] has axiomatically characterized the small inductive dimension function on separable metrizable spaces. Aarts [1971] has given a similar characterization of the large inductive dimension function for metrizable spaces. Each of these axiom systems contains an axiom of an inductive nature. Nagata [1971] has pointed out that it would be interesting, and in the spirit of Menger's work, to have an axiomatic characterization of dimension which does not contain 'inductive' axioms.

# 5

## LOCAL DIMENSION

### 1 Definitions and relations with dimension

**1.1 *Definitions*.** *The* local dimension, loc dim $X$, *of a topological space* $X$ *is defined as follows. If* $X$ *is empty, then* loc dim $X = -1$. *Otherwise* loc dim $X$ *is the least integer* $n$ *such that for every point* $x$ *of* $X$ *there is some open set* $U$ *containing* $x$ *such that* dim $\overline{U} \leqslant n$, *or if there is no such integer, then* loc dim $X = \infty$.

*The* local inductive dimension loc Ind $X$ *is defined similarly. If* $X$ *is empty, then* loc Ind $X = -1$. *Otherwise* loc Ind $X$ *is the least integer* $n$ *such that for every point* $x$ *of* $X$ *there is some open set* $U$ *containing* $x$ *such that* Ind $\overline{U} \leqslant n$, *or if there is no such integer then* loc Ind $X = \infty$.

Thus loc dim $X$ (loc Ind $X$) is the least integer $n$ such that there exists an open covering $\{U_\lambda\}$ of $X$ such that dim $\overline{U}_\lambda \leqslant n$ (Ind $\overline{U}_\lambda \leqslant n$) for each $\lambda$. It is clear that loc dim $X \leqslant$ dim $X$ and loc Ind $X \leqslant$ Ind $X$ for every topological space $X$. If loc dim $X \leqslant n$, $x \in X$ and $U$ is an open set such that $x \in U$, then there exists an open set $V$ such that $x \in V \subset U$ and dim $\overline{V} \leqslant n$. For there exists an open set $W$ such that $x \in W$ and dim $\overline{W} \leqslant n$. If $V = U \cap W$, then $x \in V \subset U$ and $\overline{V}$ is a closed subset of $\overline{W}$, so that dim $\overline{V} \leqslant n$. Hence loc dim $X \leqslant n$ if and only if every open covering of the space $X$ has an open refinement $\{U_\lambda\}$, say, such that dim $\overline{U}_\lambda \leqslant n$ for each $\lambda$. Analogous results are clearly true for local inductive dimension.

**1.2 *Remark*.** The small inductive dimension ind $X$ of a space $X$ is already a local property of $X$. For suppose that $X$ is a space with the property that each point has an open neighbourhood with closure of small inductive dimension not exceeding $n$. Then ind $X \leqslant n$. For let $x$ be a point of $X$ and let $U$ be an open set such that $x \in U$. There is some open set $W$ such that $x \in W$ and ind $\overline{W} \leqslant n$. Hence $x \in U \cap W$ and $U \cap W$ is open in $\overline{W}$. Hence there exists a set $V$ open in $\overline{W}$ such that $x \in V \subset U \cap W$ and ind $B \leqslant n-1$, where $B$ is the boundary of $V$ in $\overline{W}$. Since $V$ is open in $\overline{W}$ and $V \subset W$, it follows that $V$ is open in $X$. Also the closure of $V$ in $\overline{W}$ is its closure $\overline{V}$ in $X$ and hence

$$B = \overline{V} \backslash V = \text{bd}\,(V).$$

Thus $x \in V \subset U$, where $V$ is open and $\operatorname{ind} \operatorname{bd}(V) \leqslant n-1$. Therefore $\operatorname{ind} X \leqslant n$.

**1.3   Proposition.** *If $X$ is a space in which each open set is a union of closed sets, then*

$$\operatorname{ind} X \leqslant \operatorname{loc} \operatorname{Ind} X.$$

*Proof.* Clearly $\operatorname{loc} \operatorname{Ind} X = -1$ implies that $\operatorname{ind} X = -1$. Let us suppose that $\operatorname{loc} \operatorname{Ind} X \leqslant n$, where $n \geqslant 0$, and that $x \in U$, where $U$ is open in $X$. There is an open set $W$ such that $x \in W$ and $\operatorname{Ind} \overline{W} \leqslant n$, and there exists a closed set $F$ such that $x \in F \subset U \cap W$. Since

$$\operatorname{Ind} \overline{W} \leqslant n,$$

there is an open set $V$ of $\overline{W}$ such that $F \subset V \subset U \cap W$ and

$$\operatorname{Ind} B \leqslant n-1,$$

where $B$ is the boundary of $V$ in $\overline{W}$. Since $V$ is open in $\overline{W}$ and $V \subset W$, $V$ is open in $X$. The closure of $V$ in $\overline{W}$ is its closure $\overline{V}$ in $X$ and hence $B = \overline{V} \backslash V = \operatorname{bd}(V)$. Since $B$ has the property that each open set is a union of closed sets, it follows from a remark preceding Proposition 4.2.1 that $\operatorname{ind} B \leqslant n-1$. Thus $x \in V \subset U$, where $V$ is open and $\operatorname{ind} \operatorname{bd}(V) \leqslant n-1$. Hence $\operatorname{ind} X \leqslant n$. $\square$

**1.4   Corollary.** *If $X$ is a regular space or a $T_1$-space, then*

$$\operatorname{ind} X \leqslant \operatorname{loc} \operatorname{Ind} X. \square$$

**1.5   Proposition.** *If $X$ is a normal space, then*

$$\operatorname{loc} \dim X \leqslant \operatorname{loc} \operatorname{Ind} X.$$

*Proof.* Let $X$ be a normal space such that $\operatorname{loc} \operatorname{Ind} X \leqslant n$. Then each point $x$ of $X$ has a neighbourhood $U$ such that $\operatorname{Ind} \overline{U} \leqslant n$. The closed set $\overline{U}$ of $X$ is normal so that $\dim \overline{U} \leqslant n$. Thus $\operatorname{loc} \dim X \leqslant n$. $\square$

Summarizing the relations between the concepts of dimension and local dimension we have:

**1.6   Proposition.** *If $X$ is a normal regular space, then*

$$\operatorname{ind} X \leqslant \operatorname{loc} \operatorname{Ind} X \leqslant \operatorname{Ind} X,$$

$$\operatorname{loc} \dim X \leqslant \operatorname{loc} \operatorname{Ind} X \leqslant \operatorname{Ind} X,$$

$$\operatorname{loc} \dim X \leqslant \dim X \leqslant \operatorname{Ind} X. \square$$

Since a $T_4$-space is regular, these inequalities hold in particular for $T_4$-spaces. It will be shown by examples that no other relations hold for all normal regular spaces.

A continuous function $f: X \to Y$ is said to be a *local homeomorphism* if each point $x$ of $X$ has an open neighbourhood $U_x$ such that $f(U_x)$ is open in $Y$ and the mapping of $U_x$ onto $f(U_x)$ given by restriction of $f$ is a homeomorphism.

**1.7 Proposition.** *If $X$ and $Y$ are regular spaces and $f: X \to Y$ is a surjectuve local homeomorphism, then*

$$\operatorname{loc dim} X = \operatorname{loc dim} Y \quad and \quad \operatorname{loc Ind} X = \operatorname{loc Ind} Y.$$

*Proof.* Suppose that $\operatorname{loc dim} Y \leqslant n$, and let $x$ be a point of $X$. There exists an open neighbourhood $U$ of $x$ such that $f(U)$ is open in $Y$ and $f$ maps $U$ homeomorphically onto $f(U)$. There exists an open set $H$ of $Y$ such that $f(x) \in H \subset \bar{H} \subset f(U)$ and $\dim \bar{H} \leqslant n$. If

$$V = f^{-1}(H) \cap U,$$

then $V$ is an open neighbourhood of $x$ and $\bar{V}$ is homeomorphic with $\bar{H}$ so that $\dim \bar{V} \leqslant n$. Thus $\operatorname{loc dim} X \leqslant n$. Now suppose that

$$\operatorname{loc dim} X \leqslant n$$

and let $y$ be a point of $Y$. There exist $x$ in $X$ such that $f(x) = y$, and an open neighbourhood $U$ of $x$ such that $f$ maps $U$ homeomorphically onto $f(U)$, which is an open neighbourhood of $y$; and there exists an open set $V$ of $X$ such that $x \in V \subset \bar{V} \subset U$ and $\dim \bar{V} \leqslant n$. If $H = f(V)$, then $H$ is an open neighbourhood of $y$ and $\bar{H}$ is homeomorphic with $\bar{V}$ so that $\dim \bar{H} \leqslant n$. Thus $\operatorname{loc dim} Y \leqslant n$. Hence $\operatorname{loc dim} X = \operatorname{loc dim} Y$. The proof that $\operatorname{loc Ind} X = \operatorname{loc Ind} Y$ is similar. $\square$

## 2   Subset and sum theorems

In this section we find some subset and sum theorems for local dimension. A later example shows that the subset theorems for loc dim and loc Ind do not hold in general for $T_4$-spaces. Of course there is a closed subset theorem.

**2.1 Proposition.** *If $A$ is a closed set of a space $X$, then*

$$\operatorname{loc dim} A \leqslant \operatorname{loc dim} X \quad and \quad \operatorname{loc Ind} A \leqslant \operatorname{loc Ind} X.$$

*Proof.* Suppose that $\operatorname{loc dim} X \leqslant n$ and let $x$ be a point of $A$. Then there is an open set $U$ of $X$ such that $x \in U$ and $\dim \bar{U} \leqslant n$. Then

$U \cap A$ is an open neighbourhood of $x$ in $A$ and the closure of $U \cap A$ in $A$ is a closed set of $\overline{U}$ and hence has dimension not exceeding $n$. Hence $\operatorname{loc} \dim A \leqslant n$. Thus $\operatorname{loc} \dim A \leqslant \operatorname{loc} \dim X$ as was to be shown. An identical proof serves for local inductive dimension.□

An open subset theorem holds for local dimension of regular spaces.

**2.2　Proposition.** *If $Y$ is an open set of a regular space $X$, then* $\operatorname{loc} \dim Y \leqslant \operatorname{loc} \dim X$ *and* $\operatorname{loc} \operatorname{Ind} Y \leqslant \operatorname{loc} \operatorname{Ind} X$.

*Proof.* Suppose that $\operatorname{loc} \dim X \leqslant n$ and let $y$ be a point of $Y$. There is an open neighbourhood $U$ of $y$ in $X$ such that $\dim \overline{U} \leqslant n$. Since $X$ is a regular space there exists an open set $V$ such that

$$y \in V \subset \overline{V} \subset U \cap Y.$$

Then $V$ is an open neighbourhood of $y$ in $Y$ and $\overline{V}$ is the closure of $V$ in $Y$. Since $\overline{V}$ is a closed subset of $\overline{U}$, it follows that $\dim \overline{V} \leqslant n$. Thus $\operatorname{loc} \dim Y \leqslant n$. Hence $\operatorname{loc} \dim Y \leqslant \operatorname{loc} \dim X$. The proof that $\operatorname{loc} \operatorname{Ind} Y \leqslant \operatorname{loc} \operatorname{Ind} X$ is similar.□

The subset theorem holds for local dimension of totally normal spaces.

**2.3　Proposition.** *If $A$ is a subset of a totally normal space $X$, then* $\operatorname{loc} \dim A \leqslant \operatorname{loc} \dim X$ *and* $\operatorname{loc} \operatorname{Ind} A \leqslant \operatorname{loc} \operatorname{Ind} X$.

*Proof.* Suppose that $\operatorname{loc} \dim X \leqslant n$ and let $x$ be a point of $A$. There is a neighbourhood $U$ of $x$ in $X$ such that $\dim \overline{U} \leqslant n$. Then $U \cap A$ is an open neighbourhood of $x$ in $A$ and its closure in $A$ is a subset of the totally normal space $\overline{U}$ and hence has covering dimension not exceeding $n$ by Theorem 3.6.4. Thus $\operatorname{loc} \dim A \leqslant n$. Hence

$$\operatorname{loc} \dim A \leqslant \operatorname{loc} \dim X.$$

The proof that $\operatorname{loc} \operatorname{Ind} A \leqslant \operatorname{loc} \operatorname{Ind} X$ is similar but uses Theorem 4.4.8 instead of Theorem 3.6.4.□

We shall see that the countable sum theorem does not hold for local dimension in normal spaces. We have however:

**2.4　Proposition.** *If a normal space $X$ is the union of two closed sets $A$ and $B$ such that* $\operatorname{loc} \dim A \leqslant n$ *and* $\operatorname{loc} \dim B \leqslant n$, *then*

$$\operatorname{loc} \dim X \leqslant n.$$

*Proof.* If $x \in X \backslash A$, then $x \in B$ and there is an open set $U \cap B$ of $B$, where $U$ is open in $X$ such that $\dim (U \cap B)^- \leqslant n$. If $W = U \cap (X \backslash A)$, then $W$ is open in $X$, $x \in W$ and since $\overline{W} \subset (U \cap B)^-$, it follows that $\dim \overline{W} \leqslant n$. Similarly if $x \in X \backslash B$, there is an open neighbourhood of $x$ in $X$ which has closure of covering dimension not exceeding $n$.

If $x \in A \cap B$, then there exist open sets $U$ and $V$ containing $x$ such that $\dim (U \cap A)^- \leqslant n$ and $\dim (V \cap B)^- \leqslant n$. Let

$$W = X \backslash (A \backslash U) \backslash (B \backslash V).$$

Then $W$ is open and $x \in W$. And $W \subset (U \cap A) \cup (V \cap B)$ for if $y \in W$, then $y \notin A \backslash U$ and $y \notin B \backslash V$, but either $y \in A$ or $y \in B$, so that either $y \in A \cap U$ or $y \in B \cap V$. Thus

$$\dim \overline{W} \leqslant \dim \big( (U \cap A)^- \cup (V \cap B)^- \big) \leqslant n,$$

by the sum theorem in the normal space $(U \cap A)^- \cup (V \cap B)^-$. Thus $\operatorname{loc} \dim X \leqslant n.\square$

**2.5   Corollary.** *If $\mathscr{A}$ is a locally finite closed covering of a normal space $X$ such that $\operatorname{loc} \dim A \leqslant n$ for each $A$ in $\mathscr{A}$, then $\operatorname{loc} \dim X \leqslant n$.*

*Proof.* If $x \in X$, then $x$ belongs to only finitely many members of $\mathscr{A}$; let $B$ be their union. Then $B$ is closed in $X$, hence normal, and $\operatorname{loc} \dim B \leqslant n$, since the result of Proposition 2.4 evidently extends to any finite number of summands. If $U$ is the complement of the union of those members of $\mathscr{A}$ which do not contain $x$, then $U$ is open in $X$ and $x \in U \subset B$. It follows that there exists $V$ open in $B$ such that $x \in V \subset U$ and $\dim \overline{V} \leqslant n$, where $\overline{V}$ denotes the closure of $V$ in $B$ and hence in $X$, and $V$ is open in $X$ since $V$ is open in $U$. Thus

$$\operatorname{loc} \dim X \leqslant n.\square$$

The finite sum theorem does not hold for local inductive dimension in normal spaces or even in bicompact spaces. For the bicompact space $S$ considered in Example 4.3.4 satisfies $\operatorname{ind} S = \operatorname{Ind} S = 2$ so that $\operatorname{loc} \operatorname{Ind} S = 2$. But $S = S_1 \cup S_2$, where $S_1$ and $S_2$ are closed sets of $S$ and $\operatorname{ind} S_i = \operatorname{Ind} S_i = 1$ for $i = 1, 2$, so that $\operatorname{loc} \operatorname{Ind} S_i = 1$. We can obtain a finite sum theorem for $\operatorname{loc} \operatorname{Ind}$ by imposing a condition on the intersection of the summands.

**2.6   Proposition.** *If a normal regular space $X$ is the union of two closed sets $A$ and $B$ such that $\operatorname{loc} \operatorname{Ind} A \leqslant n$, $\operatorname{loc} \operatorname{Ind} B \leqslant n$ and*

$$\operatorname{loc} \operatorname{Ind} A \cap B \leqslant 0,$$

*then $\operatorname{loc} \operatorname{Ind} X \leqslant n$.*

*Proof.* If $x \in X \backslash A$, then $x \in B$ and there is an open set $U \cap B$ of $B$, where $U$ is open in $X$, such that $\operatorname{Ind}(U \cap B)^- \leqslant n$. If $W = U \cap (X \backslash A)$, then $W$ is open in $X$, $x \in W$ and $\operatorname{Ind} \overline{W} \leqslant n$ since $\overline{W} \subset (U \cap B)^-$. Similarly if $x \in X \backslash B$, there is an open neighbourhood of $x$ in $X$ which has closure of covering dimension not exceeding $n$. If $A \cap B$ is empty this completes the proof.

If $x \in A \cap B = C$, then there exists $U_1$ open in $A$ such that $x \in U_1$ and $\operatorname{Ind} \overline{U}_1 \leqslant n$, and there exists $U_2$ open in $B$ such that $x \in U_2$ and

$$\operatorname{Ind} \overline{U}_2 \leqslant n.$$

Since $\operatorname{loc} \operatorname{Ind} C = 0$ and $U_1 \cap U_2$ is an open set of $C$, there exists $V$ open in $C$ such that $x \in V \subset U_1 \cap U_2$ and $\operatorname{Ind} \overline{V} \leqslant 0$. Furthermore $\operatorname{ind} C = 0$ since $C$ is regular, and hence there exists an open-and-closed set $W$ of $C$ such that $x \in W \subset V$. Since $W$ is closed in $C$, it follows that $W$ is closed in $A$. Let $P = W \cup (U_1 \backslash B)$. Then

$$A \backslash P = (A \backslash W) \cap \big((A \backslash U_1) \cup C\big) = (A \backslash U_1) \cup (C \backslash W).$$

Since $U_1$ is open in $A$, it follows that $A \backslash U_1$ is closed in $A$. Since $W$ is open in $C$, it follows that $C \backslash W$ is closed in $C$ and hence in $A$. Thus $P = W \cup (U_1 \backslash B)$ is open in $A$. Since $A$ is normal, there exists an open set $W_1$ of $A$ such that

$$W \subset W_1 \subset \overline{W}_1 \subset W \cup (U_1 \backslash B).$$

Similarly there exists an open set $W_2$ of $B$ such that

$$W \subset W_2 \subset \overline{W}_2 \subset W \cup (U_2 \backslash A).$$

Let $G = W_1 \cup W_2$. Then by an argument similar to that in the proof of Proposition 4.4.13 we can see that

$$X \backslash G = (A \backslash W_1) \cup (B \backslash W_2)$$

and hence that $G$ is open in $X$. Finally $\overline{G} = \overline{W}_1 \cup \overline{W}_2$. Since $W_1 \subset U_1$, it follows that $\operatorname{Ind} \overline{W}_1 \leqslant n$. Similarly $\operatorname{Ind} \overline{W}_2 \leqslant n$. Furthermore $\overline{W}_1 \cap \overline{W}_2 = W \subset \overline{V}$ so that $\operatorname{Ind} \overline{W}_1 \cap \overline{W}_2 \leqslant 0$. It follows from Proposition 4.4.13 that $\operatorname{Ind} \overline{G} \leqslant n$. Thus $\operatorname{loc} \operatorname{Ind} X \leqslant n$. $\square$

## 3   The monotonicity of covering dimension

In this section we return to the question of the validity of the subset theorem for the covering dimension of completely normal regular spaces. We shall reduce this question to the apparently simpler problem of whether the local dimension of every completely normal regular space is equal to its dimension.

**3.1 Proposition.** *If $X$ is a normal regular space, there is a normal regular space $X^*$ containing $X$ as an open subspace such that*

$$\dim X^* \leqslant \operatorname{loc} \dim X.$$

*Proof.* If $X$ is empty, let $X^* = X$. Otherwise the points of the space $X^*$ are the points of $X$ together with one new point $x_0$. A subset $U$ of $X^*$ is said to be open if either (i) $U \subset X$ and $U$ is open in $X$, or (ii) $x_0 \in U$ and $X^* \backslash U$ is a closed set of $X$ which is contained in an open set $V$ of $X$ such that $\dim \overline{V} \leqslant \operatorname{loc} \dim X$.

It is clear that $\varnothing$ is an open set of type (i) and that $X^*$ is an open set of type (ii). The intersection of two open sets, one of which is of type (i), is an open set of type (i). If $U_1$ and $U_2$ are two open sets of type (ii), then $x_0 \in U_1 \cap U_2$ and

$$X^* \backslash (U_1 \cap U_2) = (X^* \backslash U_1) \cup (X^* \backslash U_2),$$

which is the union of two closed sets and hence is a closed set of $X$. If $X^* \backslash U_1 \subset V_1$ and $X^* \backslash U_2 \subset V_2$, where $V_1$ and $V_2$ are open in $X$ and $\dim \overline{V}_1 \leqslant \operatorname{loc} \dim X$, $\dim \overline{V}_2 \leqslant \operatorname{loc} \dim X$, then $X^* \backslash (U_1 \cap U_2) \subset V_1 \cup V_2$ and $(V_1 \cup V_2)^- = \overline{V}_1 \cup \overline{V}_2$ is a closed set of $X$ and hence is normal. Therefore $\dim (\overline{V}_1 \cup \overline{V}_2) \leqslant \operatorname{loc} \dim X$ by the sum theorem. Thus the intersection $U_1 \cap U_2$ is an open set of type (ii). The union of a collection of open sets of type (i) is an open set of type (i). If a collection of open sets contains an open set $U_1$ of type (ii), then the union $U$ of the collection contains $x_0$, and $U \backslash \{x_0\}$ is a union of open sets of $X$ and hence is open in $X$. Therefore $X^* \backslash U$ is closed in $X$, and if $X^* \backslash U_1 \subset V_1$, where $V_1$ is open in $X$ and $\dim \overline{V}_1 \leqslant \operatorname{loc} \dim X$, then $X^* \backslash U$ is also contained in $V_1$. Therefore $U$ is an open set of type (ii). Thus $X^*$ is a topological space and it is clear that $X$ is a subspace of $X^*$. The set $X$ is an open set of type (i) of $X^*$.

The space $X^*$ is normal. For if $E$ and $F$ are disjoint closed sets of $X^*$, then at least one of them, say $F$, does not contain $x_0$. Then $X^* \backslash F$ is an open set of $X^*$ containing $x_0$ and is therefore an open set of type (ii). Hence there exists an open set $V$ of $X$ such that $F \subset V$ and $\dim \overline{V} \leqslant \operatorname{loc} \dim X$. Then $V \cap (X^* \backslash E)$ is an open set of $X$ containing the closed set $F$. Since $X$ is normal, there exists an open set $W$ of $X$ such that

$$F \subset W \subset \overline{W} \subset V \cap (X^* \backslash E).$$

Let $U = X^* \backslash \overline{W}$. Then $x_0 \in U$, the set $X^* \backslash U = \overline{W}$ is closed in $X$ and $\overline{W} \subset V$, where $V$ is open in $X$ and $\dim \overline{V} \leqslant \operatorname{loc} \dim X$. Therefore $U$

is an open set of type (ii) of $X^*$, and $E \subset U$ since $\overline{W} \subset X^* \backslash E$. Furthermore $W$ is an open set of type (i) of $X^*$ which is disjoint from $U$ and $F \subset W$. Therefore $X^*$ is normal.

The space $X^*$ is regular. If $x$ is a point of $X$ and $U$ is an open set of $X^*$ containing $x$, then $U \cap X$ is an open set of $X$ and there exists an open set $V$ of $X$ such that $x \in V \subset U \cap X$ and $\dim \overline{V} \leqslant \operatorname{loc} \dim X$. Since $X$ is a regular space, there is an open set $W$ of $X$ such that $x \in W \subset \overline{W} \subset V$. Since $W$ is closed in $X$ and $\overline{W} \subset V$, where $V$ is open in $X$ and $\dim \overline{V} \leqslant \operatorname{loc} \dim X$, it follows that $X^* \backslash \overline{W}$ is an open set of type (ii) of $X^*$. Thus $x \in W \subset \overline{W} \subset U$, where $W$ is open and $\overline{W}$ is closed in $X^*$. If $U$ is an open set of $X^*$ containing $x_0$, then since $\{x_0\}$ is closed in $X^*$ and $X^*$ is a normal space, there exists an open set $W$ of $X^*$ such that $x_0 \in W \subset \overline{W} \subset U$, where $\overline{W}$ denotes the closure of $W$ in $X^*$. Therefore $X^*$ is regular.

The set $\{x_0\}$ is closed in $X^*$ and $\dim \{x_0\} = 0 \leqslant \operatorname{loc} \dim X$ since $X$ is non-empty. Let $F$ be a closed set of $X^*$ which does not contain $x_0$. Then $X^* \backslash F$ is an open set of type (ii) and hence $F$ is closed in $X$ and $F \subset V$ for some open set $V$ of $X$ such that $\dim \overline{V} \leqslant \operatorname{loc} \dim X$. Therefore $\dim F \leqslant \operatorname{loc} \dim X$. Hence $\dim X^* \leqslant \operatorname{loc} \dim X$ by Proposition 3.5.6. $\square$

**3.2**   *Remark.* If $X$ is a $T_4$-space or a completely normal space, then $X^*$ has the same property.

For let $X$ be a $T_4$-space. If $x \in X$, then $\{x\}$ is a closed set of $X$, and there exists an open set $V$ of $X$ such that $x \in V$ and

$$\dim \overline{V} \leqslant \operatorname{loc} \dim X.$$

Hence $X^* \backslash \{x\}$ is an open set of type (ii) of $X^*$, so that $\{x\}$ is a closed set of $X^*$, and $\{x_0\}$ is a closed set of $X^*$ since $X$ is open in $X^*$. Thus all one-point subsets of $X^*$ are closed. Hence $X^*$ is a $T_4$-space.

Now let $X$ be a completely normal space. If $U$ is an open set of type (i) of $X^*$, then $U \subset X$ and hence the subspace $U$ is normal. If $U$ is an open set of type (ii) of $X^*$, then $X^* \backslash U$ is closed in $X$ and $X^* \backslash U \subset V$, where $V$ is open in $X$ and $\dim \overline{V} \leqslant \operatorname{loc} \dim X$. Since $X$ is normal there is an open set $W$ of $X$ such that $X^* \backslash U \subset W \subset \overline{W} \subset V$. Then $X^* \backslash \overline{W}$ is an open set of type (ii) of $X^*$ so that $\overline{W}$ is closed in $X^*$. Since $\overline{W} \cap U$ is a subspace of the completely normal space $X$, it follows that $\overline{W} \cap U$ is normal. Since $X^* \backslash W$ is closed in the normal space $X^*$, it follows that $X^* \backslash W$ is normal. But

$$U = (\overline{W} \cap U) \cup (X^* \backslash W)$$

and $\overline{W} \cap U$ and $X^* \backslash W$ are closed sets of $U$. It follows from Proposition

1.4.3 that the subspace $U$ is normal. Since every open subspace of $X^*$ is normal, it follows from Proposition 1.4.2 that $X^*$ is completely normal.$\square$

**3.3**    *Theorem. There exists a completely normal regular space $X$ with a subspace $A$ such that* $\dim A > \dim X$ *if and only if there exists a completely normal regular space $Y$ such that* $\dim Y > \operatorname{loc dim} Y$.

*Proof.* If there exists a completely normal regular space $Y$ such that $\dim Y > \operatorname{loc dim} Y$, then by Proposition 3.1 and the remark above, there exists completely normal regular space $X$ containing $Y$ as an open subset such that $\dim X \leqslant \operatorname{loc dim} Y < \dim Y$.

Conversely if $X$ is a completely normal regular space with a subspace $A$ such that $\dim A > \dim X$, then $X$ has an open subspace $Y$ such that $\dim Y > \operatorname{loc dim} Y$. For $\dim X$ is finite since $\dim X < \dim A$. Let $\dim X = n$. Since $\dim A > n$, there is an open covering $\{G_1, \ldots, G_k\}$ of $A$ which has no open refinement of order not exceeding $n$. Let $G_i = A \cap U_i$, where $U_i$ is open in $X$, and let $Y = \bigcup_{i=1}^{k} U_i$. Then $Y$ is open in $X$ and the open covering $\{U_1, \ldots, U_k\}$ of $Y$ has no open refinement of order not exceeding $n$. Hence $\dim Y > n$. But by Proposition 2.2,

$$\operatorname{loc dim} Y \leqslant \operatorname{loc dim} X \leqslant \dim X = n.$$

Therefore $\dim Y > \operatorname{loc dim} Y$.$\square$

Next we find some conditions under which we can assert that loc dim and dim coincide. These results are of interest in narrowing down the area in which we might look for an example which would imply the failure of the subset theorem for completely normal regular spaces.

**3.4**    *Proposition. If $X$ is a weakly paracompact normal space, then* $\operatorname{loc dim} X = \dim X$.

*Proof.* Suppose that $\operatorname{loc dim} X \leqslant n$. Then each point $x$ of $X$ has an open neighbourhood $U_x$ such that $\dim \overline{U}_x \leqslant n$. Since $X$ is weakly paracompact, the open covering $\{U_x\}_{x \in X}$ has a point-finite open refinement $\{V_\lambda\}$ and since $X$ is normal, by Proposition 1.3.9, there is a closed covering $\{F_\lambda\}$ of $X$ with $F_\lambda \subset V_\lambda$ for each $\lambda$. For each $\lambda$ there exists an open $F_\sigma$-set $H_\lambda$ such that $F_\lambda \subset H_\lambda \subset V_\lambda$. Since $H_\lambda$ is an $F_\sigma$-set of $\overline{U}_x$ for some $x$, it follows that $\dim H_\lambda \leqslant n$. But $\{H_\lambda\}$ is a point-finite open covering of $X$. Thus $\dim X \leqslant n$ by Proposition 3.5.10. Thus $\dim X \leqslant \operatorname{loc dim} X$, and hence $\dim X = \operatorname{loc dim} X$.$\square$

**3.5 Corollary.** *If $X$ is a normal space which is the union of a countable family of closed weakly paracompact subsets or the union of a locally finite family of weakly paracompact subsets, all but at most one of which are closed, then* loc dim $X$ = dim $X$.

*Proof.* In the first case, $X = \bigcup_{i \in \mathbf{N}} A_i$, where the $A_i$ are closed and weakly paracompact, and since $A_i$ is normal, from Propositions 3.4 and 2.1 it follows that

$$\dim A_i = \text{loc dim } A_i \leqslant \text{loc dim } X.$$

Thus dim $X \leqslant$ loc dim $X$ by the countable sum theorem and hence dim $X$ = loc dim $X$.

In the second case let $\{A_\lambda\}$ be a locally finite covering of $X$ by weakly paracompact sets and suppose that if $\lambda \neq \lambda_0$ then $A_\lambda$ is closed. If $B = \bigcup_{\lambda \neq \lambda_0} A_\lambda$, then $B$ is a closed set of $X$. For each $\lambda \neq \lambda_0$, $A_\lambda$ is a weakly paracompact normal space and

$$\dim A_\lambda = \text{loc dim } A_\lambda \leqslant \text{loc dim } X.$$

Thus dim $B \leqslant$ loc dim $X$ by Proposition 3.5.1. If $F$ is a closed set of $X$ disjoint from $B$, then $F$ is a closed set of $A_{\lambda_0}$ and so is weakly paracompact. Hence $F$ is weakly paracompact and normal and

$$\dim F = \text{loc dim } F \leqslant \text{loc dim } X.$$

It follows from Proposition 3.5.6 that dim $X \leqslant$ loc dim $X$ and hence dim $X$ = loc dim $X$.□

For local inductive dimension we have:

**3.6 Proposition.** *If $X$ is a weakly paracompact totally normal space, then* loc Ind $X$ = Ind $X$.

*Proof.* Suppose that loc Ind $X \leqslant n$. Then each point $x$ of $X$ has an open neighbourhood $U_x$ such that Ind $\overline{U}_x \leqslant n$. The open covering $\{U_x\}_{x \in X}$ has a point-finite open refinement $\{V_\lambda\}$. Since $V_\lambda$ is an open subset of some totally normal space $U_x$, it follows from Theorem 4.4.8 that Ind $V_\lambda \leqslant n$ for each $\lambda$. Hence Ind $X \leqslant n$ by Proposition 4.4.12. Thus Ind $X \leqslant$ loc Ind $X$ and hence loc Ind $X$ = Ind $X$.□

If $x$ is a point of a space $X$, then the *dimension of $X$ at $x$*, $\dim_x X$, is defined to be the least integer $n$ such that for some open set $U$ containing $x$, dim $\overline{U} = n$, or if there is no such integer, then $\dim_x X = \infty$.

We might define local dimension of a non-empty space $X$ as follows: $\operatorname{loc dim} X$ is the least integer $n$ such that $\dim_x X \leqslant n$ for every point $x$ of $X$, or if there is no such integer, then $\operatorname{loc dim} X = \infty$.

**3.7 Proposition.** *Let $X$ be an $n$-dimensional normal space. If $X$ is weakly paracompact, or the union of a countable family of closed weakly paracompact sets, or the union of a locally finite family of weakly paracompact sets, all but at most one of which are closed, then the set of points of $X$ at which $X$ is $n$-dimensional is an $n$-dimensional closed set of $X$.*

*Proof.* Let $D$ be the set of points of $X$ at which $X$ is $n$-dimensional and suppose that $x \notin D$. Since $\operatorname{loc dim} X = \dim X = n$, it follows that $\dim_x X < n$, so that there exists an open neighbourhood $U$ of $x$ such that $\dim \overline{U} \leqslant n-1$. Then $\dim_y X \leqslant n-1$ for each point $y$ of $U$ so that $U \cap D = \varnothing$. Thus $D$ is closed.

Let $F$ be a closed set of $X$ which does not meet $D$. If $x \in F$, then $x$ has an open neighbourhood $U$ in $X$ such that $\dim \overline{U} \leqslant n-1$. Now $U \cap F$ is a neighbourhood of $x$ in $F$ and the closure of $U \cap F$ in $F$ is a closed subset of $\overline{U}$ and hence has dimension not exceeding $n-1$. Thus $\operatorname{loc dim} F \leqslant n-1$. But it follows either from Proposition 3.4 or from Proposition 3.5 that $\operatorname{loc dim} F = \dim F$. If it were true that $\dim D \leqslant n-1$, then from Proposition 3.5.6 it would follow that $\dim X \leqslant n-1$ which is false. Thus $\dim D \geqslant n = \dim X$ and hence $\dim D = n$ since $D$ is closed in $X$. $\square$

## 4 Some examples

This section will be devoted to the construction of examples. We shall show that the subset theorems for dim and Ind do not hold for all $T_4$-spaces even if the subset is required to be normal. We also show that the subset and countable sum theorems do not hold for the local dimension $\operatorname{loc dim}$ of normal spaces. The examples obtained in this section together with Example 4.3.4 show that in general there are no relations between ind, dim, Ind, $\operatorname{loc dim}$ and $\operatorname{loc Ind}$ other than those given in Proposition 1.6.

The examples of this section are obtained from a construction which we now describe. Let $T$ be the linearly ordered space of ordinals less than the first uncountable ordinal $\omega_1$ and for each $\alpha$ in $T$ let $T_\alpha = [0, \alpha]$ and $T'_\alpha = T \backslash T_\alpha$. Let $P$ be an arbitrary separable metric space. A subset $M$ of $T \times P$ is said to be *convergent* if for each $p$ in $P$ there exists some $\alpha$ in $T$ such that $T'_\alpha \times \{p\} \subset M$. We shall see that any convergent set is normal and has the same covering dimension as $P$. In the proof of this fact we make use of several lemmas which now follow. In Lemmas 4.1, 4.2 and 4.3, $P$ will be a separable metric space and $M$ will be a convergent subset of $T \times P$.

**4.1 Lemma.** *If $\{G_i\}_{i \in \mathbf{N}}$ is a family of open sets of $T \times P$ such that $M \subset \bigcup_{i \in \mathbf{N}} G_i$ and $p$ is a point of $P$, then there exist an open neighbourhood $W$ of $p$ in $P$ and $\beta$ in $T$ such that $(T'_\beta \times W) \cap M \subset G_j$ for some $j$.*

*Proof.* Since $M$ is convergent, there exists $\alpha$ in $T$ such that $T'_\alpha \times \{p\} \subset M$. It follows from Lemma 1.4.13 that there exist $\beta$ in $T$ and an integer $j$ such that $T'_\beta \times \{p\} \subset G_j$. If $\gamma \in T'_\beta$, let $H_\gamma = \{q \in P \mid (\gamma, q) \in G_j\}$. Then $H_\gamma$ is an open neighbourhood of $p$ in $P$. Let $\rho$ be the metric in $P$ and for each $\gamma$ in $T'_\beta$ let $\epsilon(\gamma) = \rho(p, P \backslash H_\gamma)$ so that $\epsilon(\gamma) > 0$ for each $\gamma$. If $r$ is the greatest lower bound of the real numbers $\epsilon(\gamma)$ for $\gamma$ in $T'_\beta$, then $r > 0$. For suppose that $r = 0$ and let $\{\gamma_k\}_{k \in \mathbf{N}}$ be an increasing sequence of ordinals in $T'_\beta$ such that $\epsilon(\gamma_k) \to 0$. If $\gamma_0 = \lim \gamma_k$ then $\gamma_0 \in T'_\beta$ and $\gamma_0 \notin G_j$ which is absurd. Thus $r > 0$ and if $W$ is the open ball in $P$ with centre $p$ and radius $r$, then $\{\gamma\} \times W \subset G_j$ if $\gamma \in T'_\beta$. $\square$

**4.2 Lemma.** *If $\{G_i\}_{i \in \mathbf{N}}$ is a family of open sets of $T \times P$ such that $M \subset \bigcup_{i \in \mathbf{N}} G_i$, then there exist an open covering $\{U_i\}_{i \in \mathbf{N}}$ of $P$ and $\beta$ in $T$ such that $T'_\beta \times U_i \subset G_i$ for each $i$.*

*Proof.* If $p \in P$, by Lemma 4.1 there exist an open neighbourhood $W(p)$ of $p$ in $P$, an element $\beta(p)$ of $T$ and an integer $i(p)$ such that $T'_{\beta(p)} \times W(p) \subset G_{i(p)}$. By Proposition 2.3.18, $P$ is a Lindelöf space, so that there exists a countable family $\{p_j\}_{j \in \mathbf{N}}$ of elements of $P$ such that $\{W(p_j)\}_{j \in \mathbf{N}}$ is a covering of $P$. For each positive integer $i$ let

$$U_i = \bigcup \{W(p_j) \mid i(p_j) = i\}$$

so that $\{U_i\}_{i \in \mathbf{N}}$ is an open covering of $P$. Let $\beta$ be an element of $T$ such that $\beta(p_j) \leqslant \beta$ if $j \in \mathbf{N}$. If $(\gamma, p) \in T'_\beta \times U_i$, then $p \in W(p_j)$, where $i(p_j) = i$ and $(\gamma, p) \in G_i$ since $\gamma > \beta(p_j)$). Thus $T'_\beta \times U_i \subset G_i$ for each $i$. $\square$

**4.3 Lemma.** *If $f$ is a continuous real-valued function defined on $M$, then there exists $\beta$ in $T$ such that $f$ is constant on $M \cap (T'_\beta \times \{p\})$ for each $p$ in $P$.*

*Proof.* If $p \in P$, there exists $\alpha(p)$ in $T$ such that $T'_{\alpha(p)} \times \{p\} \subset M$. Each continuous real-valued function defined on $T'_{\alpha(p)}$ is finally constant, so that there exists $\beta(p)$ in $T'_{\alpha(p)}$ such that $f(\gamma, p) = f(\beta(p), p)$ if

$\gamma \in T'_{\beta(p)}$. Let $D$ be a countable dense subset of $P$ and let $\beta$ be an element of $T$ such that $\beta(p) \leqslant \beta$ if $p \in D$. Clearly $f$ is constant on $T'_\beta \times \{p\}$ if $p \in D$. If $p \in P \backslash D$ let $\{p_k\}_{k \in \mathbb{N}}$ be a sequence of elements of $D$ which has $p$ as its limit. If $(\gamma_1, p)$ and $(\gamma_2, p)$ are elements of $M \cap (T'_\beta \times \{p\})$, then

$$f(\gamma_1, p) = \lim_k f(\gamma_1, p_k) = \lim_k f(\gamma_2, p_k) = f(\gamma_2, p). \square$$

The following property of a separable metric space $P$ allows us to construct convergent subsets $M$ of $T \times P$ such that ind $M = 0$. We recall that on subsets of a separable metric space, all dimension functions coincide.

**4.4  Lemma.** *If $P$ is a separable metric space, there exists a family $\{P_\alpha\}_{\alpha \in T}$ of zero-dimensional subspaces of $P$ such that $P_\alpha \subset P_\beta$ if $\alpha < \beta$ and $P = \bigcup_{\alpha \in T} P_\alpha$.*

*Proof.* First we obtain such a decomposition of the unit interval $I$. We define an equivalence relation $\sim$ on $I$ as follows: if $x, y \in I$, then $x \sim y$ if $x - y$ is a rational number. Each equivalence class is a countable dense subset of $I$ and there are $\mathbf{c}$ such classes, where $\mathbf{c}$ is the cardinal number of the set of real numbers. Let these classes $Q_\alpha$ be indexed by the ordinal numbers $\alpha$ less than $\omega(\mathbf{c})$, where $\omega(\mathbf{c}) \geqslant \omega_1$ is the least ordinal number of cardinality $\mathbf{c}$. If $\alpha \in T$ let

$$I_\alpha = \bigcup_{\beta < \alpha} Q_\beta \cup \bigcup_{\gamma \geqslant \omega_1} Q_\gamma.$$

Since $I_\alpha \subset I \backslash Q_\alpha$ and $Q_\alpha$ is a dense set, it follows from Example 4.1.2 that $I_\alpha$ is zero-dimensional for each $\alpha$. Thus $I = \bigcup_{\alpha \in T} I_\alpha$ is the required representation of $I$.

The Hilbert cube $K$ is the topological product $\Pi_{i \in \mathbb{N}} K_i$, where $K_i$ is a copy of the unit interval for each $i$. Let $\{K_\alpha^i\}_{\alpha \in T}$ be a family of zero-dimensional subspaces of $K_i$ such that $K_\alpha^i \subset K_\beta^i$ if $\alpha < \beta$ and $K_i = \bigcup_{\alpha \in T} K_\alpha^i$. If $\alpha \in T$, let

$$K_\alpha = \prod_{i \in N} K_\alpha^i.$$

By considering small inductive dimension it is easily seen that $K_\alpha$ is zero-dimensional. Also $K_\alpha \subset K_\beta$ if $\alpha < \beta$, and $K = \bigcup_{\alpha \in T} K_\alpha$. For if $x = \{x_i\} \in K$, then for each $i$ there exists $\alpha(i)$ such that $x_i \in K_{\alpha(i)}^i$. If $\alpha$ is the least upper bound of the countable family $\{\alpha(i)\}$ of elements of $T$, then $x_i \in K_\alpha^i$ for each $i$, so that $x \in K_\alpha$.

Finally let $P$ be an arbitrary separable metric space. It follows from

Proposition 1.5.16 that $P$ can be embedded in $K$. Let us regard $P$ as a subspace of $K$ and for each $\alpha$ in $T$ let $P_\alpha = P \cap K_\alpha$. Then

$$P = \bigcup_{\alpha \in T} P_\alpha$$

is the required representation of $P$. □

The next proposition states the main properties of convergent sets.

**4.5 Proposition.** *If $M$ is a convergent set in the product $T \times P$ of the linearly ordered space $T$ and a separable metric space $P$, then $M$ is a $T_4$-space such that*

$$\dim M = \operatorname{Ind} M = \dim P.$$

*Furthermore for a non-empty separable metric space $P$ there exists a convergent set $M$ in $T \times P$ such that*

$$\operatorname{ind} M = \operatorname{loc} \dim M = \operatorname{loc} \operatorname{Ind} M = 0.$$

*Proof.* We prove first by induction over the dimension of $P$ that $\operatorname{Ind} M \leqslant \dim P = \operatorname{Ind} P$. This assertion is clearly true if $\dim P = -1$. Suppose that the assertion is true for separable metric spaces of dimension less than $n$ and let $\dim P = n \geqslant 0$. Let $A_1$ and $A_2$ be disjoint closed sets in $M$. Then there exist open sets $G_1$ and $G_2$ of $T \times P$ such that $M \cap G_i = M \backslash A_i$ for $i = 1, 2$. Since $M \subset G_1 \cup G_2$, it follows from Lemma 4.2 that there exist open sets $U_1$ and $U_2$ of $P$ and $\beta$ in $T$ such that $T'_\beta \times U_i \subset G_i$ for $i = 1, 2$ and $U_1 \cup U_2 = P$. If $B_i = P \backslash U_i$ for $i = 1, 2$, then $B_1$ and $B_2$ are disjoint closed sets of $P$. It follows from Proposition 4.2.7 that there exists a closed set $F$ of $P$ which separates $B_1$ and $B_2$ such that $\dim F = \operatorname{Ind} F \leqslant n - 1$. Let

$$D = M \cap (T'_\beta \times F).$$

Then $D$ is closed in $M$ and since $D$ is a convergent set in $T \times F$, it follows from the induction hypothesis that $\operatorname{Ind} D \leqslant n - 1$. Since $T_\beta$ is countable, $T_\beta$ is a separable metrizable space and since $\operatorname{Ind} T_\beta = 0$, it follows from Proposition 4.5.5 that $\operatorname{Ind} T_\beta \times P = n$. Hence

$$\operatorname{Ind} M \cap (T_\beta \times P) \leqslant n$$

so that there exists $E$ closed in $M \cap (T_\beta \times P)$, and hence closed in $M$, separating $A_1 \cap (T_\beta \times P)$ and $A_2 \cap (T_\beta \times P)$ and such that $\operatorname{Ind} E \leqslant n - 1$. If $C = D \cup E$, then $C$ is a closed set of $M$ which separates $A_1$ and $A_2$ in $M$. Since $C$ is the topological sum of $D$ and $E$, it follows that $\operatorname{Ind} C \leqslant n - 1$. Thus $\operatorname{Ind} M \leqslant n$ as was to be shown. The above argument also establishes the normality of $M$. It is clear that $M$ is a Hausdorff space.

Next we show that $\dim P \leqslant \dim M$. Suppose that $\dim M \leqslant n$ and let $\{(A_1, B_1), \ldots, (A_{n+1}, B_{n+1})\}$ be a family of pairs of disjoint closed sets of $P$. For each $i = 1, \ldots, n+1$, $\left(M \cap (T \times A_i), M \cap (T \times B_i)\right)$ is a pair of disjoint closed sets of $M$. Hence by Proposition 3.1.9 there exist continuous functions $g_i : X \to [-1, 1]$ for $i = 1, \ldots, n+1$ such that $g_i(z) = 1$ if $z \in M \cap (T \times A_i)$, $g_i(z) = -1$ if $z \in M \cap (T \times B_i)$ and

$$\bigcap_{i=1}^{n+1} g_i^{-1}(0) = \varnothing.$$

It follows from Lemma 4.3 that if $p \in P$, there exists $\gamma(p)$ in $T$ such that $(\gamma, p) \in M$ and $g_i(\gamma, p) = g_i\left(\gamma(p), p\right)$ if $\gamma \in T$ and $\gamma \geqslant \gamma(p)$. Let us define $f_i : P \to [-1, 1]$ by $f_i(p) = g_i\left(\gamma(p), p\right)$ if $p \in P$. Then $f_i$ is continuous. For if $p \in P$ and $\{p_k\}$ is a sequence in $P$ with $p$ as its limit, then for each $k$ there exists $\beta_k$ in $T$ such that $\beta_k > \gamma(p)$ and

$$g_i(\beta_k, p_k) = f_i(p_k).$$

Let $\beta$ be the limit of the sequence $\{\beta_k\}$ in $T$. Then the sequence $\{(\beta_k, p_k)\}$ in $M$ has $(\beta, p)$ as its limit. And

$$f_i(p) = g_i(\beta, p) = \lim_k g_i(\beta_k, p_k) = \lim_k f_i(p_k).$$

Furthermore $f_i(p) = 1$ if $p \in A_i$, $f_i(p) = -1$ if $p \in B_i$ and

$$\bigcap_{i=1}^{n+1} f_i^{-1}(0) = \varnothing.$$

Hence $\dim P \leqslant n$ by Proposition 3.1.9. Since $\dim M \leqslant \operatorname{Ind} M$ we can conclude that

$$\dim M = \operatorname{Ind} M = \dim P.$$

To complete the proof we show that if $P = \bigcup_{\alpha \in T} P_\alpha$ is a decomposition of the separable metric space $P$ into zero-dimensional subsets such that $P_\beta \subset P_\alpha$ if $\beta < \alpha$, then the convergent set

$$M = \bigcup_{\alpha \in T} (T_\alpha' \times P_\alpha)$$

of $T \times P$ satisfies $\operatorname{ind} M = 0$. If $(\alpha, p) \in M$, then

$$(\alpha, p) \in M_\alpha = M \cap (T_\alpha \times P).$$

If $(\gamma, q) \in M_\alpha$, then $q \in P_\beta$ for some $\beta$ in $T$ such that $\beta < \gamma \leqslant \alpha$. Since $P_\beta \subset P_\alpha$ if $\beta < \alpha$, it follows that $M_\alpha \subset T_\alpha \times P_\alpha$. Hence $M_\alpha$ is a zero-

dimensional subspace of $M$. Since $M_\alpha$ is an open-and-closed set of $M$ it follows that
$$\operatorname{ind} M = \operatorname{loc\,dim} M = \operatorname{loc\,Ind} M = 0. \square$$

**4.6**　*Remark.* If in the above construction we take $P$ to be the $n$-cube $I^n$, where $n$ is a positive integer, then we can obtain a $T_4$-space $M$ which satisfies
$$\dim M = \operatorname{Ind} M = n > 0,$$
whilst
$$\operatorname{ind} M = \operatorname{loc\,dim} M = \operatorname{loc\,Ind} M = 0.$$

Since $\operatorname{loc\,dim} M = 0$, by Proposition 3.1 and Remark 3.2 there exists a $T_4$-space $N$ which contains $M$ as an open subspace and satisfies $\dim N = 0$ and hence $\operatorname{Ind} N = 0$. Thus the subset theorems for dim and Ind do not hold for all $T_4$-spaces even if the subset is required to be normal. Since $\operatorname{ind} M = 0$, by Proposition 4.1.3 $M$ can be embedded in a totally disconnected bicompact space. Thus totally disconnected bicompact spaces may contain subspaces of arbitrary covering dimension and large inductive dimension. Next we consider the subset theorem for local dimension.

**4.7**　*Example.* The subset theorem does not hold for local dimension of $T_4$-spaces.

Let $n$ be a positive integer. From the above results there exists a $T_4$-space $N$ such that $\dim N = \operatorname{Ind} N = 0$ which contains a normal subspace $M$ such that $\dim M = \operatorname{Ind} M = n$ and
$$\operatorname{ind} M = \operatorname{loc\,dim} M = \operatorname{loc\,Ind} M = 0.$$

Let $R$ consist of the union of countably many copies $N_i$, $i \in \mathbf{N}$, of $N$ together with a special point $y_0$, and let $Q$ be the subset of $R$ consisting of the union of the corresponding countable number of copies $M_i$ of $M$ together with the point $y_0$. A base for the topology of $R$ consists of the open sets of each $N_i$ together with the sets $\{y_0\} \cup \bigcup_{i>j} N_i$, where $j \in \mathbf{N}$. The induced topology on $Q$ has a base consisting of the open sets of each $M_i$ together with the sets $\{y_0\} \cup \bigcup_{i>j} M_i$, where $j \in \mathbf{N}$.

If $x$ and $y$ are distinct points of $N_j$, then $x$ and $y$ have disjoint neighbourhoods in $N_j$, and hence in $R$, since $N_j$ is a Hausdorff space. If $x \in N_i$ and $y \in N_j$, where $i \neq j$, then $N_i$ and $N_j$ are open and disjoint. If $y = y_0$ and $x \in N_j$, then $N_j$ and $\{y_0\} \cup \bigcup_{i>j} N_i$ are disjoint neighbourhoods of $x$ and $y$ respectively. Thus $R$ is a Hausdorff space. If $E$ and $F$ are disjoint closed sets of $R$, then one of them, say $F$, does not contain $y_0$. Then $y_0$ has a neighbourhood which does not meet $F$ and hence $F \subset \bigcup_{i \leqslant j} N_i$ for some $j$. Since each $N_i$ is normal, there exist disjoint open sets $U_i$ and $V_i$ of $N_i$ for $i \leqslant j$ such that $E \cap N_i \subset U_i$ and $F \cap N_i \subset V_i$. Let

$$U = \bigcup_{i \leqslant j} U_i \cup \{y_0\} \cup \bigcup_{i > j} N_i, \quad V = \bigcup_{i \leqslant j} V_i.$$

Then $U$ and $V$ are disjoint open sets of $R$, $E \subset U$ and $F \subset V$. Thus $R$ is a normal space. A similar argument shows that the subspace $Q$ is normal.

If $F$ is a closed set of $Q$ which does not contain $\{y_0\}$, then $F$ is a closed set of $\bigcup_{i \leqslant j} N_i$ for some $j$ and hence

$$\dim F \leqslant \dim (\bigcup_{i \leqslant j} N_i).$$

But $\dim (\bigcup_{i \leqslant j} N_i) = 0$ by the sum theorem, since each $N_i$ is closed in $R$ and $\dim N_i = 0$. Since $\dim \{y_0\} = 0$ it follows from Proposition 3.5.6 that $\dim R = 0$ and hence that $\operatorname{Ind} R = 0$ and $\operatorname{ind} R = 0$. It follows that $\operatorname{loc} \dim R = 0$ and $\operatorname{loc} \operatorname{Ind} R = 0$.

Since $Q$ is a non-empty subspace of $R$, it follows that $\operatorname{ind} Q = 0$. The point $y_0$ has a neighbourhood $U$ in $Q$ such that $\dim \overline{U} \leqslant \operatorname{loc} \dim Q$, and $U$ contains a set of the form $\{y_0\} \cup \bigcup_{i > j} M_i$ and hence contains the closed set $M_{j+1}$. Therefore $\dim \overline{U} \geqslant \dim M_{j+1} = n$. Thus $\operatorname{loc} \dim Q \geqslant n$. Now let $F$ be a closed set of $Q$ and let $V$ be an open set of $Q$ such that $F \subset V$. Then $y_0$ has a neighbourhood $\{y_0\} \cup \bigcup_{i > j} M_i$ which either does not meet $F$ or is contained in $V$. Since $\operatorname{Ind} M_i = n$, for each $i \leqslant j$ there is an open set $W_i$ such that $F \cap M_i \subset W_i \subset V \cap M_i$ and

$$\operatorname{Ind} \operatorname{bd} (W_i) \leqslant n - 1,$$

where $\operatorname{bd} (W_i)$ is the boundary of $W_i$ in $Q$, since $M_i$ is an open-and-closed set of $Q$. Let $W$ be the union of the sets $W_i$ for $i \leqslant j$ together with the open-and-closed set $\{y_0\} \cup \bigcup_{i > j} M_i$ in the case that this set meets $F$. Then $W$ is open in $Q$ and $F \subset W \subset V$, and $\operatorname{bd} (W)$ is the topological sum of the subspaces $\operatorname{bd} (W_i)$ so that $\operatorname{Ind} \operatorname{bd} (W) \leqslant n - 1$. Therefore $\operatorname{Ind} Q \leqslant n$. Since $\operatorname{loc} \dim Q \leqslant \operatorname{loc} \operatorname{Ind} Q \leqslant \operatorname{Ind} Q$, it follows that $\operatorname{loc} \dim Q = \operatorname{loc} \operatorname{Ind} Q = \operatorname{Ind} Q = n$, and

$$\operatorname{loc} \dim Q \leqslant \dim Q \leqslant \operatorname{Ind} Q$$

so that $\dim Q = n$.

Thus the $T_4$-space $R$ which satisfies $\operatorname{loc} \dim R = \operatorname{loc} \operatorname{Ind} R = 0$ has a normal subspace $Q$ which satisfies $\operatorname{loc} \dim Q = \operatorname{loc} \operatorname{Ind} Q = n > 0$. Thus the subset theorem does not hold for the local dimension of $T_4$-spaces even if the subspace is required to be normal. Also although the finite sum theorem holds for the local dimension ($\operatorname{loc} \dim$) of normal spaces, the countable sum theorem does not hold. For the normal space $Q$ is the union of countably many closed sets $\{y_0\}, M_1, M_2, \ldots,$ such that $\operatorname{loc\,dim}\{y_0\} = 0$ and $\operatorname{loc\,dim} M_i = 0$ for each $i$, but

$$\operatorname{loc\,dim} Q = n > 0.$$

We have already noted that Example 4.3.4 shows that the finite sum theorem does not hold for $\operatorname{loc\,Ind}$ even for bicompact spaces.

**4.8** *Remark.* It can now be seen that there are no relations between ind, $\operatorname{loc\,dim}$, $\operatorname{loc\,Ind}$, dim and Ind for normal regular spaces other than those given in Proposition 1.6. For consider the spaces $M$ and $Q$ discussed above. We observe that $\operatorname{loc\,Ind} M < \operatorname{Ind} M$ and

$$\operatorname{loc\,dim} M < \dim M.$$

Also $\operatorname{ind} Q < \operatorname{loc\,Ind} Q = \operatorname{Ind} Q$ and $\operatorname{ind} Q < \operatorname{loc\,dim} Q = \dim Q$. For the space $S$ studied in Example 4.3.4 we note that

$$\dim S = \operatorname{loc\,dim} S < \operatorname{ind} S = \operatorname{loc\,Ind} S = \operatorname{Ind} S.$$

Let $p$ and $q$ be positive integers such that $p < q$. The topological sum $A_{pq}$ of the $p$-cube $I^p$ and a $T_4$-space $M$, which satisfies

$$\operatorname{ind} M = \operatorname{loc\,dim} M = \operatorname{loc\,Ind} M = 0$$

and $\dim M = \operatorname{Ind} M = q$, is a $T_4$-space which satisfies

$$\operatorname{ind} A_{pq} = \operatorname{loc\,dim} A_{pq} = \operatorname{loc\,Ind} A_{pq} = p,$$

$$\dim A_{pq} = \operatorname{Ind} A_{pq} = q.$$

The topological sum $B_{pq}$ of the $p$-cube $I^p$ and a $T_4$-space $Q$ such that $\operatorname{ind} Q = 0$ and $\operatorname{loc\,dim} Q = \operatorname{Ind} Q = q$ is a $T_4$-space such that $\operatorname{ind} B_{pq} = 0$ and

$$\operatorname{loc\,dim} B_{pq} = \operatorname{loc\,Ind} B_{pq} = \dim B_{pq} = \operatorname{Ind} B_{pq} = q.$$

In Chapter 8 we shall show that if $p$ and $q$ are positive integers such that $p < q$, then there exist bicompact spaces $C_{pq}, D_{pq}$ such that $\dim C_{pq} = \dim D_{pq} = p$ and $\operatorname{ind} C_{pq} = \operatorname{Ind} D_{pq} = q$.

We shall conclude this section by studying Nagami's example of a $T_4$-space for which all three principal dimension functions differ. For the construction and analysis of this example, two results are required which will also be used in the study of examples in Chapter 8.

**4.9** *Proposition.* *Let $X$ be a non-empty normal space and let $A$ be a closed subspace of $X$ which is a retract of $X$ and has the property that for each open set $M$ such that $A \subset M$ there exists an open-and-closed set $N$*

*such that $A \subset N \subset M$. If* $\operatorname{Ind} A \leqslant m$ *and* $\operatorname{Ind} B \leqslant n$ *for each closed set $B$ disjoint from $A$, then* $\operatorname{Ind} X \leqslant m+n$.

*Proof.* For each non-negative integer $m$ let $(\alpha_m)$ be the statement: if $A$ is a closed subspace of a normal space $X$ which is a retract of $X$ and has the property that for each open set $M$ such that $A \subset M$ there exists an open-and-closed set $N$ such that $A \subset N \subset M$, and

$$\operatorname{Ind} A \leqslant m,$$

then $\operatorname{Ind} X \leqslant m+n$, where $n$ is an integer such that $\operatorname{Ind} B \leqslant n$ for every closed set $B$ disjoint from $A$. We shall prove that if $m > 0$ and $(\alpha_{m-1})$ is true, then $(\alpha_m)$ is true. The latter half of this proof can be trivially modified to give a proof of $(\alpha_0)$. Thus the proposition will be proved by induction.

Let $E$ be a closed set of $X$ and let $G$ be an open set such that $E \subset G$. There exists $U$ open in $A$ such that

$$E \cap A \subset U \subset \overline{U} \subset G \cap A \quad \text{and} \quad \operatorname{Ind} \operatorname{bd}_A (U) \leqslant m-1.$$

Let $\phi: X \to A$ be a retraction, let $A' = \operatorname{bd}_A (U)$ and let $X' = \phi^{-1}(A')$. Then $A'$ is closed in $A$ and so closed in $X$. Hence $X'$ is closed in $X$ so that $X'$ is normal and $A'$ is closed in $X'$. Clearly $A'$ is a retract of $X'$. Let $M' = M \cap X'$, where $M$ is open in $X$, and suppose that $A' \subset M'$. Then $M \cup (X \backslash X')$ is an open set of $X$ which contains $A$. Hence there exists an open-and-closed set $N$ of $X$ such that $A \subset N \subset M \cup (X \backslash X')$. If $N' = N \cap X'$, then $N'$ is an open-and-closed set of $X'$ and

$$A' \subset N' \subset M'.$$

Finally $\operatorname{Ind} A' \leqslant m-1$ and if $B'$ is a closed set of $X'$ disjoint from $A'$ then $B'$ is a closed set of $X$ disjoint from $A$ so that $\operatorname{Ind} B' \leqslant n$. Since $(\alpha_{m-1})$ is true it follows that $\operatorname{Ind} X' \leqslant m+n-1$.

Let $P = \phi^{-1}(U)$. Then $P$ is open in $X$ and

$$\operatorname{bd}(P) = \overline{P} \backslash P \subset \phi^{-1}(\overline{U}) \backslash \phi^{-1}(U) = X'.$$

Hence $\operatorname{Ind} \operatorname{bd}(P) \leqslant m+n-1$. Now $A \cap P \subset A \cap \phi^{-1}(\overline{U}) = \overline{U} \subset G$ and $A \cap E \subset U \subset P$. Hence $(\overline{P} \backslash G) \cup (E \backslash P)$ is a closed set of $X$ which is disjoint from $A$. It follows that there exists an open-and-closed set $V$ such that $A \subset V$ and $V$ is disjoint from $(\overline{P} \backslash G) \cup (E \backslash P)$. Since $X \backslash V$ is a closed set of $X$ which is disjoint from $A$, it follows that $\operatorname{Ind}(X \backslash V) \leqslant n$. Hence there exists $H$ open in $X \backslash V$, and hence open in $X$, such that

$$E \backslash V \subset H \subset G \backslash V \quad \text{and} \quad \operatorname{Ind} \operatorname{bd}(H) \leqslant n-1.$$

Let $W = (P \cap V) \cup H$. Since $V$ is disjoint from $E \backslash P$ it follows that

$E \cap V \subset P$. But $E \backslash V \subset H$ so that $E \subset W$. Since $V$ is disjoint from $\bar{P} \backslash G$ it follows that $P \cap V \subset G$. Clearly $H \subset G$ so that $W \subset G$. Finally bd $(W)$ is the topological sum of bd $(P \cap V)$ and bd $(H)$. Also

$$\mathrm{bd}\,(P \cap V) \subset \mathrm{bd}\,(P) \cap V,$$

therefore Ind bd $(P \cap V) \leqslant m+n-1$. Since Ind bd $(H) \leqslant n-1$ it follows that
$$\mathrm{Ind\,bd}\,(W) \leqslant m+n-1.$$
Thus Ind $X \leqslant m+n$. $\square$

If $Z$ and $X$ are topological spaces, a function $f: Z \rightarrow X$ is said to be *almost-open* if for each point $x$ of $f(Z)$ there exists some point $z$ in $Z$ such that $f(z) = x$ and $f(U)$ is a neighbourhood of $x$ in $f(Z)$ if $U$ is a neighbourhood of $z$.

**4.10 Proposition.** *If $X$ is an infinite bicompact space, there exists a totally disconnected bicompact space $Z$ of weight not exceeding the weight of $X$ and an almost-open continuous surjection $f: Z \rightarrow X$.*

*Proof.* Let $w(X) = \tau$ and let $\Lambda$ be a set such that $|\Lambda| = \tau$. Let $\{V_\lambda\}_{\lambda \in \Lambda}$ and $\{W_\lambda\}_{\lambda \in \Lambda}$ be families of open sets of $X$ such that for each point $x$ of $X$ and each open set $U$ such that $x \in U$ there exists some $\lambda$ in $\Lambda$ such that $x \in V_\lambda \subset \bar{V}_\lambda \subset W_\lambda \subset U$. If $\lambda \in \Lambda$ let $G_\lambda(1) = W_\lambda$ and $G_\lambda(2) = X \backslash \bar{V}_\lambda$ and let $K_\lambda(i) = [G_\lambda(i)]^-$ for $i = 1, 2$. For each $\lambda$ in $\Lambda$, let $D_\lambda = \{1, 2\}$ with the discrete topology and as usual let $D^\tau$ denote the topological product $\Pi_{\lambda \in \Lambda} D_\lambda$. If $z = \{z_\lambda\}_{\lambda \in \Lambda} \in D^\tau$ let

$$F(z) = \bigcap_{\lambda \in \Lambda} K_\lambda(z_\lambda).$$

For each $z$ in $D$, either $F(z)$ is empty or $F(z)$ is a one-point set. For if $x$ and $x'$ are distinct points of $X$, then since $X$ is a Hausdorff space there exists an open set $U$ such that $x \in U$ and $x' \notin \bar{U}$. There exists $\lambda$ such that $x \in V_\lambda \subset \bar{V}_\lambda \subset W_\lambda \subset U$. Then $x' \notin K_\lambda(1)$ and $x \notin K_\lambda(2)$. Let

$$Z = \{z \in D^\tau \mid F(z) \text{ is not empty}\}.$$

If $y = \{y_\lambda\} \in D^\tau$ and $y \notin Z$, then since $X$ is compact there exists a finite subset $M$ of $\Lambda$ such that $\bigcap_{\lambda \in M} K_\lambda(y_\lambda) = \varnothing$. If $V$ is the subset of $D^\tau$ consisting of those points $v = \{v_\lambda\}$ for which $v_\lambda = y_\lambda$ if $\lambda \in M$, then $V$ is an open neighbourhood of $y$ in $D^\tau$ which is disjoint from $Z$. Thus $Z$ is a closed set of the totally disconnected bicompact space $D^\tau$. Hence $Z$ is a totally disconnected bicompact space.

Let $f: Z \rightarrow X$ be defined as follows: if $z = \{z_\lambda\} \in Z$, then $f(z)$ is the unique point of $X$ in $\bigcap_{\lambda \in \Lambda} K_\lambda(z_\lambda)$. Let $U$ be an open set of $X$ and suppose that $z \in f^{-1}(U)$, where $z = \{z_\lambda\}_{\lambda \in \Lambda}$. Then $\bigcap_{\lambda \in \Lambda} K_\lambda(z_\lambda) \subset U$. Since $X$ is compact it follows that there exists a finite subset $M$ of $\Lambda$ such that $\bigcap_{\lambda \in M} K_\lambda(z_\lambda) \subset U$. Let $V$ be the subset of $Z$ consisting of those elements $\{v_\lambda\}_{\lambda \in \Lambda}$ for which $v_\lambda = z_\lambda$ if $\lambda \in M$. Then $V$ is open in $Z$ and $z \in V \subset f^{-1}(U)$. Thus $f^{-1}(U)$ is open in $Z$. Hence $f$ is continuous. Furthermore $f$ is an almost-open surjection. For let $x$ be a point of $X$. For each $\lambda$, choose $z_\lambda$ in $\{1, 2\}$ such that $x \in G_\lambda(z_\lambda)$. If $z = \{z_\lambda\}_{\lambda \in \Lambda}$, then $z \in Z$ and $f(z) = x$. Moreover let $U$ be an open neighbourhood of $z$. There exists a finite subset $M$ of $\Lambda$ such that $z \in V \subset U$, where $V$ consists of those points $\{v_\lambda\}_{\lambda \in \Lambda}$ of $Z$ such that $v_\lambda = z_\lambda$ if $\lambda \in M$. Let $W = \bigcap_{\lambda \in M} G_\lambda(z_\lambda)$. Then $W$ is open in $X$ and $x \in W$. If $x' \in W$, then for each $\lambda$ in $\Lambda \setminus M$ choose $y_\lambda$ such that $x' \in G_\lambda(y_\lambda)$ whilst $y_\lambda = z_\lambda$ if $\lambda \in M$. If $y = \{y_\lambda\}_{\lambda \in \Lambda}$, then $y \in V$ and $f(y) = x'$. Thus $x \in W \subset f(V) \subset f(U)$. Hence $f(U)$ is a neighbourhood of $x$. $\square$

**4.11** *Example.* There exists a $T_4$-space $Z$ such that

$$\text{ind } Z = 0, \quad \dim Z = 1 \quad \text{and} \quad \text{Ind } Z = 2.$$

By Proposition 4.10, there exists a closed set $C$ of the Cantor set and a continuous almost-open surjection $f: C \to I$, where $I$ is the unit interval. Let

$$P = I \cup (C \times \Lambda \times I),$$

where $\Lambda$ is a set of cardinality greater than $\mathbf{c}$, the cardinal of the set $\mathbf{R}$ of real numbers. Let a base for the topology of $P$ consist of all sets open in the topological product $C \times \Lambda \times I$, where $\Lambda$ has the discrete topology, together with all sets of the form

$$U \cup \left(f^{-1}(U) \times (\Lambda \setminus K) \times I\right),$$

where $U$ is open in $I$ and $K$ is a finite subset of $\Lambda$. It was established in the proof of Lemma 4.4 that for each ordinal $\alpha < \omega_1$ there exists a zero-dimensional dense subset $I_\alpha$ of $I$ such that $I_\alpha \subset I_\beta$ if $\alpha < \beta$ and $I = \bigcup_{\alpha < \omega_1} I_\alpha$. Let $T$ be the linearly ordered space of ordinals $\alpha$ such that $\alpha < \omega_1$ and let $T_\alpha = [0, \alpha]$ and $T'_\alpha = T \setminus T_\alpha$ if $\alpha \in T$. If $\alpha \in T$ let $P_\alpha$ be the subset of $P$ given by

$$P_\alpha = I_\alpha \cup \left(f^{-1}(I_\alpha) \times \Lambda \times I_\alpha\right).$$

We define $Z$ to be the subspace of the topological product $T \times P$ given by

$$Z = \bigcup_{\alpha \in T} (T'_\alpha \times P_\alpha).$$

We note that $Z \cap (T \times I) = M$, where $M$ is a convergent set in $T \times I$ such that $\operatorname{ind} M = 0$ and $\dim M = \operatorname{Ind} M = 1$.

Let $Y = Z \cap \big(T \times (C \times \Lambda \times I)\big)$. Then the subspace $Y$ of $Z$ is the topological sum of a family $\{A_\lambda\}_{\lambda \in \Lambda}$ of subspaces, where

$$A_\lambda = Z \cap \big(T \times (C \times \{\lambda\} \times I)\big)$$

if $\lambda \in \Lambda$. We observe that $C \times \{\lambda\} \times I$ is a separable metric space and that $A_\lambda$ is a convergent set in $T \times (C \times \{\lambda\} \times I)$. It is now easily seen that $\operatorname{ind} A_\lambda = 0$ and $\dim A_\lambda = \operatorname{Ind} A_\lambda = 1$. It follows that $Y$ is a normal space such that $\operatorname{ind} Y = 0$ and $\dim Y = \operatorname{Ind} Y = 1$.

If $\alpha \in T$, then the subspace $Z_\alpha = Z \cap (T_\alpha \times P)$ of $Z$ is a Lindelöf space. For let $\mathcal{U}$ be an open covering of $Z_\alpha$. It is not difficult to show that if $\beta \in T_\alpha$, then $Z \cap (\{\beta\} \times P)$ is contained in the union of countably many members of $\mathcal{U}$. Since $T_\alpha$ is countable, it follows that $Z_\alpha$ is covered by countably many members of $\mathcal{U}$. Since $Z_\alpha$ is clearly regular, it follows that $Z_\alpha$ is a normal space. If $\alpha \in T$, the subspace

$$Y_\alpha = Z_\alpha \cap Y$$

of $Z$ is the topological sum of a family $\{A_\lambda^\alpha\}_{\lambda \in \Lambda}$ of subspaces, where

$$A_\lambda^\alpha = Z \cap \big(T_\alpha \times (C \times \{\lambda\} \times I)\big)$$

if $\lambda \in \Lambda$. Since $A_\lambda^\alpha$ is zero-dimensional in all senses for each $\lambda$, it follows that $\operatorname{ind} Y = \dim Y = \operatorname{Ind} Y = 0$. The further analysis of this example is given in the four propositions $A$, $B$, $C$ and $D$ which follow.

**Proposition A.** *The space $Z$ is normal.*

*Proof.* Let $E$ and $F$ be disjoint closed sets of $Z$. Then for each point $x$ of $I$ there exist $\alpha(x)$ in $T$, an open neighbourhood $V_x$ of $x$ in $I$ and a finite subset $K_x$ of $\Lambda$ such that

$$Z \cap \big(T'_{\alpha(x)} \times (V_x \cup (f^{-1}(V_x) \times (\Lambda \backslash K_x) \times I))\big)$$

is contained either in $Z \backslash E$ or in $Z \backslash F$. There exists a finite subset $B$ of $I$ such that $\bigcup_{x \in B} V_x = I$. Let $K = \bigcup_{x \in B} K_x$ so that $K$ is a finite subset of $\Lambda$ and let $\beta = \max_{x \in B} \alpha(x)$. Now let

$$Z' = Z \cap [T'_\beta \times \big(I \cup (C \times (\Lambda \backslash K) \times I)\big)]$$

and

$$Z'' = Z \cap [T'_\beta \times C \times K \times I].$$

Then $Z = Z_\beta \cup Z' \cup Z''$. Furthermore $Z_\beta$, $Z'$ and $Z''$ are disjoint and open-and-closed in $Z$. The subspace $Z_\beta$ is normal so that there exist disjoint open sets $G_0$ and $H_0$ such that $E \cap Z_\beta \subset G_0$ and $F \cap Z_\beta \subset H_0$.

The subspace $Z''$ is closed in $Y$ and hence is normal, so that there exist disjoint open sets $G''$ and $H''$ such that $E \cap Z'' \subset G''$ and

$$F \cap Z'' \subset H''.$$

Finally let $\{W_1, \ldots, W_s\}$ be a finite open covering of $I$ which is a star-refinement of the covering $\{V_x\}_{x \in B}$, and for each $i = 1, \ldots, s$ let

$$U_i = Z \cap [T'_\beta \times \left(W_i \cup (f^{-1}(W_i) \times (\Lambda \backslash K) \times I)\right)].$$

If $\mathscr{U} = \{U_1, \ldots, U_s\}$, then $\mathscr{U}$ is an open covering of $Z'$ which is a star-refinement of the covering $\{Z' \backslash E, Z' \backslash F\}$. If $G' = \text{St}(E \cap Z', \mathscr{U})$ and $H' = \text{St}(F \cap Z', \mathscr{U})$, then $G'$ and $H'$ are disjoint open sets such that $E \cap Z' \subset G'$ and $F \cap Z' \subset H'$. If $G = G_0 \cup G' \cup G''$ and $H = H_0 \cup H' \cup H''$, then $G$ and $H$ are disjoint open sets such that $E \subset G$ and $F \subset H$. Thus $Z$ is a normal space. $\square$

**Proposition B.** $\text{ind } Z = \text{loc dim } Z = \text{loc Ind } Z = 0.$

*Proof.* If $\alpha \in T$, then $Z_\alpha$ is a normal space. If $M_\alpha = Z_\alpha \cap (T \times I)$, then $M_\alpha$ is closed in $Z_\alpha$ and $\dim M_\alpha = 0$. Furthermore $Z_\alpha \backslash M_\alpha = Y_\alpha$ and $\dim Y_\alpha = 0$. It follows from Proposition 3.5.7 that $\dim Z_\alpha = 0$ and hence

$$\dim Z_\alpha = \text{ind } Z_\alpha = \text{Ind } Z_\alpha = 0.$$

For each $\alpha$, $Z_\alpha$ is an open-and-closed set in $Z$ and if $(\alpha, p) \in Z$ then $(\alpha, p) \in Z_\alpha$. Thus $\text{ind } Z = \text{loc dim } Z = \text{loc Ind } Z = 0.$ $\square$

**Proposition C.** $\dim Z = 1.$

*Proof.* The space $Z$ is normal and $M = Z \cap (T \times I)$ is a closed set such that $\dim M = 1$. Furthermore $Z \backslash M = Y$ and $\dim Y = 1$. It follows from Proposition 3.5.7 that $\dim Z = 1.$ $\square$

**Proposition D.** $\text{Ind } Z = 2.$

*Proof.* First we obtain an upper bound for $\text{Ind } Z$ by an application of Proposition 4.9. Let us define $\phi: P \to I$ as follows: if $(c, \lambda, t) \in C \times \Lambda \times I$, then $\phi(c, \lambda, t) = f(c)$ and if $u \in I$, then $\phi(u) = u$. It is clear that $\phi$ is continuous and hence $\phi$ is a retraction of $P$ onto the subspace $I$ of $P$. If we define $\psi: T \times P \to T \times I$ by $\psi(\alpha, p) = \left(\alpha, \phi(p)\right)$ if $(\alpha, p) \in T \times P$, then the restriction of $\psi$ to $Z$ is a retraction of $Z$ onto $M$. Let $U$ be an open set of $Z$ such that $M \subset U$. Then there exist $\xi$ in $T$ and a finite subset $K$ of $\Lambda$ such that

$$V_1 = Z \cap [T'_\xi \times \left(I \cup (C \times (\Lambda \backslash K) \times I)\right)] \subset U,$$

and $V_1$ is an open-and-closed set of $Z$. Since $\operatorname{Ind} Z_\xi = 0$, there exists an open-and-closed set $V_2$ of $Z_\xi$ such that $Z_\xi \cap M \subset V_2 \subset Z_\xi \cap U$. If $V = V_1 \cup V_2$, then $V$ is an open-and-closed set of $Z$ and $M \subset V \subset U$. If $B$ is a closed set of $Z$ disjoint from $M$, then $B \subset Y$ so that $\operatorname{Ind} B \leqslant 1$. Since $\operatorname{Ind} M = 1$, it follows from Proposition 4.9 that $\operatorname{Ind} Z \leqslant 2$.

Now we see that the reverse inequality holds. Consider the disjoint closed sets $E$ and $F$ of $M$ given by

$$E = Z \cap (T \times \{0\}) \quad \text{and} \quad F = Z \cap (T \times \{1\}).$$

Then $E$ and $F$ are disjoint closed sets of $Z$. Let $U$ be an open set of $Z$ such that $E \subset U \subset \bar{U} \subset Z \backslash F$, let $B = \operatorname{bd}(U)$ and let $V = Z \backslash \bar{U}$. Let $G$ be the subset of $I$ consisting of those points $x$ such that

$$\{\alpha \in T \mid (\alpha, x) \in U\}$$

is cofinal in $T$. If $x \in G$, then $0 \leqslant x < 1$, $G$ is an open set of $I$, and if $x \in G$ and $x > 0$, then there exists an integer $i(x)$ such that

$$1/i(x) < \min\{x, 1-x\}$$

and the set

$$\{\alpha \in T \mid Z \cap [\{\alpha\} \times (x - 1/i(x), x + 1/i(x))] \subset U\}$$

is cofinal in $T$. Similarly if $H$ is the set of points $x$ of $I$ such that $\{\alpha \in T \mid (\alpha, x) \in V\}$ is cofinal in $T$, then $H$ is an open set of $I$, $0 < x \leqslant 1$ if $x \in H$, and if $x \in H$ and $x < 1$ then there exists an integer $j(x)$ such that $1/j(x) < \min\{x, 1-x\}$ and the set

$$\{\alpha \in T \mid Z \cap [\{\alpha\} \times (x - 1/j(x), x + 1/j(x))] \subset V\}$$

is cofinal in $T$.

We shall consider three cases: (i) $G \cap H \neq \varnothing$; (ii) $G \cap H = \varnothing$ and $G \cup H$ is not dense in $I$; (iii) $G \cap H = \varnothing$ and $G \cup H$ is dense in $I$. We shall see that in each case $\operatorname{Ind} B \geqslant 1$.

(i) If $G \cap H \neq \varnothing$ and $x \in G \cap H$, then $0 < x < 1$. Let

$$k = \max\{i(x), j(x)\}$$

and let

$$T_1 = \{\alpha \in T \mid Z \cap [\{\alpha\} \times (x - 1/k, x + 1/k)] \subset B\}.$$

Since $\bar{U} \cap \bar{V} = B$, the set $T_1$ is cofinal in $T$. And $T_1$ is a closed set of $T$. Thus if

$$B_1 = Z \cap (T_1 \times [x - 1/2k, x + 1/2k]),$$

then $B_1$ is a closed subset of $B$. Since $T_1$ is homeomorphic with $T$, it follows from Proposition 4.5 that $\operatorname{Ind} B_1 = 1$. Hence $\operatorname{Ind} B \geqslant 1$.

(ii) If $G \cap H = \varnothing$ and $G \cup H$ is not dense in $I$, then $I \backslash (G \cup H)$ contains a non-empty closed subinterval $I_1$ of $I$. Then $B \cap (T \times I_1)$ is a convergent set in $T \times I_1$. It follows from Proposition 4.5 that $\operatorname{Ind} B \cap (T \times I_1) = 1$. Since $B \cap (T \times I_1)$ is a closed subset of $B$ it follows that $\operatorname{Ind} B \geqslant 1$.

(iii) If $G \cap H = \varnothing$ and $G \cup H$ is dense in $I$, let $a$ be the least upper bound of $G$. Since $G$ is open, it follows that $a \notin G$ and since $H$ is open and disjoint from $G$ it follows that $a \notin H$. Thus $0 < a < 1$. Let

$$S = \{\alpha \in T \mid (\alpha, a) \in (M \cap U)^- \cap (M \cap V)^-\}.$$

Since $G \cup H$ is dense in $I$ it is not difficult to see that $S$ is cofinal in $T$, and it is clear that $S$ is closed in $T$. If $z = (\beta, x) \in M \cap U$, there exists a finite subset $K_z$ of $\Lambda$ such that

$$Z \cap [\{\beta\} \times f^{-1}(x) \times (\Lambda \backslash K_z) \times I] \subset U.$$

If $w = (\gamma, y) \in M \cap V$, there exists a finite subset $K_w$ of $\Lambda$ such that

$$Z \cap [\{\gamma\} \times f^{-1}(y) \times (\Lambda \backslash K_w) \times I] \subset V.$$

Let $\Lambda_1 = \bigcup \{K_z \mid z \in M \cap U\}$ and $\Lambda_2 = \bigcup \{K_w \mid w \in M \cap V\}$. Since $|M| = \mathbf{c}$ it follows that $|\Lambda_1 \cup \Lambda_2| \leqslant \mathbf{c}$ and hence $\Lambda_1 \cup \Lambda_2 \neq \Lambda$. Let $\nu$ be an element of $\Lambda \backslash (\Lambda_1 \cup \Lambda_2)$. Since $f$ is an almost-open mapping, there exists $c$ in $C$ such that $f(c) = a$ and $f(N)$ is a neighbourhood of $a$ in $I$ for each neighbourhood $N$ of $c$ in $C$. We shall show that if $\alpha \in S$ and $t \in \bigcup_{\gamma < \alpha} I_\gamma$ then $z = (\alpha, c, \nu, t) \in B$. Each neighbourhood of $z$ in $Z$ contains a neighbourhood of the form

$$W = Z \cap [(\xi, \alpha] \times N \times \{\nu\} \times J],$$

where $\xi$ is an ordinal such that $\xi < \alpha$, $N$ is an open neighbourhood of $c$ in $C$ and $J$ is an open interval containing $t$. If $W' = Z \cap [(\xi, \alpha] \times f(N)]$, then $W'$ is a neighbourhood of $(\alpha, a)$ in $M$. Since $\alpha \in S$ it follows that $W' \cap U \neq \varnothing$. If $(\beta, x) \in W' \cap U$, then $x \in I_\gamma$ for some $\gamma < \beta$ and there exists $d$ in $N$ such that $f(d) = x$. Since $I_\gamma$ is dense in $I$, there exists $s$ such that $s \in J \cap I_\gamma$. Then $(\beta, d, \nu, s) \in W$. But

$$Z \cap [\{\beta\} \times f^{-1}(x) \times \{\nu\} \times I] \subset U,$$

by choice of $\nu$, so that $(\beta, d, \nu, s) \in U$. Thus $z \in \overline{U}$. Similarly $z \in \overline{V}$ so that $z \in B$. It follows that

$$B' = Z \cap [S \times \{c\} \times \{\nu\} \times I] \subset B.$$

Since $S$ is cofinal in $T$, it follows from Proposition 4.5 that $\operatorname{Ind} B' = 1$. But $B'$ is a closed set and again $\operatorname{Ind} B \geqslant 1$.

Thus we have found disjoint closed sets $E$ and $F$ such that $\operatorname{Ind} B \geqslant 1$

for every closed set $B$ which separates $E$ and $F$. Thus $\operatorname{Ind} Z \geqslant 2$. This completes the proof that $\operatorname{Ind} Z = 2.\;\square$

Summarizing, there exists a $T_4$-space $Z$ such that

$$\operatorname{ind} Z = 0, \quad \dim Z = 1, \quad \operatorname{Ind} Z = 2.$$

## Notes

The definitions of local dimension and local inductive dimension were made by Dowker [1955]. He established the facts about these dimension functions given in this chapter. Proposition 3.4 is an extension by Zarelua [1963] of a result due to Dowker. The finite sum theorem for local inductive dimension was obtained by Pears [1970].

The question of whether the subset theorem holds for completely normal Hausdorff spaces was posed by Čech [1948]. Filippov announced at a recent conference that he has a negative answer to this question. The reduction of the problem to a question about local dimension is due to Dowker [1955].

The space described in Remark 4.6 in the case $n = 1$ is Dowker's 'Example M'. The generalizations described in Proposition 4.5 are due to Smirnov [1958]. Taking $P$ to be the Hilbert cube, say, we find that there exists a $T_4$-space $M$ such that $\operatorname{ind} M = 0$ and $\dim M = \infty$. By Proposition 4.1.3, a $T_4$-space with small inductive dimension equal to zero can be embedded in a bicompact space which is zero-dimensional in all senses. Thus there exist bicompact spaces which are zero-dimensional in all senses and have normal subspaces of infinite covering dimension. Example 4.7 was given by Dowker. Example 4.11 was constructed by Nagami [1966]. In addition to the methods of Dowker and Smirnov this construction employs an idea due to Vopěnka [1958] which will be studied in Chapter 8.

# 6

# IMAGES OF ZERO-DIMENSIONAL SPACES

## 1 Canonical coverings and dimension

In this section two new dimension functions $\delta$ and $\Delta$ are introduced. The definitions are in terms of families of locally finite closed coverings of a special type.

**1.1 Definition.** *A locally finite closed covering $\{F_\lambda\}_{\lambda \in \Lambda}$ of a space $X$ is called a* canonical covering *if $F_\lambda = \bar{G}_\lambda$ for each $\lambda$ in $\Lambda$, where $\{G_\lambda\}_{\lambda \in \Lambda}$ is a disjoint family of open sets.*

**1.2 Proposition.** *Each locally finite closed covering of a space has a refinement which is a canonical covering.*

*Proof.* Let $\{A_\lambda\}_{\lambda \in \Lambda}$ be a locally finite closed covering of a space $X$ and let $\Lambda$ be well-ordered. We show first that for each $\lambda$ there exists an open set $V_\lambda$ such that $\bar{V}_\lambda \subset A_\lambda$ and $\{\bar{V}_\lambda\}_{\lambda \in \Lambda}$ is a covering of $X$. The proof is by transfinite induction. Suppose that $\mu \in \Lambda$ and that for each $\lambda < \mu$ there exists an open set $V_\lambda$ such that $\bar{V}_\lambda \subset A_\lambda$ and

$$\bigcup_{\lambda \leqslant \nu} \bar{V}_\lambda \cup \bigcup_{\lambda > \nu} A_\lambda = X$$

for each $\nu < \mu$. If $x \in X$, the set $\{\lambda \in \Lambda \mid x \in A_\lambda\}$ is finite. Let $\xi$ be its greatest element. If $\xi \geqslant \mu$ then $x \in A_\xi \subset \bigcup_{\lambda \geqslant \mu} A_\lambda$. If $\xi < \mu$, then since $\bigcup_{\lambda \leqslant \xi} \bar{V}_\lambda \cup \bigcup_{\lambda > \xi} A_\lambda = X$, it follows that $x \in \bigcup_{\lambda \leqslant \xi} \bar{V}_\lambda \subset \bigcup_{\lambda < \mu} \bar{V}_\lambda$. Thus

$$\bigcup_{\lambda < \mu} \bar{V}_\lambda \cup \bigcup_{\lambda \geqslant \mu} A_\lambda = X.$$

Now let

$$V_\mu = X \setminus (\bigcup_{\lambda < \mu} \bar{V}_\lambda \cup \bigcup_{\lambda > \mu} A_\lambda).$$

Since $\{\bar{V}_\lambda\}_{\lambda < \mu}$ and $\{A_\lambda\}_{\lambda > \mu}$ are locally finite families, it follows that $V_\mu$ is an open set. Furthermore $\bar{V}_\mu \subset A_\mu$ and $\bigcup_{\lambda \leqslant \mu} \bar{V}_\lambda \cup \bigcup_{\lambda > \mu} A_\lambda = X$. The existence of the required covering follows by transfinite induction. Now if $\lambda \in \Lambda$ let

$$G_\lambda = V_\lambda \setminus \bigcup_{\mu < \lambda} \bar{V}_\mu.$$

Then $G_\lambda$ is an open set for each $\lambda$. If $\lambda \neq \mu$, $G_\lambda$ and $G_\mu$ are disjoint, and if $F_\lambda = \bar{G}_\lambda$, then $F_\lambda \subset \bar{V}_\lambda \subset A_\lambda$ so that the family $\{F_\lambda\}_{\lambda \in \Lambda}$ is locally finite. If it is a covering of $X$, it will be the required canonical covering. If $x \in X$, there exists $\mu$ such that $x \in \overline{V}_\mu$ and $x \notin \overline{V}_\lambda$ if $\lambda < \mu$. The set $\bigcup_{\lambda < \mu} \overline{V}_\lambda$ is closed and $x \notin \bigcup_{\lambda < \mu} \overline{V}_\lambda$. Thus if $W$ is any neighbourhood of $x$, then $W \backslash \bigcup_{\lambda < \mu} \overline{V}_\lambda$ is a neighbourhood of $x$ and therefore has a non-empty intersection with $V_\mu$. Thus $W$ intersects $V_\mu \backslash \bigcup_{\lambda < \mu} \overline{V}_\lambda = G_\mu$ so that $x \in \overline{G}_\mu = F_\mu$. Hence $\{F_\lambda\}_{\lambda \in \Lambda}$ is a covering of $X$. $\square$

**1.3  Remark.** It follows from the above result that a topological space is normal if and only if every locally finite open covering has a refinement which is a canonical covering and that a topological space is paracompact and normal if and only if every open covering has a refinement which is a canonical covering.

If one canonical covering is a refinement of another then we have more information than is usually the case when one covering is a refinement of another.

**1.4  Proposition.** *Let $\{E_\rho\}_{\rho \in R}$ and $\{F_\lambda\}_{\lambda \in \Lambda}$ be canonical coverings of a space $X$ such that $\{E_\rho\}$ is a refinement of $\{F_\lambda\}$. Then for each $\rho$ in $R$, there is a unique element $\tau(\rho)$ of $\Lambda$ such that $E_\rho \subset F_{\tau(\rho)}$, and if $\lambda \in \Lambda$ then*

$$F_\lambda = \bigcup_{\tau(\rho) = \lambda} E_\rho.$$

*Proof.* For each $\rho$ in $R$ let $E_\rho = \overline{H}_\rho$ and for each $\lambda$ in $\Lambda$ let $F_\lambda = \overline{G}_\lambda$, where $\{H_\rho\}_{\rho \in R}$, $\{G_\lambda\}_{\lambda \in \Lambda}$ are families of disjoint open sets. For each $\rho$ in $R$, there exists $\tau(\rho)$ such that $E_\rho \subset F_{\tau(\rho)}$. If $\lambda \neq \tau(\rho)$, $E_\rho$ is not contained in $F_\lambda$, for $G_\lambda \cap F_{\tau(\rho)} = \varnothing$ so that $G_\lambda \cap H_\rho = \varnothing$. It follows that $H_\rho \cap F_\lambda = \varnothing$ so that $E_\rho$ is not contained in $F_\lambda$. Finally let $\lambda \in \Lambda$ and $x \in G_\lambda$. There exists $\rho$ in $R$ such that $x \in E_\rho$. Since $G_\lambda \cap H_\rho \neq \varnothing$, it follows that $\tau(\rho) = \lambda$ and

$$G_\lambda \subset \bigcup_{\tau(\rho) = \lambda} E_\rho \subset F_\lambda.$$

Since $\{E_\rho\}$ is locally finite, it follows that $\bigcup_{\tau(\rho) = \lambda} E_\rho = F_\lambda$. $\square$

**1.5  Definition.** *A family $\{\mathscr{F}_\alpha\}_{\alpha \in \Omega}$ of canonical coverings of a space $X$ is said to be* directed *if $\Omega$ is directed by the following relation: if $\alpha, \beta \in \Omega$, then $\alpha \leqslant \beta$ if $\mathscr{F}_\beta$ is a refinement of $\mathscr{F}_\alpha$.*

Let us call a canonical covering of a space $X$ which is a subset of the set of closed sets of $X$ and is indexed by itself a *proper canonical covering* of $X$. The proper canonical coverings of $X$ form a set and the next proposition shows that, regarded as a family indexed by itself, the set of proper canonical coverings is directed.

**1.6 Proposition.** *Any two canonical coverings of a space have a common refinement which is a canonical covering.*

*Proof.* Let $\{\bar{G}_\lambda\}, \{\bar{H}_\rho\}$ be canonical coverings of a space $X$, where $\{G_\lambda\}, \{H_\rho\}$ are disjoint families of open sets. Then the collection of all open sets $G_\lambda \cap H_\rho$ is disjoint and $\{(G_\lambda \cap H_\rho)^-\}$ is locally finite. Since $\{\bar{H}_\rho\}$ is a covering of $X$, it follows that

$$G_\lambda = \bigcup_\rho G_\lambda \cap \bar{H}_\rho \subset \bigcup_\rho (G_\lambda \cap H_\rho)^-$$

so that $\bar{G}_\lambda \subset \bigcup_\rho (G_\lambda \cap H_\rho)^-$. But $\{\bar{G}_\lambda\}$ is a covering of $X$ so that

$$X = \bigcup_{\lambda, \rho} (G_\lambda \cap H_\rho)^-.$$

Thus $\{(G_\lambda \cap H_\rho)^-\}_{\lambda, \rho}$ is a canonical covering of $X$ which is a refinement of the given coverings. $\square$

**1.7 Definition.** *A family $\{\mathscr{F}_\alpha\}_{\alpha \in \Omega}$ of coverings of a space $X$ is said to be* regular *if for each point $x$ of $X$ and each open set $U$ such that $x \in U$ there exists $\alpha$ in $\Omega$ such that $\mathrm{St}(x, \mathscr{F}_\alpha) \subset U$. The family $\{\mathscr{F}_\alpha\}_{\alpha \in \Omega}$ of coverings of $X$ is said to be* fine *if each open covering of $X$ has a refinement $\mathscr{F}_\alpha$ for some $\alpha$ in $\Omega$.*

**1.8 Remark.** If there exists a regular family of locally finite closed coverings of a space $X$, then $X$ is a regular space. For let $\{\mathscr{F}_\alpha\}_{\alpha \in \Omega}$ be a regular family of locally finite closed coverings of $X$. If $U$ is an open set of $X$ and $x$ is a point of $U$, then there exists $\alpha$ in $\Omega$ such that $\mathrm{St}(x, \mathscr{F}_\alpha) \subset U$. If

$$V = X \setminus \bigcup \{F \in \mathscr{F}_\alpha \mid x \notin F\},$$

then $V$ is an open neighbourhood of $x$ such that $V \subset \mathrm{St}(x, \mathscr{F}_\alpha) \subset U$. Since $\mathrm{St}(x, \mathscr{F}_\alpha)$ is a closed set we see that $x \in V \subset \bar{V} \subset U$. Thus $X$ is regular. It is clear that if there exists a fine family of locally finite closed coverings of a space $X$, then $X$ is a paracompact normal space.

**1.9 Proposition.** *If $X$ is a regular space or a $T_1$-space then each fine family of coverings of $X$ is regular.*

*Proof.* Let $X$ be a space in which each open set is the union of closed sets and let $\{\mathscr{F}_\alpha\}_{\alpha \in \Omega}$ be a fine family of coverings of $X$. Let $U$ be open in $X$ and let $x$ be a point of $U$. Then there exists a closed set $F$ of $X$ such that $x \in F \subset U$. The open covering $\{U, X \setminus F\}$ of $X$ has a refinement $\mathscr{F}_\alpha$ for some $\alpha$ in $\Omega$. And $\mathrm{St}(x, \mathscr{F}_\alpha) \subset U$. $\square$

The new dimension functions can now be introduced.

**1.10   Definition.** *For a topological space* $X$ *we define*

$$\delta(X) = \Delta(X) = -1$$

*if and only if* $X$ *is empty. For a non-empty topological space we define* $\delta(X)$ *to be the least integer* $n$ *for which there exists a regular directed family* $\{\mathscr{F}_\alpha\}_{\alpha \in \Omega}$ *of canonical coverings of* $X$ *such that the order of* $\mathscr{F}_\alpha$ *does not exceed* $n$ *for each* $\alpha$ *in* $\Omega$, *and* $\delta(X) = \infty$ *if there exists no such integer. We define* $\Delta(X)$ *to be the least integer* $n$ *for which there exists a fine directed family* $\{\mathscr{F}_\alpha\}_{\alpha \in \Omega}$ *of canonical coverings of* $X$ *such that the order of* $\mathscr{F}_\alpha$ *does not exceed* $n$ *for each* $\alpha$ *in* $\Omega$, *and* $\Delta(X) = \infty$ *if there exists no such integer.*

**1.11   Proposition.** *If* $X$ *is a regular space or a* $T_1$-*space, then*

$$\delta(X) \leqslant \Delta(X).$$

*Proof.* This follows immediately from Proposition 1.9. $\square$

**1.12   Proposition.** *If* $X$ *is a topological space, then*

$$\operatorname{ind} X = 0 \quad \text{if and only if} \quad \delta(X) = 0.$$

*Proof.* If $\operatorname{ind} X = 0$, then $X$ has a base for its topology which consists of open-and-closed sets. It follows that the set of all finite disjoint open coverings, which is clearly directed when considered as a family indexed by itself, is also regular. A finite disjoint open covering is a canonical covering. Thus $\delta(X) = 0$.

Conversely let $\delta(X) = 0$. Then there exists a regular family of disjoint locally finite closed coverings. But a disjoint locally finite closed covering consists of open-and-closed sets. It follows that $X$ has a base for its topology which consists of open-and-closed sets so that $\operatorname{ind} X = 0$. $\square$

**1.13   Definition.** *A topological space* $X$ *with a base of open-and-closed sets such that each open covering of* $X$ *has a disjoint open refinement is called a* perfectly zero-dimensional *space.*

**1.14   Proposition.** *The following statements about a topological space* $X$ *are equivalent*:
   (a) $X$ *is a perfectly zero-dimensional space*;
   (b) $X$ *is a strongly paracompact regular space and*

$$\operatorname{ind} X = \dim X = \operatorname{Ind} X = 0;$$

(c) $X$ is a paracompact regular space such that $\dim X = 0$;

(d) $X$ is a regular space such that $\Delta(X) = 0$.

*Proof.* A space $X$ has a base of open-and-closed sets if and only if $\operatorname{ind} X = 0$ and such a space is regular. Since a disjoint open covering is star-finite, it follows that $(a) \Rightarrow (b)$. It is clear that $(b) \Rightarrow (c)$. Since a regular space $X$ such that $\dim X = 0$ satisfies $\operatorname{ind} X = 0$, it is clear that $(c) \Rightarrow (a)$. If $X$ is a perfectly zero-dimensional space, then the set of disjoint open coverings of $X$ is a directed fine set of canonical coverings so that $\Delta(X) = 0$. Conversely if $X$ is a regular space and $\Delta(X) = 0$, then $\operatorname{ind} X = 0$ by Proposition 1.12, and since a canonical covering of order zero is a disjoint open covering, it follows that $X$ is a perfectly zero-dimensional space. Thus $(a) \Leftrightarrow (d)$. $\square$

**1.15**   *Remark.* In §3 the dimension functions $\delta$ and $\Delta$ will be characterized. These characterizations will enable us to see that $\operatorname{ind} X \leqslant \delta(X)$ for all topological spaces $X$ and that $\operatorname{Ind} X \leqslant \Delta(X)$ for all regular spaces $X$. For a normal regular space $X$ therefore

$$\dim X \leqslant \operatorname{Ind} X \leqslant \Delta(X),$$

so that there is some interest in finding classes of spaces $X$ for which $\dim X = \Delta(X)$. First we make a digression to study irreducible mappings which will be needed to establish the characterizations.


## 2   Irreducible mappings

In this section we introduce a class of continuous surjections which take canonical coverings of the domain into canonical coverings of the range. We have already seen in Chapter 2 that perfect mappings preserve local finiteness.

**2.1**   *Definition.* A continuous surjection $f: X \to Y$ is called an irreducible mapping if $f(A) \neq Y$ for each proper closed set $A$ of $X$.

If $f: X \to Y$ is a surjection and $B \subset X$, let us define

$$f^*(B) = \{y \in Y \mid f^{-1}(y) \subset B\}.$$

**2.2**   *Lemma.* A continuous surjection $f: X \to Y$ is an irreducible mapping if and only if $f^*(U)$ is non-empty for each non-empty open set $U$ of $X$ and is a closed mapping if and only if $f^*(U)$ is open for each non-empty open set $U$ of $X$.

*Proof.* Both assertions follow at once on observing that

$$Y \backslash f(A) = f^*(X \backslash A)$$

for every subset $A$ of $X$. $\square$

**2.3 Lemma.** *If* $f: X \to Y$ *is a closed irreducible mapping, then for each open set* $U$ *of* $X$

$$f(\overline{U}) = \left(f^*(U)\right)^-.$$

*Proof.* Clearly $f^*(U) \subset f(U) \subset f(\overline{U})$ so that $\left(f^*(U)\right)^- \subset f(\overline{U})$ since $f$ is a closed mapping. To establish the reverse inclusion, consider the set $W = f^{-1}f^*(U)$. Then $W$ is an open set and $W \subset U$. We shall prove that $\overline{W} = \overline{U}$. Let $x$ be a point of $\overline{U}$ and let $V$ be an open set such that $x \in V$. Then $U \cap V$ is a non-empty open set of $X$ so that since $f$ is an irreducible mapping there exists a point $y_0$ of $Y$ such that

$$f^{-1}(y_0) \subset U \cap V.$$

Since $f^{-1}(y_0) \subset U$, it follows that $f^{-1}(y_0) \subset W$ so that $V \cap W \neq \varnothing$. Thus $x \in \overline{W}$ so that $\overline{W} = \overline{U}$. Clearly $f(W) = f^*(U)$. Thus

$$f(\overline{U}) = f(\overline{W}) = \left(f(W)\right)^- = \left(f^*(U)\right)^-. \square$$

**2.4 Proposition.** *If* $f: X \to Y$ *is a perfect irreducible mapping and* $\{F_\lambda\}_{\lambda \in \Lambda}$ *is a canonical covering of* $X$, *then* $\{f(F_\lambda)\}_{\lambda \in \Lambda}$ *is a canonical covering of* $Y$.

*Proof.* It follows from Lemma 2.5.6 that $\{f(F_\lambda)\}_{\lambda \in \Lambda}$ is a locally finite closed covering of $Y$. If $\lambda \in \Lambda$, let $G_\lambda = X \backslash \bigcup_{\mu \neq \lambda} F_\mu$ so that $\overline{G}_\lambda = F_\lambda$. By Lemma 2.3, $f(F_\lambda) = f(\overline{G}_\lambda) = \left(f^*(G_\lambda)\right)^-$ and $f^*(G_\lambda)$ is an open set by Lemma 2.2. Since the family $\{G_\lambda\}_{\lambda \in \Lambda}$ is disjoint, the family

$$\{f^*(G_\lambda)\}_{\lambda \in \Lambda}$$

is disjoint. $\square$

There is a plentiful supply of perfect irreducible mappings.

**2.5 Proposition.** *If* $f: X \to Y$ *is a perfect mapping, then there is a closed set* $A$ *of* $X$ *such that* $f | A : A \to Y$ *is a perfect irreducible mapping.*

*Proof.* Let $\mathscr{U}$ be the set of open subsets $U$ of $X$ such that $f^*(U)$ is empty. Consider $\mathscr{U}$ to be partially ordered by inclusion and let $\mathscr{V}$ be a chain in $\mathscr{U}$. Let $V$ be the union of the members of $\mathscr{V}$ and suppose that $f^*(V) \neq \varnothing$. If $y_0 \in f^*(V)$, then the compact set $f^{-1}(y_0)$ is contained in the union of the members of $\mathscr{V}$, and hence since $\mathscr{V}$ is a chain, $f^{-1}(y_0)$ is contained in some member of $\mathscr{V}$ which is absurd. Hence $V \in \mathscr{U}$. It follows from Zorn's lemma that $\mathscr{U}$ has a maximal element $U$, say. Let $A = X \backslash U$. Then $A$ is a closed set of $X$, and $f^{-1}(y) \cap A \neq \varnothing$ if $y \in Y$ so that $f | A : A \to Y$ is surjective. Thus

$$f | A : A \to Y$$

is a perfect mapping. Let $B$ be a proper closed subset of $A$. Then $X \backslash B$ is an open set of $X$ which properly contains $U$. Thus

$$f^*(X \backslash B) \neq \varnothing$$

so that $f(B) \neq Y$. Thus $f | A$ is a perfect irreducible mapping. $\square$

We conclude this section with two easily obtained properties of irreducible mappings which will be needed in Chapter 8.

**2.6  Proposition.** *If $f : X \to Y$ and $g : Y \to Z$ are continuous surjections such that $g \circ f : X \to Z$ is an irreducible mapping, then $f$ and $g$ are irreducible mappings.*

*Proof.* Let $U$ be a non-empty open set of $X$. Then $(g \circ f)^*(U)$ is non-empty. If $z \in (g \circ f)^*(U)$ and $y \in g^{-1}(z)$, then $f^{-1}(y) \subset f^{-1}g^{-1}(z) \subset U$ so that $y \in f^*(U)$. Thus $f^*(U)$ is non-empty and it follows that $f$ is an irreducible mapping. Next let $V$ be a non-empty open set of $Y$. Then $(g \circ f)^*\big(f^{-1}(V)\big)$ is non-empty. If $z \in (g \circ f)^*\big(f^{-1}(V)\big)$, then

$$f^{-1}g^{-1}(z) \subset f^{-1}(V)$$

so that $g^{-1}(z) \subset V$. Thus $g^*(V)$ is non-empty and it follows that $g$ is an irreducible mapping. $\square$

**2.7  Proposition.** *If $f_1 : X_1 \to Y_1$ and $f_2 : X_2 \to Y_2$ are irreducible mappings, then*

$$f_1 \times f_2 : X_1 \times X_2 \to Y_1 \times Y_2$$

*is an irreducible mapping.*

*Proof.* Let $U$ be a non-empty open set of $X_1 \times X_2$ and let $(x_1, x_2) \in U$. There exist open sets $V_1$ and $V_2$ of $X_1$ and $X_2$ respectively such that $(x_1, x_2) \in V_1 \times V_2 \subset U$. Since $f_1$ and $f_2$ are irreducible mappings, $f_1^*(V_1)$ and $f_2^*(V_2)$ are non-empty. If $y_1 \in f_1^*(V_1)$ and $y_2 \in f_2^*(V_2)$, then

$$(y_1, y_2) \in Y_1 \times Y_2$$

and

$$(f_1 \times f_2)^{-1}(y_1, y_2) = f_1^{-1}(y_1) \times f_2^{-1}(y_2) \subset V_1 \times V_2 \subset U.$$

Thus $(f_1 \times f_2)^*(U)$ is non-empty. It follows that $f_1 \times f_2$ is an irreducible mapping. $\square$

## 3  Images of zero-dimensional spaces

In this section we shall characterize the dimension functions $\delta$ and $\Delta$ in terms of images of zero-dimensional spaces. First the notion of a directed family of canonical coverings is generalized.

**3.1  Definition.** *Let $\{\mathscr{F}_\alpha\}_{\alpha \in \Omega}$ be a family of locally finite closed coverings of a space, where for each $\alpha$ in $\Omega$, $\mathscr{F}_\alpha$ is the indexed covering $\{F_\lambda^\alpha\}_{\lambda \in \Lambda(\alpha)}$. The family $\{\mathscr{F}_\alpha\}_{\alpha \in \Omega}$ is said to be strongly directed if $\Omega$ is a directed set and for each pair $\alpha, \beta$ in $\Omega$ such that $\alpha \leqslant \beta$ there exists a function $\rho_{\alpha\beta} \colon \Lambda(\beta) \to \Lambda(\alpha)$ such that $\{\Lambda(\alpha), \rho_{\alpha\beta}\}_{\alpha, \beta \in \Omega}$ is an inverse system of sets over $\Omega$, and furthermore for each $\lambda$ in $\Lambda(\alpha)$*

$$F_\lambda^\alpha = \cup\{F_\mu^\beta \,|\, \rho_{\alpha\beta}(\mu) = \lambda\}.$$

Let $\Phi = \{\mathscr{F}_\alpha\}_{\alpha \in \Omega}$ be a strongly directed family of locally finite closed coverings of a space $X$. We associate with $\Phi$ an inverse system $\{X_\alpha, \pi_{\alpha\beta}\}_{\alpha, \beta \in \Omega}$ of topological spaces, over the directed set $\Omega$. For each $\alpha$ in $\Omega$ let $X_\alpha = \cup_{\lambda \in \Lambda(\alpha)}\big(F_\lambda^\alpha \times (\lambda)\big)$, where $(\lambda)$ denotes the set with one element $\lambda$ (this notation being used here to avoid confusion with the indexed families $\{\lambda_\alpha\}$ which occur subsequently). Let the closed sets of $X_\alpha$ be the empty set and all subsets of the form $\cup_{\lambda \in \Lambda(\alpha)}\big(B_\lambda \times (\lambda)\big)$, where $B_\lambda$ is a closed set of $F_\lambda^\alpha$. Thus $X_\alpha$ is the disjoint topological sum of the subspaces $F_\lambda^\alpha$ of $X$. Let us define $\sigma_\alpha \colon X_\alpha \to X$ by $\sigma_\alpha(x, \lambda) = x$ if $(x, \lambda) \in X_\alpha$. Then clearly $\sigma_\alpha$ is a continuous closed surjection and $\sigma_\alpha^{-1}(x)$ is finite for each $x$ in $X$ since $\mathscr{F}_\alpha$ is locally finite. If $\alpha, \beta \in \Omega$ and $\alpha \leqslant \beta$ let $\pi_{\alpha\beta} \colon X_\beta \to X_\alpha$ be defined by putting $\pi_{\alpha\beta}(x, \mu) = \big(x, \rho_{\alpha\beta}(\mu)\big)$ if $(x, \mu) \in X_\beta$. Clearly $\pi_{\alpha\beta}$ is surjective. The function $\pi_{\alpha\beta}$ is continuous. For if $B$ is closed in $X_\alpha$, then $B = \cup_{\lambda \in \Lambda(\alpha)}\big(B_\lambda \times (\lambda)\big)$, where $B_\lambda$ is closed in $F_\lambda^\alpha$, so that

$$\pi_{\alpha\beta}^{-1}(B) = \bigcup_{\mu \in \Lambda(\beta)} \big((B_{\rho(\mu)} \cap F_\mu^\beta) \times (\mu)\big),$$

where $\rho = \rho_{\alpha\beta}$. The function $\pi_{\alpha\beta}$ is closed. For if $A$ is closed in $X_\beta$, then $A = \cup_{\mu \in \Lambda(\beta)}\big(A_\mu \times (\mu)\big)$, where $A_\mu$ is closed in $F_\mu^\beta$, so that

$$\pi_{\alpha\beta}(A) = \bigcup_{\lambda \in \Lambda(\alpha)} \big(B_\lambda \times (\lambda)\big),$$

where
$$B_\lambda = \bigcup \{A_\mu \mid \rho_{\alpha\beta}(\mu) = \lambda\}$$

if $\lambda \in \Lambda(\alpha)$. Finally if $(x, \lambda) \in X_\alpha$, then $\pi_{\alpha\beta}^{-1}(x, \lambda)$ consists of those elements $(x, \mu)$ of $X_\beta$ such that $x \in F_\mu^\beta$ and $\rho_{\alpha\beta}(\mu) = \lambda$. Thus $\pi_{\alpha\beta}^{-1}(x, \lambda)$ is a finite discrete space. Since $\{\Lambda(\alpha), \rho_{\alpha\beta}\}_{\alpha, \beta \in \Omega}$ is an inverse system of sets over $\Omega$, it is evident that $\{X_\alpha, \pi_{\alpha\beta}\}_{\alpha, \beta \in \Omega}$ is an inverse system of spaces over $\Omega$. Let $Z(\Phi)$ be the inverse limit of this inverse system and for each $\alpha$ in $\Omega$ let $\pi_\alpha : Z(\Phi) \to X_\alpha$ be the canonical mapping. It follows from Proposition 1.7.5 that for each $\alpha$ in $\Omega$, $\pi_\alpha$ is a continuous closed surjection such that $\pi_\alpha^{-1}(x, \lambda)$ is bicompact if $(x, \lambda) \in X_\alpha$. Evidently the space $Z(\Phi)$ can be taken to consist of elements $(x, \xi)$, where $\xi = \{\lambda(\alpha)\}_{\alpha \in \Omega}$ is an element of the inverse limit of the inverse system $\{\Lambda(\alpha), \rho_{\alpha\beta}\}_{\alpha, \beta \in \Omega}$ and $x \in \bigcap_{\alpha \in \Omega} F_{\lambda(\alpha)}^\alpha$. If $(x, \xi) \in Z(\Phi)$, where $\xi = \{\lambda(\alpha)\}_{\alpha \in \Omega}$, then $\pi_\alpha(x, \xi) = \big(x, \lambda(\alpha)\big)$. For each $\alpha$ in $\Omega$ and each $\lambda$ in $\Lambda(\alpha)$ let
$$V_{\alpha\lambda} = \pi_\alpha^{-1}\big(F_\lambda^\alpha \times (\lambda)\big).$$

If $\mathscr{V}_\alpha = \{V_{\alpha\lambda}\}_{\lambda \in \Lambda(\alpha)}$, then $\mathscr{V}_\alpha$ is a disjoint covering of $Z(\Phi)$ by open-and-closed sets. If $\alpha, \beta \in \Omega$ and $\alpha \leqslant \beta$, then $\mathscr{V}_\beta$ is a refinement of $\mathscr{V}_\alpha$ and each member of $\mathscr{V}_\alpha$ is the union of those members of $\mathscr{V}_\beta$ which it contains. Let us put $\mathscr{V}(\Phi) = \{V_{\alpha\lambda}\}_{\lambda \in \Lambda(\alpha), \alpha \in \Omega}$. Finally we define

$$f : Z(\Phi) \to X$$

by $f(x, \xi) = x$ if $(x, \xi) \in Z(\Phi)$. Since $f = \sigma_\alpha \circ \pi_\alpha$ for each $\alpha$ in $\Omega$, it is clear that $f$ is continuous and closed and since $\sigma_\alpha^{-1}(x)$ is finite if $x \in X$, we see that $f^{-1}(x)$ is bicompact if $x \in X$. Thus given a strongly directed family $\Phi$ of locally finite closed coverings of a space $X$, we have constructed a space $Z(\Phi)$ and a perfect mapping $f : Z(\Phi) \to X$. The properties of the space $Z(\Phi)$ and the mapping $f$ will be established in the next four lemmas. The notation introduced above will be used throughout this section.

**3.2   Lemma.** *If $\Phi$ is a regular strongly directed family of locally finite closed coverings of a topological space, then $\mathscr{V}(\Phi)$ is a base for the topology of $Z(\Phi)$.*

*Proof.* Let $z \in Z(\Phi)$ and let $U$ be an open set of $Z(\Phi)$ such that $z \in U$. Then for some $\alpha$ in $\Omega$, there exists $W_\alpha$ open in $X_\alpha$ such that

$$z \in \pi_\alpha^{-1}(W_\alpha) \subset U.$$

If $\pi_\alpha(z) = (x, \lambda)$, then there exists $W$ open in $X$ such that $x \in F_\lambda^\alpha \cap W$ and $(F_\lambda^\alpha \cap W) \times (\lambda) \subset W_\alpha$. Since $\Phi$ is regular, there exists $\beta$ such that

$\alpha \leqslant \beta$ and $\mathrm{St}\,(x, \mathscr{F}_\beta) \subset W$. If $\pi_\beta(z) = (x, \mu)$, then $x \in F_\mu^\beta \subset F_\lambda^\alpha \cap W$ so that $\pi_{\alpha\beta}\big(F_\mu^\beta \times (\mu)\big) \subset W_\alpha$. Hence

$$z \in V_{\beta\mu} \subset \pi_\alpha^{-1}(W_\alpha) \subset U.$$

Thus $\mathscr{V}(\Phi)$ is a base for the topology of $Z(\Phi)$. $\square$

**3.3 Corollary.** *If $\Phi$ is a regular strongly directed family of locally finite closed coverings of a topological space, then*

$$\mathrm{ind}\, Z(\Phi) = 0.$$

*Proof.* By Lemma 3.2, $Z(\Phi)$ has a base for its topology which consists of open-and-closed sets. $\square$

**3.4 Lemma.** *If $\Phi$ is a regular strongly directed family of locally finite closed coverings of a $T_1$-space, then $Z(\Phi)$ is a Hausdorff space.*

*Proof.* If $(x, \xi) \in Z(\Phi)$ and $\xi = \{\lambda(\alpha)\}_{\alpha\in\Omega}$, then $x \in \bigcap_{\alpha\in\Omega} F_{\lambda(\alpha)}^\alpha$. In fact $\bigcap_{\alpha\in\Omega} F_{\lambda(\alpha)}^\alpha = \{x\}$. For if $y \in X$ and $y \neq x$, then since $\Phi$ is regular and $X$ is a $T_1$-space, there exists $\alpha$ in $\Omega$ such that $y \notin \mathrm{St}\,(x, \mathscr{F}_\alpha)$ and hence $y \notin F_{\lambda(\alpha)}^\alpha$. Now suppose that $z = (x, \xi)$ and $z' = (x', \xi')$ are distinct points of $Z(\Phi)$, where $\xi = \{\lambda(\alpha)\}_{\alpha\in\Omega}$ and $\xi' = \{\lambda'(\alpha)\}_{\alpha\in\Omega}$. It follows from the first remark that $\xi \neq \xi'$ so that $\lambda(\alpha) \neq \lambda'(\alpha)$ for some $\alpha$ in $\Omega$. Then $V_{\alpha\lambda(\alpha)}$ and $V_{\alpha\lambda'(\alpha)}$ are disjoint open neighbourhoods of $z$ and $z'$ respectively. $\square$

**3.5 Lemma.** *If $\Phi$ is a fine strongly directed family of locally finite closed coverings of a regular space, then $Z(\Phi)$ is a perfectly zero-dimensional space.*

*Proof.* If $\Phi$ is a fine family of coverings of a regular space $X$, it follows from Proposition 1.9 that $\Phi$ is a regular family and hence by Lemma 3.2, $Z(\Phi)$ has a base for its topology which consists of open-and-closed sets. It remains to show that each open covering of $Z(\Phi)$ has a disjoint open refinement. Let $\mathscr{U}$ be an open covering of $Z(\Phi)$. If $x \in X$, then $f^{-1}(x)$ is a bicompact subset of $Z(\Phi)$. Hence there exists a finite subset $K$ of $\Omega$ and a member $W_\alpha$ of $\mathscr{V}_\alpha$ for each $\alpha$ in $K$ such that $f^{-1}(x) \subset \bigcup_{\alpha\in K} W_\alpha$ and each set $W_\alpha$ is contained in some member of $\mathscr{U}$. Choose $\beta$ in $\Omega$ such that $\alpha \leqslant \beta$ if $\alpha \in K$. Since each set $W_\alpha$ is the union of members of $\mathscr{V}_\beta$, it follows that there exists a set $\mathscr{V}_x$ of disjoint open sets of $Z(\Phi)$, each of which is contained in some member of $\mathscr{U}$, such that the union $V_x$ of $\mathscr{V}_x$ contains $f^{-1}(x)$. Since $f$ is a closed mapping, $V_x$ is an open set of $Z(\Phi)$ and $f^{-1}(x) \subset V_x$, there exists an open set

$W_x$ of $X$ such that $x \in W_x$ and $f^{-1}(W_x) \subset V_x$. Since the family $\Phi$ is fine, the open covering $\{W_x\}_{x \in X}$ of $X$ has a refinement $\mathscr{F}_\gamma$ for some $\gamma$ in $\Omega$. If $\lambda \in \Lambda(\gamma)$, choose $x(\lambda)$ in $X$ such that $F_\lambda^\gamma \subset W_{x(\lambda)}$. Then

$$V_{\gamma\lambda} = \cup \{V_{\gamma\lambda} \cap V \mid V \in \mathscr{V}_{x(\lambda)}\}.$$

Thus $\{V_{\gamma\lambda} \cap V \mid V \in \mathscr{V}_{x(\lambda)}, \lambda \in \Lambda(\gamma)\}$ is a disjoint open covering of $Z(\Phi)$ which is a refinement of $\mathscr{U}$. Hence $Z(\Phi)$ is a perfectly zero-dimensional space. $\square$

The final lemma requires a preliminary remark. If $z_0$ and $z_1$ are distinct points of $Z(\Phi)$, then there exists $\alpha$ in $\Omega$ such that $\pi_\alpha(z_0) \neq \pi_\alpha(z_1)$. An easy induction shows that if $B$ is a finite subset of $Z(\Phi)$, then there exists $\beta$ in $\Omega$ such that $\pi_\beta|B$ is injective.

**3.6    Lemma.** *If $\Phi = \{\mathscr{F}_\alpha\}_{\alpha \in \Omega}$ is a strongly directed family of locally finite closed coverings of a space $X$ such that for each $\alpha$ in $\Omega$, the order of $\mathscr{F}_\alpha$ does not exceed $n$, then*

$$f: Z(\Phi) \to X$$

*satisfies the condition that $f^{-1}(x)$ contains at most $n+1$ points if $x \in X$.*

*Proof.* Suppose there exists a point $x$ of $X$ such that $f^{-1}(x)$ contains $n+2$ distinct points $z_0, \ldots, z_{n+1}$. Let $B = \{z_0, \ldots, z_{n+1}\}$. As pointed out above there exists $\beta$ in $\Omega$ such that $\pi_\beta|B$ is injective. Since

$$\pi_\beta(B) \subset \sigma_\beta^{-1}(x)$$

it follows that $\sigma_\beta^{-1}(x)$ contains $n+2$ distinct points. But this is absurd for $\sigma_\beta^{-1}(x)$ contains at most $n+1$ points since the order of $\mathscr{F}_\beta$ does not exceed $n$. $\square$

We can now characterize the dimension functions $\delta$ and $\Delta$.

**3.7    Proposition.** *The following statements about a topological space $X$ are equivalent:*

*(a) $\delta(X) \leqslant n$;*

*(b) there exists a regular strongly directed family $\{\mathscr{F}_\alpha\}_{\alpha \in \Omega}$ of locally finite closed coverings of $X$ such that the order of $\mathscr{F}_\alpha$ does not exceed $n$ for each $\alpha$ in $\Omega$;*

*(c) there exists a space $Z$ such that $\operatorname{ind} Z = 0$, and a continuous closed surjection $f: Z \to X$ such that $f^{-1}(x)$ contains at most $n+1$ points if $x \in X$.*

*Proof.* (a) $\Rightarrow$ (b). If $\delta(X) \leqslant n$, then $X$ has a regular directed family $\Phi = \{\mathscr{F}_\alpha\}_{\alpha \in \Omega}$ of canonical coverings such that the order of $\mathscr{F}_\alpha$ does not exceed $n$ for each $\alpha$ in $\Omega$. For each $\alpha$ in $\Omega$ let $\mathscr{F}_\alpha = \{F_\lambda^\alpha\}_{\lambda \in \Lambda(\alpha)}$, where $F_\lambda^\alpha$ is non-empty if $\lambda \in \Lambda(\alpha)$. If $\alpha, \beta \in \Omega$ and $\mathscr{F}_\beta$ is a refinement of $\mathscr{F}_\alpha$, then by Proposition 1.4 we can define a function

$$\rho_{\alpha\beta}\colon \Lambda(\beta) \to \Lambda(\alpha)$$

by taking for each $\mu$ in $\Lambda(\beta)$, $\rho_{\alpha\beta}(\mu)$ to be the unique element $\lambda$ of $\Lambda(\alpha)$ such that $F_\mu^\beta \subset F_\lambda^\alpha$. It also follows from Proposition 1.4 that for each $\lambda$ in $\Lambda(\alpha)$ we have

$$F_\lambda^\alpha = \bigcup\{F_\mu^\beta \,|\, \rho_{\alpha\beta}(\mu) = \lambda\}.$$

If additionally, $\mathscr{F}_\gamma$ is a refinement of $\mathscr{F}_\beta$, then by uniqueness we have $\rho_{\alpha\gamma} = \rho_{\alpha\beta} \circ \rho_{\beta\gamma}$. Thus the regular directed family $\Phi$ of canonical coverings can be given the structure of a strongly directed family of coverings and we see that (b) holds.

(b) $\Rightarrow$ (c). Let $\Phi = \{\mathscr{F}_\alpha\}_{\alpha \in \Omega}$ be a regular strongly directed family of locally finite closed coverings such that the order of $\mathscr{F}_\alpha$ does not exceed $n$ for each $\alpha$ in $\Omega$. Taking $Z = Z(\Phi)$, then by Corollary 3.3 we have a space $Z$ such that $\operatorname{ind} Z = 0$ and by Lemma 3.6, we have a continuous closed surjection $f\colon Z \to X$ such that $f^{-1}(x)$ contains at most $n+1$ points if $x \in X$. Thus (c) holds.

(c) $\Rightarrow$ (a). Let $f\colon Z \to X$ be a continuous closed surjection, where $\operatorname{ind} Z = 0$ and $f^{-1}(x)$ contains at most $n+1$ points if $x \in X$. By Proposition 2.5 we can suppose that $f$ is irreducible. Let $\Omega$ be the set of all disjoint open coverings of $Z$. Since a disjoint open covering is locally finite and consists of open-and-closed sets, $\Omega$ is a set of canonical coverings of $Z$, which is clearly directed. For each $\alpha$ in $\Omega$ let

$$\mathscr{F}_\alpha = \{f(F) \,|\, F \in \alpha\}.$$

Since $f$ is a perfect irreducible mapping, $\mathscr{F}_\alpha$ is a canonical covering of $X$ by Proposition 2.4. Thus $\{\mathscr{F}_\alpha\}_{\alpha \in \Omega}$ is a directed family of canonical coverings of $X$ and clearly for each $\alpha$ in $\Omega$ the order of $\mathscr{F}_\alpha$ does not exceed $n$. If it is shown that this family is regular, it will follow that $\delta(X) \leqslant n$. If $U$ is open in $X$ and $x \in U$, then $f^{-1}(x) \subset f^{-1}(U)$. Since $f^{-1}(x)$ is finite and $\operatorname{ind} Z = 0$, there exists an open-and-closed set $W$ such that $f^{-1}(x) \subset W \subset f^{-1}(U)$. Let $\alpha = \{W, Z\backslash W\}$ so that

$$\mathscr{F}_\alpha = \{f(W), f(Z\backslash W)\}.$$

Since $f^{-1}(x) \subset W$, it follows that $x \notin f(Z\backslash W)$, so that

$$\operatorname{St}(x, \mathscr{F}_\alpha) = f(W) \subset U.\,\square$$

**3.8    Proposition.** *The following statements about a regular space X are equivalent:*

(a) $\Delta(X) \leqslant n$;

(b) *there exists a fine strongly directed family* $\{\mathscr{F}_\alpha\}_{\alpha \in \Omega}$ *of locally finite closed coverings of X such that the order of* $\mathscr{F}_\alpha$ *does not exceed n for each* $\alpha$ *in* $\Omega$;

(c) *there exists a perfectly zero-dimensional space Z and a continuous closed surjection* $f : Z \to X$ *such that* $f^{-1}(x)$ *contains at most* $n+1$ *points if* $x \in X$.

*Proof.* (a) $\Rightarrow$ (b). If $\Delta(X) \leqslant n$, then there exists a fine directed family $\{\mathscr{F}_\alpha\}_{\alpha \in \Omega}$ of canonical coverings of $X$ such that the order of $\mathscr{F}_\alpha$ does not exceed $n$ for each $\alpha$ in $\Omega$. Therefore (b) holds since a directed family of canonical coverings can be given the structure of a strongly directed family.

(b) $\Rightarrow$ (c). Let $\Phi = \{\mathscr{F}_\alpha\}_{\alpha \in \Omega}$ be a fine strongly directed family of locally finite closed coverings of the regular space $X$ such that the order of $\mathscr{F}_\alpha$ does not exceed $n$ for each $\alpha$ in $\Omega$. If $Z = Z(\Phi)$, then $Z$ is a perfectly zero-dimensional space by Lemma 3.5, and by Lemma 3.6 there exists a continuous closed surjection $f : Z \to X$ such that $f^{-1}(x)$ contains at most $n+1$ points if $x \in X$. Thus (c) holds.

(c) $\Rightarrow$ (a). Let $f : Z \to X$ be a continuous closed surjection, where $Z$ is a perfectly zero-dimensional space and $f^{-1}(x)$ contains at most $n+1$ points if $x \in X$. By Proposition 2.5 we can suppose that $f$ is irreducible. As in the corresponding part of the proof of Proposition 3.7 we have a directed family $\Phi = \{\mathscr{F}_\alpha\}_{\alpha \in \Omega}$ of canonical coverings of $X$ such that each $\mathscr{F}_\alpha$ has order not exceeding $n$, where $\Omega$ is the set of disjoint open coverings of $X$ and $\mathscr{F}_\alpha = \{f(F) \mid F \in \alpha\}$ if $\alpha \in \Omega$. It remains to show that the family $\Phi$ is fine. But if $\mathscr{U}$ is an open covering of $X$, then $\mathscr{V} = \{f^{-1}(U) \mid U \in \mathscr{U}\}$ is an open covering of $Z$. Hence there exists $\alpha$ in $\Omega$ such that $\alpha$ is a refinement of $\mathscr{V}$ and it is clear that $\mathscr{F}_\alpha$ is a refinement of $\mathscr{U}$. $\square$

We shall exploit the above characterizations to relate the dimension functions $\delta$ and $\Delta$ to the small and large inductive dimension functions. First we study continuous closed surjections $f : X \to Y$ such that $f^{-1}(y)$ is finite for all $y$ in $Y$. Lemma 3.9 will be used in Chapter 9 also.

**3.9    Lemma.** *Let X and Y be topological spaces and let* $f : X \to Y$ *be a continuous closed surjection such that* $f^{-1}(y)$ *contains at most* $n+1$

*points if $y \in Y$. If $H$ is a non-empty open set of $X$ and $B$ is a closed set of $Y$ such that $B \subset f(H) \backslash f^*(H)$, then the function*

$$g : (X \backslash H) \cap f^{-1}(B) \to B$$

*given by restriction of $f$ is a continuous closed surjection such that $g^{-1}(y)$ contains at most $n$ points if $y \in B$.*

*Proof.* If $y \in B$, then $y \notin f^*(H)$ so that there exists $x$ in $X \backslash H$ such that $f(x) = y$. Thus $g$ is surjective. The mapping $f^{-1}(B) \to B$ given by restriction of $f$ is closed, so that $g$ is closed. If $y \in B$, then

$$g^{-1}(y) = f^{-1}(y) \backslash H.$$

But since $y \in f(H)$, the set $f^{-1}(y) \cap H$ is non-empty and hence $g^{-1}(y)$ contains at most $n$ points. $\square$

**3.10  Proposition.** *If $X$ is a topological space such that $\operatorname{ind} X = 0$ and $f : X \to Y$ is a continuous closed surjection such that $f^{-1}(y)$ contains at most $n + 1$ points if $y \in Y$, then $Y$ is a regular space such that*

$$\operatorname{ind} Y \leqslant n.$$

*Proof.* Since $X$ is a regular space and $f$ is a perfect mapping, it follows from Proposition 2.5.5 that $Y$ is a regular space. If $n = 0$ then $f$ is a homeomorphism, so that $\operatorname{ind} Y = 0$ and the proposition holds. Suppose that the proposition holds for surjections such that the inverse image of each point contains at most $n$ points. Let $y$ be a point of $Y$ and let $G$ be an open set of $Y$ such that $y \in G$. If $x \in f^{-1}(y)$, then there exists an open-and-closed set $H_x$ such that $x \in H_x \subset f^{-1}(G)$. Therefore since $f^{-1}(y)$ is finite there exists an open-and-closed set $H$ such that

$$f^{-1}(y) \subset H \subset f^{-1}(G).$$

If $V = f^*(H)$, then $V$ is open in $Y$ and $y \in V \subset G$. Since

$$\bar{V} \subset f(\bar{H}) = f(H)$$

it follows that $\operatorname{bd}(V) \subset f(H) \backslash V$. By Lemma 3.9 there exists a closed subspace $A$ of $X$ and a continuous closed surjection $g : A \to \operatorname{bd}(V)$ such that $g^{-1}(y)$ contains at most $n$ points if $y \in \operatorname{bd}(V)$. Hence

$$\operatorname{ind} \operatorname{bd}(V) \leqslant n - 1$$

by the induction hypothesis. Thus $\operatorname{ind} X \leqslant n$ and the proposition is established by induction. $\square$

**3.11  Proposition.** *If $X$ is a topological space such that $\operatorname{Ind} X = 0$ and $f: X \to Y$ is a continuous closed surjection such that $f^{-1}(y)$ contains at most $n + 1$ points if $y \in Y$, then $Y$ is a normal space such that*

$$\operatorname{Ind} Y \leqslant n.$$

*Proof.* The proof, which closely resembles that of Proposition 3.10, is left to the reader.□

**3.12  Proposition.** (a) *If $X$ is a topological space, then*

$$\operatorname{ind} X \leqslant \delta(X).$$

(b) *If $X$ is a regular space, then*

$$\operatorname{Ind} X \leqslant \Delta(X).$$

*Proof.* Part (a) follows at once from Propositions 3.7 and 3.10. Similarly part (b) follows from Propositions 3.8 and 3.11.□

We can also obtain subset theorems for $\delta$ and $\Delta$ as follows.

**3.13  Proposition.** (a) *If $A$ is a subspace of a space $X$, then*

$$\delta(A) \leqslant \delta(X).$$

(b) *If $A$ is a closed subspace of a regular space $X$, then*

$$\Delta(A) \leqslant \Delta(X).$$

*Proof.* If $f: Z \to X$ is a continuous closed surjection and $A$ is a subspace of $X$, then the mapping $f^{-1}(A) \to A$ given by restriction of $f$ is closed. Assertions (a) and (b) are now seen to follow at once from Propositions 3.7 and 3.8 respectively.□

We can exploit the characterization theorems to obtain a sufficient condition for the coincidence of $\delta$ and $\Delta$.

**3.14  Proposition.** *If $X$ is a completely paracompact regular space, then*

$$\delta(X) = \Delta(X).$$

*Proof.* It suffices to show that $\Delta(X) \leqslant \delta(X)$. Suppose that $\delta(X) \leqslant n$. Then by Proposition 3.7 there exists a space $Z$ such that $\operatorname{ind} Z = 0$, and a continuous closed surjection $f: Z \to X$ such that $f^{-1}(x)$ contains at most $n + 1$ points if $x \in X$. Since $f$ is a perfect mapping and $X$ is a completely paracompact space, $Z$ is a completely paracompact space. Hence $\dim Z = 0$ by Proposition 4.5.8. Thus $Z$ is a perfectly zero-dimensional space and it follows from Proposition 3.8 that

$$\Delta(X) \leqslant n. \square$$

It follows from Proposition 3.12 that if $X$ is a normal regular space then

$$\dim X \leqslant \operatorname{Ind} X \leqslant \Delta(X).$$

Since there exist bicompact spaces for which $\dim X < \operatorname{Ind} X$ (see Example 4.3.4 and §2 of Chapter 8), there exist bicompact spaces for which $\dim X < \Delta(X)$. We can however characterize the paracompact regular spaces $X$ for which $\dim X \leqslant n$ in terms of certain perfect images of perfectly zero-dimensional spaces. First we have:

**3.15  Proposition.** *A topological space $X$ is a paracompact regular space if and only if there exists a perfectly zero-dimensional space $Z$ and a perfect mapping $f \colon Z \to X$.*

*Proof.* Since by Proposition 2.1.25 the closed image of a paracompact normal space is paracompact, and by Proposition 2.5.5 the perfect image of a regular space is regular, it is clear that the perfect image of a perfectly zero-dimensional space is a paracompact regular space. Conversely suppose that $X$ is a paracompact regular space, and let $\Phi$ be the strongly directed family of all proper canonical coverings of $X$. There exists a perfect mapping $f \colon Z(\Phi) \to X$. Since $X$ is a paracompact normal space, $\Phi$ is a fine family. Since $X$ is a regular space, it follows from Lemma 3.5 that $Z(\Phi)$ is a perfectly zero-dimensional space. $\square$

**3.16  Definition.** *Let $n$ be a non-negative integer. A continuous surjection $f \colon X \to Y$ is said to have the property $(\mathcal{N}_n)$ if for each finite open covering $\mathcal{U}$ of $Y$ there exists an open covering $\{V_\lambda\}_{\lambda \in \Lambda}$ of $X$ such that (i) $\{f(V_\lambda)\}_{\lambda \in \Lambda}$ is a refinement of $\mathcal{U}$ and (ii) for each point $y$ of $Y$ the set*

$$\{\lambda \in \Lambda \mid f^{-1}(y) \cap V_\lambda \neq \varnothing\}$$

*has at most $n+1$ elements.*

**3.17  Lemma.** *Let $f \colon X \to Y$ be a continuous closed surjection with the property $(\mathcal{N}_n)$ of a normal space $X$ onto a space $Y$. Then $Y$ is a normal space such that $\dim Y \leqslant n$.*

*Proof.* Since $X$ is a normal space and $f$ is a closed mapping, $Y$ is a normal space. Let $\mathcal{U} = \{U_1, \ldots, U_k\}$ be a finite open covering of $Y$.

Since $f$ has the property $(\mathcal{N}_n)$, there exists an open covering $\{V_\lambda\}_{\lambda \in \Lambda}$ of $X$ such that $\{f(V_\lambda)\}_{\lambda \in \Lambda}$ is a refinement of $\mathcal{U}$ and

$$\{\lambda \in \Lambda \mid f^{-1}(y) \cap V_\lambda \neq \varnothing\}$$

has at most $n+1$ elements if $y \in Y$. For each $\lambda$ in $\Lambda$ choose $i(\lambda)$ such that $f(V_\lambda) \subset U_{i(\lambda)}$ and let $W_i = \bigcup_{i(\lambda)=i} V_\lambda$. Then $\{W_1, \ldots, W_k\}$ is an open covering of $X$, $f(W_i) \subset U_i$ for each $i$, and if $y \in Y$ then $f^{-1}(y) \cap W_i \neq \varnothing$ for at most $n+1$ integers $i$. Since $X$ is a normal space, there exists a closed covering $\{F_1, \ldots, F_k\}$ of $X$ such that $F_i \subset W_i$ for each $i$. If $\mathscr{F} = \{f(F_1), \ldots, f(F_k)\}$, then $\mathscr{F}$ is a closed covering of $Y$ of order not exceeding $n$ and $f(F_i) \subset U_i$ for each $i$. Hence $\dim Y \leqslant n$.☐

**3.18  Proposition.** *A topological space $X$ is a paracompact regular space such that $\dim X \leqslant n$ if and only if there exists a perfectly zero-dimensional space $Z$ and a perfect mapping $f: Z \to X$ which has the property $(\mathcal{N}_n)$.*

*Proof.* If there exists a perfect mapping $f: Z \to X$, where $Z$ is a perfectly zero-dimensional space and $f$ has the property $(\mathcal{N}_n)$, then $X$ is a paracompact regular space by Proposition 3.15 and $\dim X \leqslant n$ by Lemma 3.17. Conversely let $X$ be a paracompact regular space such that $\dim X \leqslant n$. Let $\Phi$ be the fine strongly directed family of proper canonical coverings of $X$ and let us apply to $\Phi$ the notation introduced at the beginning of the section. If $Z = Z(\Phi)$, then $Z$ is a perfectly zero-dimensional space by Lemma 3.5, and there exists a perfect mapping $f: Z \to X$, defined above. To complete the proof we show that $f$ has the property $(\mathcal{N}_n)$. If $\mathcal{U}$ is an open covering of $X$, then since $X$ is a paracompact normal space such that $\dim X \leqslant n$, it follows from Proposition 1.2 that there exists $\alpha$ in $\Omega$ such that $\mathscr{F}_\alpha$ is a refinement of $\mathcal{U}$ and the order of $\mathscr{F}_\alpha$ does not exceed $n$. If $V_{\alpha\lambda} = \pi_{\alpha\lambda}^{-1}\big(F_\lambda^\alpha \times (\lambda)\big)$ then $\mathscr{V}_\alpha = \{V_{\alpha\lambda}\}_{\lambda \in \Lambda(\alpha)}$ is a disjoint open covering of $Z$. Since

$$f(V_{\alpha\lambda}) = F_\lambda^\alpha$$

if $\lambda \in \Lambda(\alpha)$, it follows that $\{f(V_{\alpha\lambda})\}_{\lambda \in \Lambda(\alpha)}$ is a refinement of $\mathcal{U}$. If $x \in X$, then $V_{\alpha\lambda} \cap f^{-1}(x) \neq \varnothing$ implies that $x \in F_\lambda^\alpha$ so that

$$\{\lambda \in \Lambda(\alpha) \mid V_{\alpha\lambda} \cap f^{-1}(x) \neq \varnothing\}$$

has at most $n+1$ members.☐

## 4   Dimension and bicompactification

In this section we examine the relation between $\Delta(X)$ and $\Delta(\beta X)$ for a Tihonov space $X$, where $\beta X$ denotes the Stone–Čech bicompactification of $X$. We begin by establishing that if $X$ is a $T_4$-space, then $\dim X = \dim \beta X$ and $\operatorname{Ind} X = \operatorname{Ind} \beta X$. Only special cases are required in this section, but these results are clearly of independent interest.

We shall regard a Tihonov space $X$ as a subspace of its Stone–Čech bicompactification $\beta X$. If $A$ is a subset of $X$ we shall reserve the notation $\bar{A}$ and $\operatorname{bd}(A)$ for the closure and boundary respectively of $A$ in $X$. For any subset $B$ of $\beta X$ we shall write $\operatorname{cl}_{\beta X}(B)$ and $\operatorname{bd}_{\beta X}(B)$ for the closure and boundary respectively of $B$ in $\beta X$.

For each open set $U$ of a Tihonov space $X$, let

$$\hat{U} = \beta X \backslash \operatorname{cl}_{\beta X}(X \backslash U).$$

Then $\hat{U}$ is open in $\beta X$ and since $X \backslash U$ is a closed set of $X$ it follows that $X \cap \operatorname{cl}_{\beta X}(X \backslash U) = X \backslash U$ so that $\hat{U} \cap X = U$. Furthermore $\hat{U}$ is the largest open set of $\beta X$ with this property. For if $V$ is open in $\beta X$ and $X \cap V = U$, then

$$\operatorname{cl}_{\beta X}(X \backslash U) = \operatorname{cl}_{\beta X}(X \backslash V) \subset \beta X \backslash V,$$

so that $V \subset \beta X \backslash \operatorname{cl}_{\beta X}(X \backslash U) = \hat{U}$. Also since $X$ is dense in $\beta X$ it follows that

$$\operatorname{cl}_{\beta X}(U) = \operatorname{cl}_{\beta X}(\hat{U}).$$

If $U_1, \ldots, U_k$ are open sets of $X$, then

$$\beta X \backslash \left( \bigcap_{i=1}^{k} U_i \right) = \operatorname{cl}_{\beta X} \left( \bigcup_{i=1}^{k} (X \backslash U_i) \right) = \bigcup_{i=1}^{k} \operatorname{cl}_{\beta X}(X \backslash U_i)$$

so that

$$\left( \bigcap_{i=1}^{k} U_i \right)^{\widehat{}} = \bigcap_{i=1}^{k} \hat{U}_i.$$

It is evident that if $U$ and $V$ are open in $X$ and $U \subset V$, then $\hat{U} \subset \hat{V}$. Thus if $U_1, \ldots, U_k$ are open sets of $X$, then

$$\bigcup_{i=1}^{k} \hat{U}_i \subset \left( \bigcup_{i=1}^{k} U_i \right)^{\widehat{}}.$$

**4.1   Lemma.** *If $U_1, \ldots, U_k$ are open sets of a $T_4$-space, then*

$$\bigcup_{i=1}^{k} \hat{U}_i = \left( \bigcup_{i=1}^{k} U_i \right)^{\widehat{}}.$$

*Proof.* We have

$$\beta X \backslash \bigcup_{i=1}^{k} \hat{U}_i = \bigcap_{i=1}^{k} (\beta X \backslash \hat{U}_i) = \bigcap_{i=1}^{k} \mathrm{cl}_{\beta X} (X \backslash U_i)$$

and by Proposition 1.6.10

$$\bigcap_{i=1}^{k} \mathrm{cl}_{\beta X} (X \backslash U_i) = \mathrm{cl}_{\beta X} \left( X \backslash \bigcup_{i=1}^{k} U_i \right). \square$$

**4.2   Corollary.** *If $\mathcal{U} = \{U_1, \ldots, U_k\}$ is an open covering of a $T_4$-space $X$, then $\{\hat{U}_1, \ldots, \hat{U}_k\}$ is an open covering of $\beta X$ of the same order as $\mathcal{U}$.*

*Proof.* Since $\hat{X} = \beta X$, it follows from Lemma 4.1 that $\{\hat{U}_1, \ldots, \hat{U}_k\}$ is an open covering of $\beta X$. If $1 \leqslant i(0) < \ldots < i(n) \leqslant k$, then

$$\bigcap_{j=0}^{n} \hat{U}_{i(j)} = \left( \bigcap_{j=0}^{n} U_{i(j)} \right)^{\hat{}},$$

so that the order of $\{\hat{U}_1, \ldots, \hat{U}_k\}$ is clearly the same as the order of $\mathcal{U}$. $\square$

**4.3   Proposition.** *If $X$ is a $T_4$-space, then*

$$\dim X = \dim \beta X.$$

*Proof.* Suppose first that $\dim \beta X \leqslant n$. If $\{U_1, \ldots, U_k\}$ is an open covering of $X$ then, by Corollary 4.2, $\{\hat{U}_1, \ldots, \hat{U}_k\}$ is an open covering of $\beta X$. Hence there exists an open covering $\{H_1, \ldots, H_k\}$ of $\beta X$ of order not exceeding $n$ such that $H_i \subset U_i$ for each $i$. If $V_i = H_i \cap X$, then

$$\{V_1, \ldots, V_k\}$$

is an open covering of $X$ of order not exceeding $n$ such that $V_i \subset U_i$ for each $i$.

Now suppose that $\dim X \leqslant n$, and let $\{G_1, \ldots, G_k\}$ be an open covering of $\beta X$. There exists an open covering $\{H_1, \ldots, H_k\}$ of $\beta X$ such that $\mathrm{cl}_{\beta X} (H_i) \subset G_i$ for each $i$, and there exists an open covering $\{U_1, \ldots, U_k\}$ of $X$ of order not exceeding $n$ such that $U_i \subset H_i \cap X$ for each $i$. By Corollary 4.2, $\{\hat{U}_1, \ldots, \hat{U}_k\}$ is an open covering of $\beta X$ of order not exceeding $n$, and $\hat{U}_i \subset \mathrm{cl}_{\beta X} (U_i) \subset \mathrm{cl}_{\beta X} (H_i) \subset G_i$ for each $i$. Thus $\dim \beta X \leqslant n$. $\square$

The next lemma allows us to make inductive arguments about the dimension of the Stone–Čech bicompactification.

**4.4   Lemma.** *If $U$ is an open set in a $T_4$-space $X$, then*

$$\mathrm{bd}_{\beta X} (\hat{U}) = \mathrm{cl}_{\beta X} \big( \mathrm{bd} (U) \big) = \beta (\mathrm{bd} (U)).$$

*Proof.* Since $\mathrm{cl}_{\beta X}(U) = \mathrm{cl}_{\beta X}(\hat{U})$, it follows that $\overline{U} = X \cap \mathrm{cl}_{\beta X}(\hat{U})$ and hence $\mathrm{cl}_{\beta X}(\overline{U}) = \mathrm{cl}_{\beta X}(\hat{U})$. Thus

$$\mathrm{cl}_{\beta X}\big(\mathrm{bd}\,(U)\big) = \mathrm{cl}_{\beta X}\big(\overline{U} \cap (X\backslash U)\big) \subset \mathrm{cl}_{\beta X}(\overline{U}) \cap \mathrm{cl}_{\beta X}(X\backslash U)$$
$$= \mathrm{cl}_{\beta X}(\hat{U}) \cap (\beta X\backslash\hat{U}) = \mathrm{bd}_{\beta X}(\hat{U}).$$

To see that the reverse inclusion holds, let $V = X\backslash\overline{U}$. Since $U$ and $V$ are disjoint open sets of $X$, it follows that $\hat{U}$ and $\hat{V}$ are disjoint open sets of $\beta X$. Hence $\mathrm{cl}_{\beta X}(\hat{U}) \subset \beta X\backslash\hat{V}$ so that $\mathrm{bd}_{\beta X}(\hat{U}) \subset \beta X\backslash\hat{V}$. Thus

$$\mathrm{bd}_{\beta X}(\hat{U}) \subset (\beta X\backslash\hat{U}) \cap (\beta X\backslash\hat{V}) = \beta X\backslash(\hat{U} \cup \hat{V}).$$

But $\hat{U} \cup \hat{V} = (U \cup V)^\wedge$ by Lemma 4.1 so that

$$\mathrm{bd}_{\beta X}(\hat{U}) \subset \beta X\backslash(U \cup V)^\wedge = \mathrm{cl}_{\beta X}\big(X\backslash(U \cup V)\big)$$
$$= \mathrm{cl}_{\beta X}\big(\overline{U} \cap (X\backslash U)\big) = \mathrm{cl}_{\beta X}\big(\mathrm{bd}\,(U)\big).$$

Finally since $\mathrm{bd}\,(U)$ is a closed set in the $T_4$-space $X$, it follows from Proposition 1.6.11 that $\mathrm{cl}_{\beta X}\big(\mathrm{bd}\,(U)\big) = \beta\big(\mathrm{bd}\,(U)\big).\ \square$

**4.5  Proposition.** *If $X$ is a $T_4$-space, then*

$$\mathrm{Ind}\,X = \mathrm{Ind}\,\beta X.$$

*Proof.* We prove first, by induction, that $\mathrm{Ind}\,X \leqslant \mathrm{Ind}\,\beta X$. This is clearly true if $\mathrm{Ind}\,\beta X = -1$ and we suppose it is true for spaces with Stone–Čech bicompactification of large inductive dimension not exceeding $n-1$, where $n$ is a non-negative integer. Let $X$ be a $T_4$-space such that $\mathrm{Ind}\,\beta X \leqslant n$, let $E$ be a closed set of $X$ and let $G$ be an open set of $X$ such that $E \subset G$. Since $E$ and $X\backslash G$ are disjoint closed sets of $X$, it follows from Proposition 1.6.10 that $K = \mathrm{cl}_{\beta X}(E)$ and $\mathrm{cl}_{\beta X}(X\backslash G)$ are disjoint closed sets of $\beta X$. Thus $K \subset \hat{G}$ and since $\mathrm{Ind}\,\beta X \leqslant n$ there exists $V$ open in $\beta X$ such that

$$K \subset V \subset \hat{G} \quad \text{and} \quad \mathrm{Ind}\,\mathrm{bd}_{\beta X}(V) \leqslant n-1.$$

If $U = V \cap X$, then $U$ is open in $X$ and $E \subset U \subset G$. Also $U \subset V \subset \hat{U}$, so that $\mathrm{cl}_{\beta X}(\hat{U}) = \mathrm{cl}_{\beta X}(U) = \mathrm{cl}_{\beta X}(V)$. Thus

$$\mathrm{bd}_{\beta X}(\hat{U}) = \big(\mathrm{cl}_{\beta X}(\hat{U})\big)\backslash\hat{U} \subset \big(\mathrm{cl}_{\beta X}(V)\big)\backslash V = \mathrm{bd}_{\beta X}(V).$$

Hence $\mathrm{Ind}\,\mathrm{bd}_{\beta X}(\hat{U}) \leqslant n-1$. It follows from Lemma 4.4 that

$$\mathrm{Ind}\,\beta\big(\mathrm{bd}\,(U)\big) \leqslant n-1$$

so that $\mathrm{Ind}\,\mathrm{bd}\,(U) \leqslant n-1$ by the induction hypothesis. Thus $\mathrm{Ind}\,X \leqslant n$. It follows by induction that $\mathrm{Ind}\,X \leqslant \mathrm{Ind}\,\beta X$ if $X$ is a $T_4$-space.

Now we show, by induction, that $\operatorname{Ind}\beta X \leqslant \operatorname{Ind} X$. This is clearly true if $\operatorname{Ind} X = -1$ and we suppose it is true for spaces of large inductive dimension not exceeding $n-1$, where $n$ is a non-negative integer. Let $X$ be a $T_4$-space such that $\operatorname{Ind} X \leqslant n$, let $E$ be a closed set of $\beta X$ and let $G$ be an open set of $\beta X$ such that $E \subset G$. Then there exist $V$ and $W$ open in $\beta X$ such that

$$E \subset V \subset \operatorname{cl}_{\beta X}(V) \subset W \subset \operatorname{cl}_{\beta X}(W) \subset G.$$

Since $\operatorname{Ind} X \leqslant n$, there exists $U$ open in $X$ such that

$$X \cap \operatorname{cl}_{\beta X}(V) \subset U \subset X \cap W \quad \text{and} \quad \operatorname{Ind}\operatorname{bd}(U) \leqslant n-1.$$

Then $\hat{U}$ is open in $\beta X$ and $E \subset V \subset (V \cap X)^\frown \subset \hat{U}$. Furthermore $\operatorname{cl}_{\beta X}(\hat{U}) = \operatorname{cl}_{\beta X}(U) \subset \operatorname{cl}_{\beta X}(W) \subset G$. Thus $E \subset \hat{U} \subset G$. By the induction hypothesis $\operatorname{Ind}\beta\big(\operatorname{bd}(U)\big) \leqslant n-1$. Thus $\operatorname{Ind}\operatorname{bd}_{\beta X}(U) \leqslant n-1$ by Lemma 4.4. Hence $\operatorname{Ind}\beta X \leqslant n$. It follows by induction that

$$\operatorname{Ind}\beta X \leqslant \operatorname{Ind} X$$

if $X$ is a $T_4$-space. $\square$

**4.6    Remarks.** (1) Propositions 4.3 and 4.5 are not true if the hypothesis that $X$ is a normal space is omitted. For if $X$ is the Tihonov plank studied in Example 4.3.1, and $Y$ is the subspace $X \backslash \{(\omega_1, \omega_0)\}$, then $Y$ is a non-normal Tihonov space and $\dim Y = \operatorname{Ind} Y = 1$ whilst

$$\dim X = \operatorname{Ind} X = 0.$$

But it is not difficult to see that $X$ is the Stone–Čech bicompactification $\beta Y$ of $Y$.

(2) There is no result analogous to Propositions 4.3 and 4.5 for small inductive dimension. For as noted in Remark 5.4.8, there exists a $T_4$-space $X$ such that $\operatorname{ind} X = 1$ and $\operatorname{Ind} X = 2$. By Proposition 4.5, $\operatorname{Ind}\beta X = 2$ so that $\operatorname{ind}\beta X \leqslant 2$. But by Proposition 4.2.4, $\operatorname{ind}\beta X = 1$ would imply that $\operatorname{Ind}\beta X = 1$. Thus we have $\operatorname{ind} X = 1$ and $\operatorname{ind}\beta X = 2$. We can show that for the Stone–Čech bicompactification of a totally normal space, the small and large inductive dimensions coincide.

**4.7    Proposition.** *If $X$ is a totally normal Hausdorff space, then*

$$\operatorname{ind}\beta X = \operatorname{Ind}\beta X.$$

*Proof.* We know that $\operatorname{ind}\beta X \leqslant \operatorname{Ind}\beta X$ with equality if $\operatorname{ind}\beta X \leqslant 1$. We prove by induction that $\operatorname{Ind}\beta X \leqslant \operatorname{ind}\beta X$. Let $n \geqslant 2$ be an integer and suppose that the result is known to be true for spaces with Stone–Čech bicompactification of small inductive dimension not exceeding

$n-1$. Let $X$ be a totally normal Hausdorff space such that $\operatorname{ind}\beta X \leqslant n$, let $E$ be a closed set of $\beta X$ and let $G$ be an open set of $\beta X$ such that $E \subset G$. If $z \in E$, then since $\operatorname{ind}\beta X \leqslant n$, there exists an open set $V$ of $\beta X$ such that $z \in V \subset \operatorname{cl}_{\beta X}(V) \subset G$ and $\operatorname{ind}\operatorname{bd}_{\beta X}(V) \leqslant n-1$. Since $E$ is a compact set in $\beta X$, it follows that there exist open sets $V_1, \ldots, V_m$ of $\beta X$ such that $\operatorname{cl}_{\beta X}(V_i) \subset G$ and $\operatorname{ind}\operatorname{bd}_{\beta X}(V_i) \leqslant n-1$ for each $i = 1, \ldots, m$, and $E \subset \bigcup_{i=1}^m V_i$. For each $i$ let $U_i = V_i \cap X$ and let $U = \bigcup_{i=1}^m U_i$. Then $\hat U$ is open in $\beta X$ and $E \subset \hat U$ since $V_i \subset \hat U_i$ for each $i$ and $\hat U = \bigcup_{i=1}^m \hat U_i$ by Lemma 4.1. Furthermore

$$\operatorname{cl}_{\beta X}(\hat U_i) = \operatorname{cl}_{\beta X}(U_i) = \operatorname{cl}_{\beta X}(V_i)$$

for each $i$, so that $\operatorname{cl}_{\beta X}(\hat U) \subset G$. Since $\operatorname{cl}_{\beta X}(\hat U_i) = \operatorname{cl}_{\beta X}(V_i)$ and $V_i \subset \hat U_i$, it follows that $\operatorname{bd}_{\beta X}(\hat U_i) \subset \operatorname{bd}_{\beta X}(V_i)$. Hence $\operatorname{ind}\operatorname{bd}_{\beta X}(\hat U_i) \leqslant n-1$ so that $\operatorname{ind}\beta\big(\operatorname{bd}(U_i)\big) \leqslant n-1$ by Lemma 4.4. It follows from the induction hypothesis that $\operatorname{Ind}\beta\big(\operatorname{bd}(U_i)\big) \leqslant n-1$. Thus $\operatorname{Ind}\operatorname{bd}(U_i) \leqslant n-1$ by Proposition 4.5. Since $\operatorname{bd}(U) \subset \bigcup_{i=1}^m \operatorname{bd}(U_i)$, it follows from the finite sum theorem for Ind that $\operatorname{Ind}\operatorname{bd}(U) \leqslant n-1$. Thus

$$\operatorname{Ind}\beta\big(\operatorname{bd}(U)\big) \leqslant n-1$$

by Proposition 4.5 and hence $\operatorname{Ind}\operatorname{bd}_{\beta X}(\hat U) \leqslant n-1$ by Lemma 4.4. Since $\hat U$ is open in $\beta X$ and $E \subset \hat U \subset G$, it follows that $\operatorname{Ind}\beta X \leqslant n$. The proof is now completed by induction. $\square$

We next examine the relation of $\Delta(X)$ to $\Delta(\beta X)$ for a Tihonov space $X$.

**4.8   Lemma.** *Let $X$ be a Tihonov space and let $Y$ be a $T_4$-space. If $f: X \to Y$ is a continuous closed surjection such that $f^{-1}(y)$ contains at most $n+1$ points if $y \in Y$, then $(\beta f)^{-1}(z)$ contains at most $n+1$ points if $z \in \beta Y$, where $\beta f: \beta X \to \beta Y$ is the continuous extension of $f$.*

*Proof.* Suppose that there exists a point $z_0$ of $\beta Y$ such that $(\beta f)^{-1}(z_0)$ contains $n+2$ distinct points $u_0, \ldots, u_n, u_{n+1}$ of $\beta X$. Then there exist open sets $V_0, \ldots, V_n, V_{n+1}$ of $\beta X$ such that $u_i \in V_i$ and

$$\operatorname{cl}_{\beta X}(V_i) \cap \operatorname{cl}_{\beta X}(V_j) = \varnothing$$

if $0 \leqslant i < j \leqslant n+1$. For each $i$, let $A_i$ be the closure of $V_i \cap X$ in $X$. Then $A_0, \ldots, A_n, A_{n+1}$ are $n+2$ disjoint closed sets of $X$. By hypothesis $\bigcap_{i=0}^{n+1} f(A_i) = \varnothing$. Since $f$ is a closed mapping, $f(A_i)$ is a closed set of $Y$ for each $i$. Since $Y$ is a $T_4$-space it follows from Proposition 1.6.10 that

$$\bigcap_{i=0}^{n+1} \operatorname{cl}_{\beta Y} f(A_i) = \varnothing.$$

But $(\beta f)\big(\mathrm{cl}_{\beta X}(A_i)\big) = \mathrm{cl}_{\beta Y} f(A_i)$ since $\beta f$ is closed so that

$$\bigcap_{i=0}^{n+1} (\beta f)\big(\mathrm{cl}_{\beta X}(A_i)\big) = \varnothing .$$

Since $\mathrm{cl}_{\beta X}(A_i) = \mathrm{cl}_{\beta X}(V_i)$, it follows that $u_i \in \mathrm{cl}_{\beta X}(A_i)$ so that

$$z_0 \in (\beta f)\big(\mathrm{cl}_{\beta X}(A_i)\big)$$

for $i = 0, \ldots, n+1$, which is a contradiction. It follows that if $z \in \beta Y$, then $(\beta f)^{-1}(z)$ contains at most $n+1$ elements. $\square$

**4.9    Proposition.** *If $X$ is a Tihonov space, then*

$$\Delta(\beta X) \leqslant \Delta(X).$$

*Proof.* If $\Delta(X) \leqslant n$, then by Proposition 3.8 there exists a perfectly zero-dimensional space $Z$ and a continuous closed surjection $f: Z \to X$ such that $f^{-1}(x)$ contains at most $n+1$ points if $x \in X$. And it follows from Lemma 3.4 that we can suppose that $Z$ is a Hausdorff space. Since $\Delta(X)$ is finite, $X$ is paracompact and hence $X$ is a $T_4$-space. It follows from Lemma 4.8 that $\beta f: \beta Z \to \beta X$ is a continuous closed surjection such that $(\beta f)^{-1}(z)$ contains at most $n+1$ points if $z \in \beta X$. Since $\dim Z = 0$, it follows from Proposition 4.3 that $\dim \beta Z = 0$. Thus $\beta Z$ is a perfectly zero-dimensional space. It follows from Proposition 3.8 that $\Delta(\beta X) \leqslant n$. $\square$

**4.10    Remark.** The space $T$ of ordinals less than the first uncountable ordinal $\omega_1$ is a non-paracompact space. It follows that $\Delta(T) = \infty$. Since every continuous function on $T$ is finally constant, it is easily seen that the Stone–Čech bicompactification $\beta T$ of $T$ is the space of ordinals not exceeding $\omega_1$. Thus $\beta T$ is a bicompact space such that $\dim \beta T = 0$ so that $\beta T$ is a perfectly zero-dimensional space. Thus we have $\Delta(T) = \infty$ and $\Delta(\beta T) = 0$. We shall see that provided $\Delta(X)$ is finite then we have $\Delta(X) = \Delta(\beta X)$ for a Hausdorff space $X$. For the proof of this fact we use a special case of the following proposition. More results of this type will be established in Chapter 9.

**4.11    Proposition.** *Let $X$ be a normal space, let $Y$ be a paracompact Hausdorff space and let $f: X \to Y$ be a continuous closed surjection such that $\dim f^{-1}(y) = 0$ if $y \in Y$. Then $\dim X \leqslant \dim Y$.*

*Proof.* We shall show that if $\dim Y \leqslant n$ then $\dim X \leqslant n$. Let

$$\{U_1, \ldots, U_k\}$$

be an open covering of $X$. If $y \in Y$, then since $Y$ is a $T_1$-space, $f^{-1}(y)$ is a closed subspace of $X$ and thus is a normal space. Since

$$\dim f^{-1}(y) = 0,$$

there exists a disjoint closed covering $\{F_{1y}, \ldots, F_{ky}\}$ of $f^{-1}(y)$ such that $F_{iy} \subset U_i$ for each $i$. Since $X$ is a normal space and the sets $F_{iy}$ are closed in $X$, there exist disjoint open sets $G_{iy}$ such that

$$F_{iy} \subset G_{iy} \subset U_i \quad \text{for} \quad i = 1, \ldots, k.$$

Since $f^{-1}(y) \subset \bigcup_{i=1}^{k} G_{iy}$ and $f$ is a closed mapping, there exists an open neighbourhood $W_y$ of $y$ in $Y$ such that $f^{-1}(W_y) \subset \bigcup_{i=1}^{k} G_{iy}$. Since $Y$ is a paracompact normal space such that $\dim Y \leqslant n$, it follows from Corollary 3.4.4 that there exists an open covering $\{V_y\}_{y \in Y}$ of $Y$ of order not exceeding $n$ such that $V_y \subset W_y$ for each $y$. If $y \in Y$ and $i = 1, \ldots, k$ let $H_{iy} = G_{iy} \cap f^{-1}(V_y)$. Then $\{H_{iy}\}_{y \in Y, i=1, \ldots, k}$ is an open covering of $X$ of order not exceeding $n$ which is a refinement of $\{U_1, \ldots, U_k\}$. Thus $\dim X \leqslant n$. $\square$

**4.12 Proposition.** *If $X$ is a paracompact Hausdorff space, then*

$$\Delta(X) = \Delta(\beta X).$$

*Proof.* Suppose that $\Delta(\beta X) \leqslant n$. Then by Proposition 3.8 there exists a perfectly zero-dimensional Hausdorff space $Q$ and a continuous closed surjection $g : Q \to \beta X$ such that the set $g^{-1}(y)$ contains at most $n+1$ points if $y \in \beta X$. Since $g$ is a perfect mapping, it follows from Proposition 2.5.9 that $Q$ is a bicompact space. Let $P = g^{-1}(X)$ and let $f : P \to X$ be given by restriction of $g$. Then $f$ is a continuous closed surjection and $f^{-1}(x)$ contains at most $n+1$ points if $x \in X$. Let $h : \beta P \to Q$ be the extension of the inclusion of $P$ in $Q$. Then by uniqueness, $g \circ h = \beta f : \beta P \to \beta X$. It follows from Lemma 4.8 that the inverse image of each point of $\beta X$ under $\beta f$ is finite and hence the inverse image of each point of $Q$ under $h$ is finite. Thus since $h$ maps the bicompact space $\beta P$ onto a closed subspace of $Q$ and since $Q$ is a bicompact space such that $\dim Q = 0$, we can apply Proposition 4.11 to conclude that $\dim \beta P = 0$. It follows that $\dim P = 0$. Since $X$ is a paracompact space and $f : P \to X$ is a perfect mapping, we see that $P$ is a paracompact Hausdorff space. Hence $P$ is a perfectly zero-dimensional space. Thus $\Delta(X) \leqslant n$, by Proposition 3.8. Thus

$$\Delta(X) \leqslant \Delta(\beta X)$$

and the result now follows from Proposition 4.9. $\square$

**4.13    Corollary.** *If $X$ is a Hausdorff space such that $\Delta(X)$ is finite, then*

$$\Delta(X) = \Delta(\beta X).$$

*Proof.* A space $X$ for which $\Delta(X)$ is finite is paracompact. $\square$


**Notes**

The first application of directed families of canonical coverings in dimension theory was made by Aleksandrov and Ponomarev [1960]. By modifying a method of proof due to Proskuryakov [1951] that covering dimension and inductive dimension coincide for separable metric spaces, they found a sufficient condition for coincidence of covering dimension and inductive dimension for bicompact spaces. As a development of this work and an application of his work on projective spectra, Ponomarev [1962, 1963] introduced the study of the dimension functions $\delta$ and $\Delta$. He called $\delta(X)$ the *approximation dimension* of a space $X$, and $\Delta(X)$ the *cofinal approximation dimension* of $X$. Motivated by a classical result (which will be discussed in the next chapter) in dimension theory, Nagami [1962] sought to characterize the images of zero-dimensional spaces under certain continuous closed mappings and found results closely related to those of Ponomarev. The concept of a strongly directed family of coverings is due to Nagami. Most of the work in § 3 is a synthesis of the independent results of Ponomarev [1963] and Nagami [1962]. The method of presentation here is due to A. T. Al-Ani. Mack and Pears [1974] have considered dimension functions resembling $\delta$ and $\Delta$. Nagami [1962] defined the property $(\mathcal{N}_n)$ for mappings, saying that mappings satisfying $(\mathcal{N}_n)$ have *vague order* not exceeding $n+1$. He proved Lemma 3.17 and Proposition 3.18.

Wallman [1938] proved that the dimension of a space is equal to the dimension of its Wallman compactification. Since the Stone–Čech bicompactification of a $T_4$-space coincides with its Wallman compactification, this was the first proof of Proposition 4.3. A different approach will be found in Chapter 10. The equality $\operatorname{Ind} X = \operatorname{Ind} \beta X$ if $X$ is a $T_4$-space was found by Vedenisov [1941]. Proposition 4.7 was established by Smirnov [1951]. Mack and Pears [1974] proved that $\Delta(X) = \Delta(\beta X)$ if $X$ is a paracompact Hausdorff space.

# 7

# THE DIMENSION OF PSEUDO-METRIZABLE
# AND METRIZABLE SPACES

## 1  Images of zero-dimensional spaces

This section begins with an important characterization of the dimension of a pseudo-metrizable space. We show that the pseudo-metrizable spaces satisfy a decomposition theorem for dimension. We shall exploit this result in proving that each finite-dimensional subspace of a pseudo-metrizable space has a '$G_\delta$-envelope' of the same dimension and in proving that if $X$ is a pseudo-metrizable space then $\dim X = \Delta(X)$.

**1.1  Theorem.** *If $X$ is a pseudo-metrizable space, then $\dim X \leqslant n$ if and only if $X = \bigcup_{i=1}^{n+1} A_i$, where $\dim A_i \leqslant 0$ for $i = 1, \ldots, n+1$.*

*Proof.* Suppose that $X$ is a pseudo-metrizable space such that

$$X = \bigcup_{i=1}^{n+1} A_i,$$

where $\dim A_i \leqslant 0$ for $i = 1, \ldots, n+1$. Then $\dim X \leqslant n$ by Proposition 3.5.11. We establish the converse by induction. The statement is evidently true for spaces of dimension zero. Suppose that it is true for pseudo-metrizable spaces of dimension not exceeding $n-1$, where $n$ is a positive integer. Let $X$ be a pseudo-metrizable space such that $\dim X \leqslant n$. Then by Proposition 4.5.3, $X$ has a $\sigma$-locally finite base $\mathscr{V}$ for its topology such that $\operatorname{Ind} \operatorname{bd}(V) \leqslant n-1$ if $V \in \mathscr{V}$. The proof of Lemma 4.5.2 shows that $X = A \cup B$, where $\dim A \leqslant n-1$ and $\dim B \leqslant 0$. By the induction hypothesis, $A = \bigcup_{i=1}^{n} A_i$, where

$$\dim A_i \leqslant 0$$

for $i = 1, \ldots, n$. Taking $A_{n+1} = B$ we obtain the required decomposition of $X$. The proof is now completed by induction. $\square$

Next a lemma is proved which will be used to prove the existence of $G_\delta$-envelopes. We obtain a more general result than is required for the immediate application, but it will also be used later in the section.

**1.2 Lemma.** *Let $\{U_\lambda\}_{\lambda \in \Lambda}$ be a locally finite family of open sets of a perfectly normal space $X$, let $\{F_\lambda\}_{\lambda \in \Lambda}$ be a family of closed sets such that $F_\lambda \subset U_\lambda$ for each $\lambda$, and let $\{A_n\}_{n \in \mathbf{N}}$ be a countable family of subsets of $X$ such that $\dim A_n \leqslant 0$ for each $n$. Then there exists a family $\{V_\lambda\}_{\lambda \in \Lambda}$ of open sets of $X$ such that $F_\lambda \subset V_\lambda \subset \overline{V}_\lambda \subset U_\lambda$ for each $\lambda$, and for each $n$, if $x \in A_n$ then*

$$\{\lambda \in \Lambda \mid x \in \mathrm{bd}\,(V_\lambda)\}$$

*has at most $n - 1$ members.*

*Proof.* Let $\Lambda$ be well-ordered. The proof is by transfinite induction. Let $\mu$ be an element of $\Lambda$ and suppose that for each $\lambda < \mu$ we have constructed an open set $V_\lambda$ such that $F_\lambda \subset V_\lambda \subset \overline{V}_\lambda \subset U_\lambda$ and for each $\nu < \mu$ and each $n$, if $x \in A_n$ then

$$\{\lambda \in \Lambda \mid \lambda \leqslant \nu, x \in \mathrm{bd}\,(V_\lambda)\}$$

has at most $n - 1$ members. For each $x$ in $X$, let $n(x)$ denote the number of elements in

$$M_x = \{\lambda \in \Lambda \mid \lambda < \mu, x \in \mathrm{bd}\,(V_\lambda)\}.$$

Suppose that $x \in A_n$. The set $M_x$ is finite and if $\xi$ is the largest member of $M_x$, then $\xi < \mu$ and

$$M_x = \{\lambda \in \Lambda \mid \lambda \leqslant \xi, x \in \mathrm{bd}\,(V_\lambda)\},$$

so that $M_x$ has at most $n - 1$ members. Thus if $x \in A_n$, then

$$n(x) \leqslant n - 1.$$

For each positive integer $n$, let $H_n = \{x \in A_n \mid n(x) = n - 1\}$ and let $K = \bigcup_{i \in \mathbf{N}} H_i$. Then for each $n$, $\bigcup_{i=1}^n H_i$ is open in $K$. For if $x \in \bigcup_{i=1}^n H_i$, then $n(x) \leqslant n - 1$. Let

$$U = X \backslash \bigcup \{\mathrm{bd}\,(V_\lambda) \mid \lambda < \mu, x \notin \mathrm{bd}\,(V_\lambda)\}.$$

Then since $\{\mathrm{bd}\,(V_\lambda)\}_{\lambda \in \Lambda}$ is locally finite, $U$ is open in $X$, and $x \in U$. If $y \in U$ and $y \in \mathrm{bd}\,(V_\lambda)$ for some $\lambda < \mu$, then $x \in \mathrm{bd}\,(V_\lambda)$ so that if $y \in U$ then $n(y) \leqslant n(x) \leqslant n - 1$. It follows that $U \cap (\bigcup_{i \geqslant n+1} H_i) = \varnothing$ so that $U \cap K$ is an open neighbourhood of $x$ in $K$ which is contained in $\bigcup_{i=1}^n H_i$. Since $H_i \subset A_i$, we have $\dim H_i \leqslant 0$ for each $i$. It follows from Proposition 3.5.5 that $\dim K \leqslant 0$. Hence from Lemma 4.4.16 we see, on recalling that $\dim K \leqslant 0$ is equivalent to $\mathrm{Ind}\,K \leqslant 0$, that there exists an open set $V_\mu$ such that $F_\mu \subset V_\mu \subset \overline{V}_\mu \subset U_\mu$ and $\mathrm{bd}\,(V_\mu) \cap K = \varnothing$. For each $x$ in $X$, let $m(x)$ denote the number of elements in the set

$$\{\lambda \in \Lambda \mid \lambda \leqslant \mu, x \in \mathrm{bd}\,(V_\lambda)\}.$$

We must show that if $x \in A_n$, then $m(x) \leqslant n-1$. If $x \in A_n$ and $x \in K$, then $m(x) = n(x) \leqslant n-1$. If $x \in A_n$ and $x \notin K$, then

$$n(x) \leqslant m(x) \leqslant n(x) + 1.$$

But since $x \notin H_n$, it follows that $n(x) < n-1$ so that

$$m(x) \leqslant n(x) + 1 \leqslant n-1.$$

Thus $\{V_\lambda\}_{\lambda \leqslant \mu}$ has the required properties. The proof is completed by transfinite induction. $\square$

**1.3    Corollary.** *If $X$ is a pseudo-metrizable space and $A$ is a subspace of $X$ such that* $\dim A \leqslant 0$, *then there exists a $\sigma$-locally finite base $\mathscr{V}$ for the topology of $X$ such that* $\mathrm{bd}\,(V)$ *is disjoint from $A$ if $V \in \mathscr{V}$.*

*Proof.* By Proposition 2.3.4 there exists a sequence $\{\mathscr{U}_i\}_{i \in \mathbf{N}}$ of locally finite open coverings of $X$ such that for each point $x$ of $X$ and each open set $U$ such that $x \in U$, there exists some integer $i$ such that $\mathrm{St}\,(x, \mathscr{U}_i) \subset U$. For each $i$, let $\mathscr{U}_i = \{U_{i\lambda}\}_{\lambda \in \Lambda}$ and choose a closed covering $\{F_{i\lambda}\}_{\lambda \in \Lambda}$ such that $F_{i\lambda} \subset U_{i\lambda}$ for each $\lambda$. Applying Lemma 1.2 with the countable family $\{A_n\}_{n \in \mathbf{N}}$ of subspaces of $X$, where $A_1 = A$ and $A_n = \varnothing$ if $n > 1$, we obtain a family $\mathscr{V} = \{V_{i\lambda}\}_{i \in \mathbf{N}, \lambda \in \Lambda}$ of open sets of $X$ such that $F_{i\lambda} \subset V_{i\lambda} \subset \overline{V}_{i\lambda} \subset U_{i\lambda}$ for each $i$ and $\lambda$, and

$$\mathrm{bd}\,(V_{i\lambda}) \cap A = \varnothing.$$

Then $\mathscr{V}$ is the required $\sigma$-locally finite base for the topology of $X$. $\square$

**1.4    Theorem.** *Let $X$ be a pseudo-metrizable space and let $A$ be a non-empty subspace of $X$ such that* $\dim A \leqslant n$. *Then there exists a $G_\delta$-set $H$ of $X$ such that $A \subset H$ and* $\dim H \leqslant n$.

*Proof.* Suppose first that $\dim A = 0$. Then by Corollary 1.3 there exists a $\sigma$-locally finite base $\mathscr{V}$ for the topology of $X$ such that $\mathrm{bd}\,(V)$ is disjoint from $A$ if $V \in \mathscr{V}$. Since $\{\mathrm{bd}\,(V) \mid V \in \mathscr{V}\}$ is a $\sigma$-locally finite family of closed sets of $X$, the set

$$H = X \backslash \bigcup_{V \in \mathscr{V}} \mathrm{bd}\,(V)$$

is a $G_\delta$-set of $X$. Clearly $A \subset H$. If $\mathscr{W} = \{H \cap V \mid V \in \mathscr{V}\}$, then $\mathscr{W}$ is a $\sigma$-locally finite base for the topology of $H$. If $V \in \mathscr{V}$, then

$$\mathrm{bd}_H\,(H \cap V) \subset \mathrm{bd}\,(V) \cap H,$$

so that $\mathrm{bd}_H\,(H \cap V)$ is empty. Hence $\mathscr{W}$ is a base for the topology of $H$ which consists of open-and-closed sets. By Lemma 4.5.1,

$$\dim H \leqslant 0.$$

Now suppose that $\dim A \leqslant n$. Then by Theorem 1.1, $A = \bigcup_{i=1}^{n+1} A_i$, where $\dim A_i \leqslant 0$ for $i = 1, \ldots, n+1$. Thus for each $i$ there exists a $G_\delta$-set $H_i$ such that $A_i \subset H_i$ and $\dim H_i \leqslant 0$. If $H = \bigcup_{i=1}^{n+1} H_i$, then $H$ is a $G_\delta$-set such that $A \subset H$ and $\dim H \leqslant n$ by Theorem 1.1.□

The following consequence of Theorems 1.1 and 1.4 will be useful later.

**1.5  Corollary.** *If $X$ is a pseudo-metrizable space such that $\dim X \leqslant n$ and $F$ is an $F_\sigma$-set of $X$ such that $\dim F \leqslant m < n$, then there exists an $F_\sigma$-set $K$ of $X$ which is disjoint from $F$ and satisfies*

$$\dim K \leqslant n-m-1 \quad and \quad \dim (X \backslash K) \leqslant m.$$

*Proof.* It follows from Theorem 1.1 that $X = Y \cup Z$, where $\dim Y \leqslant m$ and $\dim Z \leqslant n-m-1$. By Theorem 1.4 there exists a $G_\delta$-set $Q$ such that $Z \subset Q$ and $\dim Q \leqslant n-m-1$. If $P = X \backslash Q$, then $P \subset Y$ so that $\dim P \leqslant m$. Since $F$ and $P$ are $F_\sigma$-sets, it follows that $\dim F \cup P \leqslant m$. By Theorem 1.4 there exists a $G_\delta$-set $H$ such that $F \cup P \subset H$ and $\dim H \leqslant m$. If $K = X \backslash H$, then $K$ is an $F_\sigma$-set which is disjoint from $F$ and $\dim (X \backslash K) \leqslant m$. Since $K \subset Z$, it follows that $\dim K \leqslant n-m-1$.□

Next there is a characterization of the spaces $X$, which are pseudo-metrizable and satisfy $\dim X \leqslant n$, which should be compared with the characterization of pseudo-metrizability given in Proposition 2.3.9. Two lemmas precede this characterization. To shorten the statements we make the following definition: if $\mathscr{F} = \{F_\gamma\}_{\gamma \in \Gamma}$ is a covering of a space $X$ and $\{A_n\}_{n \in \mathbb{N}}$ is a family of subsets of $X$ such that $\dim A_n \leqslant 0$ for each $n$, then $\mathscr{F}$ is said to be *compatible* with $\{A_n\}_{n \in \mathbb{N}}$ if for each point $x$ of $X$, the set $\{\gamma \in \Gamma \mid x \in F_\gamma\}$ has at most $n$ elements if $x \in A_n$.

**1.6  Lemma.** *Let $X$ be a perfectly normal space, let $\{A_n\}_{n \in \mathbb{N}}$ be a countable family of subsets of $X$ such that $\dim A_n \leqslant 0$ for each $n$, and let $\{U_\lambda\}_{\lambda \in \Lambda}$ be a locally finite open covering of $X$. Then there exists a closed covering $\{F_\lambda\}_{\lambda \in \Lambda}$ compatible with $\{A_n\}_{n \in \mathbb{N}}$ such that $F_\lambda \subset U_\lambda$ for each $\lambda$.*

*Proof.* There exists a closed covering $\{E_\lambda\}_{\lambda \in \Lambda}$ of $X$ such that $E_\lambda \subset U_\lambda$ for each $\lambda$. By Lemma 1.2 there exists a family $\{V_\lambda\}_{\lambda \in \Lambda}$ of open sets such that $E_\lambda \subset V_\lambda \subset \overline{V}_\lambda \subset U_\lambda$ for each $\lambda$ and for each $x$ the set

$$\{\lambda \in \Lambda \mid x \in \mathrm{bd}\,(V_\lambda)\}$$

contains at most $n-1$ elements if $x \in A_n$. Let $\Lambda$ be well-ordered and for each $\lambda$ in $\Lambda$, let

$$G_\lambda = V_\lambda \backslash \bigcup_{\mu < \lambda} \bar{V}_\mu$$

and let $F_\lambda = \bar{G}_\lambda$. Then $\mathscr{F} = \{F_\lambda\}_{\lambda \in \Lambda}$ is a locally finite closed covering of $X$, as established in the proof of Proposition 6.1.2, and clearly $F_\lambda \subset U_\lambda$ for each $\lambda$. It remains to show that $\mathscr{F}$ is compatible with $\{A_n\}_{n \in \Lambda}$. Let $x \in A_n$ and let $\mu$ be the first element of $\Lambda$ such that $x \in V_\mu$. Then $x \notin F_\lambda$ if $\lambda > \mu$. If $\lambda < \mu$ and $x \in F_\lambda$, then $x \in \bar{V}_\lambda$ and $x \notin V_\lambda$ so that $x \in \mathrm{bd}\,(V_\lambda)$. Since $\{\lambda \in \Lambda \mid x \in \mathrm{bd}\,(V_\lambda)\}$ has at most $n-1$ elements, it follows that $\{\lambda \in \Lambda \mid x \in F_\lambda\}$ has at most $n$ elements. $\square$

**1.7   Lemma.** *Let $X$ be a perfectly normal space, let $\{U_\lambda\}_{\lambda \in \Lambda}$ be a locally finite open covering of $X$, let $\mathscr{A} = \{A_n\}_{n \in \mathbb{N}}$ be a countable family of subsets of $X$ such that $\dim A_n \leqslant 0$ for each $n$ and let $\{F_\gamma\}_{\gamma \in \Gamma}$ be a locally finite closed covering of $X$ compatible with $\mathscr{A}$. Then there exists a locally finite closed covering $\{F_{(\gamma,\lambda)}\}_{(\gamma,\lambda) \in \Gamma \times \Lambda}$ compatible with $\mathscr{A}$ such that*

$$F_{(\gamma,\lambda)} \subset U_\lambda$$

*for each $\gamma$ and $\lambda$, and for each $\gamma$ in $\Gamma$,*

$$F_\gamma = \bigcup_{\lambda \in \Lambda} F_{(\gamma,\lambda)}.$$

*Proof.* We begin by establishing the statement $(A)$: for given $\beta$ in $\Gamma$ there exist closed sets $F_{(\beta,\lambda)}$ for $\lambda$ in $\Lambda$ such that $F_{(\beta,\lambda)} \subset U_\lambda$ and

$$\bigcup_{\lambda \in \Lambda} F_{(\beta,\lambda)} = F_\beta,$$

and the family indexed by $\Lambda \cup (\Gamma \backslash \{\beta\})$ which consists of the sets $F_{(\beta,\lambda)}$ for $\lambda$ in $\Lambda$ together with the $F_\gamma$ for $\gamma$ in $\Gamma \backslash \{\beta\}$ is locally finite and compatible with $\mathscr{A}$.

For each point $x$ of $X$, let $n(x)$ denote the number of elements in $\{\gamma \in \Gamma \mid x \in F_\gamma\}$, and for each positive integer $n$ let

$$H_n = \{x \in X \mid n(x) = n\}.$$

Then $\bigcup_{n \in \mathbb{N}} H_n = X$ and $\bigcup_{i=1}^n H_i$ is an open set for each $n$. For if $x \in \bigcup_{i=1}^n H_i$ and $V_x = X \backslash \bigcup \{F_\gamma \mid x \notin F_\gamma\}$, then $V_x$ is an open set such that $x \in V_x \subset \bigcup_{i=1}^n H_i$. By hypothesis $A_n \subset \bigcup_{i=1}^n H_i$ for each $n$. If $m, i \in \mathbb{N}$ and $i \leqslant m$ let

$$G_{mi} = F_\beta \cap A_m \cap H_{m+1-i}.$$

Then $G_{mi} \subset A_m$ for each $m$ and each $i \leqslant m$ so that $\dim G_{mi} \leqslant 0$. Now for each positive integer $i$ let $B_i = \bigcup_{m \geqslant i} G_{mi}$. If $i, m \in \mathbb{N}$ and $m \geqslant i$, then $B_i \cap H_{m+1-i} = G_{mi}$ so that

$$\bigcup_{n=i}^{m} G_{ni} = B_i \cap \left( \bigcup_{n=i}^{m} H_{n+1-i} \right) = B_i \cap \left( \bigcup_{s=1}^{m+1-i} H_s \right).$$

Thus $\{G_{ni}\}_{n \geqslant i}$ is a countable covering of $B_i$ such that $\bigcup_{n=i}^{m} G_{ni}$ is open in $B_i$ and $\dim G_{mi} \leqslant 0$ for each $m \geqslant i$. It follows from Proposition 3.5.5 that $\dim B_i \leqslant 0$. We have a locally finite open covering

$$\{U_\lambda \cap F_\beta\}_{\lambda \in \Lambda}$$

of $F_\beta$. It follows from Lemma 1.6 that there exists a locally finite closed covering $\{F_{(\beta, \lambda)}\}_{\lambda \in \Lambda}$ of $F_\beta$ such that $F_{(\beta, \lambda)} \subset U_\lambda \cap F_\beta$ for each $\lambda$ and $\{F_{(\beta, \lambda)}\}_{\lambda \in \Lambda}$ is compatible with the countable family $\{B_i\}_{i \in \mathbb{N}}$ of subsets of $F_\beta$. The family $\{F_{(\beta, \lambda)}\}_{\lambda \in \Lambda}$ is locally finite in $X$ and consists of closed sets of $X$. Thus the family $\mathscr{F}$ indexed by $\Lambda \cup (\Gamma \backslash \{\beta\})$ which consists of the sets $F_{(\beta, \lambda)}$ for $\lambda$ in $\Lambda$ and the sets $F_\gamma$ for $\gamma$ in $\Gamma \backslash \{\beta\}$ is a locally finite closed covering of $X$. To complete the proof of $(A)$ we must show that $\mathscr{F}$ is consistent with $\mathscr{A}$. If $x \in A_m \backslash F_\beta$ then

$$\{\gamma \in \Gamma \backslash \{\beta\} \mid x \in F_\gamma\}$$

has at most $m$ elements. If $x \in A_m \cap F_\beta$, then $x \in H_{m+1-i}$ for some $i$ so that $x \in G_{mi} \subset B_i$. Since $x \in H_{m+1-i}$ and $x \in F_\beta$ the set

$$\{\gamma \in \Gamma \backslash \{\beta\} \mid x \in F_\gamma\}$$

has $m-i$ elements. Since $x \in B_i$ the set $\{\lambda \in \Lambda \mid x \in F_{(\beta, \lambda)}\}$ has at most $i$ elements. Thus $\mathscr{F}$ is consistent with $\mathscr{A}$. This completes the proof of statement $(A)$.

We now proceed with the proof of the lemma, by transfinite induction. Suppose that $\Gamma$ is well-ordered and if $\alpha \in \Gamma$ let

$$\Gamma(\alpha) = \{\gamma \in \Gamma \mid \gamma < \alpha\}, \quad \Delta(\alpha) = \{\gamma \in \Gamma \mid \gamma \leqslant \alpha\},$$

and let $\Gamma'(\alpha) = \Gamma \backslash \Gamma(\alpha)$ and $\Delta'(\alpha) = \Gamma \backslash \Delta(\alpha)$. Suppose that $\beta \in \Gamma$ and that we have a family $\{F_{(\gamma, \lambda)}\}_{(\gamma, \lambda) \in \Gamma(\beta) \times \Lambda}$ of closed sets of $X$ such that $F_{(\gamma, \lambda)} \subset U_\lambda$ if $(\gamma, \lambda) \in \Gamma(\beta) \times \Lambda$ and $\bigcup_{\lambda \in \Lambda} F_{(\gamma, \lambda)} = F_\gamma$ if $\gamma \in \Gamma(\beta)$ and suppose furthermore that if $\alpha \in \Gamma(\beta)$, then the family $\mathscr{F}_\alpha$ indexed by

$$(\Delta(\alpha) \times \Lambda) \cup \Delta'(\alpha),$$

which consists of the sets $F_{(\gamma, \lambda)}$ for $(\gamma, \lambda)$ in $\Delta(\alpha) \times \Lambda$ and the sets $F_\gamma$ for $\gamma$ in $\Delta'(\alpha)$, is a locally finite closed covering of $X$ compatible with $\mathscr{A}$. Now consider the family $\mathscr{F}$ indexed by $(\Gamma(\beta) \times \Lambda) \cup \Gamma'(\beta)$ which consists of the sets $F_{(\gamma, \lambda)}$ for $(\gamma, \lambda)$ in $\Gamma(\beta) \times \Lambda$ and the sets $F_\gamma$ for $\gamma$ in $\Gamma'(\beta)$. Then $\mathscr{F}$ is a closed covering of $X$. If $x \in X$, let

$$\Gamma_x = \{\gamma \in \Gamma \mid x \in F_\gamma\}.$$

If $\Gamma_x \subset \Gamma'(\beta)$, then since $\{F_\gamma\}_{\gamma \in \Gamma}$ is locally finite it is clear that there exists a neighbourhood of $x$ meeting only finitely many members of $\mathscr{F}$. Furthermore if $x \in A_n$, then $\Gamma_x$ has at most $n$ members so that $x$ belongs to at most $n$ members of $\mathscr{F}$. If $\Gamma_x \cap \Gamma(\beta) \neq \varnothing$, then since $\Gamma_x$ is finite there exists $\alpha$ in $\Gamma_x \cap \Gamma(\beta)$ such that $x \notin F_\gamma$ if $\alpha < \gamma < \beta$. Since $\mathscr{F}_\alpha$ is locally finite there exists a neighbourhood $W$ of $x$ such that $W$ meets only finitely many members of $\mathscr{F}_\alpha$ and $W \cap F_\gamma = \varnothing$ if $\alpha < \gamma < \beta$. Evidently $W$ meets only finitely many members of $\mathscr{F}$. Furthermore if $x \in A_n$, then since $x$ belongs to at most $n$ members of $\mathscr{F}_\alpha$, it follows that $x$ belongs to at most $n$ members of $\mathscr{F}$. Thus $\mathscr{F}$ is a locally finite closed covering of $X$ which is compatible with $\mathscr{A}$. We now apply $(A)$ to obtain a family of closed sets $\{F_{(\beta, \lambda)}\}_{\lambda \in \Lambda}$ such that $F_\beta = \bigcup_{\lambda \in \Lambda} F_{(\beta, \lambda)}$ and $F_{(\beta, \lambda)} \subset U_\lambda$ for each $\lambda$ and such that furthermore the family $\mathscr{F}_\beta$ indexed by $(\Delta(\beta) \times \Lambda) \cup \Delta'(\beta)$ which consists of the sets $F_{(\gamma, \lambda)}$ in $(\gamma, \lambda)$ in $\Delta(\beta) \times \Lambda$ and the sets $F_\gamma$ for $\gamma$ in $\Delta'(\beta)$ is a locally finite closed covering of $X$ compatible with $\mathscr{A}$. The proof is now completed by transfinite induction. $\square$

A strongly directed family $\{\mathscr{F}_i\}_{i \in \mathbf{N}}$ of locally finite coverings of a space $X$, indexed by the set $\mathbf{N}$ of positive integers directed by its usual order, will be called a strongly directed sequence of coverings of $X$.

**1.8   Theorem.** *The following statements about a topological space $X$ are equivalent:*

(a) *$X$ is a pseudo-metrizable space such that* $\dim X \leqslant n$;

(b) *there exists a regular strongly directed sequence $\{\mathscr{F}_i\}_{i \in \mathbf{N}}$ of locally finite closed coverings of $X$ such that the order of $\mathscr{F}_i$ does not exceed $n$ if $i \in \mathbf{N}$;*

(c) *there exists a pseudo-metrizable space $Z$ such that $\dim Z = 0$ and a continuous closed surjection $f: Z \to X$ such that $f^{-1}(x)$ contains at most $n+1$ points if $x \in X$.*

*Proof.* $(a) \Rightarrow (b)$. Let $X$ be a pseudo-metrizable space such that

$$\dim X \leqslant n.$$

By Proposition 2.3.4 there exists a sequence $\{\mathcal{U}_i\}_{i\in\mathbb{N}}$ of locally finite open coverings of $X$ such that for each point $x$ of $X$ and each open set $U$ such that $x \in U$, there exists some integer $i$ such that

$$\operatorname{St}(x, \mathcal{U}_i) \subset U.$$

For each $i$, let $\mathcal{U}_i = \{U_{i\lambda}\}_{\lambda\in\Lambda}$. In order to construct the required strongly directed sequence of coverings, we first define the inverse system of indexing sets $\{\Lambda(i), \rho_{ij}\}_{i,j\in\mathbb{N}}$ as follows. Let $\Lambda(1) = \Lambda$ and for each $i$ let $\Lambda(i+1) = \Lambda(i) \times \Lambda$. Let $\rho_{i,i+1}: \Lambda(i+1) \to \Lambda(i)$ be defined by putting $\rho_{i,i+1}(\xi, \lambda) = \xi$ if $\xi \in \Lambda(i)$ and $\lambda \in \Lambda$. If $i, j \in \mathbb{N}$ and $i < j$ let $\rho_{ij} = \rho_{i,i+1} \circ \ldots \circ \rho_{j-1,j}$. Since $\dim X \leqslant n$, it follows from Theorem 1.1 that $X = \bigcup_{i=1}^{n+1} A_i$, where $\dim A_i \leqslant 0$ for $i = 1, \ldots, n+1$. If we put $A_i = \varnothing$ if $i > n+1$, then it is clear that the order of a covering of $X$ compatible with $\mathscr{A} = \{A_i\}_{i\in\mathbb{N}}$ does not exceed $n$. By Lemma 1.6 there exists a locally finite closed covering $\mathscr{F}_1 = \{F_\lambda\}_{\lambda\in\Lambda(1)}$ compatible with $\mathscr{A}$ such that $F_\lambda \subset U_{1\lambda}$ for each $\lambda$. Suppose that we have a locally finite closed covering $\mathscr{F}_i = \{F_\xi\}_{\xi\in\Lambda(i)}$ of $X$ which is compatible with $\mathscr{A}$. Then by Lemma 1.7 there exists a locally finite closed covering $\mathscr{F}_{i+1} = \{F_{(\xi,\lambda)}\}_{(\xi,\lambda)\in\Lambda(i)\times\Lambda}$, compatible with $\mathscr{A}$, such that $F_{(\xi,\lambda)} \subset U_{i+1,\lambda}$ if $\lambda \in \Lambda$, and $F_\xi = \bigcup_{\lambda\in\Lambda} F_{(\xi,\lambda)}$ if $\xi \in \Lambda(i)$. It is clear that in this way we can construct a strongly directed sequence $\Phi = \{\mathscr{F}_i\}_{i\in\mathbb{N}}$ of locally finite closed coverings of $X$ of order not exceeding $n$ with $\{\Lambda(i), \rho_{ij}\}$ as associated inverse system of indexing sets. Since $\mathscr{F}_i$ is a refinement of $\mathcal{U}_i$ for each $i$, it is clear that the strongly directed sequence $\Phi$ is regular.

(b) $\Rightarrow$ (c). Let $\Phi = \{\mathscr{F}_i\}_{i\in\mathbb{N}}$ be a regular strongly directed sequence of locally finite closed coverings of a space $X$ such that the order of $\mathscr{F}_i$ does not exceed $n$ if $i \in \mathbb{N}$. Let $Z$ be the space $Z(\Phi)$ defined in §3 of Chapter 6 and let $f: Z \to X$ be the continuous closed surjection defined there. By Lemma 6.3.2, $Z$ has a base for its topology which consists of countably many disjoint open coverings of $Z$. Thus $Z$ is a regular space and it follows from Proposition 4.5.3 that $Z$ is a pseudo-metrizable space such that $\dim Z = 0$. By Lemma 6.3.6, $f^{-1}(x)$ contains at most $n+1$ points if $x \in X$.

(c) $\Rightarrow$ (a). Suppose that $X$ is a topological space for which (c) holds. Since $X$ is the perfect image of a pseudo-metrizable space, it follows from Proposition 2.5.7 that $X$ is a pseudo-metrizable space. Since a pseudo-metrizable space $Z$ such that $\dim Z = 0$ is perfectly zero-dimensional, it follows from Proposition 6.3.8 that $\Delta(X) \leqslant n$ and hence $\dim X \leqslant n$ by Proposition 6.3.12. $\square$

**1.9  Corollary.** *If $X$ is a pseudo-metrizable space, then*

$$\dim X = \operatorname{Ind} X = \Delta(X).$$

*Proof.* If $X$ is a normal regular space, then it follows from Proposition 6.3.12 that

$$\dim X \leqslant \operatorname{Ind} X \leqslant \Delta(X).$$

But it follows from Proposition 6.3.8 and Theorem 1.8 that

$$\Delta(X) \leqslant \dim X$$

if $X$ is a pseudo-metrizable space.$\square$

We observe that Theorem 1.8 gives more information than the equality $\dim X = \Delta(X)$ for a pseudo-metrizable space $X$. It was proved that if $X$ is a pseudo-metrizable space such that $\dim X \leqslant n$, then there exists a perfectly zero-dimensional *pseudo-metrizable* space $Z$ and a continuous closed surjection $f\colon Z \to X$ such that $f^{-1}(x)$ contains at most $n+1$ points if $x \in X$. The following proposition gives more supplementary information about the dimension of metrizable spaces. It was shown in Example 2.3.15 that each open covering of the Baire space $B(\tau)$ of weight $\tau$, where $\tau$ is an infinite cardinal number, has a disjoint open refinement. Thus $\dim B(\tau) = 0$.

**1.10  Proposition.** *A topological space $X$ is a metrizable space such that $\dim X \leqslant n$ and $w(X) \leqslant \tau$, where $\tau$ is an infinite cardinal number, if and only if there exists a subspace $Z$ of the Baire space $B(\tau)$ and a continuous closed surjection $f\colon Z \to X$ such that $f^{-1}(x)$ contains at most $n+1$ points if $x \in X$.*

*Proof.* Let $X$ be a metrizable space such that $\dim X \leqslant n$ and $w(X) \leqslant \tau$. Let $\Phi = \{\mathscr{F}_i\}_{i \in \mathbf{N}}$ be a strongly directed sequence of locally finite closed coverings of order not exceeding $n$, constructed as in the proof that $(a)$ implies $(b)$ in Theorem 1.8. Then $\mathscr{F}_i = \{F_\xi\}_{\xi \in \Lambda(i)}$, where $\Lambda(i)$ denotes the cartesian product of $i$ copies of a set $\Lambda$ and the inverse system of indexing sets associated with $\Phi$ is $\{\Lambda(i), \rho_{ij}\}_{i,j \in \mathbf{N}}$, where $\rho_{ij}\colon \Lambda(j) \to \Lambda(i)$ is the natural projection if $i < j$. We can suppose that $|\Lambda| = \tau$ since $w(X) \leqslant \tau$. There exists a continuous closed surjection $f\colon Z(\Phi) \to X$ such that $f^{-1}(x)$ contains at most $n+1$ points if $x \in X$. If $\boldsymbol{\lambda} = \{\lambda_i\}_{i \in \mathbf{N}}$ is a sequence of elements of $\Lambda$ and $k \in \mathbf{N}$ let $\langle \boldsymbol{\lambda}, k \rangle = (\lambda_1, \ldots, \lambda_k)$. It is easily seen that the elements of $Z(\Phi)$ can be taken to be the pairs $(x, \boldsymbol{\lambda})$, where $\boldsymbol{\lambda}$ is a sequence of elements of

$\Lambda$ and $x \in \bigcap_{k \in \mathbb{N}} F_{\langle \lambda, k \rangle}$. Since $X$ is a $T_1$-space, it follows from Lemma 6.3.4 that if $(x, \lambda) \in Z(\Phi)$ then $\bigcap_{k \in \mathbb{N}} F_{\langle \lambda, k \rangle} = \{x\}$. Thus we can define an injection $\theta : Z(\Phi) \to B(\tau)$, where $B(\tau)$ is the Baire space based on $\Lambda$, by putting $\theta(x, \lambda) = \lambda$ if $(x, \lambda) \in Z(\Phi)$. A base for the topology of $Z(\Phi)$ consists of all sets of the form

$$\{(x, \lambda) \in Z(\Phi) \mid \langle \lambda, k \rangle = \xi\},$$

where $k \in \mathbb{N}$ and $\xi \in \Lambda(k)$. Thus it is clear that $\theta$ is continuous and that the image of each open set of $Z = Z(\Phi)$ is open in $\theta(Z)$. Hence $\theta$ is an embedding and the necessity of the condition is established. Sufficiency of the condition follows at once from Theorem 1.8 on noting that by Corollary 2.5.8 and Proposition 2.5.5, the perfect image of a metrizable space of weight not exceeding $\tau$ is a metrizable space of weight not exceeding $\tau$. $\square$

Finally we obtain a condition for pseudo-metrizability and $n$-dimensionality which is related to Theorem 1.8 and which will be applied in the next section.

**1.11 Proposition.** *A topological space $X$ is a pseudo-metrizable space such that* $\dim X \leqslant n$ *if and only if there exists a sequence* $\{\mathscr{B}_i\}_{i \in \mathbb{N}}$ *of locally finite coverings of $X$ of order not exceeding $n$, such that for each $i$ the closure of each member of $\mathscr{B}_{i+1}$ is contained in some member of $\mathscr{B}_i$, and for each point $x$ of $X$ and each open set $U$ such that $x \in U$, there exists some integer $i$ such that* $\mathrm{St}\,(x, \mathscr{B}_i) \subset U$.

*Proof.* It follows from Theorem 1.8 that the condition is necessary. Now let $X$ be a space which satisfies this condition. For each positive integer $i$ let $\mathscr{B}_i = \{B_\lambda\}_{\lambda \in \Lambda(i)}$, where $\Lambda(i)$ and $\Lambda(j)$ are disjoint if $i$ and $j$ are distinct positive integers. Let us define $\tau_i : \Lambda(i+1) \to \Lambda(i)$ such that $B_\lambda \subset B_{\tau_i(\lambda)}$ if $\lambda \in \Lambda(i+1)$.

For each positive integer $i$, let $X_i = \bigcap_{\lambda \in \Lambda(i)} (B_\lambda \times (\lambda))$ and let the closed sets of $X_i$ be the empty set and all subsets of $X_i$ of the form $\bigcup_{\lambda \in \Lambda(i)} (C_\lambda \times (\lambda))$, where $C_\lambda$ is a closed set of the subspace $B_\lambda$ if $\lambda \in \Lambda(i)$. Now define $\sigma_i : X_i \to X$ by $\sigma_i(x, \lambda) = x$ if $(x, \lambda) \in X_i$. Then $\sigma_i$ is continuous. Since the order of $\mathscr{B}_i$ does not exceed $n$, it follows that $\sigma_i^{-1}(x)$ contains at most $n+1$ points if $x \in X$, and the subspace $\sigma_i^{-1}(x)$ of $X_i$ is discrete if $x \in X$. For each positive integer $i$ let

$$\pi_{i, i+1} : X_{i+1} \to X_i$$

be defined by

$$\pi_{i, i+1}(x, \lambda) = \big(x, \tau_i(\lambda)\big) \quad \text{if} \quad (x, \lambda) \in X_{i+1}.$$

The function $\pi_{i,i+1}$ is continuous. For each $i$, let $\pi_{ii}$ be the identity function on $X_i$. If $i,j \in N$ and $i < j$ let $\pi_{ij}: X_j \to X_i$ be the composite $\pi_{i,i+1} \circ \ldots \circ \pi_{j-1,j}$. Then we have an inverse system $X = \{X_i, \pi_{ij}\}$ of spaces over the directed set $N$ of positive integers. Let $Z$ be the inverse limit of $X$ and let $\pi_i: Z \to X_i$ be the canonical mapping if $i \in N$. Then $Z$ consists of the pairs $(x, \xi)$, where $\xi = \{\lambda(i)\}_{i \in N}$ such that $\lambda(i) \in \Lambda(i)$ and $\tau_i(\lambda(i+1)) = \lambda(i)$ for each $i$ and $x \in \bigcap_{i \in N} B_{\lambda(i)}$. Furthermore $\pi_i(x, \xi) = (x, \lambda(i))$ if $(x, \xi) \in Z$ and $\xi = \{\lambda(i)\}$. A base for the topology of $Z$ consists of the sets $V_\lambda = \pi_i^{-1}(B_\lambda \times (\lambda))$, where $\lambda \in \Lambda(i)$ and $i \in N$. For each $i$, $\{V_\lambda\}_{\lambda \in \Lambda(i)}$ is a disjoint covering of $Z$ by open-and-closed sets. It follows from Proposition 4.5.3 that $Z$ is a pseudo-metrizable space such that $\dim Z \leqslant 0$. We note that if $\lambda \in \Lambda(i)$ and $\mu \in \Lambda(i+1)$, then $V_\lambda \cap V_\mu = \varnothing$ if $\tau_i(\mu) \neq \lambda$.

Let $x$ be a point of $X$. For each $i$, $\sigma_i^{-1}(x)$ is a non-empty finite discrete subspace of $X_i$, and if $i,j \in N$ and $i \leqslant j$ then

$$\pi_{ij}(\sigma_j^{-1}(x)) \subset \sigma_i^{-1}(x).$$

Thus the subspaces $\sigma_i^{-1}(x)$ form a subsystem of $X$. By Proposition 1.7.2, the inverse limit of this subsystem is non-empty. It follows that there exists $\xi$ such that $(x, \xi) \in Z$. Thus, in particular, $Z$ is non-empty and if we define $f: Z \to X$ by putting $f(x, \xi) = x$ if $(x, \xi) \in Z$, then $f$ is a surjection. Since $f = \sigma_i \circ \pi_i$ for each $i$, it follows that $f$ is continuous. Suppose that there exists some point $x$ of $X$ such that $f^{-1}(x)$ contains a set $B$ with $n+2$ elements. Then there exists some positive integer $k$ such that $\pi_k|B$ is injective. Since $\pi_k(B) \subset \sigma_k^{-1}(x)$ and $\sigma_k^{-1}(x)$ contains at most $n+1$ points, this is absurd. Hence $f^{-1}(x)$ contains at most $n+1$ points if $x \in X$. Finally we show that the continuous surjection $f: Z \to X$ is closed. Let $E$ be a closed set of $Z$ and let $x_0 \in [f(E)]^-$. Since $Z = \bigcup_{\lambda \in \Lambda(1)} V_\lambda$, it follows that

$$f(E) = \bigcup_{\lambda \in \Lambda(1)} f(E \cap V_\lambda).$$

But $f(E \cap V_\lambda) \subset B_\lambda$ and the family $\{B_\lambda\}_{\lambda \in \Lambda(1)}$ is locally finite. Hence

$$[f(E)]^- = \bigcup_{\lambda \in \Lambda(1)} [f(E \cap V_\lambda)]^-.$$

Thus we can choose $\lambda(1)$ in $\Lambda(1)$ such that $x_0 \in [f(E \cap V_{\lambda(1)})]^-$. Now suppose that we have found $\lambda(i)$ in $\Lambda(i)$ for $i = 1, \ldots, k$ such that $x_0 \in [f(E \cap \bigcap_{i=1}^k V_{\lambda(i)})]^-$. Since $Z = \bigcup_{\lambda \in \Lambda(k+1)} V_\lambda$ we can employ the above argument again to find $\lambda(k+1)$ in $\Lambda(k+1)$ such that $x_0 \in [f(E \cap \bigcap_{i=1}^{k+1} V_{\lambda(i)})]^-$. By induction we can find $\xi = \{\lambda(i)\}_{i \in N}$ such that $\lambda(i) \in \Lambda(i)$ for each $i$ and $x_0 \in [f(E \cap \bigcap_{i=1}^k V_{\lambda(i)})]^-$ for each positive integer $k$. Since $V_{\lambda(i)} \cap V_{\lambda(i+1)} \neq \varnothing$, it follows that $\lambda(i) = \tau_i\big(\lambda(i+1)\big)$ for each positive integer $i$. Since $f(V_{\lambda(i)}) \subset B_{\lambda(i)}$, it follows that

$$x_0 \in \bigcap_{i \in \mathbf{N}} \bar{B}_{\lambda(i)}.$$

But $\bar{B}_{\lambda(i+1)} \subset B_{\lambda(i)}$ for each $i$ and hence $x_0 \in \bigcap_{i \in \mathbf{N}} B_{\lambda(i)}$. Thus if

$$z_0 = (x_0, \xi),$$

then $z_0 \in Z$ and $f(z_0) = x_0$. But since $V_{\lambda(i)} \cap E \neq \varnothing$ for each $i$, it follows that every neighbourhood of $z_0$ meets $E$ and hence $z_0 \in E$ since $E$ is a closed set. Hence $x_0 \in f(E)$. Thus $f$ is a closed mapping. It follows from Theorem 1.8 that $X$ is a pseudo-metrizable space such that $\dim X \leqslant n$. $\square$

## 2  Sequences of open coverings and inverse limits

In this section we determine the dimension of a pseudo-metrizable space in terms of the existence of a sequence of open coverings with certain properties. We give a characterization of the topological spaces $X$ which are pseudo-metrizable and satisfy $\dim X \leqslant n$ which is analogous to the pseudo-metrization condition of Proposition 2.3.8. In the latter part of the section we consider inverse limits of metrizable spaces.

**2.1  Proposition.** *A space $X$ is pseudo-metrizable and satisfies $\dim X \leqslant n$ if and only if there exists a sequence $\{\mathscr{U}_i\}_{i \in \mathbf{N}}$ of open coverings of $X$ such that (a) $\mathscr{U}_i$ has order not exceeding $n$ and $\mathscr{U}_{i+1}$ is a refinement of $\mathscr{U}_i$ for each positive integer $i$, and (b) for each point $x$ of $X$ and each open set $U$ such that $x \in U$ there exists an integer $i$ and an open set $V$ such that $x \in V$ and $\mathrm{St}(V, \mathscr{U}_i) \subset U$.*

*Proof.* Let $X$ be a pseudo-metric space such that $\dim X \leqslant n$. For each positive integer $i$ let $\mathscr{W}_i$ be the open covering $\{B_{1/2^i}(x)\}_{x \in X}$. Let $\mathscr{U}_1$ be an open covering of $X$ of order not exceeding $n$ which is a refinement of $\mathscr{W}_1$. By induction we can construct a sequence $\{\mathscr{U}_i\}_{i \in \mathbf{N}}$ of open coverings of $X$ such that for each $i$, $\mathscr{U}_i$ is a refinement of $\mathscr{W}_i$ of order not exceeding $n$ and $\mathscr{U}_{i+1}$ is a refinement of $\mathscr{U}_i$. This sequence of coverings evidently satisfies the condition of the proposition.

Now suppose that $\{\mathscr{U}_i\}_{i \in \mathbf{N}}$ is a sequence of open coverings of a space $X$ such that the condition of the proposition is satisfied. It follows from Proposition 2.3.8 that the space $X$ is pseudo-metrizable and it remains to show that $\dim X \leqslant n$. If $i \in \mathbf{N}$ let $\mathscr{U}_i = \{U_\lambda\}_{\lambda \in \Lambda(i)}$, where we suppose that $\Lambda(i)$ and $\Lambda(j)$ are disjoint if $i \neq j$. We remark first since $\mathcal{U}_{i+1}$ is a refinement of $\mathcal{U}_i$ for each $i$, it follows that if $x$ is a point of $X$ and $U$ is an open set of $X$ such that $x \in U$, then for some $i$ there exists $\lambda$ in $\Lambda(i)$ such that $x \in U_\lambda$ and $\operatorname{St}(U_\lambda, \mathcal{U}_i) \subset U$. Now let $\mathcal{G}$ be an open covering of $X$. For each positive integer $i$ let $\Omega(i)$ be the subset of $\Lambda(i)$ consisting of those $\lambda$ for which $\operatorname{St}(U_\lambda, \mathcal{U}_i)$ is contained in some member of $\mathcal{G}$, and let

$$X_i = \bigcup_{\lambda \in \Omega(i)} U_\lambda.$$

Clearly $\{X_i\}_{i \in \mathbf{N}}$ is a countable open covering of $X$. For each $i$, let

$$L(i) = \{\lambda \in \Lambda(i) \mid U_\lambda \cap X_i \neq \varnothing\},$$

$$M(i) = \{\lambda \in L(i) \mid U_\lambda \cap (\bigcup_{j<i} X_j) = \varnothing\},$$

$$N(i) = \{\lambda \in L(i) \mid U_\lambda \cap (\bigcup_{j<i} X_j) \neq \varnothing\}.$$

Then $L(1) = M(1)$ and $L(i)$ is the disjoint union of $M(i)$ and $N(i)$ for each $i$. For each positive integer $i$, let us define

$$\tau_{i,i+1} : \Lambda(i+1) \to \Lambda(i)$$

so that if $\mu \in \Lambda(i+1)$ and $\tau_{i,i+1}(\mu) = \lambda$, then $U_\mu \subset U_\lambda$. If $i, j \in N$ and $i < j$ let

$$\tau_{ij} = \tau_{i,i+1} \circ \ldots \circ \tau_{j-1,j}.$$

Clearly $\tau_{ij} \circ \tau_{jk} = \tau_{ik}$ if $i < j < k$. If $i, j \in \mathbf{N}$ and $i < j$ and $\lambda \in M(i)$, let $N(i,j,\lambda) = \{\mu \in N(j) \mid \tau_{ij}(\mu) = \lambda, \tau_{kj}(\mu) \notin M(k) \text{ if } i < k < j\}$. Then

$$N(j) = \bigcup \{N(i,j,\lambda) \mid \lambda \in M(i), i < j\}.$$

For if $\mu \in N(j)$, then $U_\mu \cap X_r \neq \varnothing$ for some $r < j$, so that $U_\nu \cap X_r \neq \varnothing$, where $\nu = \tau_{rj}(\mu)$. If $r$ is the least integer with this property, then $\tau_{rj}(\mu) \in M(r)$. Now let $i$ be the largest integer such that $i < j$ and $\tau_{ij}(\mu) \in M(i)$. If $\tau_{ij}(\mu) = \lambda$, then $\mu \in N(i,j,\lambda)$ as required.

If $i \in N$ and $\lambda \in M(i)$, let

$$V_\lambda = (U_\lambda \cap X_i) \cup \bigcup \{U_\mu \cap X_j \mid \mu \in N(i,j,\lambda), j > i\},$$

and let $\mathcal{V} = \{V_\lambda\}_{\lambda \in M(i), i \in \mathbf{N}}$. Then $\mathcal{V}$ is a family of open sets of $X$ and $V_\lambda \subset U_\lambda$ if $\lambda \in M(i)$. We shall complete the proof by showing that $\mathcal{V}$ is a covering of $X$ of order not exceeding $n$ which is a refinement of $\mathcal{G}$.

Let $x$ be a point of $X$. Then there exists $j$ such that $x \in X_j$ and $x \in U_\mu$ for some $\mu$ in $\Lambda(j)$. It follows that $\mu \in L(j)$. If $\mu \in M(j)$, then $x \in U_\mu \cap X_j \subset V_\mu$. If $\mu \notin M(j)$, then $\mu \in N(j)$ so that there exist $i < j$ and

$\lambda$ in $M(i)$ such that $\mu \in N(i, j, \lambda)$ and in this case $x \in U_\mu \cap X_j \subset V_\lambda$. Thus $\mathscr{V}$ is an open covering of $X$. Furthermore $\mathscr{V}$ is a refinement of $\mathscr{G}$. For if $\lambda \in M(i)$, then $U_\lambda \cap X_i \neq \varnothing$ so that there exists $\nu$ in $\Omega(i)$ such that $U_\lambda \cap U_\nu \neq \varnothing$. Since $\mathrm{St}\,(U_\nu, \mathscr{U}_i) \subset G$ for some member $G$ of $\mathscr{G}$, it follows that $U_\lambda \subset G$. But $V_\lambda \subset U_\lambda$ so that $V_\lambda \subset G$.

Finally the order of $\mathscr{V}$ does not exceed $n$. For suppose that this is not the case. Then there exist a point $x$ of $X$ and $n+2$ distinct elements $\lambda(0), \ldots, \lambda(n+1)$ of $\bigcup_{i \in \mathbb{N}} M(i)$ such that

$$x \in \bigcap_{r=0}^{n+1} V_{\lambda(r)}.$$

Let $\lambda(r) \in M\big(k(r)\big)$ for each $r$. There exists an integer $s$ such that $x \in X_s \backslash \bigcup_{j<s} X_j$. Since $V_{\lambda(r)} \cap (\bigcup_{j<k(r)} X_j) = \varnothing$, it follows that $k(r) \leqslant s$ for $r = 0, \ldots, n+1$. Let

$$A = \{r \mid k(r) < s\}, B = \{r \mid k(r) = s\},$$

so that $A$ and $B$ are disjoint sets such that $A \cup B = \{0, \ldots, n+1\}$. If $r \in A$, then since $x \notin X_{k(r)}$ there exist $m(r) \geqslant s$ and $\mu(r)$ in

$$N\big(k(r), m(r), \lambda(r)\big)$$

such that $x \in U_{\mu(r)} \cap X_{m(r)}$, and we put $\nu(r) = \tau_{s, m(r)}\big(\mu(r)\big)$. If $r \in B$ we put $\nu(r) = \lambda(r)$. Then $\nu(0), \ldots, \nu(n+1)$ are distinct elements of $\Lambda(s)$. For it is evident that if $r, t \in B$ and $r \neq t$, then $\nu(r) \neq \nu(t)$. If $r \in A$ and $t \in B$, then $\nu(t) \in M(s)$ whilst $\nu(r) \notin M(s)$ so that $\nu(r) \neq \nu(t)$. Lastly suppose that $r, t \in A$ and $r \neq t$. If $k(r) = k(t) = k$, then

$$\tau_{ks}\big(\nu(r)\big) = \lambda(r) \neq \lambda(t) = \tau_{ks}\big(\nu(t)\big),$$

so that $\nu(r) \neq \nu(t)$. If $k(r) \neq k(t)$ let us suppose that

$$k(r) < k(t) < s \leqslant m(r).$$

Then $\tau_{k(t)s}\big(\nu(r)\big) = \tau_{k(t)\,m(r)}\big(\mu(r)\big) \notin M\big(k(t)\big)$. Since

$$\tau_{k(t)s}\big(\nu(t)\big) = \lambda(t) \in M\big(k(t)\big)$$

it follows that $\nu(r) \neq \nu(t)$. Thus $\nu(0), \ldots, \nu(n+1)$ are distinct elements of $\Lambda(s)$ as asserted. But

$$x \in \bigcap_{r=0}^{n+1} U_{\nu(r)}.$$

Since the order of $\mathscr{U}_s$ does not exceed $n$, this is absurd. It follows that the order of $\mathscr{V}$ does not exceed $n$ as was to be shown. $\square$

If $X$ is a pseudo-metric space and $\mathcal{U} = \{U_\lambda\}_{\lambda \in \Lambda}$ is a covering of $X$, then we define

$$\text{mesh}\,\mathcal{U} = \sup_{\lambda \in \Lambda} \text{diam}\, U_\lambda.$$

**2.2   Corollary.** *If $X$ is a pseudo-metric space, then* $\dim X \leqslant n$ *if and only if there exists a sequence* $\{\mathcal{U}_i\}_{i \in \mathbf{N}}$ *of open coverings of $X$ such that the order of* $\mathcal{U}_i$ *does not exceed $n$ and* $\mathcal{U}_{i+1}$ *is a refinement of* $\mathcal{U}_i$ *if* $i \in \mathbf{N}$ *and*

$$\lim_{i \to \infty} \text{mesh}\,\mathcal{U}_i = 0.$$

*Proof.* The proof of the necessity of this condition is the same as the proof of the necessity of the condition given in Proposition 2.1. Now suppose that $X$ is a pseudo-metric space and that $\{\mathcal{U}_i\}_{i \in \mathbf{N}}$ is a sequence of open coverings satisfying the condition. Let $U$ be an open set of $X$ and let $x$ be a point of $U$. Then there exists a positive integer $j$ such that $B_{1/2^j}(x) \subset U$. There exists a positive integer $i$ such that $\text{mesh}\,\mathcal{U}_i < 1/2^{j+1}$. If $V = B_{1/2^{j+1}}(x)$, then $V$ is open, $x \in V$ and $\text{St}\,(V, \mathcal{U}_i) \subset U$. It follows from Proposition 2.1 that $\dim X \leqslant n$. $\square$

Next we have another application of Proposition 2.1. Let us say that a space $X$ is the inverse limit of a sequence of spaces if $X$ is the inverse limit of an inverse system $\{X_i, \pi_{ij}\}$ of spaces over the set $\mathbf{N}$ of positive integers, directed by its usual order.

**2.3   Proposition.** *If a space $X$ is the inverse limit of a sequence* $\{X_i\}_{i \in \mathbf{N}}$ *of pseudo-metrizable spaces such that* $\dim X_i \leqslant n$ *if* $i \in \mathbf{N}$, *then $X$ is a pseudo-metrizable space such that* $\dim X \leqslant n$.

*Proof.* If $i, j \in \mathbf{N}$ and $i \leqslant j$, let $\pi_{ij} : X_j \to X_i$ be the connecting mapping, and let $\pi_i : X \to X_i$ be the canonical mapping if $i \in \mathbf{N}$. If $i \in \mathbf{N}$, let $\{\mathcal{U}_{ij}\}_{j \in \mathbf{N}}$ be a sequence of open coverings of $X_i$ such that $\mathcal{U}_{i,j+1}$ is a refinement of $\mathcal{U}_{ij}$ for each $j$, and for each point $x_i$ of $X_i$ and each open neighbourhood $V_i$ of $x_i$ in $X_i$ there exist $W_i$ open in $X_i$ and an integer $j$ such that $x_i \in W_i$ and $\text{St}\,(W_i, \mathcal{U}_{ij}) \subset V_i$. Let $\mathcal{V}_1$ be an open covering of $X_1$ of order not exceeding $n$ which is a refinement of $\mathcal{U}_{11}$. Suppose that $i$ is a positive integer and that $\mathcal{V}_i$ is an open covering of $X_i$. Consider the finite set of open coverings of $X_{i+1}$ which consists of $\pi_{i,i+1}^{-1}(\mathcal{V}_i)$ and $\pi_{j,i+1}^{-1}(\mathcal{U}_{j,i+1})$ for $j = 1, \ldots, i+1$. These coverings have a common open refinement $\mathcal{V}_{i+1}$ of order not exceeding $n$. Thus by induction we have an open covering $\mathcal{V}_j$ of $X_j$ for each positive integer $j$ such that the order of $\mathcal{V}_j$ does not exceed $n$ and $\mathcal{V}_j$ is a refinement of each of the coverings $\pi_{ij}^{-1}(\mathscr{U}_{ij})$, where $i \leqslant j$, and of $\pi_{j-1,j}^{-1}(\mathscr{V}_{j-1})$ if $j > 1$. Now for each $j$ let $\mathscr{U}_j = \pi_j^{-1}(\mathscr{V}_j)$. Then $\{\mathscr{U}_j\}_{j \in \mathbf{N}}$ is a sequence of open coverings of $X$ of order not exceeding $n$ and $\mathscr{U}_{j+1}$ is a refinement of $\mathscr{U}_j$ for each $j$. Let $U$ be open in $X$ and let $x$ be a point of $U$. Then there exist an integer $i$ and $V_i$ open in $X_i$ such that $x \in \pi_i^{-1}(V_i) \subset U$. If $x_i = \pi_i(x)$, then $x_i \in V_i$ so that there exist an open set $W_i$ of $X_i$ and an integer $j$ such that $x_i \in W_i$ and $\mathrm{St}\,(W_i, \mathscr{U}_{ij}) \subset V_i$. If $W = \pi_i^{-1}(W_i)$, then $W$ is open in $X$ and $x \in W$. If $j \leqslant i$, then $\mathrm{St}\,(W_i, \mathscr{V}_i) \subset V_i$ so that $\mathrm{St}\,(W, \mathscr{U}_i) \subset U$. If $j > i$, then $\mathrm{St}\big(\pi_{ij}^{-1}(W_i), \mathscr{V}_j\big) \subset \pi_{ij}^{-1}(V_i)$ so that $\mathrm{St}\,(W, \mathscr{U}_j) \subset U$. It now follows from Proposition 2.1 that $X$ is a pseudo-metrizable space such that $\dim X \leqslant n$. $\square$

This section concludes with another result involving inverse limits of metrizable spaces. In §5 of Chapter 2 the paracompact $M$-spaces were characterized as the inverse limits of inverse systems of metrizable spaces with perfect connecting mappings. We shall show that if $X$ is a paracompact $M$-space such that $\dim X \leqslant n$, then $X$ is the inverse limit of an inverse system of metrizable spaces $Y_\xi$, with perfect connecting mappings, such that $\dim Y_\xi \leqslant n$ for all $\xi$. Some preliminary results must be obtained.

**2.4  Proposition.** *If $X$ is a paracompact normal space such that $\dim X \leqslant n$ and $\{\mathscr{V}_i\}_{i \in \mathbf{N}}$ is a sequence of open coverings of $X$, then there exists a sequence $\{\mathscr{U}_i\}_{i \in \mathbf{N}}$ of open coverings of $X$ of order not exceeding $n$ such that for each $i$, $\mathscr{U}_i$ is a refinement of $\mathscr{V}_i$, $\mathscr{U}_{i+1}$ is a strong star-refinement of $\mathscr{U}_i$ and each member of $\mathscr{U}_{i+1}$ meets at most $n + 1$ members of $\mathscr{U}_i$.*

*Proof.* Since $X$ is a paracompact normal space such that $\dim X \leqslant n$, there exists a locally finite open refinement $\mathscr{U}_1$ of $\mathscr{V}_1$ such that the order of $\{\bar{U} \mid U \in \mathscr{U}_1\}$ does not exceed $n$. Suppose that $\mathscr{U}_k$ is a locally finite open covering of $X$ such that the order of $\{\bar{U} \mid U \in \mathscr{U}_k\}$ does not exceed $n$. If $x \in X$, let

$$G_x = X \backslash \bigcup \{\bar{U} \mid U \in \mathscr{U}_k \quad \text{and} \quad x \notin \bar{U}\}$$

and let $\mathscr{G}_{k+1} = \{G_x\}_{x \in X}$. By Proposition 2.1.27 there exists an open strong star-refinement $\mathscr{W}_{k+1}$ of $\mathscr{U}_k$. Now since $X$ is a paracompact normal space such that $\dim X \leqslant n$, we can choose a locally finite open covering $\mathscr{U}_{k+1}$ of $X$, which is a refinement of the open coverings $\mathscr{G}_{k+1}$, $\mathscr{V}_{k+1}$ and $\mathscr{W}_{k+1}$, such that the order of $\{\bar{U} \mid U \in \mathscr{U}_{k+1}\}$ does not exceed $n$. Clearly $\mathscr{U}_{k+1}$ is a strong star-refinement of $\mathscr{U}_k$. If $U \in \mathscr{U}_{k+1}$, then $U \subset G_x$ for some $x$ in $X$. Thus if $V \in \mathscr{U}_k$ and $U \cap V \neq \varnothing$ then $x \in \bar{V}$.

Since the order of $\{\bar{V} \mid V \in \mathcal{U}_k\}$ does not exceed $n$, it follows that $U$ meets at most $n+1$ members of $\mathcal{U}_k$. It now follows by induction that we can construct the required sequence $\{\mathcal{U}_i\}_{i \in \mathbf{N}}$ of open coverings of $X$. $\square$

We recall from §5 of Chapter 2 that if $\Psi = \{\mathcal{U}_i\}_{i \in \mathbf{N}}$ is a sequence of open coverings of a space $X$ such that $\mathcal{U}_{i+1}$ is a strong star-refinement of $\mathcal{U}_i$ for each $i$, then there exist a metric space $W(\Psi)$ and a continuous surjection $g: X \to W(\Psi)$ which is a $\mathcal{U}_i$-mapping for every $i$.

**2.5 Proposition.** *If* $\Psi = \{\mathcal{U}_i\}_{i \in \mathbf{N}}$ *is a sequence of open coverings, each of order not exceeding $n$, of a space such that for each positive integer $i$, $\mathcal{U}_{i+1}$ is a strong star-refinement of $\mathcal{U}_i$ and each member of $\mathcal{U}_{i+1}$ meets finitely many members of $\mathcal{U}_i$, then* dim $W(\Psi) \leqslant n$.

*Proof.* Let $\Psi = \{\mathcal{U}_i\}_{i \in \mathbf{N}}$ be a sequence of open coverings of a space $X$ with the properties stated above. For each $i$, let $\mathcal{U}_i = \{U_\lambda\}_{\lambda \in \Lambda(i)}$, where we suppose that $\Lambda(i)$ and $\Lambda(j)$ are disjoint if $i$ and $j$ are distinct positive integers. Let $W = W(\Psi)$. The notation of §5 of Chapter 2 will be used throughout this proof. We recall that if $w \in W$ and $x \in g^{-1}(w)$, then
$$g^{-1}(w) = \{y \in X \mid d(x,y) = 0\}.$$
If $i \in \mathbf{N}$ and $\lambda \in \Lambda(i)$ let
$$B_\lambda = \{w \in W \mid g^{-1}(w) \subset U_\lambda\}.$$
Then $\mathcal{B}_i = \{B_\lambda\}_{\lambda \in \Lambda(i)}$ is a covering of $W$ since $g$ is a $\mathcal{U}_i$-mapping. Since the order of $\mathcal{U}_i$ does not exceed $n$, it is clear that the order of $\mathcal{B}_i$ does not exceed $n$. Furthermore each member of $\mathcal{B}_{i+1}$ meets only finitely many members of $\mathcal{B}_i$. Now let $w$ be a point of $W$ and suppose that $x \in g^{-1}(w)$. There exists $\lambda$ in $\Lambda(i+1)$ such that St $(x, \mathcal{U}_{i+2}) \subset U_\lambda$. Let $G$ be the open ball in $W$ with centre $w$ and radius $1/2^{i+3}$ and suppose that $v \in G$. If $y \in g^{-1}(v)$, then $d(x,y) < 1/2^{i+3}$ so that $D(x,y) < 1/2^{i+1}$. Since $D(x,y) \leqslant 1/2^{i+2}$ it follows that $y \in$ St $(x, \mathcal{U}_{i+2}) \subset U_\lambda$. Hence
$$g^{-1}(v) \subset U_\lambda$$
so that $v \in B_\lambda$. Thus $G \subset B_\lambda$ so that the open neighbourhood $G$ of $w$ meets only finitely many members of $\mathcal{B}_i$. Thus each covering $\mathcal{B}_i$ of $W$ is locally finite.

If $\lambda \in \Lambda(i+1)$ there exists $\mu$ in $\Lambda(i)$ such that St $(U_\lambda, \mathcal{U}_{i+1}) \subset U_\mu$. Suppose that $w \in W$ and $w \notin B_\mu$. Then $g^{-1}(w) \not\subset U_\mu$ so that there exists $x$ in $g^{-1}(w)$ such that $x \notin U_\mu$. Hence St $(x, \mathcal{U}_{i+1})$ is disjoint from $U_\lambda$.

Now let $v$ be a point of the open ball $P$ in $W$ with centre $w$ and radius $1/2^{i+2}$. If $y \in g^{-1}(v)$, then $d(x, y) < 1/2^{i+2}$ so that $D(x, y) \leqslant 1/2^{i+1}$, from which it follows that $y \in \mathrm{St}\,(x, \mathscr{U}_{i+1})$ and hence $y \notin U_\lambda$. Thus if $v \in P$, then $g^{-1}(v)$ is not contained in $U_\lambda$ so that $v \notin B_\lambda$. Hence $w \notin \bar{B}_\lambda$. It follows that $\bar{B}_\lambda \subset B_\mu$.

Finally let $H$ be open in $W$ and let $w$ be a point of $H$. There exists a positive integer $i$ such that $Q \subset H$, where $Q$ is the open ball in $W$ of radius $1/2^{i-1}$ and centre $w$. Suppose that $\lambda \in \Lambda(i)$ and that $w \in B_\lambda$. If $v \in B_\lambda$, then $g^{-1}(v) \subset U_\lambda$. Thus if $x \in g^{-1}(w)$ and $y \in g^{-1}(v)$ it follows that $y \in \mathrm{St}\,(x, \mathscr{U}_i)$ so that $D(x, y) \leqslant 1/2^i < 1/2^{i-1}$ and hence $d(x, y) < 1/2^{i-1}$. Hence $v \in Q$ and it follows that $\mathrm{St}\,(w, \mathscr{B}_i) \subset H$. It follows from Proposition 1.11 that $\dim W(\Psi) \leqslant n$.$\square$

**2.6   Proposition.** *Let $X$ be a paracompact $M$-space such that*

$$\dim X \leqslant n$$

*and let $f \colon X \to Z$ be a perfect mapping, where $Z$ is a metric space. If $\mathscr{U}$ is an open covering of $X$, then there exists a metric space $W$ such that $\dim W \leqslant n$ and perfect mappings $g \colon X \to W$ and $h \colon W \to Z$ such that $g$ is a $\mathscr{U}$-mapping and $f = h \circ g$.*

*Proof.* For each positive integer $i$, let $\mathscr{G}_i$ be the open covering of $Z$ by the open $\frac12^i$-balls and let $\mathscr{V}_i = f^{-1}(\mathscr{G}_i)$. From the proof of Proposition 2.5.20 we see that if $\Psi = \{\mathscr{U}_i\}_{i \in \mathbf{N}}$ is a sequence of open coverings of $X$ such that $\mathscr{U}_1$ is a refinement of $\mathscr{U}$ and $\mathscr{U}_i$ is a refinement of $\mathscr{V}_i$ and $\mathscr{U}_{i+1}$ is a strong star-refinement of $\mathscr{U}_i$ for each $i$, then $g \colon X \to W(\Psi)$ is a perfect $\mathscr{U}$-mapping and there exists a perfect mapping $h \colon W(\Psi) \to Z$ such that $f = h \circ g$. Since $\dim X \leqslant n$, it follows from Proposition 2.4 that we can choose the sequence $\Psi = \{\mathscr{U}_i\}_{i \in \mathbf{N}}$ to satisfy the above conditions and additionally to satisfy the conditions that for each $i$ the order of $\mathscr{U}_i$ does not exceed $n$ and each member of $\mathscr{U}_{i+1}$ meets at most $n + 1$ members of $\mathscr{U}_i$. It then follows from Proposition 2.5 that $\dim W(\Psi) \leqslant n$.$\square$

**2.7   Proposition.** *If $X$ is a paracompact $M$-space such that $\dim X \leqslant n$, then $X$ is a homeomorphic with the inverse limit of an inverse system $\{Y_\xi, \pi_{\xi\eta}\}_{\xi, \eta \in \Omega}$, where each space $Y_\xi$ is a metrizable space such that*

$$\dim Y_\xi \leqslant n$$

*and $\pi_{\xi\eta}$ is a perfect mapping if $\xi, \eta \in \Omega$ and $\xi \leqslant \eta$.*

*Proof.* Let $\Lambda$ be the set of open coverings of $X$, let $\Omega_0 = \Lambda$ and for each non-negative integer $n$ let $\Omega_{n+1} = \Omega_n \times \Omega_n$. Let $\Omega = \bigcup_{n \geqslant 0} \Omega_n$

be given the structure of a directed set as in the proof of Proposition 2.5.21. If $\lambda \in \Lambda = \Omega_0$, then by Proposition 2.6 there exists a metric space $Y_\lambda$ such that $\dim Y_\lambda \leqslant n$ and a perfect $\lambda$-mapping $g_\lambda : X \to Y_\lambda$. Now suppose that $n \geqslant 0$ and that for each $\alpha$ in $\bigcup_{k \leqslant n} \Omega_k$ we have found a metrizable space $Y_\alpha$ such that $\dim Y_\alpha \leqslant n$ and a perfect mapping $g_\alpha : X \to Y_\alpha$. Suppose furthermore that for each pair $\alpha, \beta$ in $\bigcup_{k \leqslant n} \Omega_k$ such that $\alpha \leqslant \beta$ we have found a perfect mapping

$$\pi_{\alpha\beta} : Y_\beta \to Y_\alpha$$

such that $\pi_{\alpha\beta} \circ g_\beta = g_\alpha$, $\pi_{\alpha\alpha}$ is the identity, and if $\alpha, \beta, \gamma \in \bigcup_{k \leqslant n} \Omega_k$ and $\alpha \leqslant \beta \leqslant \gamma$ then $\pi_{\alpha\beta} \circ \pi_{\beta\gamma} = \pi_{\alpha\gamma}$. Let $\xi = (\alpha, \beta) \in \Omega_{n+1}$. By Proposition 2.5.15 there exist a metrizable space $Z$ and perfect mappings $f : X \to Z$, $\sigma : Z \to X_\alpha$ and $\rho : Z \to X_\beta$ such that $\sigma \circ f = g_\alpha$ and $\rho \circ f = g_\beta$. By Proposition 2.6 there exist a metrizable space $Y_\xi$ such that

$$\dim Y_\xi \leqslant n,$$

and perfect mappings $g_\xi : X \to Y_\xi$ and $h : Y_\xi \to Z$ such that $h \circ g_\xi = f$. If $\pi_{\alpha\xi} = \sigma \circ h$ and $\pi_{\beta\xi} = \rho \circ h$, then $\pi_{\alpha\xi} : Y_\xi \to Y_\alpha$ and $\pi_{\beta\xi} : Y_\xi \to Y_\beta$ are perfect mappings such that $\pi_{\alpha\xi} \circ g_\xi = g_\alpha$ and $\pi_{\beta\xi} \circ g_\xi = g_\beta$. As in the proof of Proposition 2.5.21, it follows by induction that if $\xi \in \Omega$ then there exist a metrizable space $Y_\xi$ such that $\dim Y_\xi \leqslant n$, and a perfect mapping $g_\xi : X \to Y_\xi$, and if $\xi, \eta \in \Omega$ and $\xi \leqslant \eta$ there exists a perfect mapping $\pi_{\xi\eta} : Y_\eta \to Y_\xi$ such that $\pi_{\xi\eta} \circ g_\eta = g_\xi$, $\pi_{\xi\xi}$ is the identity mapping, and if $\xi, \eta, \zeta \in \Omega$ and $\xi \leqslant \eta \leqslant \zeta$ then $\pi_{\xi\eta} \circ \pi_{\eta\zeta} = \pi_{\xi\zeta}$. Thus $\{Y_\xi, \pi_{\xi\eta}\}_{\xi,\eta \in \Omega}$ is an inverse system of spaces over $\Omega$. Let $Y$ be the inverse limit of this inverse system with canonical mappings $\pi_\xi : Y \to Y_\xi$ if $\xi \in \Omega$. Then, as in the proof of Proposition 2.5.21, the unique continuous function $g : X \to Y$, such that $\pi_\xi \circ g = g_\xi$ if $\xi \in \Omega$, is a homeomorphism.$\square$

## 3  Universal spaces for metrizable spaces of given weight and dimension

We have seen that if $\tau$ is an infinite cardinal number then the Baire space $B(\tau)$ is a zero-dimensional metrizable space of weight equal to $\tau$. We prove now that $B(\tau)$ is a universal space for such spaces.

**3.1  Theorem.** *If $\tau$ is an infinite cardinal number, then the Baire space $B(\tau)$ is a universl space for zero-dimensional metrizable spaces of weight not exceeding $\tau$.*

*Proof.* Let $X$ be a metrizable space such that $\dim X = 0$ and $w(X) \leqslant \tau$. Then by Proposition 1.10 there exist a subspace $Z$ of the Baire space

$B(\tau)$ and a continuous closed surjection $f\colon Z \to X$ such that $f^{-1}(x)$ is a one-point set if $x \in X$. Thus $f$ is a continuous closed bijection and hence a homeomorphism. Thus $X$ is homeomorphic with a subspace of $B(\tau)$. $\square$

**3.2 Corollary.** *If $X$ is the topological product of a countable family $\{X_i\}_{i \in \mathbb{N}}$ of metrizable spaces such that $\dim X_i = 0$ if $i \in \mathbb{N}$, then*

$$\dim X = 0.$$

*Proof.* Let $\tau$ be an infinite cardinal number such that $w(X_i) \leqslant \tau$ for each $i$. Then each $X_i$ can be embedded in $B(\tau)$ so that $X$ can be embedded in a countable product of copies of $B(\tau)$. But a countable product of copies of $B(\tau)$ is homeomorphic with $B(\tau)$. Thus $X$ can be embedded in $B(\tau)$ and hence $\dim X = 0$. $\square$

We recall that each metrizable space of weight not exceeding $\tau$ can be embedded in the countable topological product $P = \Pi_{i \in \mathbb{N}} P_i$, where for each $i$, $P_i$ is a copy of the hedgehog $J(\tau)$ based on a set $\Lambda$ of cardinality equal to $\tau$. We recall that the points of $J(\tau)$ are equivalence classes of $I \times \Lambda$ under the relation $\sim$, where $(t, \lambda) \sim (u, \mu)$ if $(t, \lambda) = (u, \mu)$ or if $t = u = 0$; we denote the equivalence class containing $(t, \lambda)$ by $[t, \lambda]$ and we put $[0, \lambda] = \mathbf{0}$. If $p = \{p_i\} \in P$ and $p_i = [t_i, \lambda_i]$ for each $i$, let $N(p)$ denote the cardinality of the set

$$\{i \in \mathbb{N} \mid t_i \text{ is rational and } t_i > 0\}.$$

Now let

$$M_n(\tau) = \{p \in P \mid N(p) \leqslant n\}.$$

Then $M_n(\tau)$ is a metrizable space and evidently $w\big(M_n(\tau)\big) \leqslant \tau$. Furthermore $\dim M_n(\tau) \leqslant n$. For $M_n(\tau) = \bigcup_{k=0}^{n} L_k$, where

$$L_k = \{p \in P \mid N(p) = k\}$$

and we shall show that $\dim L_k = 0$ for each $k$. Let

$$K = \{[t, \lambda] \in J(\tau) \mid t = 0 \text{ or } t \text{ is irrational}\}.$$

The subspace $K \backslash \{\mathbf{0}\}$ of $K$ is the topological sum of subspaces each homeomorphic with the space of irrationals in the unit interval so that $\dim (K \backslash \{\mathbf{0}\}) = 0$. It follows that $\dim K = 0$. Since $L_0$ is the topological product of countably many copies of $K$, it follows from Corollary 3.2 that $\dim L_0 = 0$. Now suppose that $k > 0$. Let

$$C = \{m_1, \ldots, m_k\}$$

be a subset of $\mathbf{N}$ with $k$ elements, let $\mathbf{r} = (r_1, \ldots, r_k)$ be a $k$-tuple of rational numbers such that $0 < r_i \leqslant 1$ for each $i$, and let $\mathbf{\mu} = (\mu_1, \ldots, \mu_k)$ be a $k$-tuple of elements of $\Lambda$. Let $L_C(\mathbf{r}, \mathbf{\mu})$ be the subset of $L_k$ consisting of those points $\{[t(i), \lambda(i)]\}_{i \in \mathbf{N}}$ such that $t(m_i) = r_i$ and $\lambda(m_i) = \mu_i$ for $i = 1, \ldots, k$. Then $L_C(\mathbf{r}, \mathbf{\mu})$ is homeomorphic with $L_0$ so that $\dim L_C(\mathbf{r}, \mathbf{\mu}) = 0$. Now let

$$L_C(\mathbf{r}) = \bigcup \{L_C(\mathbf{r}, \mathbf{\mu}) \mid \mathbf{\mu} \in \Lambda^k\}.$$

Then $L_C(\mathbf{r})$ is the topological sum of the subspaces $L_C(\mathbf{r}, \mathbf{\mu})$ so that $\dim L_C(\mathbf{r}) = 0$. But $L_C(\mathbf{r})$ is a closed subset of $L_k$ and the collection of all such sets is a countable closed covering of $L_k$. Thus $\dim L_k = 0$. It follows from Theorem 1.1 that $\dim M_n(\tau) \leqslant n$.

We shall see that in fact $\dim M_n(\tau) = n$ and $w(M_n(\tau)) = \tau$ and that $M_n(\tau)$ is a universal space for metrizable spaces of dimension not exceeding $n$ and weight not exceeding $\tau$.

**3.3 Lemma.** *Let $X$ be a pseudo-metrizable space such that $\dim X \leqslant n$, let $\{W_m\}_{m \in \mathbf{N}}$ be a countable family of open sets, let $\{F_m\}_{m \in \mathbf{N}}$ be a family of closed sets of $X$ such that $F_m \subset W_m$ for each $m$, and let $D$ be the set of rational numbers in the open interval $(0, 1/\sqrt{2})$. Then there exists a family $\{U(m, r) \mid (m, r) \in \mathbf{N} \times D\}$ of open sets of $X$ such that the order of the family $\{\mathrm{bd}\, U(m, r) \mid (m, r) \in \mathbf{N} \times D\}$ does not exceed $n - 1$, and if $(m, r) \in \mathbf{N} \times D$ then* $F_m \subset U(m, r) \subset K(m, r) \subset W_m$
*and*

$$K(m, r) = \bigcap_{q < r} U(m, q), \quad U(m, r) = \bigcup_{q > r} K(m, q),$$

*where $K(m, r)$ is the closure of $U(m, r)$ in $X$.*

*Proof.* Since $\dim X \leqslant n$, it follows from Theorem 1.1 that $X = \bigcup_{k=0}^{n} A_k$, where $\dim A_k \leqslant 0$ for each $k$. We choose a pseudo-metric $d$ which induces the topology of $X$.

Let the set $D$ of rational numbers in the open interval $(0, 1/\sqrt{2})$ be enumerated as $r_i$ for $i$ in $\mathbf{N}$ such that

$$r_2 < r_1 < r_3, \quad r_4 < r_2 < r_5 < r_1 < r_6 < r_3 < r_7$$

and so on. For each non-negative integer $s$ let $D_s = \{r_i \mid 2^s \leqslant i < 2^{s+1}\}$. If $r_i \in D_s$ and $2^s < i < 2^{s+1} - 1$, then in $\bigcup_{t \leqslant s} D_t$ ordered by magnitude $r_i$ has an immediate predecessor $r_a$ and an immediate successor $r_b$, and $r_a$ and $r_b$ are members of $\bigcup_{t < s} D_t$. We must construct open sets $U(m, r_i)$ for all pairs $(m, i)$ in $\mathbf{N} \times \mathbf{N}$. The construction will be by induction over $\mathbf{N} \times \mathbf{N}$ which we linearly order by the relation

$$(p, j) < (m, i)$$

if $p+j < m+i$ or if $p+j = m+i$ and $p < m$. By Lemma 4.4.16, there exists an open set $U(1, r_1)$ such that

$$F_1 \subset U(1, r_1) \subset K(1, r_1) \subset W$$

and $A_0 \cap \operatorname{bd} U(1, r_1) = \varnothing$, where $K(1, r_1)$ is the closure of $U(1, r_1)$. If $(m, i) \in \mathbf{N} \times \mathbf{N}$ let $L_{mi} = \{(p, j) \in \mathbf{N} \times \mathbf{N} \mid (p, j) < (m, i)\}$ and suppose that we have constructed an open set $U(p, r_j)$ with closure $K(p, r_j)$ if $(p, j) \in L_{mi}$ such that

(a) $F_p \subset U(p, r_j) \subset K(p, r_j) \subset W_p$;

(b) if $(p, h), (p, j)$ are members of $L_{mi}$ and $r_h < r_j$, then

$$K(p, r_j) \subset U(p, r_h);$$

(c) for $k = 0, \ldots, n$, if $x \in A_k$ then the set

$$\{(p, j) \in L_{mi} \mid x \in \operatorname{bd} U(p, r_j)\}$$

contains at most $k$ elements.

We show how to construct $U(m, r_i)$. Suppose that $i = 2^s + q$, where $s \geqslant 1$ and $0 \leqslant q < 2^s$. If $0 < q < 2^s - 1$, let $r_a$ and $r_b$ be the immediate predecessor and successor of $r_i$ in $\bigcup_{t \leqslant s} D_t$ ordered by magnitude. Then $(m, a) \in L_{mi}$ and $(m, b) \in L_{mi}$. If $s$ is odd let

$$E = K(m, r_b) \quad \text{and} \quad G = U(m, r_a) \cap B_{1/s}\big(K(m, r_b)\big),$$

and if $s$ is even let

$$E = K(m, r_b) \cup \big(X \backslash B_{1/s}\big(X \backslash K(m, r_a)\big)\big) \quad \text{and} \quad G = U(m, r_a).$$

The modifications which should be made in the definitions of $E$ and $G$ in the cases $q = 0$ and $q = 2^s - 1$ will be obvious. In all cases $E$ is a closed set of $X$, $G$ is an open set and $E \subset G$. For each point $x$ of $X$, let $k(x)$ denote the number of elements in the set

$$\{(p, j) \in L_{mi} \mid x \in \operatorname{bd} U(p, r_j)\}$$

and let

$$H = \{x \in X \mid x \in A_k \quad \text{for some } k \text{ and } \quad k(x) = k\}.$$

Then $\dim H \leqslant 0$ by an argument similar to one employed in the proof of Lemma 1.2. It follows from Lemma 4.4.16 that there exists an open set $U(m, r_i)$ such that

$$E \subset U(m, r_i) \subset K(m, r_i) \subset G$$

and $H \cap \operatorname{bd} U(m, r_i) = \varnothing$, where $K(m, r_i)$ is the closure of $U(m, r_i)$. If $x \in A_k$ and $x \in \operatorname{bd} U(m, r_i)$, then $k(x) \leqslant k-1$. Thus if $x \in A_k$ then there are at most $k$ pairs $(p, j)$ such that $(p, j) \leqslant (m, i)$ and $x \in \operatorname{bd} U(p, r_j)$. By induction we construct an open set $U(m, r_i)$ for each pair $(m, i)$.

Thus we obtain a family of open sets $\{U(m,r) \mid (m,r) \in \mathbf{N} \times D\}$ such that the order of the family $\{\operatorname{bd} U(m,r) \mid (m,r) \in \mathbf{N} \times D\}$ does not exceed $n-1$ and

$$F_m \subset U(m,r) \subset K(m,r) \subset W_m$$

if $(m,r) \in \mathbf{N} \times D$, where $K(m,r)$ is the closure of $U(m,r)$. It remains to verify the final conditions. If $r, q \in D$ and $q > r$, then

$$K(m,r) \subset U(m,q)$$

for each $m$. Let $(m,r) \in \mathbf{N} \times D$ and suppose that $r \in D_s$. If $x \in K(m,r)$, let $\delta = d(x, K(m,r)) > 0$ and choose an odd integer $t$ such that

$$t > \max\{s, 1/\delta\}.$$

If $q$ is the immediate precedessor of $r$ in $\bigcup_{u \leqslant t} D_u$, then

$$U(m,q) \subset B_{1/t}(K(m,r)).$$

It follows that $x \notin U(m,q)$. Thus

$$K(m,r) = \bigcap_{q<r} U(m,q).$$

If $x \in U(m,r)$, let $\delta = d(x, X \backslash K(m,r))$ and choose an even integer $t$ such that $t > \max\{s, 1/\delta\}$. If $q$ is the immediate successor of $r$ in $\bigcup_{u \leqslant t} D_u$, then

$$K(m,q) \supset X \backslash B_{1/t}(X \backslash K(m,r)).$$

Since $d(x, X \backslash K(m,r)) > 1/t$, it follows that $x \in K(m,q)$. Thus

$$U(m,r) = \bigcup_{r<q} K(m,q). \square$$

**3.4   Theorem.** *The space $M_n(\tau)$ is a universal space for metrizable spaces of weight not exceeding $\tau$ and covering dimension not exceeding $n$.*

*Proof.* We have seen that $M_n(\tau)$ is a metrizable space of weight not exceeding $\tau$ and that $\dim M_n(\tau) \leqslant n$. The proof will be complete if we show that if $X$ is a metrizable space such that $w(X) \leqslant \tau$ and $\dim X \leqslant n$, then $X$ can be embedded in $M_n(\tau)$. By Proposition 2.3.6 there exist a family $\{F_{m\lambda}\}_{m \in \mathbf{N}, \lambda \in \Lambda}$ of closed sets and a family

$$\{W_{m\lambda}\}_{m \in \mathbf{N}, \lambda \in \Lambda}$$

of open sets such that $F_{m\lambda} \subset W_{m\lambda}$ for each $m$ and $\lambda$, the family

$$\{W_{m\lambda}\}_{\lambda \in \Lambda}$$

is discrete for each $m$, and if $x$ is a point of $X$ and $U$ is an open set such that $x \in U$, then there exist $m$ and $\lambda$ such that

$$x \in F_{m\lambda} \subset W_{m\lambda} \subset U.$$

Since $w(X) \leqslant \tau$, we can suppose that $|\Lambda| = \tau$ and we shall construct an embedding of $X$ in the countable product of copies of $J(\tau)$ based on $\Lambda$.

For each positive integer $m$ let $F_m = \bigcup_{\lambda \in \Lambda} F_{m\lambda}$ and $W_m = \bigcup_{\lambda \in \Lambda} W_{m\lambda}$. Then $F_m$ is a closed set, $W_m$ is an open set of $X$ and $F_m \subset W_m$. Since $\dim X \leqslant n$, it follows from Lemma 3.3 that there exists a family $\{U(m, r) \mid (m, r) \in \mathbf{N} \times D\}$ of open sets of $X$, where $D$ is the set of rational numbers in the open interval $(0, 1/\sqrt{2})$, such that the order of the family $\{\operatorname{bd} U(m, r) \mid (m, r) \in \mathbf{N} \times D\}$ does not exceed $n-1$, and if $(m, r) \in \mathbf{N} \times D$ then

$$F_m \subset U(m, r) \subset K(m, r) \subset W_m$$

and

$$K(m, r) = \bigcap_{q < r} U(m, q), \quad U(m, r) = \bigcup_{q > r} K(m, q),$$

where $K(m, r)$ is the closure of $U(m, r)$ in $X$. If $(m, r) \in \mathbf{N} \times D$ and $\lambda \in \Lambda$ let

$$U_\lambda(m, r) = U(m, r) \cap W_{m\lambda},$$

and let $K_\lambda(m, r)$ be the closure of $U_\lambda(m, r)$ in $X$. Then $K_\lambda(m, r) \subset W_m$ and $K_\lambda(m, r) \subset X \backslash W_{m\mu}$ if $\lambda \neq \mu$, so that $K_\lambda(m, r) \subset W_{m\lambda}$. Since

$$K(m, r) = \bigcup_{\lambda \in \Lambda} K_\lambda(m, r)$$

it follows that $K(m, r) \cap W_{m\lambda} = K_\lambda(m, r)$.

If $m \in \mathbf{N}$ and $\lambda \in \Lambda$, let us define a continuous function

$$f_{m\lambda} : \overline{W}_{m\lambda} \to I$$

by putting $f_{m\lambda}(x) = \sup \{r \in D \mid x \in U_\lambda(m, r)\}$ if there exists $r$ in $D$ such that $x \in U_\lambda(m, r)$, and $f_{m\lambda}(x) = 0$ otherwise. Let $A_m = \bigcup_{\lambda \in \Lambda} \overline{W}_{m\lambda}$ so that $A_m$ is the topological sum of the subspaces $\overline{W}_{m\lambda}$, and define $f_m : A_m \to J(\tau)$ by putting

$$f_m(x) = [f_{m\lambda}(x), \lambda] \quad \text{if} \quad x \in \overline{W}_{m\lambda}.$$

If $x \in A_m \backslash \bigcup_{\lambda \in \Lambda} W_{m\lambda}$ then $f_m(x) = \mathbf{0}$. It follows that we can define a continuous function $g_m : X \to J(\tau)$ such that $g_m \mid A_m = f_m$ and $g_m(x) = \mathbf{0}$ if $x \in X \backslash \bigcup_{\lambda \in \Lambda} W_{m\lambda}$. The countable family $\{g_m\}_{m \in \mathbf{N}}$ of continuous functions on $X$ separates points from closed sets. For let $A$ be a closed set of $X$ and let $x$ be a point of $X \backslash A$. There exist $m$ and $\lambda$ such that

$x \in F_{m\lambda} \subset W_{m\lambda} \subset X \backslash A$, so that $g_m(x) = [1/\sqrt{2}, \lambda]$ and $V_\lambda = \{[t, \lambda] \mid t > 0\}$ is a neighbourhood of $g_m(x)$ in $J(\tau)$ which is disjoint from $g_m(A)$. Since $X$ is a $T_1$-space, the family $\{g_m\}_{m \in \mathbb{N}}$ also separates points of $X$. Hence the continuous function $g$ given by $g(x) = \{g_m(x)\}_{m \in \mathbb{N}}$ if $x \in X$, mapping $X$ into a countable product of copies of $J(\tau)$, is an embedding. We complete the proof by showing that $g(X) \subset M_n(\tau)$. Let $x$ be a point of $X$. If $g_m(x) = [r, \lambda]$, where $r$ is rational and $r > 0$, then $r \in D$ and $x \in W_{m\lambda}$. Since $f_m(x) = r$, it follows that $x \in U(m, q)$ if $q \in D$ and $q < r$, so that $x \in K(m, r)$. If $q \in D$ and $q > r$, then there exists $c$ in $D$ such that $q > c > r$. Since $x \notin U_\lambda(m, c)$, it follows that $x \notin K_\lambda(m, q)$ so that $x \notin K(m, q)$. Hence $x \notin U(m, r)$ and it follows that $x \in \operatorname{bd} U(m, r)$. Since there are at most $n$ pairs $(m, r)$ such that $x \in \operatorname{bd} U(m, r)$, it follows that $g(x) \in M_n(\tau)$. Thus $g(X) \subset M_n(\tau)$ and hence there exists an embedding of $X$ in $M_n(\tau)$. $\square$

**3.5** *Remark.* Let $X$ be a metrizable space such that $w(X) \geqslant \aleph_0$ and let us define $w^*(X)$ to be the least cardinal of a set $\Lambda$ for which there exist a family $\{F_{m\lambda}\}_{m \in \mathbb{N}, \lambda \in \Lambda}$ of closed sets and a family $\{W_{m\lambda}\}_{m \in \mathbb{N}, \lambda \in \Lambda}$ of open sets such that $F_{m\lambda} \subset W_{m\lambda}$ for each $m, \lambda$, the family $\{W_{m\lambda}\}_{\lambda \in \Lambda}$ is discrete for each $m$, and for each point $x$ of $X$ and each open set $U$ such that $x \in U$ there exist $m, \lambda$ such that

$$x \in F_{m\lambda} \subset W_{m\lambda} \subset U.$$

The above proof shows that if $X$ is a metrizable space such that $w^*(X) = \tau$ and $\dim X \leqslant n$, then $X$ can be embedded in $M_n(\tau)$. If $w(X) > \aleph_0$, then $w^*(X) = w(X)$. But if $w(X) = \aleph_0$, then $w^*(X) = 1$. For let $X$ be a separable metrizable space and let $\mathscr{W}$ be a countable base for the topology of $X$. The set of pairs $(V, W)$, such that $V, W \in \mathscr{W}$ and $\bar{V} \subset W$, is countable, and we enumerate such pairs as $(V_m, W_m)$ for $m$ in $\mathbb{N}$. The families $\{\bar{V}_m\}_{m \in \mathbb{N}}$ and $\{W_m\}_{m \in \mathbb{N}}$ satisfy the above conditions, for if $U$ is open and $x \in U$, then there exists some $m$ such that $x \in \bar{V}_m \subset W_m \subset U$. Thus, since $J(1) = I$, we see that a universal space for separable metrizable spaces of covering dimension equal to $n$ is the subspace of the cube $I^{\aleph_0}$ consisting of those points at most $n$ of whose non-zero coordinates are rational. We can obtain a universal space for $n$-dimensional separable metrizable spaces which is a subspace of a finite dimensional space. We find in fact a universal space for strongly metrizable spaces of given weight and dimension.

Let $n$ and $m$ be integers such that $0 \leqslant n \leqslant m$ and let $Q_n^m$ denote the set of points in the cube $I^m$ of which at most $n$ coordinates are rational. The mapping which takes the point $(x_1, \ldots, x_m)$ of $I^m$ to the point

$(x_1, \ldots, x_m, 0, 0, \ldots)$ of $I^{\aleph_0}$ is an embedding of $Q^m_{n_i}$ in the subspace of $I^{\aleph_0}$ consisting of those points of which at most $n$ non-zero coordinates are rational. It follows that $\dim Q^m_n \leqslant n$. We shall see that if

$$m \geqslant 2n+1$$

then $\dim Q^m_n = n$, that $Q^{2n+1}_n$ is a universal space for $n$-dimensional separable metrizable spaces and that $B(\tau) \times Q^{2n+1}_n$ is a universal space for $n$-dimensional strongly metrizable spaces of weight not exceeding $\tau$.

The following embedding lemma will be used.

**3.6   Lemma.** *Let $X$ be a $T_1$-space and let $\{\mathscr{U}_i\}_{i \in \mathbb{N}}$ be a sequence of open coverings of $X$ with the property that for each point $x$ of $X$ and each open set $U$ such that $x \in U$ there exists some integer $i$ such that $\mathrm{St}\,(x, \mathscr{U}_i) \subset U$. If $f \colon X \to Z$ is a continuous function which is a $\mathscr{U}_i$-mapping for every positive integer $i$, then $f$ is an embedding.*

*Proof.* Suppose that $x$ is a point of $X$ and that $U$ is an open set of $X$ such that $x \in U$. There exists some $i$ such that $\mathrm{St}\,(x, \mathscr{U}_i) \subset U$. Since $f$ is a $\mathscr{U}_i$-mapping, there exists an open set $W$ of $Z$ such that $f(x) \in W$ and $f^{-1}(W) \subset \mathrm{St}\,(x, \mathscr{U}_i)$. Thus $W \cap f(X) \subset f(U)$ and hence $f(U)$ is open in $f(X)$. If $y$ is a point of $X$ and $y \neq x$, then the above argument with $U = X \backslash \{y\}$ shows that $f(x) \neq f(y)$. Hence $f$ is an injection and it follows that $f$ is an embedding. $\square$

The following property of strongly pseudo-metrizable spaces is also needed.

**3.7   Lemma.** *If $X$ is a strongly pseudo-metrizable space, then for each positive integer $m$ there exist a disjoint open covering $\mathscr{W}_m$ and a star-finite open covering $\mathscr{U}_m$ of $X$ such that*

   (a) *$\mathscr{W}_{m+1}$ is a refinement of $\mathscr{W}_m$ for each $m$;*
   (b) *$\mathscr{U}_m$ is a refinement of $\mathscr{W}_m$ for each $m$ and each member of $\mathscr{W}_m$ contains only finitely many members of $\mathscr{U}_m$;*
   (c) *for each point $x$ of $X$ and each open set $G$ such that $x \in G$, there exists an integer $m$ such that $\mathrm{St}\,(x, \mathscr{U}_m) \subset G$.*

*Proof.* Since $X$ is a strongly pseudo-metrizable space, by Proposition 2.3.14 there exists a sequence $\{\mathscr{V}_m\}_{m \in \mathbb{N}}$ of star-finite open coverings of $X$ such that for each point $x$ of $X$ and each open set $G$ of $X$ such that $x \in G$, there exists some integer $j$ such that $\mathrm{St}\,(x, \mathscr{V}_j) \subset G$. There is no loss of generality in assuming that $\mathscr{V}_{m+1}$ is a refinement of $\mathscr{V}_m$ for each $m$. If $m \in \mathbb{N}$, let $\mathscr{V}_m = \{V_\alpha^i\}_{i \in \mathbb{N}, \alpha \in \Omega(m)}$, where $V_\alpha^i \cap V_\beta^j = \varnothing$ if $\alpha, \beta \in \Omega(m)$ and $\alpha \neq \beta$. If $\alpha \in \Omega(m)$, let $W_\alpha = \bigcup_{i \in \mathbb{N}} V_\alpha^i$. Then

$$W_m = \{W_\alpha\}_{\alpha \in \Omega(m)}$$

is a disjoint open covering of $X$ and $\mathscr{W}_{m+1}$ is a refinement of $\mathscr{W}_m$ for each $m$ since $\mathscr{V}_{m+1}$ is a refinement of $\mathscr{V}_m$.

For each $m$ there exists a closed covering $\{F_\alpha^i\}_{i \in \mathbb{N}, \alpha \in \Omega(m)}$ of $X$ such that $F_\alpha^i \subset V_\alpha^i$ for each $i$ and $\alpha$. Let $m$ be a positive integer and let $\alpha$ be an element of $\Omega(m)$. For each positive integer $j$ such that $j \leqslant m$ let $\alpha(j)$ be the unique element of $\Omega(j)$ such that $W_\alpha \subset W_{\alpha(j)}$, and let

$$\mathscr{H}_j(\alpha) = \{W_\alpha \cap V_{\alpha(j)}^{m+1-j}, W_\alpha \backslash F_{\alpha(j)}^{m+1-j}\}.$$

Now let $\mathscr{U}(\alpha)$ be a finite open covering of $W_\alpha$, which is a refinement of each of the coverings $\mathscr{H}_j(\alpha), j = 1, \ldots, m$, of $W_\alpha$. If $\mathscr{U}_m = \bigcup_{\alpha \in \Omega(m)} \mathscr{U}(\alpha)$, then $\mathscr{U}_m$ is an open covering of $X$ which is a refinement of $\mathscr{W}_m$ and each member of $\mathscr{W}_m$ contains only finitely many members of $\mathscr{U}_m$. Let $G$ be an open set of $X$ and let $x$ be a point of $G$. There exists a positive integer $j$ such that $\operatorname{St}(x, \mathscr{V}_j) \subset G$. Suppose that $x \in F_\alpha^r$ and let $m$ be a positive integer such that $m + 1 - j = r$. Then each member of $\mathscr{U}_m$ which contains $x$ is contained in $V_\alpha^r$. Thus

$$\operatorname{St}(x, \mathscr{U}_m) \subset V_\alpha^r \subset \operatorname{St}(x, \mathscr{V}_j) \subset G. \ \square$$

**3.8   Theorem.** *If $\tau$ is an infinite cardinal number, then the topological product $B(\tau) \times Q_n^{2n+1}$ is a universal space for strongly metrizable spaces of weight not exceeding $\tau$ and dimension (in all senses) equal to $n$.*

*Proof.* Since $B(\tau) \times Q_n^{2n+1}$ is a subspace of the strongly metrizable space $B(\tau) \times I^{2n+1}$, it follows that $B(\tau) \times Q_n^{2n+1}$ is strongly metrizable and it is clear that the weight of $B(\tau) \times Q_n^{2n+1}$ is equal to $\tau$. Since $B(\tau)$ is zero-dimensional, it follows from Proposition 4.5.5 that

$$\dim B(\tau) \times Q_n^{2n+1} = \dim Q_n^{2n+1} \leqslant n.$$

We shall complete the proof by showing that each $n$-dimensional strongly metrizable space of weight not exceeding $\tau$ can be embedded in $B(\tau) \times Q_n^{2n+1}$.

If $L$ is the subspace of $I^{2n+1}$ consisting of those points for which $n+1$ given coordinates have given rational values, then $L \cap Q_n^{2n+1} = \varnothing$. There are countably many such subspaces and we enumerate them to form a family $\{L_k\}_{k \in \mathbb{N}}$ of subspaces of $I^{2n+1}$. Clearly

$$Q_n^{2n+1} = I^{2n+1} \backslash \bigcup_{k \in \mathbb{N}} L_k.$$

Let $X$ be a strongly metrizable space such that $w(X) \leqslant \tau$ and $\dim X \leqslant n$. By Lemma 3.7, there exist a sequence $\{\mathscr{W}_m\}_{m \in \mathbf{N}}$ of disjoint open coverings of $X$ such that $\mathscr{W}_{m+1}$ is a refinement of $\mathscr{W}_m$ for each $m$, and a sequence $\{\mathscr{U}_m\}_{m \in \mathbf{N}}$ of open coverings of $X$ such that $\mathscr{U}_m$ is a refinement of $\mathscr{W}_m$ and each member of $\mathscr{W}_m$ contains only finitely many members of $\mathscr{U}_m$, and furthermore if $G$ is an open set and $x$ is a point of $G$, then there exists an integer $m$ such that $\mathrm{St}\,(x, \mathscr{U}_m) \subset G$. Let $\mathscr{W}_m = \{W_\alpha\}_{\alpha \in \Omega(m)}$, where $\Omega(m)$ and $\Omega(k)$ are disjoint if $k \neq m$ and $W_\alpha$ is non-empty if $\alpha \in \Omega(m)$. Since $w(X) \leqslant \tau$ it follows that $|\Omega(m)| \leqslant \tau$. Let $\Omega$ be a set of cardinality equal to $\tau$ which contains the disjoint union of the sets $\Omega(m)$ for $m$ in $\mathbf{N}$ and let $B(\tau)$ be the Baire space based on $\Omega$. Let $\phi : X \to B(\tau)$ be the continuous function defined as follows: if $x \in X$, then

$$\phi(x) = (\alpha_1, \alpha_2, \ldots, \alpha_k, \ldots),$$

where $\alpha_k$ is the unique element of $\Omega(k)$ such that $x \in W_{\alpha_k}$. If $\alpha \in \Omega(m)$, let $T_\alpha = \phi(W_\alpha)$ so that $T_\alpha$ is open in the subspace $\phi(X)$ of $B(\tau)$.

Now consider the topological product $B(\tau) \times I^{2n+1}$ and let $\pi_1$ and $\pi_2$ be the projections of $B(\tau) \times I^{2n+1}$ onto $B(\tau)$ and $I^{2n+1}$ respectively. Let $M(X)$ be the set of continuous functions

$$f : X \to B(\tau) \times I^{2n+1},$$

such that $\pi_1 \circ f = \phi$. If $\mathscr{C}$ be the set of subsets $C$ of $\bigcup_{m \in \mathbf{N}} \Omega(m)$ such that $\{W_\alpha\}_{\alpha \in C}$ is a disjoint open covering of $X$. If $C \in \mathscr{C}$ and $f \in M(X)$, let $N_C(f)$ be the subset of $M(X)$ consisting of those functions $g$ such that

$$\sup_{x \in W_\alpha} \rho\big(\pi_2 f(x), \pi_2 g(x)\big) < 1/m$$

if $\alpha \in C \cap \Omega(m)$, where $\rho$ is the Euclidean metric in $I^{2n+1}$. If $\alpha \in \Omega(m)$, let $\mathscr{T}_\alpha$ be the open covering of $T_\alpha \times I^{2n+1}$ which consists of all sets of the form $T_\alpha \times G$, where $G$ is an open ball in $I^{2n+1}$ of radius $1/m$. For each positive integer $k$ let $M_k(X)$ be the subset of $M(X)$ consisting of those functions $f$ for which there exists some $C$ in $\mathscr{C}$ with the property that if $\alpha \in C$, then the covering $f^{-1}(\mathscr{T}_\alpha)$ of $W_\alpha$ is a refinement of $\mathscr{U}_k$ and the closure of $\pi_2 f(W_\alpha)$ is disjoint from $L_k$. We observe that if $f \in M_k(X)$, then $f$ is a $\mathscr{U}_k$-mapping and $f(X) \subset B(\tau) \times (I^{2n+1} \backslash L_k)$.

Three facts $(A)$, $(B)$ and $(C)$ supply the essential steps in the proof of the theorem.

$(A)$ If $k \in \mathbf{N}$, $f \in M(X)$ and $C \in \mathscr{C}$, then

$$N_C(f) \cap M_k(X) \neq \varnothing.$$

If $\alpha \in C \cap \Omega(m)$, let $p(\alpha) = \max\{6m, k\}$ and let

$$D_\alpha = \{\beta \in \Omega(p(\alpha)) \mid W_\beta \subset W_\alpha\}.$$

Let $\beta \in D_\alpha$. Since $I^{2n+1}$ is compact, there exists a finite subcovering of the open covering $f^{-1}(\mathcal{T}_\beta)$ of $W_\beta$. Finitely many members of $\mathcal{U}_k$ meet $W_\beta$ since $p(\alpha) \geqslant k$. Since $\dim W_\beta \leqslant n$ it follows from Proposition 3.1.10 that there exists a finite open covering $\mathcal{V}_\beta = \{V_1, \ldots, V_s\}$ of $W_\beta$ of order not exceeding $n$ which is a star-refinement of $f^{-1}(\mathcal{T}_\beta)$ and of $\mathcal{U}_k$. For each $i = 1, \ldots, s$ let

$$G_i = B_{1/3m}\big(\pi_2 f(V_i)\big).$$

Choose points $w_0, \ldots, w_n$ of $L_k$ and a point $z_i$ in $G_i$ for $i = 1, \ldots, s$ such that $w_0, \ldots, w_n, z_1, \ldots, z_s$ are in general position (no $m+2$ of these points lie in an $m$-dimensional affine subspace of $\mathbf{R}^{2n+1}$ if $m \leqslant 2n$). For $i = 1, \ldots, s$, there exists a continuous function $\lambda_i: W_\beta \to I$ such that $\Sigma_{i=1}^s \lambda_i(x) = 1$ if $x \in W_\beta$ and $\lambda_i(x) = 0$ if $x \notin V_i$. Define

$$\psi_\beta: W_\beta \to I$$

by putting

$$\psi_\beta(x) = \sum_{i=1}^s \lambda_i(x) z_i \quad \text{if} \quad x \in W_\beta.$$

The function $\psi_\beta$ has three important properties.

(i) *If $x \in W_\beta$, then $\rho\big(\pi_2 f(x), \psi_\beta(x)\big) < 2/3m$.*

Suppose that $x \in W_\beta$. For $i = 1, \ldots, s$ there exists $u_i$ in $\pi_2 f(V_i)$ such that $\rho(z_i, u_i) < 1/3m$. If $x \in V_i$, then $f(V_i) \subset T_\beta \times G$, where $G$ is an open ball in $I^{2n+1}$ of radius $1/p(\alpha)$. Thus $\pi_2 f(V_i) \subset G$ and it follows that

$$\rho\big(\pi_2 f(x), u_i\big) < 2/p(\alpha) < 1/3m.$$

Thus if $x \in V_i$, then $\rho\big(\pi_2 f(x), z_i\big) < 2/3m$. Since $\lambda_i(x) = 0$ for those values of $i$ such that $x \notin V_i$, it follows that $\rho\big(\pi_2 f(x), \psi_\beta(x)\big) < 2/3m$.

(ii) *The closure of $\psi_\beta(W_\beta)$ is disjoint from $L_k$.*

If $\Gamma$ is a subset of $\{1, \ldots, s\}$ with $t$ elements, where $t \leqslant n+1$, let $K_\Gamma$ be the closed affine subspace spanned by $\{z_i\}_{i \in \Gamma}$. In view of the general position hypothesis, $K_\Gamma \cap L_k = \varnothing$ for each such $\Gamma$. If $K$ is the union of the finite collection of subspaces $K_\Gamma$, then $K$ is a closed set disjoint from $L_k$. If $x \in W_\beta$ and $\Gamma(x) = \{i \mid x \in V_i\}$, then $\Gamma(x)$ has at most $n+1$ elements and $\psi_\beta(x) \in K_{\Gamma(x)} \subset K$. It follows that the closure of $\psi_\beta(W_\beta)$ is disjoint from $L_k$.

(iii) *There exists an integer $q(\beta)$ such that $q(\beta) > p(\alpha)$ and*

$$x \in \mathrm{St}\,(x_0, \mathscr{V}_\beta)$$

*if* $x, x_0 \in W_\beta$ *and* $\rho\big(\psi_\beta(x_0), \psi_\beta(x)\big) < 2/q_{(\beta)}.$

There are finitely many subspaces $K_\Gamma$ as defined in (ii). Hence there exists an integer $q(\beta)$ such that $q(\beta) > p(\alpha)$ and

$$\rho(K_\Gamma, K_{\Gamma'}) \geqslant 2/q(\beta)$$

if $K_\Gamma$ and $K_{\Gamma'}$ are disjoint. Suppose that $x_0, x \in W_\beta$ and

$$\rho\big(\psi_\beta(x_0), \psi_\beta(x)\big) < 2/q(\beta).$$

If $\Gamma = \{i \mid x_0 \in V_i\}$ and $\Gamma' = \{i \mid x \in V_i\}$, then $\psi_\beta(x_0) \in K_\Gamma$ and $\psi_\beta(x) \in K_{\Gamma'}$. It follows that $K_\Gamma \cap K_{\Gamma'} \neq \varnothing$ so that $\Gamma \cap \Gamma' \neq \varnothing$ in view of the general position hypothesis. Thus $x \in \mathrm{St}\,(x_0, \mathscr{V}_\beta)$.

There exists a continuous function $\psi \colon X \to I^{2n+1}$ such that $\psi \mid W_\beta = \psi_\beta$ if $\beta \in D_\alpha$ and $\alpha \in C$. Define $g$ in $M(X)$ by putting

$$g(x) = \big(\phi(x), \psi(x)\big) \quad \text{if} \quad x \in X.$$

Then $g \in N_C(f)$. For if $\alpha \in C \cap \Omega(m)$ and $\beta \in D_\alpha$, then

$$\rho\big(\pi_2 f(x), \psi_\beta(x)\big) < 2/3m$$

if $x \in W_\beta$ by (i). Hence

$$\sup_{x \in W_\alpha} \rho\big(\pi_2 f(x), \psi(x)\big) \leqslant 2/3m < 1/m.$$

Also $g \in M_k(X)$, for if $\alpha \in C$ and $\beta \in D_\alpha$ let $\beta = p(\alpha)$ and $q = q(\beta) > p$. Let

$$E_\beta = \{\gamma \in \Omega(q) \mid W_\gamma \subset W_\beta\}.$$

Let $E = \bigcup_{\alpha \in C} \bigcup \{E_\beta \mid \beta \in D_\alpha\}$ so that $E \in \mathscr{C}$. If $\gamma \in E_\beta$, where $\beta \in D_\alpha$ and $\alpha \in C$, then

$$\pi_2 g(W_\gamma) \subset \pi_2 g(W_\beta) = \psi_\beta(W_\beta).$$

Thus the closure of $\pi_2 g(W_\gamma)$ is disjoint from $L_k$, by (ii). Let

$$x_0 \in g^{-1}(T_\gamma \times G),$$

where $G$ is an open ball in $I^{2n+1}$ of radius $1/q$. If $x \in g^{-1}(T_\gamma \times G)$, then $\rho\big(\psi_\beta(x), \psi_\beta(x_0)\big) < 2/q$ and hence $x \in \mathrm{St}\,(x_0, \mathscr{V}_\beta)$. Thus

$$g^{-1}(T_\gamma \times G) \subset \mathrm{St}\,(x_0, \mathscr{V}_\beta).$$

Since $\mathscr{V}_\beta$ is a star-refinement of $\mathscr{U}_k$, it follows that the covering $g^{-1}(\mathscr{T}_\gamma)$ of $W_\gamma$ is a refinement of $\mathscr{U}_k$. Thus $(A)$ is established.

(B) *If* $k \in \mathbb{N}$ *and* $g \in M_k(X)$, *then there exists* $D$ *in* $\mathscr{C}$ *such that*

$$N_D(g) \subset M_k(X).$$

There exists $C$ in $\mathscr{C}$ such that if $\alpha \in C$ then the closure $R_\alpha$ of $\pi_2 g(W_\alpha)$ in $I^{2n+1}$ is disjoint from $L_k$ and $g^{-1}(\mathscr{T}_\alpha)$ is a refinement of $\mathscr{U}_k$. If

$$\alpha \in C \cap \Omega(m),$$

let

$$b(\alpha) = \rho(R_\alpha, L_k) > 0$$

and let $r(\alpha)$ be an integer such that

$$r(\alpha) \geqslant \max\{2m, 1/b(\alpha)\}.$$

Let $D$ be the subset of $\Omega$ consisting of those elements $\gamma$ such that $\gamma \in \Omega(r(\alpha))$ for some $\alpha$ in $C$ and $W_\gamma \subset W_\alpha$. Then $D \in \mathscr{C}$. Let $h$ be a member of $N_D(g)$. Suppose that $\gamma \in D$ and $W_\gamma \subset W_\alpha$, where $\alpha \in C \cap \Omega(m)$. If $G_1$ and $G_2$ are the open balls in $I^{2n+1}$ with centre a point $x$ and radii $1/m$ and $1/r(\alpha)$ respectively, then

$$h^{-1}(T_\gamma \times G_2) \subset g^{-1}(T_\alpha \times G_1).$$

For if not, there exists $x$ in $W_\gamma$ such that $h(x) \in T_\gamma \times G_2$ and $g(x) \notin T_\alpha \times G_1$. It follows that $\pi_2 h(x) \in G_2$ and $\pi_2 g(x) \notin G_1$. Thus

$$\rho(\pi_2 g(x), \pi_2 h(x)) > 1/m - 1/r(\alpha) \geqslant 1/r(\alpha).$$

This is a contradiction of the fact that $h \in N_D(g)$. Since $g^{-1}(\mathscr{T}_\alpha)$ is a refinement of $\mathscr{U}_k$, it follows that $h^{-1}(\mathscr{T}_\gamma)$ is a refinement of $\mathscr{U}_k$. Again suppose that $\gamma \in D$ and that $W_\gamma \subset W_\alpha$, where $\alpha \in C$. Then

$$\sup_{x \in W_\gamma} \rho(\pi_2 g(x), \pi_2 h(x)) < 1/r(\alpha) \leqslant b(\alpha),$$

so that there exists $c(\gamma) > 0$ such that $\rho(\pi_2 g(x), \pi_2 h(x)) \leqslant b(\alpha) - c(\gamma)$ if $x \in W_\gamma$. Thus if $x \in W_\gamma$ and $z \in L_k$, it follows that

$$b(\alpha) \leqslant \rho(\pi_2 g(x), z) \leqslant \rho(\pi_2 g(x), \pi_2 h(x)) + \rho(\pi_2 h(x), z).$$

Hence $\rho(\pi_2 h(x), z) \geqslant c(\gamma)$. Thus $\rho(z, \pi_2 h(W_\gamma)) \geqslant c(\gamma) > 0$ if $z \in L_k$. It follows that the closure of $\pi_2 h(W_\gamma)$ in $I^{2n+1}$ is disjoint from $L_k$. Thus $h \in M_k(X)$ and hence $N_D(g) \subset M_k(X)$. This completes the proof of (B).

(C) *If* $k \in \mathbb{N}$, $f \in M(X)$, $C \in \mathscr{C}$ *and* $g \in N_C(f) \cap M_k(X)$, *then there exists* $D$ *in* $\mathscr{C}$ *such that*

$$N_D(g) \subset N_C(f) \cap M_k(X).$$

Let $\alpha \in C \cap \Omega(m)$ and let

$$a(\alpha) = \sup_{x \in W_\alpha} \rho(\pi_2 f(x), \pi_2 g(x))$$

so that $a(\alpha) < 1/m$. Now choose an integer $d(\alpha)$ such that

$$1/d(\alpha) < 1/m - a(\alpha).$$

Then $d(\alpha) > m$. Let

$$A_\alpha = \{\beta \in \Omega\big(d(\alpha)\big) \mid W_\beta \subset W_\alpha\},$$

and let $C' = \bigcup_{\alpha \in C} A_\alpha$. Then $C' \in \mathscr{C}$. Let $h$ be a member of $N_{C'}(g)$. If $\alpha \in C \cap \Omega(m)$ and $x \in W_\alpha$, then there exists $\beta$ in $A_\alpha$ such that $x \in W_\beta$. Thus $\rho\big(\pi_2 f(x), \pi_2 h(x)\big) \leqslant a(\alpha) + 1/d(\alpha)$ and hence

$$\sup_{x \in W_\alpha} \rho\big(\pi_2 f(x), \pi_2 h(x)\big) \leqslant a(\alpha) + 1/d(\alpha) < 1/m.$$

Thus $h \in N_C(f)$ and therefore $N_{C'}(g) \subset N_C(f)$. By $(B)$, there exists $C''$ in $\mathscr{C}$ such that $N_{C''}(g) \subset M_k(X)$. Now let $D' = \{\beta \in C' \mid W_\beta \subset W_\gamma$ for some $\gamma$ in $C''\}$ and let $D'' = \{\gamma \in C'' \mid W_\gamma \subset W_\beta$ for some $\beta$ in $C'\}$. If $D = D' \cup D''$, then $D \in \mathscr{C}$ and it is clear that $N_D(g) \subset N_C(f) \cap M_k(X)$.

We can now complete the proof of the theorem. It follows from $(A)$ that $M_1(X)$ is not empty, and if we choose $f_1$ in $M_1(X)$, then it follows from $(B)$ that there exists $C_1$ in $\mathscr{C}$ such that

$$N_{C_1}(f_1) \subset M_1(X).$$

Now suppose that $k$ is an integer, $k > 1$, and we have $f_{k-1}$ in $M_{k-1}(X)$ and $C_{k-1}$ in $\mathscr{C}$. If $\alpha \in C_{k-1} \cap \Omega(m)$ let

$$D_\alpha = \{\beta \in \Omega(2m) \mid W_\beta \subset W_\alpha\}.$$

If $D_{k-1} = \bigcup\{D_\alpha \mid \alpha \in C_{k-1}\}$, then $D_{k-1} \in \mathscr{C}$. By $(A)$ and $(C)$ there exist $f_k$ in $M_k(X)$ and $C_k$ in $\mathscr{C}$ such that

$$N_{C_k}(f_k) \subset N_{D_{k-1}}(f_{k-1}) \cap M_k(X).$$

We can suppose that if $\alpha \in C_k \cap \Omega(m)$, then $m > k$. In this way we can construct by induction a sequence $\{f_k\}$ of members of $M(X)$ and two sequences $\{C_k\}$ and $\{D_k\}$ of members of $\mathscr{C}$.

Now let $k$ be a positive integer. If $j \geqslant k$, then $f_j \in N_{C_k}(f_k)$. If $x \in X$, then $x \in W_\alpha$ for some $\alpha$ in $C_k$ and if $\alpha \in \Omega(m)$, then

$$\rho\big(\pi_2 f_j(x), \pi_2 f_k(x)\big) < 1/m < 1/k.$$

It follows that the sequence of functions $\{\pi_2 f_k\}$ converges uniformly to a continuous function $\Psi: X \to I^{2n+1}$. Let $h: X \to B(\tau) \times I^{2n+1}$ be defined by putting

$$h(x) = \big(\phi(x), \Psi(x)\big) \quad \text{if} \quad x \in X.$$

Then $h \in M(X)$ and in fact $h \in \bigcap_{k \in \mathbf{N}} M_k(X)$. For suppose that

$$\alpha \in C_k \cap \Omega(m)$$

and let $x$ be a point of $W_\alpha$. There exists $\beta$ in $D_k$ such that $x \in W_\beta$. If $j \geqslant k$ then $f_j \in N_{D_k}(f_k)$ so that $\rho\big(\pi_2 f_k(x), \pi_2 f_j(x)\big) < 1/2m$. Hence

$$\rho\big(\pi_2 f_k(x), \Psi(x)\big) \leqslant 1/2m.$$

Thus if $\alpha \in C_k \cap \Omega(m)$, then

$$\sup_{x \in W_\alpha} \rho\big(\pi_2 f(x), \pi_2 h(x)\big) \leqslant 1/2m < 1/m.$$

Thus $h \in N_{C_k}(f_k)$. But $N_{C_k}(f_k) \subset M_k(X)$. Hence

$$h \in \bigcap_{k \in \mathbf{N}} M_k(X).$$

It follows that $h$ is a $\mathcal{U}_k$-mapping for each $k$ and hence $h$ is an embedding by Lemma 3.6. Furthermore $h(X) \subset B(\tau) \times (I^{2n+1} \backslash L_k)$ for each $k$ so that

$$h(X) \subset B(\tau) \times (I^{2n+1} \backslash \bigcup_{k \in \mathbf{N}} L_k) = B(\tau) \times Q_n^{2n+1}. \ \square$$

In the case of separable metrizable spaces the above result can be improved.

**3.9  Proposition.** *The space $Q_n^{2n+1}$ is a universal space for $n$-dimensional (in all senses) separable metrizable spaces.*

*Proof.* Let $X$ be a separable metrizable space and let $\mathscr{B}$ be a countable base for the topology of $X$. There are countably many pairs $(U, V)$ of members of $\mathscr{B}$ such that $\overline{V} \subset U$ and we enumerate them as $(U_m, V_m)$ for $m$ in $\mathbf{N}$. Let $\mathcal{U}_m = \{U_m, X \backslash \overline{V}_m\}$ and let $\mathcal{W}_m = \{X\}$ for each $m$. If $x$ is a point of $X$ and $U$ is an open neighbourhood of $x$, then there exists an integer $m$ such that $x \in \overline{V}_m \subset U_m \subset U$. Thus

$$\mathrm{St}\,(x, \mathcal{U}_m) = U_m \subset U.$$

Employing the method of Theorem 3.8 with the sequences $\{\mathcal{W}_m\}_{m \in \mathbf{N}}$ and $\{\mathcal{U}_m\}_{m \in \mathbf{N}}$ of open coverings of $X$, we obtain an embedding

$$h : X \to B(\aleph_0) \times Q_n^{2n+1},$$

such that $\pi_1 \circ h$ is a constant mapping, where $\pi_1$ is the projection of the product onto $B(\aleph_0)$. It follows that $X$ can be embedded in $Q_n^{2n+1}$. $\square$

## 4  Prabir Roy's example

We have seen that if $X$ is a pseudo-metrizable space, then

$$\dim X = \mathrm{Ind}\,X,$$

whilst if $X$ is a strongly pseudo-metrizable space, then

$$\dim X = \operatorname{ind} X = \operatorname{Ind} X.$$

In this section we shall see that the three dimensions do not coincide for all pseudo-metrizable spaces or even for all metrizable spaces. There follows the wonderful example due to Prabir Roy of a metrizable space $P$ such that

$$\operatorname{ind} P = 0 \quad \text{and} \quad \dim P = \operatorname{Ind} P = 1.$$

To describe the set $P$ we must make a number of definitions and we introduce at the same time some notation which will be used throughout the section.

(1) For each non-negative integer $n$, let $[n] = \{0, \dots, n\}$ and let $X_n$ denote the set of functions $x \colon [n] \to \mathbf{R}$ such that $x(0) = 0$ and $x(i) \neq 0$ if $i > 0$. Let $X = \bigcup_{n \geqslant 0} X_n$. For each $x$ in $X$, write $|x| = n$ if $x \in X_n$.

(2) Let $Y$ be the set of all injective functions $y \colon \mathbf{N} \to \mathbf{R}^+$, where $\mathbf{R}^+ = \{t \in \mathbf{R} \mid t > 0\}$. If $t \in \mathbf{R}^+$ let

$$Y_t = \{y \in Y \mid y(i) = t \quad \text{for some } i\}.$$

The set $Y_t$ has cardinality $\mathbf{c}$ so that we can choose a bijection

$$\phi_t \colon \mathbf{R}^+ \to Y_t.$$

Let $\theta_t \colon Y_t \to \mathbf{R}^+$ be the inverse of $\phi_t$.

(3) Let $Z$ be the set of all functions $z \colon \mathbf{N} \to \mathbf{R}^+$.

(4) Let $P_1$ be the set of all functions $p \colon \mathbf{N} \to \mathbf{R} \backslash \{0\}$.

(5) Let $P_2 = X \times Y \times Z$ and if $p \in P_2$ let $X(p)$, $Y(p)$, $Z(p)$ denote the coordinates of $p$.

Now let $$P = P_1 \cup P_2.$$

A topology is defined on $P$ by specifying a base. As a preliminary to the construction of the base, one more definition is required. If $p \in P_2$ and $n, i \in \mathbf{N}$, let $\gamma(p, n, i, +)$ be the element of $X$ defined as follows: $|\gamma(p, n, i, +)| = |X(p)| + n + 1$, and

$$\gamma(p, n, i, +)(j) = X(p)(j) \quad \text{for} \quad j = 0, \dots, |X(p)|,$$

$$\gamma(p, n, i, +)\big(|X(p)| + 1\big) = Y(p)(n + i - 1),$$

$$\gamma(p, n, i, +)\big(|X(p)| + 2\big) = -\theta_{Y(p)(n+i-1)}\big(Y(p)\big),$$

and if $n > 1$,

$$\gamma(p, n, i, +)\big(|X(p)| + 2 + j\big) = -Z(p)(j) \quad \text{for} \quad j = 1, \dots, n-1.$$

We define $\gamma(p, n, i, -)$ in $X$ as follows: $|\gamma(p, n, i, -)| = |X(p)| + n + 1$ and

$$\gamma(p, n, i, -)(j) = \begin{cases} \gamma(p, n, i, +)(j) & \text{if } j \leqslant |X(p)|, \\ -\gamma(p, n, i, +)(j) & \text{if } j > |X(p)|. \end{cases}$$

In an expression such as $\gamma(p, n, i, \epsilon)$, it is to be understood that $\epsilon$ denotes $+$ or $-$. Since $Y(p)$ is injective, if $i \neq j$ then

$$\gamma(p, n, i, +) \neq \gamma(p, n, j, +)$$

since $\gamma(p, n, i, +)\big(|X(p)| + 1\big) \neq \gamma(p, n, j, +)\big(|X(p)| + 1\big)$. Similarly

$$\gamma(p, n, i, -) \neq \gamma(p, n, j, -),$$

and if $i, j \in \mathbf{N}$ then $\gamma(p, n, i, +) \neq \gamma(p, n, j, -)$ since

$$\gamma(p, n, i, +)\big(|X(p)| + 1\big) > 0$$

and

$$\gamma(p, n, j, -)\big(|X(p)| + 1\big) < 0.$$

The members of the base for the topology of $P$ are of two types. If $x \in X$ and $|x| > 0$ let

$$V_x^1 = \{p \in P_1 \mid p(i) = x(i) \quad \text{for} \quad i = 1, \ldots, |x|\},$$

$$V_x^2 = \{p \in P_2 \mid |X(p)| \geqslant |x| \quad \text{and} \quad X(p)(i) = x(i) \quad \text{for} \quad i = 1, \ldots, |x|\}.$$

Let $V_x = V_x^1 \cup V_x^2$ and let $\mathscr{V}_1$ be the set of subsets of $P$ of the form $V_x$ for some $x$ in $X$ such that $|x| > 0$. If $p \in P_2$ and $n \in \mathbf{N}$ let

$$T_{(p, n)}^+ = \bigcup_{i \in \mathbf{N}} V_{\gamma(p, n, i, +)}, \quad T_{(p, n)}^- = \bigcup_{i \in \mathbf{N}} V_{\gamma(p, n, i, -)},$$

let $T_{(p, n)} = T_{(p, n)}^+ \cup T_{(p, n)}^-$ and let

$$V_{(p, n)} = R_{(p, n)} \cup T_{(p, n)},$$

where

$$R_{(p, n)} = \{q \in P_2 \mid X(q) = X(p), Y(q) = Y(p) \quad \text{and}$$

$$Z(q)(i) = Z(p)(i) \quad \text{for} \quad i = 1, \ldots, n - 1\}.$$

When $T_{(p, n)}^\epsilon$ is written, it is to be understood that $\epsilon$ denotes $+$ or $-$. Let $\mathscr{V}_2$ be the set of subsets of $P$ of the form $V_{(p, n)}$ for some $p$ in $P_2$ and some positive integer $n$. We shall see that $\mathscr{V} = \mathscr{V}_1 \cup \mathscr{V}_2$ is the base for a topology on $P$. The main properties of the members of $\mathscr{V}$ are established in the following three lemmas.

**4.1 Lemma.** *If* $x, x' \in X$ *and* $|x| > 0, |x'| > 0$, *then the following statements are equivalent*: (a) $V_x \supset V_{x'}$; (b) $V_x \cap V_{x'} \neq \varnothing$ *and* $|x| \leqslant |x'|$; (c) $|x| \leqslant |x'|$ *and* $x(i) = x'(i)$ *for* $i = 1, \ldots, |x|$.

*Proof.* (a) $\Rightarrow$ (b). Suppose that $V_x \supset V_{x'}$. Then $V_x \cap V_{x'} \neq \varnothing$. Choose $p$ in $V_{x'}^2$ such that $|X(p)| = |x'|$. Then $p \in V_x^2$ so that $|X(p)| \geqslant |x|$. Thus $|x'| \geqslant |x|$.

(b) $\Rightarrow$ (c). Suppose that (b) holds and let $p \in V_x \cap V_{x'}$. If $p \in P_1$, then $p(i) = x(i) = x'(i)$ for $i = 1, \ldots, |x|$. Similarly if $p \in P_2$, then

$$X(p)(i) = x(i) = x'(i) \quad \text{for} \quad i = 1, \ldots, |x|.$$

(c) $\Rightarrow$ (a). Obvious. $\square$

**4.2 Lemma.** *If* $x \in X$, $|x| > 0$, $p \in P_2$ *and* $n \in \mathbf{N}$, *then the following statements are equivalent*: (a) $V_x \supset V_{(p,n)}$; (b) $V_x \cap R_{(p,n)} \neq \varnothing$; (c) $|X(p)| > 0$ *and* $V_x \supset V_{X(p)}$.

*Proof.* (a) $\Rightarrow$ (b). Obvious.

(b) $\Rightarrow$ (c). Let $q \in V_x \cap R_{(p,n)}$. Then $|X(p)| = |X(q)| \geqslant |x| > 0$ and furthermore $q \in V_{X(p)}$ since $X(p) = X(q)$. Hence $V_x \cap V_{X(p)} \neq \varnothing$ and it follows from Lemma 4.1 that $V_{X(p)} \subset V_x$.

(c) $\Rightarrow$ (a). If $|X(p)| > 0$, then $V_{(p,n)} \subset V_{X(p)}$. For it is clear that $R_{(p,n)} \subset V_{X(p)}$. If $i \in \mathbf{N}$, then $|\gamma(p, n, i, \epsilon)| > |X(p)|$ and

$$\gamma(p, n, i, \epsilon)(j) = X(p)(j) \quad \text{for} \quad j = 0, \ldots, |X(p)|.$$

Thus $V_{\gamma(p,n,i,\epsilon)} \subset V_{X(p)}$ by Lemma 4.1 and it follows that $T_{(p,n)} \subset V_{X(p)}$. $\square$

**4.3 Lemma.** *If* $p, q \in P_2$ *and* $n, m \in \mathbf{N}$, *then the following statements are equivalent*: (a) $V_{(p,n)} \supset V_{(q,m)}$ *and* $|X(p)| \geqslant |X(q)|$; (b) $V_{(p,n)} \cap V_{(q,m)} \neq \varnothing$, $|X(p)| = |X(q)|$ *and* $n \leqslant m$; (c) $n \leqslant m$ *and* $R_{(p,n)} \cap R_{(q,m)} \neq \varnothing$.

*Proof.* (a) $\Rightarrow$ (b). Suppose that (a) holds. Since $q \in V_{(p,n)}$ and

$$|X(q)| \leqslant |X(p)|,$$

it follows that $|X(q)| = |X(p)|$. Suppose that $n > m$ and choose $q'$ in $P_2$ such that $X(q') = X(q)$, $Y(q') = Y(q)$, $Z(q')(m) \neq Z(p)(m)$ and $Z(q')(i) = Z(q)(i)$ for $i = 1, \ldots, m-1$ if $m > 1$. Then $q' \in R_{(q,m)} \subset V_{(q,m)}$, but $q' \notin R_{(p,n)}$ since $m \leqslant n-1$ and $Z(q')(m) \neq Z(p)(m)$ and it follows that $q' \notin V_{(p,n)}$ which is absurd. Thus $n \leqslant m$.

(b) $\Rightarrow$ (c). Suppose that (b) holds. Since $|X(p)| = |X(q)|$, it follows that $R_{(p,n)} \cap T_{(q,m)} = \varnothing$ and $R_{(q,m)} \cap T_{(p,n)} = \varnothing$. Thus either

$$R_{(p,n)} \cap R_{(q,m)} \neq \varnothing \quad \text{or} \quad T_{(p,n)} \cap T_{(q,m)} \neq \varnothing.$$

In the second case it follows from Lemma 4.1 that $T^{\epsilon}_{(p,n)} \cap T^{\epsilon}_{(q,m)} \neq \varnothing$. Thus there exist integers $i$ and $j$ such that $V_{\gamma(p,n,i,\epsilon)} \cap V_{\gamma(q,m,j,\epsilon)} \neq \varnothing$. Since $|\gamma(p,n,i,\epsilon)| = |X(p)| + n + 1 \leqslant |X(q)| + m + 1 = |\gamma(q,m,j,\epsilon)|$, it follows from Lemma 4.1 that

$$\gamma(q,m,j,\epsilon)(s) = \gamma(p,n,i,\epsilon)(s) \text{ for } s = 1, \ldots, |X(p)| + n + 1.$$

Hence $X(p) = X(q)$. Furthermore $Y(p)(n+i-1) = Y(q)(m+j-1) = t$ and $\theta_t(Y(p)) = \theta_t(Y(q))$ so that $Y(p) = Y(q)$. Finally if $n > 1$, then $Z(p)(j) = Z(q)(j)$ for $j = 1, \ldots, n-1$. Thus $q \in R_{(p,n)}$ so that

$$R_{(p,n)} \cap R_{(q,m)} \neq \varnothing.$$

(c) $\Rightarrow$ (a). Suppose that (c) holds. If $q' \in R_{(p,n)} \cap R_{(q,m)}$, then

$$X(p) = X(q') = X(q), \, Y(p) = Y(q') = Y(q)$$

and $Z(p)(i) = Z(q')(i) = Z(q)(i)$ for $i = 1, \ldots, n-1$, if $n > 1$. Thus $R_{(q,m)} \subset R_{(p,n)}$. Moreover since $|X(q)| + m + 1 \geqslant |X(p)| + n + 1$ and

$$\gamma(q,m,i,\epsilon)(j) = \gamma(p,n,m-n+i,\epsilon)(j)$$

for $j = 1, \ldots, |X(p)| + n + 1$, it follows from Lemma 4.1 that

$$T_{(q,m)} \subset T_{(p,n)}.$$

Hence $V_{(q,m)} \subset V_{(p,n)}.\square$

**4.4 Proposition.** *The set $\mathscr{V}$ of subsets of $P$ is the base for a topology on $P$.*

*Proof.* We shall show that if $U$ and $V$ are members of $\mathscr{V}$ and $p_0 \in U \cap V$, then there exists a member $W$ of $\mathscr{V}$ such that $p_0 \in W \subset U \cap V$. We consider three cases:

(i) Suppose that $U = V_x$ and $V = V_{x'}$, where $x, x' \in X$ and

$$0 < |x| \leqslant |x'|.$$

Then it follows from Lemma 4.1 that $p_0 \in V_{x'} = V_x \cap V_{x'}$.

(ii) Suppose that $U = V_x$ and $V = V_{(q,m)}$, where $x \in X$, $|x| > 0$, $q \in P_2$ and $m \in \mathbf{N}$. If $V_x \cap R_{(q,m)} \neq \varnothing$, then $p_0 \in V_{(q,m)} = V_x \cap V_{(q,m)}$ by Lemma 4.2. If $V_x \cap R_{(q,m)} = \varnothing$, then $p_0 \in V_x \cap V_{(q,m)} = V_x \cap T_{(q,m)}$. Since $T_{(q,m)}$ is a union of members of $\mathscr{V}_1$, it follows from part (i) that there exists $W$ in $\mathscr{V}_1$ such that $p_0 \in W \subset V_x \cap V_{(q,m)}$.

(iii) Suppose that $U = V_{(p,n)}$ and $V = V_{(q,m)}$, where $p, q \in P_2$, $n, m \in \mathbf{N}$ and $n \leqslant m$. If $R_{(p,n)} \cap R_{(q,m)} \neq \varnothing$, then $p_0 \in V_{(q,m)} = V_{(p,n)} \cap V_{(q,m)}$ by Lemma 4.3. If $R_{(p,n)} \cap R_{(q,m)} = \varnothing$, then

$$p_0 \in (V_{(p,n)} \cap T_{(q,m)}) \cup (V_{(q,m)} \cap T_{(p,n)}) = V_{(p,n)} \cap V_{(q,m)}.$$

Since $T_{(p,n)}$ and $T_{(q,m)}$ are unions of members of $\mathscr{V}_1$, it follows from (ii) that there exists a member $W$ of $\mathscr{V}$ such that

$$p_0 \in W \subset V_{(p,n)} \cap V_{(q,m)}. \;\square$$

Henceforth we shall regard $P$ as a topological space having the topology which has $\mathscr{V}$ as a base. To prove that $P$ is metrizable, we introduce a sequence $\{\mathscr{H}_n\}_{n\in\mathbf{N}}$ of open coverings of $P$. For each positive integer $n$ let $\mathscr{H}_n^1$ be the subset of $\mathscr{V}_1$ defined as follows:

$$\mathscr{H}_n^1 = \{V \in \mathscr{V}_1 \mid V = V_x, \quad \text{where} \quad |x| = n\}.$$

Let $\mathscr{H}_n^2$ be the subset of $\mathscr{V}_2$ defined as follows:

$$\mathscr{H}_n^2 = \{V \in \mathscr{V}_2 \mid V = V_{(q,m)}, \quad \text{where} \quad |X(q)| + m = n\}.$$

And let $\mathscr{H}_n = \mathscr{H}_n^1 \cup \mathscr{H}_n^2$.

**4.5 Proposition.** *For each positive integer $n$, $\mathscr{H}_n$ is an open covering of $X$ and $\mathscr{H}_{n+1}$ is a refinement of $\mathscr{H}_n$.*

*Proof.* Let $p$ be a point of $P$ and let $n$ be a positive integer. If $p \in P_1$ let $x$ be the element of $X$ such that $|x| = n$ and $x(i) = p(i)$ for $i = 1,\ldots,n$. Then $V_x \in \mathscr{H}_n^1$ and $p \in V_x$. If $p \in P_2$ and $|X(p)| \geqslant n$ then $p \in V_x$, where $x$ is the element of $X$ such that $|x| = n$ and $x(i) = X(p)(i)$ for $i = 1,\ldots,n$. If $p \in P_2$ and $|X(p)| < n$, let $m = n - |X(p)|$. Then $V_{(p,m)} \in \mathscr{H}_n^2$ and $p \in V_{(p,m)}$. Thus $\mathscr{H}_n$ is an open covering of $X$. Furthermore $\mathscr{H}_{n+1}$ is a refinement of $\mathscr{H}_n$. For if $x \in X$ and $V_x \in \mathscr{H}_{n+1}$ then $V_x \subset V_{x'}$, where $|x'| = n$ and $x'(i) = x(i)$ for $i = 1,\ldots,n$. If $V_{(q,m)} \in \mathscr{H}_{n+1}$, where $m > 1$, then $V_{(q,m)} \subset V_{(q,m-1)} \in \mathscr{H}_n$. If $V_{(q,1)} \in \mathscr{H}_{n+1}$, then $|X(q)| = n$ and

$$V_{(q,1)} \subset V_{X(q)} \in \mathscr{H}_n. \;\square$$

We also consider a sequence $\{\mathscr{G}_n\}_{n\in\mathbf{N}}$ of open coverings of $P$. For each positive integer $n$ let $\mathscr{G}_n^1$ be the subset of $\mathscr{V}_1$ defined as follows:

$$\mathscr{G}_n^1 = \{V \in \mathscr{V}_1 \mid V = V_x, \quad \text{where} \quad |x| \geqslant n\}.$$

Let $\mathscr{G}_n^2$ be the subset of $\mathscr{V}_2$ defined as follows:

$$\mathscr{G}_n^2 = \{V \in \mathscr{V}_2 \mid V = V_{(q,m)}, \quad \text{where} \quad |X(q)| + m \geqslant n\}.$$

And let $\mathscr{G}_n = \mathscr{G}_n^1 \cup \mathscr{G}_n^2$. Clearly $\mathscr{G}_{n+1} \subset \mathscr{G}_n$. Since $\mathscr{H}_n \subset \mathscr{G}_n$ it follows that $\mathscr{G}_n$ is an open covering of $P$. The covering $\mathscr{G}_n$ has the additional property, which will be exploited, that $P_1$ is contained in the union of members of $\mathscr{G}_n^1$ and $P_2$ is contained in the union of members of $\mathscr{G}_n^2$. This follows from the proof of Proposition 4.5 and the observation that if $p \in P_2$ and we choose $m$ such that $|X(p)| + m \geqslant n$, then $p \in V_{(p,m)}$ and $V_{(p,m)} \in \mathscr{G}_n^2$. In order to get more information about the coverings $\mathscr{G}_n$ we must take account of certain 'exceptional' subsets of $\mathscr{V}_2$. If $x \in X$ and $|x| > 0$, then $\mathscr{E}_x$ is the subset of $\mathscr{V}_2$ which consists of the sets of the form $V_{(q,m)}$, where $|X(q)| + 1 = |x|$, $X(q)(i) = x(i)$ for $i = 0, \ldots, |X(q)|$ and $Y(q) \in Y_{|x(|x|)|}$. If $p \in P_2$ and $n \in \mathbf{N}$, then

$$\mathscr{E}_{(p,n)} = \bigcup_{i \in \mathbf{N}} (\mathscr{E}_{\gamma(p,n,i,+)} \cup \mathscr{E}_{\gamma(p,n,i,-)}).$$

If $V_{(q,m)} \in \mathscr{E}_x$ then $V_{(q,j)} \in \mathscr{E}_x$ for all positive integers $j$. A similar observation can be made about $\mathscr{E}_{(p,n)}$.

**4.6 Lemma.** Let $x$ be an element of $X$ such that $|x| > 0$. There exists a positive integer $N$ such that $V \subset V_x$ if $V \in \mathscr{G}_N$, $V \notin \mathscr{E}_x$ and $V_x \cap V \neq \varnothing$.

*Proof.* If $|x| = 1$, let $N = 1$. If $|x| > 1$, then for each integer $j$ such that $1 \leqslant j \leqslant |x| - 1$ let

$$y_j = \phi_{|x(j)|}(|x(j+1)|)$$

and let $m_j$ be the unique positive integer such that $y_j(m_j) = |x(j)|$. Now define

$$N = \left( \sum_{j=1}^{|x|-1} m_j \right) + |x|.$$

Suppose that $V_{x'} \in \mathscr{G}_N^1$. Then $|x'| \geqslant N \geqslant |x|$ and it follows from Lemma 4.1 that $V_x \supset V_{x'}$ if $V_x \cap V_{x'} \neq \varnothing$. Next suppose that $V_{(p,n)} \in \mathscr{G}_N^2$ and $|X(p)| \geqslant |x|$. Since $V_{(p,n)} \subset V_{X(p)}$, it follows that if $V_x \cap V_{(p,n)} \neq \varnothing$, then $V_x \cap V_{X(p)} \neq \varnothing$ so that $V_x \supset V_{X(p)} \supset V_{(p,n)}$ by Lemma 4.1. Finally we must consider the case in which $V_{(p,n)} \in \mathscr{G}_N^2$ and $|X(p)| < |x|$. The proof is completed by showing that in this case $V_{(p,n)} \in \mathscr{E}_x$ if $V_x \cap V_{(p,n)} \neq \varnothing$. Since $|X(p)| < |x|$ it follows from Lemma 4.2 that $V_x \cap R_{(p,n)} = \varnothing$. Hence $V_x \cap V_{x'} \neq \varnothing$, where $x' = \gamma(p,n,i,\epsilon)$ for some positive integer $i$. Since

$$|x'| = |X(p)| + n + 1 \geqslant N + 1 > |x|,$$

it follows from Lemma 4.1 that $x(j) = x'(j)$ for $j = 1, \ldots, |x|$. Thus

$$x(j) = X(p)(j) \quad \text{for} \quad j = 0, \ldots, |X(p)|. \tag{1}$$

Furthermore

$$|x(|X(p)| + 1)| = |x'(|X(p)| + 1)| = Y(p)(n + i - 1). \tag{2}$$

Now suppose that $|X(p)| + 1 < |x|$ and put $k = |X(p)| + 1$ so that $Y(p)(n + i - 1) = |x(k)|$ by (2). Then

$$|x(k+1)| = |x'(k+1)| = \theta_{|x(k)|}(Y(p)).$$

Hence $Y(p) = \phi_{|x(k)|}\big(|x(k+1)|\big) = y_k$. But $y_k(m_k) = |x(k)|$ so that

$$Y(p)(m_k) = |x(k)| = Y(p)(n+i-1).$$

Thus $m_k = n+i-1$ so that $n \leqslant m_k$. But this is absurd since

$$|x| + n > |X(p)| + n \geqslant N$$

so that $n > m_k$. It follows that

$$|X(p)| + 1 = |x|. \tag{3}$$

It follows from (2) and (3) that

$$Y(p) \in Y_{|x(|x|)|}. \tag{4}$$

It follows from (1), (3) and (4) that $V_{(p,n)} \in \mathscr{E}_x$. $\square$

**Definition.** *The least positive integer with the property described in Lemma 4.6 will be denoted by $N(x)$.*

**4.7   Lemma.** *If $x \in X$, $|x| > 0$, $p \in P_2$ and $V_{(p,1)} \in \mathscr{E}_x$, then there exists a positive integer $n$ such that $V_x \cap V_{(p,n)} = \varnothing$.*

*Proof.* Since $V_{(p,1)} \in \mathscr{E}_x$, there exists a positive integer $m$ such that $Y(q)(m) = |x(|x|)|$. Let $n = m+1$. If $i \in \mathbf{N}$ then

$$|\gamma(p,n,i,\epsilon)| = |X(p)| + n + 1 > |X(p)| + 1 = |x|.$$

But since $n+i-1 > m$ it follows that

$$|x(|x|)| \neq Y(p)(n+i-1) = |\gamma(p,n,i,\epsilon)(|x|)|.$$

It follows from Lemma 4.1 that $V_x \cap T_{(p,n)} = \varnothing$. Since $|X(p)| < |x|$ it follows from Lemma 4.2 that $V_x \cap R_{(p,n)} = \varnothing$. Hence $V_x \cap V_{(p,n)} = \varnothing$. $\square$

**4.8   Lemma.** *Suppose that $p, q \in P_2$ and $n, m \in \mathbf{N}$, where*

$$|X(q)| \leqslant |X(p)| + n, \quad |X(p)| < |X(q)| + m \quad and \quad |X(p)| \neq |X(q)|.$$

*If $V_{(p,n)} \cap V_{(q,m)} \neq \varnothing$ then $|X(q)| = |X(p)| + n$ and $V_{(q,m)} \in \mathscr{E}_{(p,n)}$.*

*Proof.* Since $|X(p)| < |X(q)| + m$ it follows that $R_{(p,n)} \cap T_{(q,m)} = \varnothing$ and since $|X(q)| < |X(p)| + n + 1$ it follows that $R_{(q,m)} \cap T_{(p,n)} = \varnothing$. Since $|X(p)| \neq |X(q)|$ it follows from Lemma 4.3 that $R_{(p,n)} \cap R_{(q,m)} = \varnothing$. Thus if $V_{(p,n)} \cap V_{(q,m)} \neq \varnothing$, then $V_x \cap V_{x'} \neq \varnothing$, where $x = \gamma(p,n,i,\epsilon)$ and $x' = \gamma(q,m,j,\epsilon')$ and $i,j \in \mathbf{N}$. It follows that

$$x(s) = x'(s) \quad \text{for} \quad s = 1, \ldots, \min\{|x|, |x'|\}.$$

But $|x| = |X(p)| + n + 1$ and $|x'| = |X(q)| + m + 1$ so that

$$|X(p)| + 2 \leqslant \min\{|x|, |x'|\}$$

and if $|X(q)| < |X(p)| + n$ then also $|X(q)| + 2 \leqslant \min\{|x|, |x'|\}$. If $|X(q)| < |X(p)|$, then

$$x'(|X(p)| + 1)\, x'(|X(p)| + 2) > 0,$$

which is absurd since $x(|X(p)| + 1)\, x(|X(p)| + 2) < 0$. If

$$|X(p)| < |X(q)| < |X(p)| + n,$$

then

$$x(|X(q)| + 1)\, x(|X(q)| + 2) > 0,$$

which is absurd since $x'(|X(q)| + 1)\, x'(|X(q)| + 2) < 0$. It follows that $|X(q)| = |X(p)| + n$. Thus $|X(q)| + 1 = |x|$, and $X(q)(s) = x'(s) = x(s)$ for $s = 0, \ldots, |X(q)|$, whilst

$$x(|x|) = x'(|x|) = x'(|X(q)| + 1) = \pm Y(q)(m+j-1),$$

so that $Y(q) \in Y_{|x(|x|)|}$. Hence $V_{(q,m)} \in \mathscr{E}_x$. Since $x = \gamma(p, n, i, \epsilon)$ it follows that $V_{(q,m)} \in \mathscr{E}_{(p,n)}.\ \square$

**4.9 Lemma.** *Suppose that $p \in P_2$ and $n \in \mathbf{N}$. There exists a positive integer $N$ such that $V \subset V_{(p,n)}$ if $V \in \mathscr{G}_N$, $V \notin \mathscr{E}_{(p,n)}$ and $V_{(p,n)} \cap V \neq \varnothing$.*

*Proof.* Let $N = |X(p)| + n + 1$. If $V_x \in \mathscr{G}_N^1$, then $|x| \geqslant N > |X(p)|$ so that $V_x \cap R_{(p,n)} = \varnothing$. Thus if $V_x \cap V_{(p,n)} \neq \varnothing$, then $V_x \cap T_{(p,n)} \neq \varnothing$ so that $V_x \subset T_{(p,n)} \subset V_{(p,n)}$, by Lemma 4.1. Next suppose that $V_{(q,m)} \in \mathscr{G}_N^2$ and $|X(q)| > |X(p)| + n$. If $V_{(q,m)} \cap V_{(p,n)} \neq \varnothing$, then $V_{X(q)} \cap V_{(p,n)} \neq \varnothing$ so that $V_{X(q)} \cap T_{(p,n)} \neq \varnothing$ and it follows that $V_{(q,m)} \subset V_{X(q)} \subset T_{(p,n)} \subset V_{(p,n)}$. Finally we must consider the case in which $V_{(q,m)} \in \mathscr{G}_N^2$, $|X(q)| \leqslant |X(p)| + n$ and $V_{(q,m)} \cap V_{(p,n)} \neq \varnothing$. If $|X(p)| = |X(q)|$, then since

$$|X(q)| + m \geqslant N > |X(p)| + n,$$

we have $m > n$, and it follows from Lemma 4.3 that $V_{(q,m)} \subset V_{(p,n)}$. If $|X(p)| \neq |X(q)|$, then since $|X(q)| \leqslant |X(p)| + n$ and

$$|X(p)| < N \leqslant |X(q)| + m$$

it follows from Lemma 4.8 that $V_{(q,m)} \in \mathscr{E}_{(p,n)}.\ \square$

**Definition.** *The least integer with the property described in Lemma* 4.9 *will be denoted by* $N(p, n)$.

We can now supplement Proposition 4.4 as follows:

**4.10  Proposition.** *Suppose that $W \in \mathscr{V}$ and that $p \in W$. If $p \in P_1$, then there exists $x$ in $X$ such that $|x| > 0$ and $p \in V_x \subset W$, and if $p \in P_2$, then there exists a positive integer $n$ such that $p \in V_{(p,n)} \subset W$.*

*Proof.* If $p \in P_1$, then we need only consider the case in which $W \in \mathscr{V}_2$ so that $W = V_{(q,m)}$, say. There exists a member $V_x$ of $\mathscr{G}^1_{N(q,m)}$ such that $p \in V_x$. Since $\mathscr{E}_{(q,m)} \subset \mathscr{V}_2$, it follows that $V_x \notin \mathscr{E}_{(q,m)}$. Since $V_x \cap V_{(q,m)} \neq \varnothing$ it follows from Lemma 4.9 that $V_x \subset V_{(q,m)}$. Now suppose that $p \in P_2$. If $W = V_x \in \mathscr{V}_1$, then $|X(p)| \geqslant |x|$ so that $V_{(p,N(x))} \notin \mathscr{E}_x$. Since

$$V_{(p,N(x))} \in \mathscr{G}_{N(x)} \quad \text{and} \quad V_x \cap V_{(p,N(x))} \neq \varnothing,$$

it follows from Lemma 4.6 that $p \in V_{(p,N(x))} \subset V_x$. If $W = V_{(q,m)} \in \mathscr{V}_2$, then since $p \in V_{(q,m)}$, either $|X(p)| = |X(q)|$ or $|X(p)| \geqslant |X(q)| + m + 1$. Thus if $N = N(q,m)$, then $V_{(p,N)} \notin \mathscr{E}_{(q,m)}$. Since $V_{(p,N)} \in \mathscr{G}_N$ and

$$V_{(q,m)} \cap V_{(p,N)} \neq \varnothing,$$

it follows from Lemma 4.9 that $p \in V_{(p,N)} \subset V_{(q,m)}$.$\square$

We can now establish the first of the main properties of $P$.

**4.11  Proposition.** *The space $P$ is metrizable and $\dim P \leqslant 1$.*

*Proof.* We shall appeal to Proposition 2.1. We prove first that if $U$ is open in $P$ and $p \in U$, then there exist a member $V$ of $\mathscr{V}$ and a positive integer $N$ such that $p \in V$ and $\mathrm{St}\,(V, \mathscr{H}_N) \subset U$. If $p \in P_1$, then it follows from Proposition 4.10 that there exists $x$ in $X$ such that $|x| > 0$ and $p \in V_x \subset U$. Define $x'$ in $X$ such that $|x'| = |x| + 1$, $x'(i) = x(i)$ for $i = 1, \ldots, |x|$ and $x'(|x'|) = p(|x'|)$. Then $p \in V_{x'} \subset V_x$. Let $N = N(x')$. If $W \in \mathscr{H}_N$, $W \notin \mathscr{E}_{x'}$ and $W \cap V_{x'} \neq \varnothing$, then it follows from Lemma 4.6 that $W \subset V_{x'} \subset U$. If $W \in \mathscr{H}_N$ and $W = V_{(q,m)} \in \mathscr{E}_{x'}$, then $q \in V_x$ so that $V_x \cap R_{(q,m)} \neq \varnothing$ and hence $V_{(q,m)} \subset V_x \subset U$ by Lemma 4.2. Thus $\mathrm{St}\,(V_{x'}, \mathscr{H}_N) \subset U$. If $p \in P_2$, then it follows from Proposition 4.10 that there exists a positive integer $n$ such that $p \in V_{(p,n)} \subset U$. Hence $p \in V_{(p,n+1)} \subset V_{(p,n)}$. Let $N = N(p, n+1)$. If $W \in \mathscr{H}_N$, $W \notin \mathscr{E}_{(p,n+1)}$ and $W \cap V_{(p,n+1)} \neq \varnothing$, then it follows from Lemma 4.9 that

$$W \subset V_{(p,n+1)} \subset U.$$

If $W \in \mathscr{H}_N$ and $W = V_{(q,m)} \in \mathscr{E}_{(p,n+1)}$, then $W \in \mathscr{E}_{\gamma(p,n+1,i,\epsilon)}$ for some positive integer $i$. Thus $|X(q)| + 1 = |X(p)| + n + 1$, and for

$$j = 1, \ldots, |X(q)|,$$

$$X(q)(j) = \gamma(p, n+1, i, \epsilon)(j) = \gamma(p, n, i+1, \epsilon)(j).$$

Thus $V_{X(q)} = V_{\gamma(p,n,i+1,\varepsilon)}$ and hence $V_{(q,m)} \subset V_{\gamma(p,n,i+1,\varepsilon)} \subset V_{(p,n)} \subset U$. Thus $\mathrm{St}(V_{(p,n+1)}, \mathscr{H}_N) \subset U$. Next we show that for each positive integer $n$, the order of $\mathscr{H}_n$ does not exceed 1. If $V_x$ and $V_{x'}$ are members of $\mathscr{H}_n^1$ and $x \neq x'$, then $V_x \cap V_{x'} = \varnothing$ by Lemma 4.1. If $V_{(p,k)}$ and $V_{(q,m)}$ are members of $\mathscr{H}_n^2$ and $V_{(p,k)} \cap V_{(q,m)} \neq \varnothing$, then since

$$|X(p)| < |X(q)| + m$$

and $|X(q)| < |X(p)| + k$, it follows from Lemma 4.8 that

$$|X(p)| = |X(q)|.$$

But in this case $k = m$ and $V_{(p,k)} = V_{(q,m)}$ by Lemma 4.3. It is now clear that at most two members of $\mathscr{H}_n$, one from $\mathscr{H}_n^1$ and one from $\mathscr{H}_n^2$, contain a given point of $P$. Since $\mathscr{H}_{n+1}$ is a refinement of $\mathscr{H}_n$, it follows from Proposition 2.1 that $P$ is a pseudo-metrizable space such that $\dim P \leqslant 1$. The proof will be completed by showing that $P$ is a $T_0$-space. Thus let $p$ and $q$ be distinct points of $P$. If $p, q \in P_1$, let $j$ be a positive integer such that $p(j) \neq q(j)$ and define $x$ in $X$ so that $|x| = j$ and $x(i) = p(i)$ for $i = 1, \ldots, |x|$. Then $p \in V_x$ and $q \notin V_x$. If $p \in P_1$ and $q \in P_2$, define $x$ in $X$ so that $|x| = |X(q)| + 1$ and $x(i) = p(i)$ for $i = 1, \ldots, |x|$. Then $p \in V_x$ and $q \notin V_x$ since $|x| > |X(q)|$. Finally suppose that $p, q \in P_2$ and that $|X(p)| \leqslant |X(q)|$. If $|X(p)| < |X(q)|$, then $q \in V_{X(q)}$ and $p \notin V_{X(q)}$. If $|X(p)| = |X(q)|$ and either $X(p) \neq X(q)$ or $Y(p) \neq Y(q)$, then $p \in V_{(p,1)}$ and $q \notin V_{(p,1)}$. If $X(p) = X(q)$ and

$$Y(p) = Y(q),$$

let $n$ be a positive integer such that $Z(p)(n) \neq Z(q)(n)$. Then

$$p \in V_{(p,n+1)} \quad \text{and} \quad q \notin V_{(p,n+1)}. \square$$

Next we show that $\mathrm{ind}\, P = 0$.

**4.12   Proposition.** *Every member of the base $\mathscr{V}$ is closed.*

*Proof.* First suppose that $x \in X$, $|x| > 0$ and $p \in P \backslash V_x$. If $p \in P_1$ or $p \in P_2$ and $V_{(p,1)} \notin \mathscr{E}_x$, there exists $V$ in $\mathscr{G}_{N(x)}$ such that $p \in V$ and $V \notin \mathscr{E}_x$. It follows from Lemma 4.6 that $V \cap V_x = \varnothing$. If $p \in P_2$ and $V_{(p,1)} \in \mathscr{E}_x$, it follows from Lemma 4.7 that there exists a positive integer $n$ such that $V_{(p,n)} \cap V_x = \varnothing$.

Suppose now that $p \in P_2$, $n \in \mathbf{N}$ and $q \notin V_{(p,n)}$. If $q \in P_1$ or $q \in P_2$ and $V_{(q,1)} \notin \mathscr{E}_{(p,n)}$, then there exists $V$ in $\mathscr{G}_{N(p,n)}$ such that $q \in V$ and

$$V \notin \mathscr{E}_{(p,n)}.$$

It follows from Lemma 4.9 that $V \cap V_{(p,n)} = \varnothing$. Finally suppose that $q \in P_2$ and $V_{(q,1)} \in \mathscr{E}_{(p,n)}$. Then $|X(q)| = |X(p)| + n$. Let us suppose that

$$X(q)\big(|X(p)| + 1\big) > 0.$$

Then there exists a positive integer $k$ such that $V_{(q,1)} \in \mathscr{E}_x$, where $x = \gamma(p, n, k, +)$. By Lemma 4.7 there exists a positive integer $m$ such that $V_{(q,m)} \cap V_x = \varnothing$. Since $V_{X(q)} \cap V_{\gamma(p,n,j,+)} = \varnothing$ if $j \neq k$ and $V_{X(q)} \cap V_{\gamma(p,n,j,-)} = \varnothing$ for all $j$, it follows that $V_{(q,m)} \cap T_{(p,n)} = \varnothing$. Since $|X(q)| > |X(p)|$, it follows that $V_{X(q)} \cap R_{(p,n)} = \varnothing$ and hence

$$V_{(q,m)} \cap R_{(p,n)} = \varnothing.$$

Thus $V_{(q,m)}$ is a neighbourhood of $q$ such that $V_{(q,m)} \cap V_{(p,n)} = \varnothing$. The argument in the case $X(q)\big(|X(p)| + 1\big) < 0$ is similar. $\square$

**4.13   Corollary.** $\operatorname{ind} P = 0$.

*Proof.* A space with a base for its topology which consists of open-and-closed sets has small inductive dimension equal to zero. $\square$

The hardest part of the analysis of this example is the proof that $\operatorname{Ind} P > 0$. For this purpose some new definitions must be made. If $n \in \mathbf{N}$, let $\langle n \rangle = \{1, \ldots, n\}$ and let $\Pi_n$ denote the set of functions $\pi \colon \langle n \rangle \to \mathbf{R}^+$. Let $\Pi = \bigcup_{n \in \mathbf{N}} \Pi_n$. If $\pi \in \Pi$ we write $|\pi| = n$ if $\pi \in \Pi_n$.

A subset $K$ of $\Pi$ is called an indicator if it has the following three properties:

(i) there exists a positive integer $n$ such that $K \subset \Pi_n$, and we then write $|K| = n$;

(ii) the set of positive numbers $r$ such that $r = \pi(1)$ for some $\pi$ in $K$ is infinite;

(iii) if $|K| > 1$, $\pi \in K$ and $j$ is an integer such that $1 < j \leqslant |K|$, then the set of positive numbers $r$ such that $r = \rho(j)$ for some $\rho$ in $K$ such that $\rho(i) = \pi(i)$ if $i < j$, is infinite.

If $x \in X$ and $K$ is an indicator, then $\mathscr{S}^+(x, K)$ is defined to be the subset of $\mathscr{V}_1$ which consists of the sets $V_{x'}$ for which $|x'| = |x| + |K|$, $x'(i) = x(i)$ for $i = 0, \ldots, |x|$ and $x'(|x| + i) = \pi(i)$ for $i = 1, \ldots, |K|$ for some $\pi$ in $K$. Similarly $\mathscr{S}^-(x, K)$ is the subset of $\mathscr{V}_1$ consisting of the sets $V_{x'}$ for which $|x'| = |x| + |K|$, $x'(i) = x(i)$ for $i = 0, \ldots, |x|$ and $x'(|x| + i) = -\pi(i)$ for $i = 1, \ldots, |K|$ for some $\pi$ in $K$. Let

$$\Sigma^\epsilon(x, K) = \bigcup \{V \mid V \in \mathscr{S}^\epsilon(x, K)\},$$

where $\epsilon$ denotes $+$ or $-$. If $|x| > 0$, then $V \subset V_x$ if $V \in \mathscr{S}^\epsilon(x, K)$. Thus

$$\Sigma^+(x, K) \cup \Sigma^-(x, K) \subset V_x \quad \text{if} \quad |x| > 0.$$

If $p \in P_2$, $n \in \mathbb{N}$ and $K$ is an indicator, then $\mathscr{S}(p, n, K)$ is defined to be the subset of $\mathscr{V}_2$ which consists of the sets $V_{(q, m)}$, where $q \in R_{(p, n)}$ $m = n + |K|$ and $Z(q)(n - 1 + i) = \pi(i)$ for $i = 1, \ldots, |K|$ for some $\pi$ in $K$. Let

$$\Sigma(p, n, K) = \bigcup \{V \mid V \in \mathscr{S}(p, n, K)\}.$$

Since $V \subset V_{(p, n)}$ if $V \in \mathscr{S}(p, n, K)$ it follows that

$$\Sigma(p, n, K) \subset V_{(p, n)}.$$

**4.14   Lemma.** *If $p \in P_2$, $n \in \mathbb{N}$ and $\mathscr{H}$ is a subset of $\mathscr{V}$ with the property that for each point $q$ of $R_{(p, n)}$ there exists an indicator $K$ such that $\mathscr{S}(q, n + 1, K) \subset \mathscr{H}$, then there exists an indicator $L$ such that*

$$\mathscr{S}(p, n, L) \subset \mathscr{H}.$$

*Proof.* If $r \in \mathbb{R}^+$, choose $q(r)$ in $R_{(p, n)}$ such that $Z(q(r))(n) = r$ and let $K_r$ be an indicator such that $\mathscr{S}(q(r), n + 1, K_r) \subset \mathscr{H}$. There exists a positive integer $N$ and an infinite set $T$ of positive numbers such that $|K_r| = N$ if $r \in T$. Let $L$ be the subset of $\Pi_{N+1}$ defined as follows: if $\pi \in \Pi_{N+1}$, then $\pi \in L$ if and only if $\pi(1) \in T$ and there exists $\rho$ in $K_{\pi(1)}$ such that $\pi(i + 1) = \rho(i)$ for $i = 1, \ldots, N$. Clearly $L$ is an indicator. If $V_{(q, m)} \in \mathscr{S}(p, n, L)$, then $q \in R_{(p, n)}$, $m = n + |L|$ and there exists some $\pi$ in $L$ such that $Z(q)(n - 1 + i) = \pi(i)$ for $i = 1, \ldots, |L|$. Let $\pi(1) = r \in T$. Since there exists $\rho$ in $K_r$ such that $\rho(i) = \pi(i + 1)$ for $i = 1, \ldots, N$, it is easy to see that $V_{(q, m)} \in \mathscr{S}(q(r), n + 1, K_r)$. It follows that

$$V_{(q, m)} \in \mathscr{H}. \square$$

**4.15   Lemma.** *Suppose that $p \in P_2$ and that $\mathscr{W}$ is a subset of $\mathscr{V}$ with the property that if $q \in R_{(p, 1)}$ then (a) $V_{(q, n)} \in \mathscr{W}$ if $V_{(q, n)}$ is a subset of a member of $\mathscr{W}$, and (b) there exists some $m$ such that $V_{(q, m)} \in \mathscr{W}$. Then there exists an indicator $K$ such that $\mathscr{S}(p, 1, K) \subset \mathscr{W}$.*

*Proof.* Suppose there exists no such indicator. Let $p_1 = p$ and by induction using Lemma 4.14 construct a sequence $\{p_n\}_{n \in \mathbb{N}}$ of elements of $P_2$ such that for each $n$, $p_{n+1} \in R_{(p_n, n)}$ and there exists no indicator $K$ such that $\mathscr{S}(p_n, n, K) \subset \mathscr{W}$. Then for each positive integer $n$, $X(p_n) = X(p_{n+1}) = X(p)$, $Y(p_n) = Y(p_{n+1}) = Y(p)$ and

$$Z(p_n)(j) = Z(p_{n+1})(j) \quad \text{for} \quad j = 1, \ldots, n - 1.$$

Now let $q$ be the element of $P_2$ such that $X(q) = X(p)$, $Y(q) = Y(p)$ and $Z(q)(n) = Z(p_{n+1})(n)$ if $n \in \mathbb{N}$. Then $q \in R_{(p, 1)}$, so that by hypothesis there exists some $m > 1$ such that $V_{(q, m-1)} \in \mathscr{W}$. Since $q \in R_{(p_m, m)}$ it follows from Lemma 4.3 that $V_{(p_m, m)} \subset V_{(q, m-1)}$. But if $K$ is an indicator and $V_{(p', k)} \in \mathscr{S}_{(p_m, m, K)}$, then $p' \in R_{(p_m, m)} \subset R_{(p, 1)}$ and

$$V_{(p', k)} \subset V_{(p_m, m)} \subset V_{(q, m-1)}.$$

It follows that $V_{(p', k)} \in \mathscr{W}$. Thus $\mathscr{S}(p_m, m, K) \subset \mathscr{W}$ which is a contradiction of the definition of $p_m$. Hence the required indicator exists. $\square$

**4.16 Lemma.** *If $p \in P_2$ and $U_1$ and $U_2$ are open sets of $P$ such that $R_{(p, 1)} \subset U_1 \cup U_2$, then there exists an indicator $K$ such that*

$$\Sigma(p, 1, K) \subset U_j$$

*for $j = 1$ or $j = 2$.*

*Proof.* For $j = 1, 2$, let $\mathscr{W}_j$ be the subset of $\mathscr{V}$ consisting of those sets $V$ such that $V \subset U_j$ and $V = V_{(q, m)}$ for some $m$, where $q \in R_{(p, 1)}$. Let $\mathscr{W} = \mathscr{W}_1 \cup \mathscr{W}_2$. It is clear that $p$ in $P_2$ and the subset $\mathscr{W}$ of $\mathscr{V}$ satisfy the hypotheses of Lemma 4.15. Hence there exists an indicator $L$ such that $\mathscr{S}(p, 1, L) \subset \mathscr{W}$. Let $N = |L|$ and if $\pi \in L$ choose $q(\pi)$ in $R_{(p, 1)}$ such that $Z(q(\pi))(i) = \pi(i)$ for $i = 1, \ldots, N$. Then

$$\mathscr{S}(p, 1, L) = \{V \in \mathscr{V} \mid V = V_{(q(\pi), N+1)} \text{ for some } \pi \text{ in } L\}.$$

For $j = 1, 2$ let

$$M_1^j = \{\pi \in L \mid V_{(q(\pi), N+1)} \in \mathscr{W}_j\}.$$

If $N > 1$, then for $j = 1, 2$ and $n = 2, \ldots, N$ let $M_n^j$ be the subset of $\Pi_{N-n+1}$ consisting of those elements $\rho$ for which there exist infinitely many members $\sigma$ of $M_{n-1}^j$ such that $\sigma(i) = \rho(i)$ for $i = 1, \ldots, |\rho|$. For $n = 1, \ldots, N$ let $M_n$ be the subset of $\Pi_{N-n+1}$ consisting of those elements $\rho$ for which there exists some $\pi$ in $L$ such that $\rho(i) = \pi(i)$ for for $i = 1, \ldots, |\rho|$. Then $M_1 = L$ and it is clear that $M_1 = M_1^1 \cup M_1^2$. It is not difficult to establish by induction that $M_n^1 \cup M_n^2 = M_n$ for $n = 1, \ldots, N$. Thus in particular

$$M_N = M_N^1 \cup M_N^2.$$

But $M_N$ is the subset of $\Pi_1$ consisting of those elements $\rho$ such that $\rho(1) = \pi(1)$ for some $\pi$ in $L$. Since $L$ is an indicator, $M_N$ is an infinite set. It follows that either $M_N^1$ or $M_N^2$ is an infinite set.

Let us choose $j (= 1 \text{ or } 2)$ such that $M_N^j$ is an infinite set. If $\pi \in L$ and $n = 1, \ldots, N$ let $\pi_n$ be the element of $\Pi_{N-n+1}$ given by

$$\pi_n = \pi \mid \langle N - n + 1 \rangle.$$

Then $\pi_1 = \pi$ and $\pi_n \in M_n$ for $n = 1, \ldots, N$. Now we define

$$K = \{\pi \in L \mid \pi_n \in M_n^j \quad \text{for} \quad n = 1, \ldots, N\}.$$

It is easy to establish that $K$ is an indicator. Since $K \subset M_1^j \subset L$, it follows that $\mathscr{S}(p, 1, K)$ consists of the sets $V_{(q(\pi), N+1)}$ for $\pi$ in $K$ and it is clear that $\Sigma(p, 1, K) \subset U_j$. $\square$

**4.17 Lemma.** *Let $K$ be an indicator and let $p$ be a point of $P_2$. If $x = \gamma(p, 1, i, -)$, where $i > |K|$, then $\Sigma^+(x, K) \subset \Sigma(p, 1, K)$, and if $x = \gamma(p, 1, i, +)$, where $i > |K|$, then $\Sigma^-(x, K) \subset \Sigma(p, 1, K)$.*

*Proof.* Let $x = \gamma(p, 1, i, -)$, where $i > |K|$, and if $\pi \in K$ let $\xi(x, \pi)$ be the element of $X$ such that $|\xi(x, \pi)| = |x| + |K|$ and

$$\xi(x, \pi)(j) = x(j) \quad \text{for} \quad j = 0, \ldots, |x|,$$

$$\xi(x, \pi)(|x| + j) = \pi(j) \quad \text{for} \quad j = 1, \ldots, |K|.$$

Then

$$\mathscr{S}^+(x, K) = \{V \in \mathscr{V}_1 \mid V = V_{\xi(x, \pi)} \text{ for some } \pi \text{ in } K\}.$$

Now let $\pi$ be a member of $K$. Let us choose $q$ in $R_{(p,1)}$ such that $Z(q)(j) = \pi(j)$ for $j = 1, \ldots, |\pi|$. Clearly $V_{(q, |K|+1)} \in \mathscr{S}(p, 1, K)$. But it is easy to verify that

$$\xi(x, \pi) = \gamma(q, |K| + 1, i - |K|, -)$$

so that $V_{\xi(x, \pi)} \subset T^-_{(q, |K|+1)} \subset V_{(q, |K|+1)}$. Thus $\Sigma^+(x, K) \subset \Sigma(p, 1, K)$ as asserted. The proof of the second assertion is similar. $\square$

**4.18 Lemma.** *Suppose that $x \in X$, that $U_1$ and $U_2$ are open sets of $P$ such that $U_1 \cup U_2 = P$ and that $K_1$ and $K_2$ are indicators such that $\Sigma^+(x, K_1) \subset U_1$ and $\Sigma^-(x, K_2) \subset U_2$. Then there exist indicators $L_1$ and $L_2$ and $x'$ in $X$ with $|x'| = |x| + 1$ and $x'(i) = x(i)$ for $i = 0, \ldots, |x|$ such that*

$$\Sigma^+(x', L_1) \subset U_1 \quad \text{and} \quad \Sigma^-(x', L_2) \subset U_2.$$

*Proof.* Let $B_1$ and $B_2$ be the subsets of $R^+$ defined by

$$B_j = \{r \mid r = \pi(1) \text{ for some } \pi \text{ in } K_j\}$$

for $j = 1, 2$, and let $\sigma_j : \mathbf{N} \to B_j$ be an injective function such that the subsets $\sigma_1(\mathbf{N})$ and $\sigma_2(\mathbf{N})$ of $\mathbf{R}^+$ are disjoint. Let us define $\sigma : \mathbf{N} \to \mathbf{R}^+$ by putting $\sigma(2n-1) = \sigma_1(n)$ and $\sigma(2n) = \sigma_2(n)$ if $n \in \mathbf{N}$. Clearly $\sigma \in Y$. Let us define a subset $S$ of $Y$ as follows: $y \in S$ if and only if there exists a strictly increasing sequence $\{k_n\}_{n \in \mathbf{N}}$ of positive integers such that $y(k_n) = \sigma(n)$ if $n \in \mathbf{N}$. The set $S$ is uncountable. If $s \in S$, choose $q_s$ in $P_2$ such that $X(q_s) = x$ and $Y(q_s) = s$. It follows from

Lemma 4.16 that if $s \in S$, then there exists an indicator $K_s$ such that

$$\Sigma(q_s, 1, K_s) \subset U_j \quad \text{for} \quad j = 1 \quad \text{or} \quad j = 2.$$

For $j = 1, 2$ let

$$S_j = \{s \in S \mid \Sigma(q_s, 1, K_s) \subset U_j\}.$$

Either $S_1$ or $S_2$ is uncountable. Let us suppose that $S_1$ is uncountable. It will be clear how the argument should be modified if $S_1$ is at most countable. There exists a positive integer $N$ and an infinite subset $T$ of $S_1$ such that $|K_t| = N$ if $t \in T$. Let $\sigma_2(N) = c$.

Now let $x'$ be the element of $X$ such that $|x'| = |x| + 1$, $x'(i) = x(i)$ for $i = 0, \dots, |x|$ and $x'(|x'|) = -c$. Let $L_1$ be the subset of $\Pi_{N+1}$ consisting of those elements $\pi$ for which there exist $t$ in $T$ and $\rho$ in $K_t$ such that $\pi(1) = \theta_c(t)$ and $\pi(i+1) = \rho(i)$ for $i = 1, \dots, N$. Since $T$ is infinite, $\theta_c$ is a bijection and $K_t$ is an indicator for each $t$, it is easy to verify that $L_1$ is an indicator. Let $L_2$ be the subset of $\Pi$ defined as follows: if $\pi \in \Pi$, then $\pi \in L_2$ if and only if $|\pi| = |K_2|$ and, in the case $|K_2| > 1$, there exists some $\rho$ in $K_2$ such that $\rho(1) = c$ and $\pi(i) = \rho(i+1)$ if $i < |K_2|$. Since $c \in B_2$ and $K_2$ is an indicator it is easy to verify that $L_2$ is an indicator.

Suppose that $u \in X$, $|u| > 0$ and $V_u \in \mathscr{S}^+(x', L_1)$. Then

$$|u| = |x'| + |L_1|, \quad u(i) = x'(i) \quad \text{for} \quad i = 1, \dots, |x'|$$

and $u(|x'| + i) = \pi(i)$ for $i = 1, \dots, |L_1|$, where $\pi \in L_1$. Thus

$$|u| = |x| + N + 2, \quad u(i) = x(i) \quad \text{for} \quad i = 0, \dots, |x|, \quad u(|x| + 1) = -c,$$

and there exist $t$ in $T$ and $\rho$ in $K_t$ such that $u(|x| + 2) = \theta_c(t)$ and $u(|x| + 2 + i) = \rho(i)$ for $i = 1, \dots, N$. Let $w$ be the element of $X$ such that $|w| = |x| + 2$ and $w(i) = u(i)$ for $i = 0, \dots, |x| + 2$. Then $V_u \subset V_w$. Since $t \in S$, there exists an integer $m$ such that $t(m) = c = \sigma(2N)$ and from the definition of $S$ it follows that $m \geqslant 2N > N$. But

$$w = \gamma(q_t, 1, m, -)$$

and it now follows from Lemma 4.17 that $V_w \subset \Sigma(q_t, 1, K_t) \subset U_1$. Hence $\Sigma^+(x', L_1) \subset U_1$.

Now suppose that $u \in X$, $|u| > 0$ and $V_u \in \mathscr{S}^-(x', L_2)$. Then

$$|u| = |x'| + |K_2|, \quad u(i) = x'(i) \quad \text{for} \quad i = 0, \dots, |x'|$$

and $u(|x'| + i) = -\pi(i)$ for $i = 1, \dots, |K_2|$, where $\pi \in L_2$. There exists $\rho$ in $K_2$ such that $\rho(1) = c = -x'(|x'|)$ and $\rho(i+1) = \pi(i)$ for

$$i = 1, \dots, |K_2| - 1.$$

Thus $u(i) = x(i)$ for $i = 0, \ldots, |x|$ and $u(|x|+i) = -\rho(i)$ for

$$i = 1, \ldots, |K_2|.$$

Thus if $w$ is the element of $X$ such that $|w| = |x| + |K_2|$ and $w(i) = u(i)$ for $i = 0, \ldots, |w|$, then $V_u \subset V_w$ and $V_w \in \mathscr{S}^-(x, K_2)$. It follows that $\Sigma^-(x', L_2) \subset \Sigma^-(x, K_2) \subset U_2. \square$

We are now able to prove:

**4.19** *Proposition.* $\operatorname{Ind} P > 0$.

*Proof.* Let $x_0$ be the unique element of $X$ such that $|x_0| = 0$ and let $C_1$ and $C_2$ be the closures in $P$ of the sets $\Sigma^+(x_0, \Pi_2)$ and $\Sigma^-(x_0, \Pi_2)$ respectively. If $x \in X$ and $|x| > 0$, then $V_x \in \mathscr{S}^+(x_0, \Pi_2)$ if and only if $|x| = 2$, $x(1) > 0$ and $x(2) > 0$. If $p \in P_1 \cap C_1$ and $p \in V_{x'}$, where $|x'| > 0$, then $V_{x'} \cap V_x \neq \varnothing$ for some $x$ such that $|x| = 2$, $x(1) > 0$ and $x(2) > 0$. Hence $x'(1) > 0$ so that $p(1) > 0$. Similarly if $p \in P_1 \cap C_2$, then $p(1) < 0$. Thus $P_1 \cap C_1 \cap C_2 = \varnothing$. Suppose that $p \in P_2$ and $|X(p)| = 0$. Then $R_{(p, 1)} \cap \Sigma^+(x_0, \Pi_2) = \varnothing$. If $x' = \gamma(p, 1, i, \epsilon)$, where $i \in \mathbf{N}$, then $|x'| = 2$ and $x'(1)$ and $x'(2)$ have opposite signs so that

$$T_{(p, 1)} \cap \Sigma^+(x_0, \Pi_2) = \varnothing.$$

Thus $V_{(p, 1)} \cap \Sigma^+(x_0, \Pi_2) = \varnothing$ and it follows that $p \notin C_1$. Thus if

$$p \in P_2 \cap C_1,$$

then $|X(p)| > 0$ and $V_{X(p)} \cap V_x \neq \varnothing$ for some $x$ such that $|x| = 2$, $x(1) > 0$ and $x(2) > 0$. It follows that $X(p)(1) > 0$. Similarly if $p \in P_2 \cap C_1$, then $|X(p)| > 0$ and $X(p)(1) < 0$. Thus $P_2 \cap C_1 \cap C_2 = \varnothing$. It follows that the closed sets $C_1$ and $C_2$ are disjoint. Let us suppose that $\operatorname{Ind} P = 0$. Then there exist disjoint open sets $U_1$ and $U_2$ such that $C_1 \subset U_1$, $C_2 \subset U_2$ and $U_1 \cup U_2 = P$. Using Lemma 4.18 we can construct by induction a sequence $\{x_n\}_{n \in \mathbf{N}}$ of elements of $X$ such that if $n \in \mathbf{N}$ then: (i) $|x_n| = n$; (ii) $x_{n+1}(i) = x_n(i)$ for $i = 1, \ldots, n$; and (iii) there exist indicators $K_n^1$ and $K_n^2$ such that $\Sigma^+(x_n, K_n^1) \subset U_1$ and $\Sigma^-(x_n, K_n^2) \subset U_2$. Let $p$ be the element of $P_1$ such that $p(n) = x_n(n)$ if $n \in \mathbf{N}$. If $U$ is a neighbourhood of $p$ in $P$, then there exists a positive integer $n$ such that $p \in V_{x_n} \subset U$. But $\Sigma^+(x_n, K_n^1) \subset V_{x_n}$ so that

$$V_{x_n} \cap U_1 \neq \varnothing.$$

Since $U_1$ is a closed set it follows that $p \in U_1$. Similarly $p \in U_2$. This is absurd. Hence $\operatorname{Ind} P > 0. \square$

Summarizing we have:

**4.20   Theorem.** *The metrizable space $P$ satisfies*

$$\text{ind}\, P = 0 \quad and \quad \dim P = \text{Ind}\, P = 1.$$

*Proof.* By Corollary 4.13, $\text{ind}\, P = 0$. Since $P$ is metrizable,

$$\dim P = \text{Ind}\, P$$

so that Propositions 4.11 and 4.19 give

$$0 < \text{Ind}\, P = \dim P \leqslant 1$$

so that $\text{Ind}\, P = \dim P = 1.\,\square$

## 5   Generalizations of metric spaces and dimension theory

We conclude this chapter by studying a class of spaces which contains the class of pseudo-metrizable spaces and for which dimension theory has most of the pleasant features of the dimension theory of pseudo-metrizable spaces.

**5.1   Definition.** *A paracompact normal space which is the union of countably many closed pseudo-metrizable subspaces is called an $N$-space.*

Clearly every pseudo-metrizable space is an $N$-space. There exist $N$-spaces which are not pseudo-metrizable. A simple example arises as follows. The metrizable space $J(\aleph_0)$ has a countable closed covering $\mathscr{F}$ (by its 'spines'), each member of which is homeomorphic with the unit interval. If the set $J(\aleph_0)$ is given the weak topology with respect to the covering $\mathscr{F}$ (this is a finer topology than the metric topology of $J(\aleph_0)$) then the resulting space is an $N$-space since $\mathscr{F}$ is a closed covering by metrizable subspaces, but it is not a metrizable space since it is not a first-countable space.

It follows from Corollary 2.3.10 that the union of finitely many closed pseudo-metrizable subspaces is a closed pseudo-metrizable subspace. Thus if $X$ is an $N$-space then $X = \bigcup_{n\in\mathbb{N}} X_n$, where $X_n$ is a closed pseudo-metrizable subspace and $X_n \subset X_{n+1}$ for each $n$. Such a family $\{X_n\}_{n\in\mathbb{N}}$ of subspaces of an $N$-space $X$ is called a *monotone scale* of $X$.

It was noted in §5 of Chapter 4 that a normal space which is the union of countably many closed pseudo-metrizable subspaces is perfectly normal and for such spaces covering dimension coincides with large inductive dimension. Thus if $X$ is an $N$-space, then $X$ is perfectly normal and dim $X = \operatorname{Ind} X$. It is clear that each subspace of an $N$-space is an $N$-space. If $X$ is an $N$-space and $A$ is a subspace of $X$, then dim $A \leqslant \dim X$ since $X$ is a perfectly normal space.

We shall obtain a product theorem for the dimension of $N$-spaces. We show first that the product of $N$-spaces is an $N$-space. For the proof we require the following fact which is established in the proof of Lemma 2.2.1: if $A$ is an $F_\sigma$-set in a paracompact space $X$, then each open covering of $A$ has an open refinement which is $\sigma$-locally finite in $X$. We shall also need:

**5.2 Lemma.** *If $X$ is an $N$-space, then there exists a $\sigma$-locally finite family $\mathscr{F}$ of closed sets of $X$ such that for each point $x$ of $X$ and each open set $V$ such that $x \in V$ there exists some member $F$ of $\mathscr{F}$ such that $x \in F \subset V$.*

*Proof.* This is an immediate consequence of Proposition 2.3.9. ☐

**5.3 Proposition.** *If $X$ is a paracompact perfectly normal space and $Y$ is an $N$-space, then the topological product $X \times Y$ is a paracompact normal space.*

*Proof.* By Lemma 5.2 there exists a $\sigma$-locally finite family $\{F_\lambda\}_{\lambda \in \Lambda}$ of closed sets of $Y$ such that for each point $y$ and each open set $W$ of $Y$ such that $y \in W$ there exists some $\lambda$ such that $y \in F_\lambda \subset W$. Let $\Lambda = \bigcup_{n \in \mathbf{N}} \Lambda(n)$, where the sets $\Lambda(n)$ are disjoint and $\{F_\lambda\}_{\lambda \in \Lambda(n)}$ is locally finite for each $n$. Let $\mathscr{U}$ be an open covering of $X \times Y$ and if $\lambda \in \Lambda$, let $G_\lambda$ be the union of the open sets $G$ of $X$ such that $G \times F_\lambda$ is contained in some member of $\mathscr{U}$. Since $X$ is a paracompact perfectly normal space and $G_\lambda$ is an open set of $X$, it follows, as noted above, that $G_\lambda = \bigcup_{\alpha \in A(\lambda)} H_{\lambda\alpha}$, where $H_{\lambda\alpha}$ is open in $X$ and $H_{\lambda\alpha} \times F_\lambda$ is contained in some member of $\mathscr{U}$ for each $\alpha$, and the family $\{H_{\lambda\alpha}\}_{\alpha \in A(\lambda)}$ is $\sigma$-locally finite in $X$. The family

$$\mathscr{A} = \{H_{\lambda\alpha} \times F_\lambda \mid \alpha \in A(\lambda),\ \lambda \in \Lambda\}$$

is a covering of $X \times Y$. For if $z = (x, y) \in X \times Y$, there exists a member $U$ of $\mathscr{U}$ and open sets $V, W$ of $X, Y$ respectively such that $z \in V \times W \subset U$, and there exists $\lambda$ such that $z \in F_\lambda \subset W$. Since $V \times F_\lambda \subset U$, it follows that $x \in G_\lambda$ and hence $z \in H_{\lambda\alpha} \times F_\lambda$ for some $\alpha$. It is clear that $\mathscr{A}$ is a refinement of $\mathscr{U}$. For each $\lambda$ in $\Lambda$ and $\alpha$ in $A(\lambda)$, choose a member $U_{\lambda\alpha}$ of $\mathscr{U}$ such that $H_{\lambda\alpha} \times F_\lambda \subset U_{\lambda\alpha}$. For each positive integer $n$, the family $\{F_\lambda\}_{\lambda \in \Lambda(n)}$ of closed sets of $Y$ is locally finite. It follows from Proposition

2.1.9 that there exists a locally finite family $\{W_\lambda\}_{\lambda \in \Lambda(n)}$ of open sets of $X$ such that $F_\lambda \subset W_\lambda$ for each $\lambda$. If

$$\mathscr{W} = \{(H_{\lambda\alpha} \times W_\lambda) \cap U_{\lambda\alpha} \mid \alpha \in A(\lambda), \quad \lambda \in \Lambda\},$$

then $\mathscr{W}$ is a $\sigma$-locally finite open refinement of the open covering $\mathscr{U}$ of $X \times Y$. Since $X \times Y$ is a regular space it follows from Proposition 2.1.6 that $X \times Y$ is a paracompact normal space. $\square$

**5.4 Corollary.** *If $X$ and $Y$ are $N$-spaces, then the topological product $X \times Y$ is an $N$-space.*

*Proof.* Since an $N$-space is paracompact and perfectly normal, it follows from Proposition 5.3 that $X \times Y$ is a paracompact normal space. If $\{X_n\}_{n\in\mathbb{N}}$ and $\{Y_n\}_{n\in\mathbb{N}}$ are monotone scales of $X$ and $Y$ respectively, then $X_n \times Y_n$ is a closed pseudo-metrizable subspace of $X \times Y$ for each $n$, and $\bigcup_{n\in\mathbb{N}}(X_n \times Y_n) = X \times Y$. $\square$

**5.5 Proposition.** *If $X$ and $Y$ are $N$-spaces, at least one of which is non-empty, then* $$\dim X \times Y \leqslant \dim X + \dim Y.$$

*Proof.* If $\{X_n\}_{n\in\mathbb{N}}$ and $\{Y_n\}_{n\in\mathbb{N}}$ are monotone scales of $X$ and $Y$ respectively, then $\{X_n \times Y_n\}_{n\in\mathbb{N}}$ is a monotone scale of $X \times Y$. By Proposition 4.5.5
$$\dim X_n \times Y_n \leqslant \dim X_n + \dim Y_n$$
and $\dim X_n \leqslant \dim X$, $\dim Y_n \leqslant \dim Y$, so that it follows from the countable sum theorem that
$$\dim X \times Y \leqslant \dim X + \dim Y. \square$$

Further progress in the dimension theory of $N$-spaces depends on associating with each $N$-space a 'replica' which is a pseudo-metrizable space.

**5.6 Definition.** *If $X$ is an $N$-space, then a replica of $X$ is a pair $(\hat{X}, \rho)$, where $\hat{X}$ is a pseudo-metrizable space and $\rho: X \to \hat{X}$ is a continuous bijection such that for some monotone scale $\{X_i\}_{i\in\mathbb{N}}$ of $X$, if $i \in \mathbb{N}$ then $\rho$ maps $X_i$ homeomorphically onto a closed subspace $\rho(X_i)$ of $\hat{X}$.*

If $X$ is an $N$-space and $(\hat{X}, \rho)$ is a replica of $X$, then it follows from the countable sum theorem for dimension that $\dim X = \dim \hat{X}$. If $B$ is a subspace of $X$ and $\sigma: B \to \rho(B)$ is given by restriction of $\rho$, then $(\rho(B), \sigma)$ is a replica of $B$. Thus in particular, $\dim B = \dim \rho(B)$.

**5.7 Proposition.** *Every N-space has a replica.*

*Proof.* Let $X$ be an $N$-space and let $\{X_i\}_{i \in \mathbf{N}}$ be a monotone scale of $X$. For each $i$, the subspace $X_i$ has a base for its topology which consists of countably many locally finite open coverings of $X_i$. It follows from Proposition 2.1.10 that for each $i$ there exists a sequence $\{\mathscr{U}_{ij}\}_{j \in \mathbf{N}}$ of locally finite open coverings of $X$ such that the collection of sets $U \cap X_i$, where $U$ is a member of $\mathscr{U}_{ij}$ for some $j$, is a base for the topology of $X_i$. Let $U$ be a member of $\mathscr{U}_{ij}$. Since $X$ is a perfectly normal space, there exists a continuous real-valued function $f_U$ on $X$ such that $f_U(x) \geqslant 0$ if $x \in X$ and $U = \{x \in X \,|\, f_U(x) > 0\}$. Let us define $d_{ij}$ on $X \times X$ by putting

$$d_{ij}(x, y) = \Sigma_{U \in \mathscr{U}_{ij}} |f_U(x) - f_U(y)|$$

if $(x, y) \in X \times X$. Then $d_{ij}$ is a pseudo-metric on $X$. It is easy to see that the topology on the set $X$ induced by $d_{ij}$ is less fine than the given topology of $X$. Thus if $Y_{ij}$ is the topological space, which consists of the set $X$ with the topology induced by the pseudo-metric $d_{ij}$ and $\tau_{ij} : X \to Y_{ij}$ is the identity function, then $\tau_{ij}$ is continuous. Let $Y$ be the topological product of the family $\{Y_{ij}\}_{i, j \in \mathbf{N}}$ and let $\pi_{ij} : Y \to Y_{ij}$ be the projection if $i, j \in \mathbf{N}$. There exists a continuous injection $\tau : X \to Y$ such that $\pi_{ij} \circ \tau = \tau_{ij}$ if $i, j \in \mathbf{N}$. Let $\hat{X}$ be the subspace $\tau(X)$ of $Y$ and let $\rho : X \to \hat{X}$ be the continuous bijection given by restriction of $\tau$. Then $\hat{X}$ is a pseudo-metrizable space. Evidently if $k \in \mathbf{N}$, then the family $\{\tau_{ij} \,|\, X_k\}_{i, j \in \mathbf{N}}$ separates points of $X_k$ from closed sets. It follows from Lemma 1.5.13 that $\rho \,|\, X_k$ is an embedding of $X_k$ in $\hat{X}$. If

$$y_0 \in \hat{X} \backslash \rho(X_k),$$

then $y_0 = \rho(x_0)$, where $x_0 \in X \backslash X_k$. Suppose that $x_0 \in X_i$. Then $i > k$ and $X_k$ is closed in $X_i$. There exists a member $U$ of $\mathscr{U}_{ij}$ for some $j$ such that $x_0 \in U$ and $U \cap X_i$ is disjoint from $X_k$. If $f_U(x_0) = \epsilon > 0$ and $W$ is the open ball in $Y_{ij}$ with centre $x_0$ and radius $\epsilon$, then $\pi_{ij}^{-1}(W) \cap \hat{X}$ is an open neighbourhood of $y_0$ in $\hat{X}$ which is disjoint from $\rho(X_k)$. Hence $\rho(X_k)$ is closed in $\hat{X}$ for each $k$. Thus $(\hat{X}, \rho)$ is a replica of $X$. $\square$

**5.8 Remark.** If $X$ is an $N$-space and $(\hat{X}, \rho)$ is a replica of $X$, then $\hat{X}$ is a metrizable space if and only if $X$ is a $T_1$-space.

**5.9 Proposition.** *If $X$ is an $N$-space, then $\dim X \leqslant n$ if and only if $X = \bigcup_{i=0}^n A_i$, where $\dim A_i \leqslant 0$ for $i = 0, \ldots, n$.*

*Proof.* Sufficiency of the condition follows from Proposition 3.5.11. Let $X$ be an $N$-space such that $\dim X \leqslant n$ and let $(\hat{X}, \rho)$ be a replica of $X$. Then $\dim \hat{X} \leqslant n$ and it follows from Theorem 1.1 that

$$\hat{X} = \bigcup_{i=0}^{n} B_i,$$

where $\dim B_i \leqslant 0$ for $i = 0, \ldots, n$. If $A_i = \rho^{-1}(B_i)$, then $\dim A_i \leqslant 0$ for $i = 0, \ldots, n$ and $X = \bigcup_{i=0}^{n} A_i$. $\square$

**5.10  Proposition.** *Let $X$ be an $N$-space and let $A$ be a non-empty subspace of $X$ such that $\dim A \leqslant n$. Then there exists a $G_\delta$-set $H$ of $X$ such that $A \subset H$ and $\dim H \leqslant n$.*

*Proof.* Let $(\hat{X}, \rho)$ be a replica of $X$. Then $\dim \rho(A) \leqslant n$ so that by Theorem 1.4 there exists a $G_\delta$-set $U$ of $\hat{X}$ such that $\rho(A) \subset U$ and $\dim U \leqslant n$. If $H = \rho^{-1}(U)$, then $H$ is a $G_\delta$-set of $X$, $A \subset H$ and $\dim H \leqslant n$. $\square$

Finally we show that the analogue of Theorem 1.8 holds for $N$-spaces. The following proposition will also be needed in Chapter 9.

**5.11  Proposition.** *If $X$ is a pseudo-metrizable space and $Y$ is a perfectly normal space, then the topological product $X \times Y$ is perfectly normal.*

*Proof.* Let $\{V_\lambda\}_{\lambda \in \Lambda}$ be a $\sigma$-locally finite base for the topology of $X$ and let $\Lambda = \bigcup_{i \in \mathbf{N}} \Lambda(i)$, where the sets $\Lambda(i)$ are disjoint and the family $\{V_\lambda\}_{\lambda \in \Lambda(i)}$ is locally finite for each $i$. Let $H$ be an open set of $X \times Y$. If $\lambda \in \Lambda$ let $G_\lambda$ be the union of the open sets $G$ of $Y$ such that $V_\lambda \times G \subset H$. Then $G_\lambda$ is open in $Y$ and

$$H = \bigcup_{\lambda \in \Lambda} (V_\lambda \times G_\lambda).$$

Since $X$ and $Y$ are perfectly normal spaces, there exist continuous functions $\phi_\lambda : X \to I$ and $\psi_\lambda : X \to I$ such that

$$V_\lambda = \{x \in X \mid \phi_\lambda(x) > 0\}, \quad G_\lambda = \{y \in Y \mid \psi_\lambda(y) > 0\}.$$

If $\lambda \in \Lambda$, the real-valued function $h_\lambda$ on $X \times Y$, given by putting $h_\lambda(x, y) = \phi_\lambda(x)\,\psi_\lambda(y)$ if $(x, y) \in X \times Y$, is continuous. Since the family $\{V_\lambda \times G_\lambda\}_{\lambda \in \Lambda(i)}$ is locally finite if $i \in \mathbf{N}$, the real-valued function $h_i$ on $X \times Y$, given by

$$h_i(x, y) = \Sigma_{\lambda \in \Lambda(i)} h_\lambda(x) \quad \text{if} \quad (x, y) \in X \times Y,$$

is continuous. Now we can define a continuous function $h: X \times Y \to I$ by putting

$$h(x,y) = \sum_{i=1}^{\infty} \left( h_i(x,y)/2^i(1 + h_i(x,y)) \right)$$

if $(x,y) \in X \times Y$. Then

$$H = \{(x,y) \in X \times Y \mid h(x,y) > 0\}.$$

Thus $X \times Y$ is a perfectly normal space. $\square$

**5.12   Corollary.** *If $X$ is a pseudo-metrizable space and $Y$ is an $N$-space, then the topological product $X \times Y$ is a paracompact perfectly normal space.*

*Proof.* Since a pseudo-metrizable space is paracompact and perfectly normal, it follows from Proposition 5.3 that $X \times Y$ is a paracompact space. And since an $N$-space is perfectly normal, it follows from Proposition 5.11 that $X \times Y$ is a perfectly normal space. $\square$

The final result in the dimension theory of $N$-spaces is given by the following 'lifting lemma'.

**5.13   Lemma.** *Let $X$ be an $N$-space, let $(\hat{X}, \rho)$ be a replica of $X$, let $W$ be a pseudo-metrizable space and let $g: W \to \hat{X}$ be a continuous function. Then there exist an $N$-space $Z$ and continuous functions $f: Z \to X$ and $\sigma: Z \to W$ such that the square*

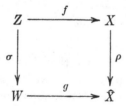

*is commutative and $(W, \sigma)$ is a replica of $Z$. If furthermore $g$ is a perfect mapping, then $f$ is a perfect mapping.*

*Proof.* Let

$$Z = \{(w,x) \in W \times X \mid g(w) = \rho(x)\}$$

and let $\sigma: Z \to W$ and $f: Z \to X$ be the restrictions to $Z$ of the projections. Then $\sigma$ and $f$ are continuous functions such that $g \circ \sigma = \rho \circ f$ and $\sigma$ is a bijection. By Corollary 5.12, $W \times X$ is a paracompact perfectly normal space so that $Z$ is a paracompact perfectly normal space by Proposition 2.2.4. Let $\{X_i\}_{i \in \mathbb{N}}$ be a monotone scale of $X$ such that if

$i \in N$ then $\rho$ maps $X_i$ homeomorphically onto $\rho(X_i)$ which is closed in $\hat{X}$. If $i \in N$, let $Z_i = f^{-1}(X_i)$ so that $Z_i$ is closed in $Z$ and $Z_i \subset Z_{i+1}$. Then $\sigma(Z_i) = g^{-1}(\rho(X_i))$ and we can define a continuous function $\theta : \sigma(Z_i) \to Z_i$ by putting $\theta(w) = (w, \rho^{-1}g(w))$ if $w \in \sigma(Z_i)$. Then $\theta$ is the inverse of the mapping of $Z_i$ onto $\sigma(Z_i)$ given by restriction of $\sigma$. Thus if $i \in N$, then $\sigma$ maps $Z_i$ homeomorphically onto $\sigma(Z_i)$ which is closed in $W$. Hence $Z$ is an $N$-space and $(W, \sigma)$ is a replica of $Z$.

Now suppose that $g$ is a perfect mapping. If $x \in X$, then

$$f^{-1}(x) = (g^{-1}\rho(x)) \times \{x\}.$$

Thus if $x \in X$, then $f^{-1}(x)$ is compact. It is clear that $f$ is surjective. It remains to show that $f$ is closed. Let $U$ be an open set of $Z$. Suppose that $x_0 \in f^*(U)$ so that $f^{-1}(x_0) \subset U$. Thus $(g^{-1}\rho(x_0)) \times \{x_0\} \subset U$. Since $g^{-1}\rho(x_0)$ is compact, it follows that there exist $G$ open in $W$ and $H$ open in $X$ such that $x_0 \in H$, $g^{-1}\rho(x_0) \subset G$ and $Z \cap (G \times H) \subset U$. Thus $\rho(x_0) \in g^*(G)$ and since $g$ is a closed mapping, $g^*(G)$ is open in $X$ by Lemma 6.2.2. If $V = H \cap \rho^{-1}g^*(G)$, then $V$ is an open neighbourhood of $x_0$. And if $x \in V$, then $x \in H$ and $g^{-1}\rho(x) \subset G$ so that

$$f^{-1}(x) \subset Z \cap (G \times H) \subset U.$$

Hence $x_0 \in V \subset f^*(U)$. Thus $f^*(U)$ is open in $X$ and it follows from Lemma 6.2.2 that $f$ is a closed mapping. $\square$

**5.14 Proposition.** *If $X$ is an $N$-space such that $\dim X \leqslant n$, then there exist a perfectly zero-dimensional $N$-space $Z$ and a continuous closed surjection $f : Z \to X$ such that $f^{-1}(x)$ contains at most $n+1$ points if $x \in X$.*

*Proof.* Let $X$ be an $N$-space such that $\dim X \leqslant n$ and let $(\hat{X}, \rho)$ be a replica of $X$. By Theorem 1.8 there exist a pseudo-metrizable space $W$ such that $\dim W = 0$ and a continuous closed surjection $g : W \to \hat{X}$ such that $g^{-1}(y)$ contains at most $n+1$ points if $y \in X$. By Lemma 5.13 there exist an $N$-space $Z$, a continuous bijection $\sigma : Z \to W$ such that $(W, \sigma)$ is a replica of $Z$ and a continuous closed surjection $f : Z \to X$ such that the diagram

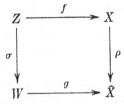

is commutative. If $x \in X$, there is a bijection between $f^{-1}(x)$ and $g^{-1}\rho(x)$ so that $f^{-1}(x)$ contains at most $n+1$ points if $x \in X$. Since the $N$-space $Z$ satisfies $\dim Z = 0$, it follows that $Z$ is a perfectly zero-dimensional space by Proposition 6.1.14. □

**5.15 Corollary.** *If $X$ is an $N$-space, then*

$$\dim X = \operatorname{Ind} X = \Delta(X).$$

*Proof.* This is an immediate consequence of Proposition 6.3.8. □

## Notes

The proofs of Theorems 1.1 and 1.4 are those of Morita [1954]. The same results were obtained by Katětov [1952] by different methods. Theorem 1.8 and Proposition 1.10 are the work of Morita [1955]. Taking $\tau = \aleph_0$ in Proposition 1.10 gives the classical theorem of Hurewicz [1930] and Kuratowski [1932]. Proposition 1.11 is due to Nagami [1960].

Proposition 2.1 was established by Nagami and Roberts [1967]. Proposition 4.5.3, Theorem 1.8 and Propositions 1.11 and 2.1 provide evidence in support of the view (Nagata [1965]) that for every metrization theorem there is an analogous theorem about dimension and metrization. Corollary 2.2 was established by Vopěnka [1959] as a generalization of a theorem of Dowker and Hurewicz [1956] which was the main part of their proof of the equality of covering dimension and large inductive dimension for metric spaces. The condition in Corollary 2.2 that $\mathcal{U}_{i+1}$ is a refinement of $\mathcal{U}_i$ cannot be omitted. There is an example due to Sitnikov [1953] of a 2-dimensional separable metric space which has a sequence of open coverings, each of order 1, with mesh tending to zero. Nagami [1960] proved Proposition 2.5. Kljušin [1964] established Proposition 2.7. Pasynkov [1963, 1965] proved that every $n$-dimensional paracompact Hausdorff space is the limit of an inverse system of $n$-dimensional metric spaces.

Katětov [1952] and Morita [1954] proved Theorem 3.1 and Corollary 3.2. Theorem 3.8 was proved by Nagata [1957]. The universal space $Q_n^{2n+1}$ for $n$-dimensional separable metrizable spaces was found by Nöbeling [1930]. A universal space for $n$-dimensional separable metrizable spaces which is compact was described by Menger [1926], but without proof. The proof is due to Lefschetz [1931]. The existence of a compact universal $n$-dimensional space for separable metrizable spaces follows from the fact that $Q_n^{2n+1}$ has a bicompactification which is $n$-dimensional and has a countable base. The proof that any Tihonov space of infinite weight has a bicompactification of the same dimension and weight will be given in Chapter 10. Hurewicz [1933a] proved that if $X$ is a compact metrizable space such that $\dim X \leqslant n$, then the set $H$ of embeddings of $X$ in $I^{2n+1}$ is a dense $G_\delta$-set in $C(X, I^{2n+1})$. Kuratowski [1937] showed that for a separable metric space $X$ such that $\dim X \leqslant n$, the set $H$ is dense in $C(X, I^{2n+1})$ but Roberts [1948] showed that it need no longer be a $G_\delta$-set. Theorem 3.4 was proved by Nagata [1963]. The countable topological product of copies of $J(\tau)$ is of course infinite-dimensional. Nagata [1966] raised the problem of finding a universal $n$-dimensional metrizable space which is a subspace of a 'simple' finite-dimensional space. Pasynkov [1967] gave alternative proofs of the existence of universal metrizable and strongly metrizable spaces of given weight and dimension.

The construction by Roy [1962, 1968] of the metrizable space $P$ such that $\operatorname{ind} P = 0$ and $\dim P = \operatorname{Ind} P = 1$ solved a problem which had been outstanding since the first attempts to construct a dimension theory for non-separable spaces. It should be noted that $P$ is a complete space. Nyikos [1973] has shown that $P$ cannot be embedded as a closed subspace in a product of infinite discrete spaces; this is an important result in the theory of zero-dimensional spaces. It follows from Proposition 4.1.3 that $P$ can be embedded in a totally disconnected bicompact space. It cannot be claimed that the relation between ind and dim for metrizable spaces is properly understood until the question of the existence of a metrizable space $X$ such that $\operatorname{ind} X = m$ and $\dim X = n$ has been settled for each pair of integers $m$ and $n$ such that $0 \leqslant m < n$. No extension of Roy's construction seems to be possible. More information about small inductive dimension for metrizable spaces is desirable. For example it would be interesting to know if for each metrizable space $X$ such that $\operatorname{ind} X \leqslant n$ there exist a metrizable space $Z$ such that $\operatorname{ind} Z = 0$ and a continuous closed surjection $f: Z \to X$ such that $f^{-1}(x)$ contains at most $n+1$ points if $x \in X$. This would be a stronger result than the equality $\operatorname{ind} X = \delta(X)$ for a metrizable space $X$. Arhangel'skiĭ [1969] has raised the question of the validity of a decomposition theorem for the small inductive dimension of metrizable spaces.

The class of Hausdorff $N$-spaces was introduced and studied by Nagami [1971]. We observe that every CW-complex is an $N$-space. For let $X$ be a CW-complex and for each non-negative integer $n$, let $X_n$ be the $n$-skeleton of $X$. Then $X$ is a paracompact perfectly normal space. Since $X_0$ is a discrete space, $X_0$ is metrizable. Suppose that $X_n$ is the union of countably many closed metrizable subspaces. The subspace $X_{n+1}\backslash X_n$ is open and homeomorphic with the topological sum of open $n$-balls and hence is metrizable. Thus since $X_{n+1}$ is perfectly normal, $X_{n+1}\backslash X_n$ is the union of countably many closed sets of $X_{n+1}$ each of which is metrizable. It follows that $X_{n+1}$ is the union of countably many closed metrizable subspaces, and by induction, $X$ is an $N$-space. It is now easily seen that

$$\operatorname{ind} X = \dim X = \operatorname{Ind} X = \Delta(X) = n,$$

where $n$ is the 'combinatorial dimension' of $X$ (Nagami [1962]). A similar inductive argument shows that $M$-spaces in the sense of Hyman are $N$-spaces, and the coincidence of covering dimension and large inductive dimension for such spaces (Pears [1971a]) follows. Nagami [1972] has shown that countable products of Hausdorff $N$-spaces have satisfactory properties with respect to dimension theory.

# 8

# THE DIMENSION OF BICOMPACT SPACES

## 1 Inverse limits

If $\mathbf{X} = \{X_\alpha, \pi_{\alpha\beta}\}_{\alpha, \beta \in \Omega}$ is an inverse system of non-empty bicompact spaces, then by Proposition 1.7.2, the inverse limit $X$ of $\mathbf{X}$ is a non-empty bicompact space. In this section we shall find a necessary and sufficient condition for the inverse limit $X$ to satisfy $\dim X \leqslant n$. The proof of this theorem depends on a sequence of five lemmas.

**1.1 Lemma.** *Let $X$ be a bicompact space and let $\mathscr{B}$ be a base for the topology of $X$ which is closed under finite unions. If $\mathscr{U} = \{U_1, ..., U_k\}$ is a finite open covering of $X$, then there exists a finite open covering $\mathscr{V} = \{V_1, ..., V_k\}$ of $X$ by members of $\mathscr{B}$ such that $V_i \subset U_i$ for each $i$ and the families $\mathscr{U}$ and $\mathscr{V}$ are similar.*

*Proof.* By Proposition 1.3.16, there exists an open covering

$$\mathscr{W} = \{W_1, ..., W_k\}$$

of $X$ such that $\overline{W_i} \subset U_i$ for each $i$, and $\mathscr{W}$ is similar to $\mathscr{U}$. Since $\overline{W_i}$ is compact, there exists an open set $V_i$ such that $\overline{W_i} \subset V_i \subset U_i$ and $V_i$ is the union of finitely many members of $\mathscr{B}$. By hypothesis $V_i$ is a member of $\mathscr{B}$ for each $i$. If $\mathscr{V} = \{V_1, ..., V_k\}$, then it is clear that $\mathscr{V}$ is the required covering. $\square$

Now let $\mathbf{X} = \{X_\alpha, \pi_{\alpha\beta}\}_{\alpha, \beta \in \Omega}$ be an inverse system of non-empty bicompact spaces over a directed set $\Omega$ and let $X$ be the inverse limit of $\mathbf{X}$ with canonical mappings $\pi_\alpha : X \to X_\alpha$ for $\alpha$ in $\Omega$. A base $\mathscr{B}$ for the topology of $X$ consists of all sets of the form $\pi_\alpha^{-1}(U_\alpha)$, where $U_\alpha$ is an open set of $X_\alpha$. We shall call $\mathscr{B}$ the *standard base* for the topology of $X$. The standard base $\mathscr{B}$ is closed under finite unions. For suppose that $K$ is a finite subset of $\Omega$ and that $U_\alpha$ is open in $X_\alpha$ if $\alpha \in K$. Choose $\beta$ in $\Omega$ such that $\alpha \leqslant \beta$ if $\alpha \in K$ and let $U_\beta = \bigcup_{\alpha \in K} \pi_{\alpha\beta}^{-1}(U_\alpha)$. Then $U_\beta$ is open in $X_\beta$ and $\bigcup_{\alpha \in K} \pi_\alpha^{-1}(U_\alpha) = \pi_\beta^{-1}(U_\beta)$. Since $X$ is a non-empty bicompact space it follows from Lemma 1.1 that if $\dim X \leqslant n$ then each open covering of $X$ has a refinement of order not exceeding $n$ which consists of members of $\mathscr{B}$. The notation which has just been introduced will be employed throughout the section.

**1.2 Lemma.** *If $\alpha \in \Omega$ and $U_\alpha$ is an open set in $X_\alpha$ such that $\pi_\alpha(X) \subset U_\alpha$, then there exists $\beta$ in $\Omega$ such that $\alpha \leqslant \beta$ and $\pi_{\alpha\beta}(X_\beta) \subset U_\alpha$.*

*Proof.* Suppose this assertion is false. If $M = \{\beta \in \Omega \mid \alpha \leqslant \beta\}$, then the subsystem $\{\pi_{\alpha\beta}^{-1}(X_\alpha \backslash U_\alpha)\}_{\beta \in M}$ over $M$ of $\mathbf{X}$ has $\pi_\alpha^{-1}(X_\alpha \backslash U_\alpha)$ as its inverse limit. But by hypothesis if $\beta \in M$, then $\pi_{\alpha\beta}^{-1}(X_\alpha \backslash U_\alpha)$ is a non-empty closed set of $X_\beta$ and thus is a non-empty bicompact space. Hence $\pi_\alpha^{-1}(X_\alpha \backslash U_\alpha)$ is non-empty. But this is absurd since $\pi_\alpha(X) \subset U_\alpha$. $\square$

**1.3 Lemma.** *Let $\mathcal{U} = \{U_1, \ldots, U_k\}$ be a finite open covering of $X$ by members of $\mathcal{B}$ and let $\alpha \in \Omega$. Then there exist $\gamma$ in $\Omega$ such that $\alpha \leqslant \gamma$ and a finite open covering $\mathcal{V}_\gamma = \{V_{\gamma 1}, \ldots, V_{\gamma k}\}$ of $X_\gamma$ such that $\pi_\gamma^{-1}(\overline{V}_{\gamma i}) \subset U_i$ for $i = 1, \ldots, k$ and the order of $\mathcal{V}_\gamma$ does not exceed the order of $\mathcal{U}$.*

*Proof.* For each $i$, $U_i = \pi_{\alpha(i)}^{-1}(U_{\alpha(i)})$, where $U_{\alpha(i)}$ is open in $X_{\alpha(i)}$. Choose $\beta$ in $\Omega$ such that $\alpha \leqslant \beta$ and $\alpha(i) \leqslant \beta$ for $i = 1, \ldots, k$ and let

$$U_{\beta i} = \pi_{\alpha(i)\beta}^{-1}(U_{\alpha(i)}).$$

Then $\pi_\beta^{-1}(U_{\beta i}) = U_i$ for each $i$, and

$$\pi_\beta(X) = \pi_\beta\left(\bigcup_{i=1}^k U_i\right) = \pi_\beta \pi_\beta^{-1}\left(\bigcup_{i=1}^k U_{\beta i}\right) \subset \bigcup_{i=1}^k U_{\beta i}.$$

Since $\pi_\beta(X)$ is a closed set in the bicompact space $X_\beta$ it follows from Proposition 1.3.16 that there exist open sets $V_{\beta i}, i = 1, \ldots, k$, of $X_\beta$ such that $\overline{V}_{\beta i} \subset U_{\beta i}$ for each $i$, $\pi_\beta(X) \subset \bigcup_{i=1}^k V_{\beta i}$ and the family $\mathcal{V}_\beta = \{V_{\beta 1}, \ldots, V_{\beta k}\}$ is similar to the family $\{U_{\beta 1} \cap \pi_\beta(X), \ldots, U_{\beta k} \cap \pi_\beta(X)\}$. If $J$ is a subset of $\{1, \ldots, k\}$ and $\bigcap_{i \in J} V_{\beta i} \neq \varnothing$, then $\pi_\beta(X) \cap \bigcap_{i \in J} U_{\beta i} \neq \varnothing$ so that $\bigcap_{i \in J} U_i = \bigcap_{i \in J} \pi_\beta^{-1}(U_{\beta i}) \neq \varnothing$. Thus the order of $\mathcal{V}_\beta$ does not exceed the order of $\mathcal{U}$. Since $\pi_\beta(X) \subset \bigcup_{i=1}^k V_{\beta i}$, by Lemma 1.2 there exists $\gamma$ in $\Omega$ such that $\beta \leqslant \gamma$ and

$$\pi_{\beta\gamma}(X_\gamma) \subset \bigcup_{i=1}^k V_{\beta i}.$$

Let $V_{\gamma i} = \pi_{\beta\gamma}^{-1}(V_{\beta i})$ for $i = 1, \ldots, k$. Then $\mathcal{V}_\gamma = \{V_{\gamma 1}, \ldots, V_{\gamma k}\}$ is an open covering of $X$, and the order of $\mathcal{V}_\gamma$ does not exceed the order of $\mathcal{V}_\beta$ and hence does not exceed the order of $\mathcal{U}$. Finally

$$\overline{V}_{\gamma i} \subset \pi_{\beta\gamma}^{-1}(\overline{V}_{\beta i}) \subset \pi_\beta^{-1}(U_{\beta i}),$$

so that

$$\pi_\gamma^{-1}(\overline{V}_{\gamma i}) \subset \pi_\gamma^{-1} \pi_{\beta\gamma}^{-1}(U_{\beta i}) = \pi_\beta^{-1}(U_{\beta i}) = U_i. \square$$

**1.4   Lemma.** *Let $\alpha$ and $\beta$ be elements of $\Omega$, let $U_\alpha$ be an open set of $X_\alpha$, let $F_\beta$ be a closed set of $X_\beta$ and suppose that $\pi_\beta^{-1}(F_\beta) \subset \pi_\alpha^{-1}(U_\alpha)$. Then there exists $\gamma$ in $\Omega$ such that $\alpha \leqslant \gamma, \beta \leqslant \gamma$ and*

$$\pi_{\beta\gamma}^{-1}(F_\beta) \subset \pi_{\alpha\gamma}^{-1}(U_\alpha).$$

*Proof.* Suppose this is not the case. Let $M = \{\gamma \in \Omega \,|\, \alpha \leqslant \gamma$ and $\beta \leqslant \gamma\}$. The subsystem $\{\pi_{\alpha\gamma}^{-1}(X_\alpha \backslash U_\alpha) \cap \pi_{\beta\gamma}^{-1}(F_\beta)\}_{\gamma \in M}$ over $M$ of $\mathbf{X}$ has $\pi_\alpha^{-1}(X_\alpha \backslash U_\alpha) \cap \pi_\beta^{-1}(F_\beta) = \pi_\beta^{-1}(F_\beta) \backslash \pi_\alpha^{-1}(U_\alpha)$ as its inverse limit. But by hypothesis, if $\gamma \in M$ then

$$\pi_{\alpha\gamma}^{-1}(X_\alpha \backslash U_\alpha) \cap \pi_{\beta\gamma}^{-1}(F_\beta) = \pi_{\beta\gamma}^{-1}(F_\beta) \backslash \pi_{\alpha\gamma}^{-1}(U_\alpha)$$

is a non-empty closed set of $X_\gamma$ and thus is a non-empty bicompact space. Since $\pi_\beta^{-1}(F_\beta) \subset \pi_\alpha^{-1}(U_\alpha)$ this is a contradiction of Proposition 1.7.2. $\square$

**1.5   Lemma.** *Let $\alpha$ and $\beta$ be elements of $\Omega$, let $\mathcal{U}_\alpha$ be an open covering of $X_\alpha$ and let $\mathcal{F}_\beta$ be a finite closed covering of $X_\beta$ such that $\pi_\beta^{-1}(\mathcal{F}_\beta)$ is a refinement of $\pi_\alpha^{-1}(\mathcal{U}_\alpha)$. Then there exists $\gamma$ in $\Omega$ such that $\pi_{\beta\gamma}^{-1}(\mathcal{F}_\beta)$ is a refinement of $\pi_{\alpha\gamma}^{-1}(\mathcal{U}_\alpha)$.*

*Proof.* Let $\mathcal{F}_\beta = \{F_1, \ldots, F_k\}$ and suppose that for each $i$,

$$\pi_\beta^{-1}(F_i) \subset \pi_\alpha^{-1}(U_i),$$

where $U_i \in \mathcal{U}_\alpha$. By Lemma 1.4, for each $i$ there exists $\gamma(i)$ in $\Omega$ such that $\alpha \leqslant \gamma(i)$, $\beta \leqslant \gamma(i)$ and $\pi_{\beta\gamma(i)}^{-1}(F_i) \subset \pi_{\alpha\gamma(i)}^{-1}(U_i)$. Choose $\gamma$ in $\Omega$ such that $\gamma(i) \leqslant \gamma$ for $i = 1, \ldots, k$. Then for each $i, \pi_{\beta\gamma}^{-1}(F_i) \subset \pi_{\alpha\gamma}^{-1}(U_i)$. Thus $\pi_{\beta\gamma}^{-1}(\mathcal{F}_\beta)$ is a refinement of $\pi_{\alpha\gamma}^{-1}(\mathcal{U}_\alpha)$. $\square$

**1.6   Proposition.** *Let $\{X_\alpha, \pi_{\alpha\beta}\}_{\alpha, \beta \in \Omega}$ be an inverse system of non-empty bicompact spaces with inverse limit $X$. Then $\dim X \leqslant n$ if and only if for each $\alpha$ in $\Omega$ and each open covering $\mathcal{U}_\alpha$ of $X_\alpha$ there exists $\beta$ in $\Omega$ such that $\alpha \leqslant \beta$ and the covering $\pi_{\alpha\beta}^{-1}(\mathcal{U}_\alpha)$ of $X_\beta$ has a refinement of order not exceeding $n$.*

*Proof.* We show first that this condition is necessary for $\dim X \leqslant n$. Suppose that $\dim X \leqslant n$ and let $\mathcal{U}_\alpha$ be an open covering of $X_\alpha$. As we noted earlier, the open covering $\pi_\alpha^{-1}(\mathcal{U}_\alpha)$ of $X$ has a refinement $\{U_1, \ldots, U_k\}$ of order not exceeding $n$ such that each $U_i$ is a member of the standard base $\mathcal{B}$ for the topology of $X$. By Lemma 1.3 there exist $\beta$ in $\Omega$ such that $\alpha \leqslant \beta$ and a finite open covering $\mathcal{V}_\beta = \{V_{\beta 1}, \ldots, V_{\beta k}\}$ of $X_\beta$ of order not exceeding $n$ such that $\pi_\beta^{-1}(\overline{V}_{\beta i}) \subset U_i$ for $i = 1, \ldots, k$.

Let $\mathcal{F}_\beta = \{\overline{V}_{\beta 1}, \ldots, \overline{V}_{\beta k}\}$. Then $\mathcal{F}_\beta$ is a finite closed covering of $X_\beta$ such that $\pi_\beta^{-1}(\mathcal{F}_\beta)$ is a refinement of $\pi_\alpha^{-1}(\mathcal{U}_\alpha)$. It follows from Lemma 1.5 that there exists $\gamma$ in $\Omega$ such that $\alpha \leqslant \gamma$, $\beta \leqslant \gamma$ and $\pi_{\beta\gamma}^{-1}(\mathcal{F}_\beta)$ is a refinement of $\pi_{\alpha\gamma}^{-1}(\mathcal{U}_\alpha)$. Thus $\pi_{\beta\gamma}^{-1}(\mathcal{V}_\beta)$ is an open covering of $X_\gamma$ of order not exceeding $n$, which is a refinement of $\pi_{\alpha\gamma}^{-1}(\mathcal{U}_\alpha)$.

Now we show that the condition is sufficient for $\dim X \leqslant n$. Let $\mathcal{U}$ be a finite open covering of $X$. We can suppose without loss of generality that the members of $\mathcal{U}$ belong to the standard base $\mathcal{B}$ for the topology of $X$. Hence by Lemma 1.3 there exist $\alpha$ in $\Omega$ and a finite open covering $\mathcal{U}_\alpha$ of $X_\alpha$ such that $\pi_\alpha^{-1}(\mathcal{U}_\alpha)$ is a refinement of $\mathcal{U}$. By hypothesis there exist $\beta$ in $\Omega$ such that $\alpha \leqslant \beta$ and a finite open covering $\mathcal{U}_\beta$ of $X_\beta$ of order not exceeding $n$ which is a refinement of $\pi_{\alpha\beta}^{-1}(\mathcal{U}_\alpha)$. Furthermore $\pi_\beta^{-1}(\mathcal{U}_\beta)$ is a finite open covering of $X$ of order not exceeding $n$ which is a refinement of $\mathcal{U}$. Hence $\dim X \leqslant n$. $\square$

**1.7 Corollary.** *If a space $X$ is the inverse limit of an inverse system $\{X_\alpha, \pi_{\alpha\beta}\}_{\alpha, \beta \in \Omega}$ of non-empty bicompact spaces such that $\dim X_\alpha \leqslant n$ for each $\alpha$ in $\Omega$, then $\dim X \leqslant n$.*

*Proof.* The condition of Proposition 1.6 is satisfied by taking $\alpha = \beta$ and applying the definition of covering dimension. $\square$

We have the following interesting characterization of the dimension of bicompact spaces.

**1.8 Proposition.** *A topological space $X$ is bicompact and satisfies $\dim X \leqslant n$ if and only if $X$ is the inverse limit of an inverse system $\{X_\alpha, \pi_{\alpha\beta}\}_{\alpha, \beta \in \Omega}$, where for each $\alpha$ in $\Omega$, $X_\alpha$ is a compact metrizable space such that $\dim X_\alpha \leqslant n$.*

*Proof.* If $X$ is the inverse limit of an inverse system $\{X_\alpha, \pi_{\alpha\beta}\}$ of compact metrizable spaces such that $\dim X_\alpha \leqslant n$ for each $\alpha$, then $X$ is a bicompact space by Proposition 1.7.2 and $\dim X \leqslant n$ by Corollary 1.7. Conversely let $X$ be a bicompact space such that $\dim X \leqslant n$. Then $X$ is a paracompact $M$-space so that by Proposition 7.2.7, $X$ is the inverse limit of an inverse system $\{X_\alpha, \pi_{\alpha\beta}\}_{\alpha, \beta \in \Omega}$, where each $X_\alpha$ is a metrizable space such that $\dim X_\alpha \leqslant n$ and each connecting mapping $\pi_{\alpha\beta}$ is perfect. It follows from Proposition 1.7.5 that the canonical mappings $\pi_\alpha$ are perfect, so that in particular $X_\alpha = \pi_\alpha(X)$ for each $\alpha$. Thus $X_\alpha$ is a compact space if $\alpha \in \Omega$. $\square$

## 2  A family of examples due to Vopenka

It has been established that if $X$ is a bicompact space then

$$\dim X \leqslant \operatorname{ind} X \leqslant \operatorname{Ind} X.$$

Example 4.3.4 shows that there exists a bicompact space $S$ such that $\dim S = 1$ and $\operatorname{ind} S = \operatorname{Ind} S = 2$. In this section we shall clarify the relation between covering dimension and the two inductive dimensions by showing that if $m$ and $n$ are positive integers such that $m \leqslant n$, then there exist bicompact spaces $X_{mn}$, $Y_{mn}$ such that

$$\dim X_{mn} = \dim Y_{mn} = m$$

and

$$\operatorname{ind} X_{mn} = \operatorname{Ind} Y_{mn} = n.$$

For the analysis of these examples some information, contained in Proposition 2.2, about the dimension of products is required.

**2.1  Lemma.** *Let $E$ be a closed set of the topological product $X \times Y$, where $Y$ is a compact space, and for each $x$ in $X$ let*

$$E_x = \{y \in Y \mid (x, y) \in E\}.$$

*If $H$ is open in $Y$, then the set*

$$\{x \in X \mid E_x \subset H\}$$

*is open in $X$.*

*Proof.* Let $U = \{x \in X \mid E_x \subset H\}$ and suppose that $z \in U$. If $E_z = Y$, then the result is trivially true. Suppose that $E_z \neq Y$, and for each $y$ in $Y \backslash E_z$ choose open sets $M_y$ and $N_y$ of $X$ and $Y$ respectively such that

$$(z, y) \in M_y \times N_y \subset (X \times Y) \backslash E.$$

Since $Y \backslash H$ is compact, there exists a finite subset $B$ of $Y \backslash E_z$ such that $Y \backslash H \subset \bigcup_{y \in B} N_y$. If $M = \bigcap_{y \in B} M_y$, then $M$ is an open neighbourhood of $z$. If $x \in M$ and $y' \in E_x$, then $y' \notin \bigcup_{y \in B} N_y$ so that $y' \in H$. Thus if $x \in M$ then $E_x \subset H$. Thus $M \subset U$. □

**2.2  Proposition.** *If $X$ and $Y$ are bicompact spaces and $\dim X = 0$, then*

$$\dim X \times Y = \dim Y \quad and \quad \operatorname{Ind} X \times Y = \operatorname{Ind} Y.$$

*Proof.* It follows from Proposition 3.2.6 that $\dim X \times Y \leqslant \dim Y$. But since $X$ is non-empty, $X \times Y$ contains a closed subspace homeomorphic with $Y$ so that $\dim X \times Y = \dim Y$.

Similarly to prove $\operatorname{Ind} X \times Y = \operatorname{Ind} Y$ it will be enough to prove the inequality $\operatorname{Ind} X \times Y \leqslant \operatorname{Ind} Y$. We shall prove by induction that if $X$ and $Y$ are bicompact spaces such that $\dim X = 0$ and $\operatorname{Ind} Y \leqslant n$, then $\operatorname{Ind} X \times Y \leqslant n$. The assertion is obviously true if $n = -1$ and we suppose that it is true in dimension $n - 1$.

Let $E$ be a closed set and $G$ an open set of $X \times Y$ such that $E \subset G$. If $x \in X$ let $E_x = \{y \in Y \mid (x, y) \in E\}$. Since $E_x$ is compact there exist open sets $P_x$ of $X$ and $V_x$ and $W_x$ of $Y$ such that

$$E_x \subset V_x \subset \bar{V}_x \subset W_x \quad \text{and} \quad P_x \times W_x \subset G.$$

By Lemma 2.1 there exists an open set $U_x$ of $X$ such that $x \in U_x \subset P_x$ and $E_z \subset V_x$ if $z \in U_x$. Thus

$$E \cap (U_x \times Y) \subset U_x \times V_x, \quad \bar{V}_x \subset W_x, \quad U_x \times W_x \subset G.$$

Since $X$ is compact and $\dim X = 0$, there exists a disjoint open covering $\{U_1, \ldots, U_s\}$ of $X$ which is a refinement of the open covering $\{U_x\}_{x \in X}$ of $X$. For each $i = 1, \ldots, s$ choose $x(i)$ in $X$ such that $U_i \subset U_{x(i)}$. Since $\operatorname{Ind} Y \leqslant n$, for each $i$ there exists $V_i$ open in $Y$ such that

$$\bar{V}_{x(i)} \subset V_i \subset W_{x(i)} \quad \text{and} \quad \operatorname{Ind} \operatorname{bd}(V_i) \leqslant n - 1.$$

Now let $H = \bigcup_{i=1}^{s}(U_i \times V_i)$. Then $H$ is open in $X \times Y$ and $E \subset H \subset G$. Since the sets $U_i$ are disjoint and open-and-closed in $X$ it follows that

$$\operatorname{bd}(H) = \bigcup_{i=1}^{s} \operatorname{bd}(U_i \times V_i) = \bigcup_{i=1}^{s} U_i \times \operatorname{bd}(V_i).$$

By the induction hypothesis, $\operatorname{Ind} U_i \times \operatorname{bd}(V_i) \leqslant n - 1$ for each $i$. But $\operatorname{bd}(H)$ is the topological sum of the subspaces $U_i \times \operatorname{bd}(V_i)$ so that $\operatorname{Ind} \operatorname{bd}(H) \leqslant n - 1$. Hence $\operatorname{Ind} X \times Y \leqslant n$. The proof is completed by induction. $\square$

The basic definition for the examples of this section will now be made. This construction has already been used in Example 5.4.11. Let $X$, $Y$ and $Z$ be topological spaces, let $f: X \to Y$ be a continuous surjection and let $\Lambda$ be an infinite discrete space. Let

$$T = X \cup (Y \times \Lambda \times Z)$$

and let a base for the topology of $T$ consist of all open sets of the topological product $X \times \Lambda \times Z$ and all sets of the form

$$U \cup \left(f^{-1}(U) \times (\Lambda \backslash K) \times Z\right),$$

where $U$ is open in $X$ and $K$ is a finite subset of $\Lambda$. This topological space is denoted by $T(X, Y, Z, f, \Lambda)$.

**2.3   Proposition.** *If $X$, $Y$ and $Z$ are bicompact spaces, then*

$$T = T(X, Y, Z, f, \Lambda)$$

*is a bicompact space. If in addition* $\dim Y = 0$, *then*

$$\dim T = \max\{\dim X, \dim Z\}$$

*and*

$$\operatorname{Ind} T \leqslant \operatorname{Ind} X + \operatorname{Ind} Z.$$

*Proof.* If $X$, $Y$ and $Z$ are Hausdorff spaces, it is clear that $T$ is a Hausdorff space. Let $X$, $Y$ and $Z$ be compact spaces and let $\mathscr{G}$ be an open covering of $T$ which consists of members of the base for the topology of $T$ described above. Since $X$ is compact we can select from $\mathscr{G}$ a finite number of sets

$$V_i = U_i \cup \left(f^{-1}(U_i) \times (\Lambda \backslash K_i) \times Z\right), \quad i = 1, \ldots, m,$$

where each $U_i$ is open in $X$, each $K_i$ is finite and $\bigcup_{i=1}^{m} U_i = X$. Then obviously

$$T \backslash \bigcup_{i=1}^{m} V_i \subset Y \times \left(\bigcup_{i=1}^{m} K_i\right) \times Z$$

and hence $T \backslash \bigcup_{i=1}^{m} V_i$ is compact. It follows that we can select from $\mathscr{G}$ a finite covering of $T$ and hence $T$ is a compact space.

Now let $X$, $Y$ and $Z$ be bicompact spaces and suppose that $\dim Y = 0$. If $W$ is an open set of $T$ such that $X \subset W$, then by an argument similar to that above there exists a finite subset $K$ of $\Lambda$ such that

$$X \cup \left(Y \times (\Lambda \backslash K) \times Z\right) \subset W.$$

Thus if $F$ is a closed set of $T$ which is disjoint from $X$, then

$$F \subset Y \times K \times Z,$$

where $K$ is a finite set. Hence $\dim F \leqslant \dim(Y \times K \times Z)$. But

$$\dim(Y \times K \times Z) = \dim Z,$$

by Proposition 2.2. Since $X$ is a closed set of $T$, it follows from Proposition 3.5.6 that

$$\dim T = \max\{\dim X, \dim Z\}.$$

Sets of the form $X \cup \left(Y \times (\Lambda \backslash K) \times Z\right)$, where $K$ is a finite subset of $\Lambda$, are open-and-closed in $T$. Hence if $W$ is an open set such that $X \subset W$, there exists an open-and-closed set $N$ such that $X \subset N \subset W$. Let us define $\phi : T \to X$ by putting $\phi(x) = x$ if $x \in X$ and $\phi(y, \lambda, z) = f(y)$ if $(y, \lambda, z) \in Y \times \Lambda \times Z$. Then $\phi$ is continuous and hence $X$ is a retract of $T$. Finally if $F$ is a closed set of $T$ disjoint from $X$ then, as shown above, $F \subset Y \times K \times Z$, where $K$ is a finite set. Thus

$$\mathrm{Ind}\, F \leqslant \mathrm{Ind}\, (Y \times K \times Z).$$

But $\mathrm{Ind}\, (Y \times K \times Z) = \mathrm{Ind}\, Z$ by Proposition 2.2. It now follows from Proposition 5.4.9 that

$$\mathrm{Ind}\, T \leqslant \mathrm{Ind}\, X + \mathrm{Ind}\, Z. \;\square$$

**2.4   Lemma.** *Let $X$, $Y$ and $Z$ be topological spaces and let $f: Y \to X$ be an almost-open continuous surjection. Let $\Lambda$ be a discrete space of cardinality greater than the cardinality of $X$ and let $T = T(X, Y, Z, f, \Lambda)$. If $U$ and $V$ are open in $T$ and $(U \cap X)^- \cap (V \cap X)^- \neq \varnothing$, then $\overline{U} \cap \overline{V}$ contains a subspace which is homeomorphic with the space $Z$.*

*Proof.* Let $U_1 = U \cap X$, $V_1 = V \cap X$ and let $x_0 \in \overline{U}_1 \cap \overline{V}_1$. There exists $y_0$ in $Y$ such that $f(y_0) = x_0$ and furthermore $f(G)$ is a neighbourhood of $x_0$ for each neighbourhood $G$ of $y_0$ in $Y$. It follows that

$$y_0 \in \big(f^{-1}(U_1)^-\big) \cap \big(f^{-1}(V_1)^-\big).$$

Since $U$ is open in $T$, for each point $x$ of $U_1$ there exists a finite subset $K_x$ of $\Lambda$ such that

$$f^{-1}(x) \times (\Lambda \backslash K_x) \times Z \subset U.$$

Similarly for each point $x$ of $V_1$ there exists a finite subset $L_x$ of $\Lambda$ such that

$$f^{-1}(x) \times (\Lambda \backslash L_x) \times Z \subset V.$$

Since $(\bigcup_{x \in U_1} K_x) \cup (\bigcup_{x \in V_1} L_x)$ is a set of cardinality not exceeding the cardinality of $X$, there exists an element $\lambda_0$ of $\Lambda$ such that

$$\lambda_0 \notin (\bigcup_{x \in U_1} K_x) \cup (\bigcup_{x \in V_1} L_x).$$

Then

$$f^{-1}(U_1) \times \{\lambda_0\} \times Z \subset U, \quad f^{-1}(V_1) \times \{\lambda_0\} \times Z \subset V$$

so that

$$\big(f^{-1}(U_1)\big)^- \times \{\lambda_0\} \times Z \subset \overline{U}, \quad \big(f^{-1}(V_1)\big)^- \times \{\lambda_0\} \times Z \subset \overline{V}.$$

But $y_0 \in \big(f^{-1}(U_1)\big)^- \cap \big(f^{-1}(V_1)\big)^-$ and we see that

$$\{y_0\} \times \{\lambda_0\} \times Z \subset \overline{U} \cap \overline{V}. \;\square$$

**2.5   Lemma.** *Let $X$ be a space and suppose that $X_2 \subset X_1 \subset X$, where $X_2$ is a non-empty connected set. Let $G$ be an open set of $X$ such that*

$G \cap X_2$ and $X_2 \backslash \bar{G}$ are non-empty. Then either (a) there exists $U$ open in $X_1$ such that $U \cap X_2 \ne \varnothing$ and $U \subset \mathrm{bd}\,(G)$, or (b)

$$(G \cap X_1)^- \cap (X_1 \backslash \bar{G})^- \cap X_2 \ne \varnothing.$$

*Proof.* Let $A = (G \cap X_1)^-$, $B = (X_1 \backslash \bar{G})^-$ and let $U = X_1 \backslash A \cup B$. If $X_2 \backslash A \cup B \ne \varnothing$ then case (a) arises. For $U$ is open in $X_1$ and

$$U \cap X_2 \ne \varnothing.$$

Furthermore since $A \supset G \cap X_1$ it follows that $X_1 \backslash A \subset X_1 \backslash G$ and since $B \supset X_1 \backslash \bar{G}$ it follows that $X_1 \backslash B \subset X_1 \cap \bar{G}$ so that

$$U \subset (X_1 \backslash G) \cap (X_1 \cap \bar{G}) = X_1 \cap \mathrm{bd}\,(G).$$

If $X_2 \subset A \cup B$, then from the relations

$$X_2 \cap A \supset X_2 \cap G \ne \varnothing, \quad X_2 \cap B \supset X_2 \backslash \bar{G} \ne \varnothing,$$

and the connectedness of $X_2$, it follows that $X_2 \cap A \cap B \ne \varnothing$ so that case (b) arises. □

The next proposition is the keystone of this section.

**2.6**   **Proposition.** *If $X$ is a bicompact space and* $\dim X > 0$, *then there exists a bicompact space $T$ such that*

$$\dim T = \dim X$$

*and*

$$\mathrm{ind}\, X + 1 \leqslant \mathrm{ind}\, T \leqslant \mathrm{Ind}\, T \leqslant 2\,\mathrm{Ind}\, X + 1.$$

*Proof.* Let $C$ be the Cantor set and let $\phi : C \to I$ be the continuous surjection defined in §3 of Chapter 4. Let $M$ be a countable discrete space and let $S = T(I, C, X, \phi, M)$. Then $S$ is a bicompact space by Proposition 2.3. It follows from Proposition 5.4.10 that there exists a bicompact space $Y$ such that $\dim Y = 0$, and an almost-open continuous surjection $f : Y \to S$. Let $\Lambda$ be a discrete space of cardinality greater than the cardinality of $S$ and let

$$T = T(S, Y, X, f, \Lambda).$$

Then by Proposition 2.3, $T$ is a bicompact space and

$$\dim T = \dim S = \dim X.$$

Furthermore

$$\mathrm{Ind}\, T \leqslant \mathrm{Ind}\, S + \mathrm{Ind}\, X.$$

But $\mathrm{Ind}\, S \leqslant 1 + \mathrm{Ind}\, X$ so that

$$\mathrm{Ind}\, T \leqslant 2\,\mathrm{Ind}\, X + 1.$$

Finally let $t_0 = 0 \in I \subset S \subset T$, and let $G$ be a neighbourhood of $t_0$ in $T$ such that $1 \notin I \cap \bar{G}$. Let $U$ be a neighbourhood of $t_0$ in $T$ such that $U \subset G$. Since $I \subset S \subset T$ and $I$ is connected, it follows from Lemma 2.5 that either (a) there exists a set $V$ open in $S$ such that $V \cap I \neq \varnothing$ and $V \subset \mathrm{bd}\,(U)$, or (b) $(U \cap S)^- \cap (S \backslash \bar{U})^- \cap I \neq \varnothing$. In case (a), $\mathrm{bd}\,(U)$ contains a subspace homeomorphic with $X$ since it is clear from the construction of $S$ that $V$ contains a subspace homeomorphic with $X$. Since the cardinality of $\Lambda$ exceeds the cardinality of $S$, it follows from Lemma 2.4 that in case (b), $\bar{U} \cap (T \backslash \bar{U})^-$ contains a subspace which is homeomorphic with $X$. Since $\bar{U} \cap (T \backslash \bar{U})^- \subset \bar{U} \cap (T \backslash U) = \mathrm{bd}\,(U)$, it follows that $\mathrm{bd}\,(U)$ contains a subspace which is homeomorphic with $X$. Thus we see that $\mathrm{ind}\,T \geqslant \mathrm{ind}\,X + 1$. $\square$

The examples can now be exhibited.

**2.7  Theorem.** *Let $m$ and $n$ be positive integers such that $m \leqslant n$. Then there exists a bicompact space $X_{mn}$ such that*

$$\dim X_{mn} = m, \quad \mathrm{ind}\,X_{mn} = n, \quad \mathrm{Ind}\,X_{mn} < \infty$$

*and there exists a bicompact space $Y_{mn}$ such that*

$$\dim Y_{mn} = m, \quad \mathrm{Ind}\,Y_{mn} = n.$$

*Proof.* We establish first the existence of the bicompact spaces $X_{1n}$ and $Y_{1n}$ by induction over $n$. We can take $X_{11} = Y_{11} = I$. Suppose that $n > 1$ and that we have found a bicompact space $X = X_{1,n-1}$ such that $\dim X = 1$, $\mathrm{ind}\,X = n-1$ and $\mathrm{Ind}\,X < \infty$. Then by Proposition 2.6, there exists a bicompact space $T$ such that $\dim T = 1$, $\mathrm{ind}\,T \geqslant n$ and $\mathrm{Ind}\,T < \infty$. From the inductive definitions of small and large inductive dimension, it follows that there exist closed subspaces $X_{1n}$ and $Y_{1n}$ of $T$ such that $\mathrm{ind}\,X_{1n} = n$ and $\mathrm{Ind}\,Y_{1n} = n$. Since $\dim X_{1n} \leqslant 1$ and $\mathrm{ind}\,X_{1n} > 0$ we must have $\dim X_{1n} = 1$ and similarly $\dim Y_{1n} = 1$ as required. By induction the construction in the case $m = 1$ is complete. Finally if $m$ and $n$ are positive integers and $m \leqslant n$, then we take $X_{mn}$ to be the topological sum of $X_{1n}$ and the $m$-cube $I^m$ and similarly we take $Y_{mn}$ to be the topological sum of $Y_{1n}$ and $I^m$. $\square$

**2.8  *Remark.*** For each positive integer $m$ there exists a bicompact space $Z_m$ such that

$$\dim Z_m = m \quad \text{and} \quad \mathrm{ind}\,Z_m = \mathrm{Ind}\,Z_m = \infty.$$

For let $Z_1$ be the disjoint union of a one-point set $\{\xi\}$ with the spaces $X_{1n}$, $n \in \mathbb{N}$. As a base for the topology of $Z_1$, take the sets $U$ such that $U \subset X_{1n}$ for some $n$ and $U$ is open in $X_{1n}$, together with the sets of the form $\{\xi\} \cup \bigcup_{n \geqslant k} X_{1n}$. Clearly $Z_1$ is a bicompact space which contains each of the spaces $X_{1n}$ as a subspace. It follows from the monotonicity of small inductive dimension that $\operatorname{ind} Z_1 = \operatorname{Ind} Z_1 = \infty$. It follows from the countable sum theorem for covering dimension that $\dim Z_1 = 1$. If $m$ is an integer greater than 1, we can take $Z_m$ to be the topological sum of $Z_1$ and the $m$-cube $I^m$.

## 3 V. V. Filippov's example

In this section we shall obtain a bicompact space with differing inductive dimensions. We begin by obtaining a decomposition of a topological space into disjoint dense subsets. First we need a set-theoretic lemma.

**3.1 Lemma.** *Let $\tau$ be an infinite cardinal number and let $\Lambda$ be a set such that $|\Lambda| = \tau$. Let $\{A_\lambda\}_{\lambda \in \Lambda}$ be a family of subsets of a set $X$ such that $|A_\lambda| \geqslant \tau$ for every $\lambda$. Then there exists a disjoint family $\{B_\lambda\}_{\lambda \in \Lambda}$ of subsets of $X$ such that $|B_\lambda| = \tau$ and $B_\lambda \subset A_\lambda$ for every $\lambda$.*

*Proof.* Let $\Lambda \times \Lambda$ be well-ordered so that the set of predecessors of $(\lambda_0, \mu_0)$ has cardinality less than $\tau$ if $(\lambda_0, \mu_0) \in \Lambda \times \Lambda$. Suppose that $(\lambda_0, \mu_0) \in \Lambda \times \Lambda$ and that for each $(\lambda, \mu)$ such that $(\lambda, \mu) < (\lambda_0, \mu_0)$ we have chosen a point $p(\lambda, \mu)$ of $A_\lambda$. Then we choose $p(\lambda_0, \mu_0)$ such that $p(\lambda_0, \mu_0) \in A_{\lambda_0}$ and $p(\lambda_0, \mu_0)$ is distinct from $p(\lambda, \mu)$ if $(\lambda, \mu) < (\lambda_0, \mu_0)$. It follows by transfinite induction that we can construct a family $\{p(\lambda, \mu)\}_{(\lambda, \mu) \in \Lambda \times \Lambda}$ of points of $X$ such that $p(\lambda, \mu) \in A_\lambda$ and

$$p(\lambda_1, \mu_1) \neq p(\lambda_2, \mu_2)$$

if $(\lambda_1, \mu_1) \neq (\lambda_2, \mu_2)$. Let

$$B_\lambda = \{x \in X \mid x = p(\lambda, \mu) \text{ for some } \mu \text{ in } \Lambda\}.$$

Then $\{B_\lambda\}_{\lambda \in \Lambda}$ is a disjoint family of subsets of $X$ such that $|B_\lambda| = \tau$ and $B_\lambda \subset A_\lambda$ for every $\lambda$. $\square$

Now we obtain the decomposition which will be used in the construction of the example later in the section.

**3.2 Proposition.** *Let $X$ be a topological space such that $w(X) = \tau$, where $\tau$ is an infinite cardinal number, and let $A$ be a subset of $X$ such* *that* $|A \cap U| \geqslant \tau$ *for each open set* $U$ *of* $X$. *Then* $A$ *is the union of* $\tau$ *disjoint dense subsets of* $X$.

*Proof.* Let $\mathscr{U} = \{U_\lambda\}_{\lambda \in \Lambda}$ be a base for the topology of $X$, where $|\Lambda| = \tau$. If $\lambda \in \Lambda$, then $|A \cap U_\lambda| \geqslant \tau$. It follows from Lemma 3.1 that there exists a disjoint family $\{B_\lambda\}_{\lambda \in \Lambda}$ of subsets of $X$ such that $|B_\lambda| = \tau$ and $B_\lambda \subset A \cap U_\lambda$ for each $\lambda$. Since $|B_\lambda| = \tau$, where $\tau$ is an infinite cardinal number, we can put $B_\lambda = \bigcup_{\mu \in \Lambda} B_{\lambda\mu}$, where $|B_{\lambda\mu}| = \tau$ and the sets $B_{\lambda\mu}$, $\mu \in \Lambda$, are disjoint. Let $\lambda_0 \in \Lambda$. If $\lambda \in \Lambda \backslash \{\lambda_0\}$, let

$$A_\lambda = \bigcup_{\nu \in \Lambda} B_{\nu\lambda}$$

and let

$$A_{\lambda_0} = (A \backslash \bigcup_{\nu \in \Lambda} B_\nu) \cup (\bigcup_{\nu \in \Lambda} B_{\nu\lambda_0}).$$

Then $A = \bigcup_{\lambda \in \Lambda} A_\lambda$ and $\{A_\lambda\}_{\lambda \in \Lambda}$ is a disjoint family of dense subsets of $X$. $\square$

We recall from the discussion in §3 of Chapter 4 that there exists a continuous order-preserving surjection $\phi_0$ of the Cantor set $C$ onto the unit interval $I$. The surjection $\phi_0$ identifies pairwise the end-points of the intervals deleted from $I$ in forming $C$ and maps them onto the set $D$ of dyadic rationals in the open unit interval $(0, 1)$. Let $D = D_1 \cup D_2$, where $D_1$ and $D_2$ are disjoint dense subsets of $I$. Since there exists an order-preserving bijection of $D$ onto $D_1$, we can construct an order-preserving surjection $\phi_1 : C \to C$ such that distinct points $s, t$ satisfy $\phi_1(s) = \phi_1(t)$ if and only if $\phi_0(s) = \phi_0(t) = d$, where $d \in D_2$. Since $\phi_1$ is an identification mapping and $\phi_1(s) = \phi_1(t)$ implies that $\phi_0(s) = \phi_0(t)$, it follows that there exists a continuous surjection $\psi_1 : C \to I$ such that $\psi_1 \circ \phi_1 = \phi_0$. Similarly there exists an identification mapping $\phi_2 : C \to C$, which identifies pairwise those points which $\phi_0$ maps into $D_1$, and there exists a continuous surjection $\psi_2 : C \to I$ such that $\psi_2 \circ \phi_2 = \phi_0$. Evidently $\psi_1$ and $\psi_2$ are identifications. For $i = 1, 2$, $\psi_i$ identifies pairwise the end-points of the intervals deleted from $I$ in forming $C$ and maps them onto $D_i$.

Several of the subsequent constructions will take the following form. Let $X$ be a space and let $\{A_\lambda\}_{\lambda \in \Lambda}$ be a disjoint family of subspaces of $X$, each of which is a copy of the Cantor set. Let

$$\Lambda = \Lambda_0 \cup \Lambda_1 \cup \Lambda_2,$$

where $\Lambda_0, \Lambda_1, \Lambda_2$ are disjoint. Consider the equivalence relation on $X$ under which distinct points $s, t$ are equivalent if and only if $s, t \in A_\lambda$, where $\lambda \in \Lambda_i$ and $\phi_i(s) = \phi_i(t)$. If $Y$ is the quotient space of $X$ with respect to this equivalence relation, then we shall say that $Y$ is obtained from $X$ by performing the identification $\phi_i$ in $A_\lambda$ if $\lambda \in \Lambda_i$, where $i = 0, 1, 2$.

Let $\xi$ be an infinite limit ordinal, and let $P_\xi, T_\xi$ be the set of ordinals not exceeding $\xi$ and the set of ordinals less than $\xi$ respectively. We shall continue to write $T_{\omega_1} = T$, $P_{\omega_1} = P$. Consider the linearly ordered set

$$W_\xi = P_\xi \cup \bigcup_{\alpha \in T_\xi} W_\xi^\alpha,$$

where $P_\xi$ has its usual order, $W_\xi^\alpha$ is the Cantor set $C$ with its usual order for each $\alpha$, and $\alpha < x < \alpha + 1$ if $x \in W_\xi^\alpha$, where $\alpha \in T_\xi$. The linearly ordered space $W_\xi$ is bicompact. Consider also the linearly ordered set

$$L_\xi = P_\xi \cup \bigcup_{\alpha \in T_\xi} L_\xi^\alpha,$$

where $P_\xi$ has its usual order, $L_\xi^\alpha$ is the unit interval $I$ with its usual order for each $\alpha$, and $\alpha < x < \alpha + 1$ if $x \in L_\xi^\alpha$, where $\alpha \in T_\xi$. The linearly ordered space $L_\xi$ is bicompact. Clearly $L_\xi$ is obtained from $W_\xi$ by performing the identification $\phi_0$ in $W_\xi^\alpha$ for each $\alpha$ in $T_\xi$. Let

$$\rho : W_\xi \to L_\xi$$

be the identification mapping. The next lemma shows that $\rho$ is an irreducible mapping.

**3.3 Lemma.** *Let $X$ and $Y$ be linearly ordered spaces and let $f : X \to Y$ be a continuous order-preserving surjection such that if $y \in Y$, then either $f^{-1}(y)$ is a one-point set or $f^{-1}(y)$ consists of two non-isolated points $x_1$ and $x_2$ such that the open interval with end-points $x_1, x_2$ is empty. Then $f$ is an irreducible mapping.*

*Proof.* Let $U$ be a non-empty open set of $X$ and let $x_0$ be a point of $U$. If $f^{-1}f(x_0) = \{x_0\}$, then $f(x_0) \in f^*(U)$. If this is not the case, then $f^{-1}f(x_0) = \{x_0, y_0\}$. Let us suppose that $x_0 < y_0$; the modifications to the argument in the case $y_0 < x_0$ will be obvious. Since $U$ is open, $x_0 \in U$ and $x_0$ is not isolated, there exists $x_1$ such that $x_1 < x_0$ and the closed interval $[x_1, x_0]$ is contained in $U$. Since $x_0$ is not isolated, there exists $x$ such that $x_1 < x < x_0$ and it is clear that

$$f^{-1}f(x) \subset [x_1, x_0] \subset U,$$

so that $f(x) \in f^*(U)$. Thus for each non-empty open set $U$ of $X$, $f^*(U)$ is non-empty and it follows from Lemma 6.2.2 that $f$ is an irreducible mapping.□

We now make the main definition of the section. Filippov's example is of the form about to be defined.

**3.4  Definition.** *Let $X$ be a bicompact space, let $\xi$ be an infinite limit ordinal and let $\Pi = (\Pi_0, \Pi_1, \Pi_2)$, where $\Pi_0, \Pi_1, \Pi_2$ are disjoint subsets of $X \times T_\xi$. We denote by*

$$F(X, \xi, \Pi)$$

*the space obtained from $X \times W_\xi$ by performing the identification $\phi_i$ in $\{x\} \times W_\xi^\alpha$ if $(x.\alpha) \in \Pi_i$, where $i = 0, 1, 2$.*

Let
$$\pi : X \times W_\xi \to F(X, \xi, \Pi)$$

be the identification mapping. The space $F(X, \xi, \Pi)$ is clearly compact. Since identifications have only been made in the sets $\{x\} \times W_\xi$ it is easy to see that $F(X, \xi, \Pi)$ is a Hausdorff space if for each $x$ and $\alpha$ distinct points of $\pi(\{x\} \times W_\xi^\alpha)$ have disjoint neighbourhoods in

$$F(X, \xi, \Pi).$$

This will be the case if for each pair $(x, \alpha)$ such that $(x, \alpha) \notin \Pi_0$ there exists a neighbourhood $N(x, \alpha)$ of $x$ in $X$ such that if $y \in N(x, \alpha)$, then $(y, \alpha) \notin \Pi_0 \cup \Pi_2$ if $(x, \alpha) \in \Pi_1$, $(y, \alpha) \notin \Pi_0 \cup \Pi_1$ if $(x, \alpha) \in \Pi_2$, and

$$(y, \alpha) \notin \Pi_0 \cup \Pi_1 \cup \Pi_2$$

if $(x, \alpha) \notin \Pi_1 \cup \Pi_2$.

If we make the identification $\psi_1$ in $\pi(\{x\} \times W_\xi^\alpha)$ if $(x, \alpha) \in \Pi_1$, the identification $\psi_2$ in $\pi(\{x\} \times W_\xi^\alpha)$ if $(x, \alpha) \in \Pi_2$, and the identification $\phi_0$ in $\pi(\{x\} \times W_\xi^\alpha)$ if $(x, \alpha) \notin \Pi_0 \cup \Pi_1 \cup \Pi_2$, then the resulting quotient space is $X \times L_\xi$. Let
$$\bar\pi : F(X, \xi, \Pi) \to X \times L_\xi$$

be the identification mapping. Since the composite

$$X \times W_\xi \xrightarrow{\pi} F(X, \xi, \Pi) \xrightarrow{\bar\pi} X \times L_\xi$$

is $1_X \times \rho$, where $1_X$ is the identity mapping on $X$, it follows from Proposition 6.2.7 that $\bar\pi \circ \pi$ is an irreducible mapping, and it then follows from Proposition 6.2.6 that $\bar\pi$ is an irreducible mapping. The continuous function $\bar\pi$ has another property which will be most important.

**3.5  Definition.** *Let $X$, $Y$ and $Z$ be bicompact spaces and let $f: Z \to X \times Y$ be an irreducible mapping. Let $z$ be a point of $Z$ and let $f(z) = (x, y)$. Then $f$ is said to be* special *at $z$ if each neighbourhood of $z$ contains a set* *of the form* $f^{-1}(V \times W)$, *where* $V$ *is open in* $X$, $W$ *is open in* $Y$ *and* $x \in V$. *The mapping* $f$ *is said to be a special irreducible mapping if it is special at every point of* $Z$.

If $f^{-1}f(z) = \{z\}$, then $f$ is special at $z$, for in that case each neighbourhood of $z$ contains a neighbourhood of $z$ of the form $f^{-1}(V \times W)$, where $V$, $W$ are open in $X$, $Y$ respectively and $f(z) \in V \times W$. In general if $f$ is special at $z$, the set of the form $f^{-1}(V \times W)$ contained in a given neighbourhood of $z$, does not contain $z$. The following sufficient condition will be useful: if $z_0 \in Z$, where $f(z_0) = (x_0, y_0)$, then $f$ is special at $z_0$ if for each open neighbourhood $U$ of $z_0$ there exists $z$ in $U$ such that $f(z) = (x_0, y)$ for some $y$ and $f^{-1}f(z) = \{z\}$. For $f^*(U)$ is a non-empty open set in $X \times Y$ and $(x_0, y) \in f^*(U)$. Hence there exist open sets $V$, $W$ of $X$, $Y$ respectively such that $x_0 \in V$, $y \in W$ and $V \times W \subset f^*(U)$. Clearly $f^{-1}(V \times W) \subset U$. Since $\bar{\pi}$ evidently satisfies this condition it follows that if $F(X, \xi, \Pi)$ is a Hausdorff space then

$$\bar{\pi} : F(X, \xi, \Pi) \to X \times L_\xi$$

is a special irreducible mapping.

The next lemma contains the key result of the section. One more definition is needed. A closed set $F$ of a space $X$ is said to be a *cut* of $X$ if $X \backslash F = G \cup H$, where $G$ and $H$ are disjoint non-empty open sets such that $\bar{G} \cap \bar{H} = F$. If $E$ is a closed set of $X$ which separates closed sets $A$ and $B$ of $X$, then $E$ contains a cut $F$ of $X$ which separates $A$ and $B$. For let $U$, $V$ be disjoint open sets such that $X \backslash E = U \cup V$ and $A \subset U$, $B \subset V$. Let $G = \bar{U}^\circ$, $F = \bar{G} \backslash G$ and $H = X \backslash \bar{G}$. Since $X \backslash \bar{H} = \bar{G}^\circ = G$, it follows that $F = \bar{G} \cap \bar{H}$, and it is clear that $A \subset G$ and $B \subset H$.

**3.6　Lemma.** *Let* $X$, $Y$ *and* $Z$ *be bicompact spaces, let* $\alpha$ *be an ordinal of cardinality exceeding the weight of* $X$, *let* $y_0$ *be a point of* $Y$ *and let* $\mathcal{U} = \{U_\beta\}_{\beta < \alpha}$ *be a family of open neighbourhoods of* $y_0$ *such that* $\overline{U_\gamma} \subset U_\beta$ *if* $\beta < \gamma$ *and every neighbourhood of* $y_0$ *contains a member of* $\mathcal{U}$. *Let* $f : Z \to X \times Y$ *be an irreducible mapping such that its restriction to* $X_0 = f^{-1}(X \times \{y_0\})$ *is an embedding, and let* $S$ *be the set of points of* $Z$ *at which* $f$ *is special. Let* $F$ *be a cut of* $Z$, *let* $F_0 = F \cap X_0$ *and let*

$$E_0 = \{x \in X \mid (x, y_0) \in f(F_0)\}.$$

*Then at least one of the following statements is true:*

(a) $F_0$ *contains a non-empty open subset of* $X_0$;

(b) *there exists an open neighbourhood $W_0$ of $y_0$ such that*

$$S \cap f^{-1}(E_0 \times W_0) \subset F.$$

*Proof.* Suppose that (b) is false. Then if $\beta$ is an ordinal less than $\alpha$, there exists a point $z_\beta$ of $S \backslash F$ such that $f(z_\beta) \in E_0 \times U_\beta$. Suppose without loss of generality that $z_\beta \in G$ and let $f(z_\beta) = (x_\beta, y_\beta)$. Then

$$f^{-1}(X \times U_\beta) \cap G$$

is an open neighbourhood of $z_\beta$. Thus since $z_\beta \in S$, there exists $V_\beta$ open in $X$ such that $x_\beta \in V_\beta$, and a non-empty open set $W_\beta$ of $Y$ such that

$$f^{-1}(V_\beta \times W_\beta) \subset f^{-1}(X \times U_\beta) \cap G.$$

Let $z'_\beta$ be the unique element of $Z$ such that $f(z'_\beta) = (x_\beta, y_0)$. Then $z'_\beta \in F$ and $f^{-1}(V_\beta \times U_\beta)$ is an open neighbourhood of $z'_\beta$. It follows that $f^{-1}(V_\beta \times U_\beta) \cap H$ is non-empty so that since $f$ is an irreducible closed mapping there exists a non-empty open set $P$ of $X \times Y$ such that

$$f^{-1}(P) \subset f^{-1}(V_\beta \times U_\beta) \cap H.$$

Let $\mathscr{B}$ be a base for the topology of $X$ such that $|\mathscr{B}| = w(X)$. There exists a member $B_\beta$ of $\mathscr{B}$ and a non-empty open set $W'_\beta$ of $Y$ such that

$$f^{-1}(B_\beta \times W'_\beta) \subset f^{-1}(V_\beta \times U_\beta) \cap H.$$

Since $B_\beta \subset V_\beta$, $W_\beta \subset U_\beta$ and $W'_\beta \subset U_\beta$, it follows that if $Q$ is a non-empty subset of $B_\beta$ then $f^{-1}(Q \times U_\beta)$ has a non-empty intersection with $G$ and with $H$. Since $\alpha$ is of cardinality exceeding $|\mathscr{B}|$, there exists a member $B$ of $\mathscr{B}$ such that $\Omega = \{\beta \mid B_\beta = B\}$ is a cofinal subset of the set of ordinals less than $\alpha$. Let $x$ be a point of $B$ and let $z$ be the unique point of $Z$ such that $f(z) = (x, y_0)$. If $N$ is an open neighbourhood of $z$, then since $f$ is a closed mapping there exists an open subset $Q$ of $B$ and a member $\beta$ of $\Omega$ such that $x \in Q$ and $f^{-1}(Q \times U_\beta) \subset N$. Hence $N$ has a non-empty intersection with $G$ and with $H$. Thus

$$z \in \bar{G} \cap \bar{H} = F.$$

Thus $f^{-1}(B \times \{y_0\}) \subset F_0$. Since $f^{-1}(B \times \{y_0\})$ is a non-empty open subset of $X_0$, we see that (a) holds.□

We make three applications now of Lemma 3.6. The results will be used in the analysis of Filippov's example and there will be a further most important application of the lemma in that analysis.

**3.7** *First application.* Let $X$ be a compact connected linearly ordered space, let $\xi$ be a limit ordinal of cardinality exceeding the weight of $X$ and let $\Pi = (\Pi_0, \Pi_1, \Pi_2)$, where $\Pi_0$, $\Pi_1$ and $\Pi_2$ are disjoint subsets of $X \times T_\xi$ such that for each point $x$ of $X$ the set

$$\{\alpha \in T_\xi \mid (x, \alpha) \in \Pi_0\}$$

is cofinal in $T_\xi$. Suppose furthermore that the space $Z = F(X, \xi, \Pi)$ is bicompact. Then $\operatorname{ind} Z \geqslant 2$.

We have a special irreducible mapping $\bar{\pi}: Z \to X \times L_\xi$ and $\bar{\pi}$ maps $X_0 = \bar{\pi}^{-1}(X \times \{\xi\})$ homeomorphically onto $X \times \{\xi\}$. Let $x_0$ and $x_1$ be the first and last elements of $X$ and let $E$ be a closed set which separates the points $z_0 = \bar{\pi}^{-1}(x_0, \xi)$ and $z_1 = \bar{\pi}^{-1}(x_1, \xi)$. Then $E$ contains a cut $F$ which separates $z_0$ and $z_1$. The conditions of Lemma 3.6 are satisfied. If case $(a)$ of the lemma arises, then $F$ contains a non-empty open set of $X_0$, and since $X_0$ is homeomorphic with $X$ it is clear that $\operatorname{ind} F \geqslant 1$. If case $(b)$ arises, then since there exists $x$ in $X$ such that $\bar{\pi}^{-1}(x, \xi) \in F$, there exists an open neighbourhood $W$ of $\xi$ in $L_\xi$ such that $\bar{\pi}^{-1}(\{x\} \times W)$ is contained in $F$. But $\bar{\pi}^{-1}(\{x\} \times W)$ contains a closed subspace homeomorphic with $I$. Again we see that $\operatorname{ind} F \geqslant 1$. Thus $\operatorname{ind} Z \geqslant 2$.

Before the second application can be made, another definition is required. Let us say that an element $x_0$ of a linearly ordered set $X$ is of countable character on the left if there exists a sequence $\{x_i\}_{i \in \mathbf{N}}$ of predecessors of $x_0$ with the property that if $x < x_0$ then there exists $i$ such that $x < x_i < x_0$. The definition of countable character on the right is analogous.

**3.8** *Second application.* Let $X$ be a compact connected linearly ordered space and let $x_0$ be a point of $X$ which is not of countable character either on the left or on the right. Let $X_1 = \{x \in X \mid x < x_0\}$ and let $X_2 = \{x \in X \mid x > x_0\}$. Let $Y$ be obtained from $C \times X$, where $C$ is the Cantor set, by performing the identification $\phi_1$ in $C \times \{x\}$ if $x \in X_1$, the identification $\phi_2$ in $C \times \{x\}$ if $x \in X_2$ and the identification $\phi_0$ in $C \times \{x_0\}$, and let

$$\theta: C \times X \to Y$$

be the identification mapping. The quotient space of $Y$ obtained by making the identification $\psi_1$ in $\theta(C \times \{x\})$ if $x \in X_1$ and the identification $\psi_2$ in $\theta(C \times \{x\})$ if $x \in X_2$ is the product space $I \times X$. Let

$$\bar{\theta}: Y \to I \times X$$

be the identification mapping. The composite $\bar{\theta} \circ \theta$ is the mapping $\phi_0 \times 1_X : C \times X \to I \times X$ which is irreducible by Proposition 6.2.7. It follows from Proposition 6.2.6 that $\bar{\theta}$ is an irreducible mapping.

Furthermore $\bar\theta$ maps $\theta(C \times \{x_0\})$ homeomorphically onto $I \times \{x_0\}$, and $\bar\theta$ is special at every point of

$$S = \theta\big[\big(C\backslash\phi_0^{-1}(D_2)\big) \times X_1\big] \cup \theta(C \times \{x_0\}) \cup \theta\big[\big(C\backslash\phi_0^{-1}(D_1)\big) \times X_2\big].$$

If $E$ is a closed set which separates the end-points $y_0 = \theta(0, x_0)$ and $y_1 = \theta(1, x_0)$ of $Y_0 = \theta(C \times \{x_0\})$, then $E$ contains a cut $F$ which separates $y_0$ and $y_1$. The conditions of Lemma 3.6 are satisfied. If case $(a)$ of the lemma arises, then $F$ contains a non-empty open subset of $Y_0$ and since $Y_0$ is homeomorphic with the unit interval $I$, it is clear that $\operatorname{ind} F \geqslant 1$. If case $(b)$ arises, then since there exists $r$ in $I$ such that $(r, x_0) \in \bar\theta(F)$, it follows that there exists an open neighbourhood $W$ of $x_0$ in $X$ such that $S \cap \bar\theta^{-1}(\{r\} \times W)$ is contained in $F$. Since $D_1$ and $D_2$ are disjoint we can suppose that $r \notin D_2$, say. Then $F$ contains $\theta(\{s\} \times [x, x_0])$, where $\phi_0(s) = r$ and $x \in X_1$. Since $F$ contains a subset homeomorphic with a non-empty closed interval in $X$, it follows that $\operatorname{ind} F \geqslant 1$. Thus $\operatorname{ind} Y \geqslant 2$. We note that this result provides a new proof that the space studied in Example 4.3.4 has both inductive dimensions not less than 2.

**3.9**  *Third application.* Let $\{Y_\alpha\}_{\alpha \in P}$ be a disjoint family of bicompact spaces indexed by the set $P$ of ordinals not exceeding $\omega_1$ and suppose that $Y_\alpha = \{a_\alpha\}$, a one-point space, if $\alpha$ is a limit ordinal, and that $\operatorname{ind} Y_\alpha = 1$ if $\alpha$ is not a limit ordinal. Let $Y = \bigcup_{\alpha \in P} Y_\alpha$. Let a base for the topology of $Y$ consist of the open sets of $Y_\alpha$ for each non-limit ordinal $\alpha$, together with the sets of the form $\bigcup_{\beta \leqslant \gamma \leqslant \alpha} Y_\gamma$, where $\alpha$ is a limit ordinal and $\beta$ is an ordinal less than $\alpha$. Then $Y$ is a bicompact space.

Let the non-limit ordinals in $P$ be partitioned into two disjoint sets $\Sigma_1$ and $\Sigma_2$ each of which is cofinal in $P$. Let $Z$ be obtained from $C \times Y$, where $C$ is the Cantor set, by performing the identification $\phi_1$ in $C \times \{y\}$ if $y \in Y_\alpha$, where $\alpha \in \Sigma_1$, the identification $\phi_2$ in $C \times \{y\}$ if $y \in Y_\alpha$, where $\alpha \in \Sigma_2$, and the identification $\phi_0$ in $C \times \{a_\alpha\}$ for each limit ordinal $\alpha$, and let

$$\theta : C \times Y \to Z$$

be the identification mapping. The quotient space of $Z$ obtained by making the identification $\psi_i$ on $\theta(C \times \{y\})$ if $y \in Y_\alpha$, where $\alpha \in \Sigma_i$ and $i = 1, 2$, is the product $I \times Y$. Let

$$\bar\theta : Z \to I \times Y$$

be the identification mapping. The composite $\bar\theta \circ \theta$ is the mapping $\phi_0 \times 1_Y : C \times Y \to I \times Y$. It follows from Propositions 6.2.7 and 6.2.6

that $\bar{\theta}$ is an irreducible mapping, and the set $S$ of points of $Z$ at which $\bar{\theta}$ is special is the union of all sets $\theta(C \times \{a_\alpha\})$, where $\alpha$ is a limit ordinal with the set

$$\theta[(C\backslash\phi_0^{-1}(D_2)) \times (\bigcup_{\alpha\in\Sigma_1} Y_\alpha)] \cup \theta[(C\backslash\phi_0^{-1}(D_1)) \times (\bigcup_{\alpha\in\Sigma_1} Y_\alpha)].$$

Let $F$ be a cut which separates the end-points $z_0 = \theta(0, a_{\omega_1})$ and $z_1 = \theta(1, a_{\omega_1})$ of $Z_0 = \theta(C \times \{a_{\omega_1}\})$, which is mapped homeomorphically by $\bar{\theta}$ onto $I \times \{a_{\omega_1}\}$. The conditions of Lemma 3.6 are satisfied. If case (a) of the lemma arises, then $F$ contains a non-empty open subset of $Z_0$ and since $Z$ is homeomorphic with the unit interval $I$, it is clear that $\operatorname{ind} F \geqslant 1$. If case (b) arises, then since there exists $r$ in $I$ such that $(r, a_{\omega_1}) \in \bar{\theta}(F)$, it follows that there exists an open neighbourhood $W$ of $a_{\omega_1}$ in $Y$ such that $S \cap \bar{\theta}^{-1}(\{r\} \times W)$ is contained in $F$. Since $D_1$ and $D_2$ are disjoint, we can suppose that $r \notin D_2$ say. Then there exists $\alpha$ in $\Sigma_1$ such that $F$ contains $\theta(\{s\} \times Y_\alpha)$, where $\phi_0(s) = r$. Since $F$ contains a subset homeomorphic with $Y_\alpha$, where $\alpha$ is a non-limit ordinal, we see that $\operatorname{ind} F \geqslant 1$. Thus $\operatorname{ind} Z \geqslant 2$.

Filippov's example is a space of the form $F(X, \xi, \Pi)$ (see Definition 3.4) in which the 'base space' $X$ is the product of a compact connected linearly ordered space with itself. The final preparatory work is concerned with the product of two compact connected linearly ordered spaces, both to establish some notation and to prove some results which will be needed in the analysis of the example.

Let us consider the topological product $X \times Y$ of two compact connected linearly ordered spaces $X$ and $Y$. By a rectangle we shall mean a subset $J_1 \times J_2$ of $X \times Y$, where $J_1$ and $J_2$ are intervals (open, half-open or closed). The boundary of a rectangle is its boundary in $X \times Y$. The vertices of a rectangle are defined in the obvious way. A horizontal segment is a subset $J \times \{y\}$ of $X \times Y$, where $J$ is a closed interval in $X$ and $y \in Y$. We shall call the horizontal segment $X \times \{y\}$ a horizontal line. Vertical lines and segments are defined analogously. By a directed horizontal segment we mean a horizontal segment $[a, b] \times \{y\}$ together with a specification of one of the points $(a, y)$ and $(b, y)$ as the initial point of the segment and the other as the terminal point. Directed vertical segments are defined analogously. By a polygonal arc we mean the union of a finite ordered set of directed segments in which the terminal point of each segment is the initial point of the following segment. The expressions 'closed polygonal arc' and 'polygonal arc without self-intersections' have the obvious meanings.

Every polygonal arc is a connected set in $X \times Y$. An open subset $U$ of $X \times Y$ is connected if and only if any two points of $U$ can be connected by a polygonal arc which lies in $U$.

Let $x_0, y_0$ be the first elements of $X$, $Y$ respectively and let $x_1, y_1$ be the last elements of $X$, $Y$ respectively. Let $L$ be a polygonal arc without self-intersections which is either closed or has its initial and terminal points in $(\{x_0\} \times Y) \cup (\{x_1\} \times Y)$. Let a function $g$ taking its values in $\{0, 1\}$ be defined on $(X \times Y) \backslash L$ by putting $g(x, y) = 0$ if $L$ crosses an odd number of times from one side of $\{x\} \times [y, y_1]$ to the other and $g(x, y) = 1$ if the number of crossings is even. Clearly $g$ is a locally constant function. It follows that if

$$U_0 = g^{-1}(0) \quad \text{and} \quad U_1 = g^{-1}(1),$$

then $U_0$ and $U_1$ are disjoint open sets such that $U_0 \cup U_1 = (X \times Y) \backslash L$. It is clear that any two points of $U_0$ can be connected by a polygonal arc which lies in $U_0$, so that $U_0$ is a connected set. Similarly $U_1$ is a connected set and thus $U_0$ and $U_1$ are the components of $(X \times Y) \backslash L$. We shall consider closed sets in $X \times Y$ which separate

$$E_0 = X \times \{y_0\} \quad \text{and} \quad E_1 = X \times \{y_1\}.$$

It is clear that if $L$ is a polygonal arc without self-intersections which connects points in $\{x_0\} \times Y$ and $\{x_1\} \times Y$ and does not intersect $E_0$ and $E_1$, then $L$ separates $E_0$ and $E_1$ for $E_0 \subset U_0$ and $E_1 \subset U_1$.

**3.10   Lemma.** *A closed set of $X \times Y$ which separates $E_0$ and $E_1$ contains a connected set $F$ which separates $E_0$ and $E_1$ and is a minimal separating set in the sense that no proper subset of $F$ separates $E_0$ and $E_1$.*

*Proof.* Let $B$ be a closed set of $X \times Y$ which separates $E_0$ and $E_1$. Since each chain in a partially ordered set is contained in a maximal chain, it follows that there is a maximal chain $\mathscr{C}$ in the set, partially ordered by inclusion, of closed sets of $B$ separating $E_0$ and $E_1$. Let $F = \bigcap_{C \in \mathscr{C}} C$. Then $F$ is a closed set of $X \times Y$. Suppose that $F$ does not separate $E_0$ and $E_1$. Then there exists a polygonal arc $L$ which connects a point in $E_0$ with a point in $E_1$ and is disjoint from $F$. But since each member of $\mathscr{C}$ separates $E_0$ and $E_1$, and $\mathscr{C}$ is a chain, we see that $\{L \cap C\}_{C \in \mathscr{C}}$ is a family of non-empty closed sets with the finite intersection property. Since $X \times Y$ is compact, it follows that

$$\bigcap_{C \in \mathscr{C}} (L \cap C) = L \cap F$$

is non-empty. This is absurd. Hence $F$ separates $E_0$ and $E_1$. Since $\mathscr{C}$ is a maximal chain it is clear that $F$ is a minimal separating set.

It remains to show that $F$ is connected. Suppose that $F$ is not connected. Then $F = F_1 \cup F_2$, where $F_1$ and $F_2$ are disjoint non-empty closed sets. Since $F_1$ does not separate $E_0$ and $E_1$, there exists an open set $V_1$ such that $F_1 \subset V_1$ and $\overline{V}_1$ does not separate $E_0$ and $E_1$. Similarly there exists an open set $V_2$ such that $F_2 \subset V_2$ and $\overline{V}_2$ does not separate $E_0$ and $E_1$. There exist open sets $W_1$ and $W_2$ each of which is the union of a finite number of rectangles such that $F_i \subset W_i \subset V_i$ for $i = 1, 2$, and $\overline{W}_1, \overline{W}_2$ are disjoint and each of them is disjoint from $E_0 \cup E_1$. Since $F$ separates $E_0$ and $E_1$, and $\overline{W}_1 \cup \overline{W}_2 \supset F$, it follows that $\overline{W}_1 \cup \overline{W}_2$ separates $E_0$ and $E_1$. Hence bd $(\overline{W}_1 \cup \overline{W}_2)$ separates $E_0$ and $E_1$. But bd $(\overline{W}_1 \cup \overline{W}_2)$ has a finite number of components $L_1, \ldots, L_k$ each of which is a polygonal arc. Since neither $\overline{W}_1$ nor $\overline{W}_2$ separates $E_0$ and $E_1$, none of these arcs connects a point in $\{x_0\} \times Y$ with a point in $\{x_1\} \times Y$. Thus each of $L_1, \ldots, L_k$ is either closed or connects points in $\{x_0\} \times Y$ or connects points in $\{x_1\} \times Y$. But it is clear that no such finite union of arcs can separate $E_0$ and $E_1$ and we have a contradiction. Thus $F$ is connected. □

**3.11  Proposition.** *If $X$ and $Y$ are compact connected linearly ordered spaces, then their topological product $X \times Y$ satisfies*

$$\operatorname{ind} X \times Y = \operatorname{Ind} X \times Y = 2.$$

*Proof.* If $E$ is a closed set of $X \times Y$ and $G$ is an open set such that $E \subset G$, then there exists an open set $W$ such that $E \subset W \subset G$ and $W$ is the union of finitely many rectangles. Then bd $(W)$ has a finite number of components each of which is a polygonal arc. Thus

$$\operatorname{Ind} \operatorname{bd}(W) = 1.$$

It follows that $\operatorname{Ind} X \times Y \leqslant 2$. But if $y_0$ and $y_1$ are the first and last elements of $Y$ and $E_0 = X \times \{y_0\}$, $E_1 = X \times \{y_1\}$, then we established in Lemma 3.10 that every closed set $F$ which separates $E_0$ and $E_1$ contains a connected closed set and so satisfies $\operatorname{Ind} F \geqslant 1$. Hence $\operatorname{Ind} X \times Y \geqslant 2$ and it follows that $\operatorname{Ind} X \times Y = 2$. Since $X \times Y$ is a bicompact space, it follows from Proposition 4.2.4 that $\operatorname{ind} X \times Y = 2$. □

Now we construct the base space for Filippov's example. It is the product $Z \times Z$, where $Z$ is the linearly ordered space we are about to study. Let $T$ be the set of ordinals less than the first uncountable ordinal $\omega_1$, and let $Z$ be the set of functions $z: T \to I$, where $I$ is the unit interval, with the property that if $z(\alpha) = 0$ for some $\alpha$ in $T$, then $z(\beta) = 0$ for all $\beta \geqslant \alpha$, and if $z(\alpha) = 1$ for some $\alpha$ in $T$, then $z(\beta) = 1$

for all $\beta \geqslant \alpha$. Let us order $Z$ as follows: if $w, z \in Z$, then $w < z$ if there exists $\alpha$ in $T$ such that $w(\gamma) = z(\gamma)$ if $\gamma < \alpha$ and $w(\alpha) < z(\alpha)$. Since $T$ is well-ordered, $Z$ is linearly ordered. The linearly ordered set $Z$ has a first element $u_0$ given by $u_0(\alpha) = 0$ if $\alpha \in T$ and a last element $u_1$ given by $u_1(\alpha) = 1$ if $\alpha \in T$.

**3.12  Lemma.** *The linearly ordered set $Z$ is order-complete.*

*Proof.* We show that each non-empty subset of $Z$ has a greatest lower bound. Let $M$ be a non-empty subset of $Z$. We shall construct an element $v$ of $Z$. Let $v(0) = \inf_{z \in M} z(0)$. Suppose that $\alpha \in T$ and that for all $\gamma$ such that $\gamma < \alpha$ we have points $v(\gamma)$ of the unit interval such that $v(\beta) = 0$ for some $\beta < \alpha$ implies that $v(\gamma) = 0$ for all $\gamma$ such that $\beta \leqslant \gamma < \alpha$, and $v(\beta) = 1$ for some $\beta < \alpha$ implies that $v(\gamma) = 1$ for all $\gamma$ such that $\beta \leqslant \gamma < \alpha$. If there exists $\beta$ such that $\beta < \alpha$ and $v(\beta) = 0$ we put $v(\alpha) = 0$, and if there exist $\beta$ such that $\beta < \alpha$ and $v(\beta) = 1$ we put $v(\alpha) = 1$. If neither of these cases arises we consider the set

$$M(\alpha) = \{z \in M \mid z(\gamma) = v(\gamma) \quad \text{if} \quad \gamma < \alpha\}.$$

If $M(\alpha)$ is empty we put $v(\alpha) = 1$ and if $M(\alpha)$ is non-empty we put

$$v(\alpha) = \inf_{z \in M(\alpha)} z(\alpha).$$

The element $v$ of $Z$ given by this transfinite construction is the greatest lower bound of $M$. Since $Z$ has a last element $u_1$ it follows that the linearly ordered set $Z$ is order-complete. $\square$

**3.13  Proposition.** *The linearly ordered space $Z$ is bicompact and connected.*

*Proof.* It follows from Lemma 3.12 and Proposition 1.5.7 that $Z$ is a bicompact space. Let $E$ and $F$ be disjoint closed sets of $Z$. Then one of these sets, say $F$, does not contain $u_0$. If $u_0 \notin E$ then $E \cup F \neq Z$. If $u_0 \in E$, let $w$ be the greatest lower bound of $F$ and let $v$ be the least upper bound of the set $\{z \in E \mid z < w\}$. Then $v \in E$, $w \in F$ and $v < w$. Let $\alpha$ be the first ordinal such that $v(\alpha) < w(\alpha)$. If $u$ is a member of $Z$ such that $u(\gamma) = v(\gamma) = w(\gamma)$ if $\gamma < \alpha$ and $v(\alpha) < u(\alpha) < w(\alpha)$ then $u \notin E \cup F$, and so $E \cup F \neq Z$. Thus $Z$ cannot be the union of two disjoint non-empty closed sets and therefore $Z$ is connected. $\square$

Let $Z_0$ be the subset of $Z$ consisting of those $z$ for which $z(\alpha) = 0$ for some $\alpha$ in $T$ or $z(\alpha) = 1$ for some $\alpha$ in $T$. If $Z_0^\alpha$ consists of those $z$ in $Z$ such that $z(\gamma) = 0$ if $\gamma > \alpha$ or $z(\gamma) = 1$ if $\gamma > \alpha$, then $Z_0 = \bigcup_{\alpha \in T} Z_0^\alpha$. Since $|Z_0^\alpha| = \mathbf{c}^{\aleph_0} = \mathbf{c}$, it follows that $|Z_0| = \mathbf{c} \cdot {}^{\aleph_0} = \mathbf{c}$. If $z_1, z_2 \in Z$ and

$z_1 < z_2$, there exists $z_0$ in $Z_0$ such that $z_1 < z_0 < z_2$. Thus $Z_0$ is dense in $Z$. Thus the linearly ordered space $Z$ has weight equal to $c$. It is easily verified that the points of $Z_0$ have countable character both on the left and on the right. If $Z_1 = Z \backslash Z_0$, then the points of $Z_1$ do not have countable character either on the left or on the right. For let $z_1$ be a point of $Z_1$ and let $\{w_i\}_{i \in \mathbf{N}}$ be a countable set of predecessors of $z_1$. Then for each $i$ there exists $\beta_i$ in $T$ such that $w_i(\gamma) = z_1(\gamma)$ if $\gamma < \beta_i$ and $w_i(\beta_i) < z_1(\beta_i)$. Let $\beta$ be the first ordinal such that $\beta_i \leqslant \beta$ for all $i$ and let us define $z$ in $Z$ by putting $z(\gamma) = z_1(\gamma)$ if $\gamma \leqslant \beta$ and $z(\gamma) = 0$ if $\gamma > \beta$. Then $z < z_1$ and $w_i \leqslant z$ for all $i$. Similarly a point of $Z_1$ does not have countable character on the right. If $z_1 \in Z_1$ then there exists an increasing sequence $\{x_\alpha\}_{\alpha \in T}$ and a decreasing sequence $\{y_\alpha\}_{\alpha \in T}$ both of which converge to $z_1$: we put $x_\alpha(\gamma) = y_\alpha(\gamma) = z_1(\gamma)$ if $\gamma \leqslant \alpha$ whilst $x_\alpha(\gamma) = 0$ and $y_\alpha(\gamma) = 1$ if $\gamma > \alpha$. It follows that if $z_1 \in Z_1$, then there exists a family $\{O_\alpha\}_{\alpha \in T}$ of neighbourhoods of $z_1$ such that $O_\beta \subset O_\alpha$ if $\alpha < \beta$ and each neighbourhood of $z_1$ contains some neighbourhood $O_\alpha$.

There is a base $\mathscr{A}$ for the topology of $Z$ which consists of all open intervals of the form $(z, z')$, where $z, z' \in Z_0$, together with all half-open intervals $[u_0, z)$ for $z$ in $Z_0$ and all half-open intervals $(z, u_1]$ for $z$ in $Z_0$. Let $\Lambda$ be the subset of $\mathscr{A} \times \mathscr{A}$ consisting of those pairs $\lambda = (U_\lambda, V_\lambda)$ such that $U_\lambda$ and $V_\lambda$ are either both open or both half-open and $\overline{V}_\lambda \subset U_\lambda$. Clearly $|\Lambda| = c$. It follows that $|\Lambda \times T| = c$. The weight of $Z$ is $c$ and if $U$ is open in $Z$, then $|U \cap Z_0| \geqslant c$ and $|U \cap Z_1| \geqslant c$. It follows from Proposition 3.2 that

$$Z_0 = \bigcup_{\lambda \in \Lambda, \, \alpha \in T} P_\alpha^\lambda,$$

where $\{P_\alpha^\lambda\}_{\lambda \in \Lambda, \, \alpha \in T}$ is a disjoint family of dense subsets of $Z$, and that

$$Z_1 = Q \cup \bigcup_{\lambda \in \Lambda, \, \alpha \in T} R_\alpha^\lambda,$$

where $\{R_\alpha^\lambda\}_{\lambda \in \Lambda, \, \alpha \in T}$ is a disjoint family of dense subsets of $Z$ and $Q$ is a dense subset of $Z$ which is disjoint from each of the sets $R_\alpha^\lambda$.

For each $\alpha$ in $T$ we construct a base $\mathscr{B}_\alpha$ for the topology of $Z$. If $\lambda \in \Lambda$, choose an interval $B_\alpha^\lambda$ which is open and has its end-points in $R_\alpha^\lambda$ if $U_\lambda$ and $V_\lambda$ are open, and which is half-open with its end-point different from $u_0$ or $u_1$ in $R_\alpha^\lambda$ if $U_\lambda$ and $V_\lambda$ are half-open, such that

$$\overline{V}_\lambda \subset B_\alpha^\lambda \subset U_\lambda.$$

It is easy to see that the family $\mathscr{B}_\alpha = \{B_\alpha^\lambda\}_{\lambda \in \Lambda}$ is a base for the topology of $Z$. The family $\mathscr{B} = \{B_\alpha^\lambda\}_{\lambda \in \Lambda, \, \alpha \in T}$ has the property that each point of $Z$ other than $u_0$ and $u_1$ is an end-point of at most one member of $\mathscr{B}$.

Now let us consider the topological product $Z^2 = Z \times Z$. We choose a point $v$ of $Z_1 \backslash Q$. There exists a family $\{O_\alpha\}_{\alpha \in T}$ of open neighbourhoods of $v$ such that $\bar{O}_\beta \subset O_\alpha$ if $\alpha < \beta$ and every neighbourhood of $v$ contains some $O_\alpha$. If $\alpha \in T$ and $\lambda \in \Lambda$, choose an interval $C_\alpha^\lambda$ with end-points in $R_\alpha^\lambda$ such that

$$\bar{O}_{\alpha+1} \subset C_\alpha^\lambda \subset \bar{C}_\alpha^\lambda \subset O_\alpha.$$

No point of $Z$ is an end-point of more than one interval belonging to the family $\mathscr{C} = \{C_\alpha^\lambda\}_{\lambda \in \Lambda, \, \alpha \in T}$. If $\alpha \in T$ and $\lambda \in \Lambda$ let

$$D_\alpha^\lambda = C_\alpha^\lambda \times B_\alpha^\lambda.$$

If $\mathscr{D} = \{D_\alpha^\lambda\}_{\lambda \in \Lambda, \, \alpha \in T}$ then each neighbourhood of each point of $\{v\} \times Z$ contains a neighbourhood of the point which is a member of $\mathscr{D}$. We note that the boundaries of two distinct rectangles in $\mathscr{D}$ either are disjoint or intersect in at most four points, none of which is a vertex of any rectangle in $\mathscr{D}$. Let $V$ consist of the vertices of rectangles in $\mathscr{D}$ neither coordinate of which is an end-point of $Z$, and let $N$ consist of those points in $\{v\} \times Z$ which do not lie on the boundary of any rectangle in $\mathscr{D}$. If $\alpha \in T$ and $\lambda \in \Lambda$ let $S_\alpha^\lambda$ be the set of points of the boundary of the rectangle $D_\alpha^\lambda$ having one coordinate in $P_\alpha^\lambda$ and let $S = \bigcup_{\lambda, \, \alpha} S_\alpha^\lambda$. If a horizontal or vertical line does not contain a non-trivial segment lying in the boundary of any rectangle in $\mathscr{D}$, then it meets $S$ in at most two points. We observe that of the coordinates of a point of $S$, exactly one belongs to $Z_0$.

Let $\mathscr{E}$ consist of sets disjoint from $\{v\} \times Z$ which are products of intervals of the form $(q, q')$, where $q, q' \in Q$, $[u_0, q)$, where $q \in Q$, or $(q, u_1]$, where $q \in Q$. Each neighbourhood of each point of $Z^2 \backslash (\{v\} \times Z)$ contains a neighbourhood of the points which is a member of $\mathscr{E}$. We note that all vertices of rectangles in $\mathscr{E}$, neither of whose coordinates is an end of $Z$, lie in $Q^2 = Q \times Q$.

The next objective is to obtain some information about the closed sets of $Z^2$ which separate

$$E_0 = Z \times \{u_0\} \quad \text{and} \quad E_1 = Z \times \{u_1\}.$$

It follows from Lemma 3.10 that each closed set which separates $E_0$ and $E_1$ contains a connected closed set which separates $E_0$ and $E_1$ and is a minimal separating set. Since $\operatorname{Ind} Z^2 = 2$, it follows that there exists a connected closed set $E$ of $Z^2$ which separates $E_0$ and $E_1$ such that $\operatorname{Ind} E = 1$. Four lemmas will precede the determination of the possible separating sets. The reader is occasionally left the task of formulating analogous results in which coordinates are interchanged, or of supplying the proofs of some special cases. Also the reader will need to decide from the context whether an expression such as $(z_1, z_2)$ denotes an open interval in $Z$ or an element of $Z^2$.

**3.14   Lemma.** *Let $F$ be a connected closed set of $Z^2$, and let $\mathbf{z} = (z', z'')$, where $z' \in Z_1$, be an accumulation point of $F \cap (Z \times \{z''\})$. Then at least one of the following two statements is true:*

(a) *$F$ contains some non-trivial segment in $Z \times \{z''\}$ ending in $\mathbf{z}$;*

(b) *$F$ contains some non-trivial segment in $\{v'\} \times Z$, where $v' \in Z_0$.*

*Proof.* The point $\mathbf{z}$ is an accumulation point of at least one of the sets $F \cap ([u_0, z'] \times \{z''\})$ and $F \cap ([z', u_1] \times \{z''\})$. Let us suppose that $z$ is an accumulation point of $F_0 = F \cap ([u_0, z'] \times \{z''\})$, and let us assume that (a) is false. Since $z' \in Z_1$ we can find sequences $\{a_\alpha\}_{\alpha \in T}$ and $\{b_\alpha\}_{\alpha \in T}$ such that $(a_\alpha, z'') \in F$, $(b_\alpha, z'') \notin F$ and $a_\alpha < b_\alpha < a_{\alpha+1} < z'$ for all $\alpha$. For each $\alpha$, let us choose open intervals $H'_\alpha$ and $H''_\alpha$ such that

$$(b_\alpha, z'') \in H'_\alpha \times H''_\alpha \subset Z \backslash F.$$

If $z'' \in Z_0$ then there exists an open interval $H$ containing $z''$ such that $H \subset H''_\alpha$ for uncountably many values of $\alpha$. If $z'' \in Z_1$, the intersection of countably many open intervals containing $z''$ contains an open interval containing $z''$. Thus in either case there exist an open interval $H$ containing $z''$ and a sequence $\{\alpha(i)\}_{i \in \mathbf{N}}$ in $T$ such that $\alpha(i) < \alpha(j)$ if $i < j$ and $H \subset H''_{\alpha(i)}$ for all $i$. The intervals $H'_{\alpha(i)}$ are disjoint. If $v'$ is the least upper bound of the set $\bigcup_{i \in \mathbf{N}} H'_{\alpha(i)}$, then $v' \in Z_0$. Let $x_i \in H'_{\alpha(i)}$ if $i \in \mathbf{N}$ so that $\{x_i\}$ is a sequence which converges to $v'$. Then for each $i$, one of the following statements is true:

(i)   $F \cap \big((x_i, x_{i+1}) \times \{y\}\big) \neq \varnothing$ if $y \in H$ and $y < z''$;

(ii)  $F \cap \big((x_i, x_{i+1}) \times \{y\}\big) \neq \varnothing$ if $y \in H$ and $y > z''$.

For suppose that there exist $y_1, y_2$ in $H$ such that $y_1 < z'' < y_2$ and $F \cap \big((x_i, x_{i+1}) \times \{y_j\}\big) = \varnothing$ for $j = 1, 2$. Then $(x_i, x_{i+1}) \times (y_1, y_2)$ is a rectangle with interior and exterior containing points of $F$ but with boundary disjoint from $F$, which is a contradiction of the connectedness of $F$. It follows that at least one of the statements (i) and (ii) is true for infinitely many values of $i$. Since $F$ is closed, it follows that $F$ contains a segment in $\{v'\} \times \big(H \cap (z'', u_1]\big)$ or a segment in

$$\{v'\} \times \big(H \cap [u_0, z'')\big).$$

Thus if (a) is false, then (b) is true. $\square$

There is one circumstance in which the hypothesis of Lemma 3.14 is satisfied.

**3.15  Lemma.** *If $F$ is a closed set in $Z^2$, and $\mathbf{z} = (z', z'')$, where $z' \in Z_1$ and $z'' \in Z_0$, is an accumulation point of $F \backslash (\{z'\} \times Z)$, then $\mathbf{z}$ is an accumulation point of $F \cap (Z \times \{z''\})$.*

*Proof.* Let us suppose without essential loss of generality that $\mathbf{z}$ is an accumulation point of $F \cap \big( [u_0, z') \times Z \big)$. Let $w$ be an element of $Z$ such that $w < z'$. If $\mathscr{V} = \{V_i\}_{i \in \mathbf{N}}$ is a countable family of neighbourhoods of $z''$ in $Z$ such that $V_{i+1} \subset V_i$ for each $i$, and every neighbourhood of $z''$ contains a member of $\mathscr{V}$, then we can construct by induction sequences $\{x_i'\}$ and $\{x_i''\}$ such that $(x_1', x_1'') \in F \cap \big( (w, z') \times V_1 \big)$ and

$$(x_{i+1}', x_{i+1}'') \in F \cap \big( (x_i', z') \times V_{i+1} \big)$$

if $i \in \mathbf{N}$. If $x'$ is the limit of the sequence $\{x_i'\}$, then $x' < z'$ since $z' \in Z_1$. Thus $w < x' < z'$ and $(x', z'') \in F$. $\square$

Before stating the next lemma, we must distinguish a subset $\mathscr{Y}$ of the set of closed sets of $Z^2$ which is needed in the study of separating sets.

Let $\{\Delta_\alpha\}_{\alpha \in P}$ be a sequence of disjoint closed intervals in $Z$ indexed by the set $P$ of ordinals not exceeding the first uncountable ordinal $\omega_1$. If $\alpha \in P$, let $\Delta_\alpha = [a_\alpha, b_\alpha]$. We call the sequence *distinguished* if

   (i) $a_\alpha = b_\alpha$ if and only if $\alpha$ is a limit ordinal;

   (ii) the sequence $\{a_\alpha\}_{\alpha \in P}$ is monotonic (either increasing or decreasing);

   (iii) the set $\bigcup_{\alpha \in P} \Delta_\alpha$ is closed in $Z$.

Let us consider a distinguished sequence $\{\Delta_\alpha\}_{\alpha \in P}$, where $\Delta_\alpha = [a_\alpha, b_\alpha]$ and $\{a_\alpha\}_{\alpha \in P}$ is increasing. Suppose that $\beta$ is a limit ordinal in $P$. It follows from (iii) that $a_\beta$ is the least upper bound of $\{a_\alpha \mid \alpha < \beta\}$. Thus if $\beta$ is a limit ordinal such that $\beta < \omega_1$, then $a_\beta \in Z_0$. But $a_{\omega_1} \in Z_1$. For let $\{z_i\}_{i \in \mathbf{N}}$ be a countable family of predecessors of $a = a_{\omega_1}$. For each $i$ there exists $\alpha(i)$ such that $\alpha(i) < \omega_1$ and $a_{\alpha(i)} > z_i$. If $\alpha$ is the first ordinal such that $\alpha \geqslant \alpha(i)$ for all $i$, then $\alpha < \omega_1$ and

$$z_i < a_\alpha < a$$

for all $i$. Thus $a$ is not of countable character on the left and it follows that $a \in Z_1$.

Let us say that a compact subset $Y$ of $Z^2$ is *distinguished* if

$$Y = \bigcup_{\alpha \in P} Y_\alpha$$

and there exist distinguished sequences $\{\Delta_\alpha'\}_{\alpha \in P}$ and $\{\Delta_\alpha''\}_{\alpha \in P}$ such that

$$Y_\alpha = Y \cap (\Delta_\alpha' \times \Delta_\alpha'') \quad \text{if} \quad \alpha \in P,$$

ind $Y_\alpha = 1$ if $\alpha$ is not a limit ordinal, and the unique element of $Y_{\omega_1}$ does not belong to $Q^2$. Let $\mathscr{Y}$ be the set of distinguished subsets of $Z^2$. For each $Y$ in $\mathscr{Y}$ we choose a fixed decomposition $Y = \bigcup_{\alpha \in P} Y_\alpha$, as above. We note that if $Y \in \mathscr{Y}$ then ind $Y = \text{Ind } Y = 1$.

To ease the proof of the next lemma, one more remark should be made. If $X$ is a compact connected space and $K$ is a closed set of $X$ with non-empty interior, then $K$ is not totally disconnected. For suppose that $K$ is a proper closed set of $X$ which is totally disconnected and let $x$ be an interior point of $K$. Then there exists $U$ open in $X$ such that $x \in U \subset K$. Since $K$ is compact and $\{x\}$ is the intersection of the open-and-closed sets of $K$ which contain $x$, there exists an open-and-closed set $H$ of $K$ such that $x \in H \subset U \subset K$. Then $H$ is a proper open-and-closed set of $X$. This is absurd. It follows that if $D$ is a closed rectangle in $Z^2$ and $F$ is a closed connected set in $Z^2$ such that $F$ contains an interior point of $D$, then since $F \cap D$ has non-empty interior in $F$ we can conclude that $F \cap D$ is not totally disconnected and hence ind $F \cap D \geqslant 1$.

**3.16** *Lemma. Let $F$ be a closed connected subset of $Z^2$ such that $\text{Ind } F = 1$ and suppose that $\mathbf{z} = (z', z'')$ is a point of $Z_1^2 \backslash Q^2$ which is an accumulation point of $F \backslash [(\{z'\} \times Z) \cup (Z \times \{z''\})]$. Then $F$ contains a member of $\mathscr{Y}$.*

*Proof.* Let us suppose without essential loss of generality that $\mathbf{z}$ is an accumulation point of $F_0 = F \cap \big([u_0, z'] \times [u_0, z'']\big)$. Then we can construct by induction sequences $\{a'_\alpha\}_{\alpha \in T}$ and $\{a''_\alpha\}_{\alpha \in T}$ which converge to $z'$ and $z''$ respectively such that: (i) if $\alpha < \beta$ then $a'_\alpha < a'_\beta$ and $a''_\alpha < a''_\beta$; (ii) if $\alpha$ is a limit ordinal then the sequences $\{a'_\gamma\}_{\gamma < \alpha}$ and $\{a''_\gamma\}_{\gamma < \alpha}$ converge to $a'_\alpha$ and $a''_\alpha$ respectively; and (iii) if $\alpha$ is not a limit ordinal then there exist $b'_\alpha$, $b''_\alpha$ such that $a'_\alpha < b'_\alpha < a'_{\alpha+1}$, $a''_\alpha < b''_\alpha < a''_{\alpha+1}$ and the open rectangle $(a'_\alpha, b'_\alpha) \times (a''_\alpha, b''_\alpha)$ contains a point of $F$. Taking

$$\Delta'_\alpha = [a'_\alpha, b'_\alpha]$$

if $\alpha$ is not a limit ordinal, $\Delta'_\alpha = \{a_\alpha\}$ if $\alpha$ is a limit ordinal and $\Delta'_{\omega_1} = \{z'\}$, we obtain a distinguished sequence of intervals in $Z$. Similarly we obtain a distinguished sequence by taking $\Delta''_\alpha = [a''_\alpha, b''_\alpha]$ if $\alpha$ is not a limit ordinal, $\Delta''_\alpha = \{a''_\alpha\}$ if $\alpha$ is a limit ordinal and $\alpha < \omega_1$ and $\Delta''_{\omega_1} = \{z''\}$. If $\alpha \in P$ let

$$Y_\alpha = F \cap (\Delta'_\alpha \times \Delta''_\alpha)$$

and let $Y = \bigcup_{\alpha \in P} Y_\alpha$. Then by the remark above ind $Y_\alpha = 1$ if $\alpha$ is not a limit ordinal, since $Y_\alpha$ contains an interior point of the closed rectangle $\Delta'_\alpha \times \Delta''_\alpha$. Moreover the unique point of $Y_{\omega_1}$ does not belong to $Q^2$. Thus $Y \in \mathscr{Y}$. $\square$

**3.17 Lemma.** *Let $F$ be a connected closed set of $Z^2$ with $\operatorname{Ind} F = 1$ such that for some rectangle $D$ in $\mathscr{D}$, $F$ contains a non-trivial segment lying on the boundary of $D$ but $F$ is not contained in the boundary of $D$. Then $F$ contains either (a) a member of $\mathscr{Y}$, or (b) a non-trivial segment $J \times \{z\}$ or $\{z\} \times J$, where $z \in Z_0$, or (c) two non-trivial segments intersecting in a common end-point, precisely one of which lies in the boundary of $D$.*

*Proof.* The intersection of $F$ with the boundary of $D$ contains a segment. Let $L$ be the maximal polygonal arc containing this segment and contained in $F$ and in the boundary of $D$. Then $L$ is a proper closed subset of $F$. Since $F$ is connected, there exists a point $\mathbf{z}$ of $L$ which is an accumulation point of $F \backslash L$. Suppose first that $\mathbf{z}$ is not a vertex of $D$ and let $\mathbf{z} = (z', z'')$, where the line $\{z'\} \times Z$ contains a segment of the boundary of $D$ so that $z' \in Z_1 \backslash Q$. In view of the maximality of $L$, $\mathbf{z}$ is an accumulation point of $F \backslash (\{z'\} \times Z)$. In the case in which $\mathbf{z}$ is an accumulation point of $F \cap (Z \times \{z''\})$, it follows from Lemma 3.14 that either (b) or (c) holds. By Lemma 3.15 this is necessarily the case if $z'' \in Z_0$. Thus there remains the case in which $z'' \in Z_1$ and $\mathbf{z}$ is not an accumulation point of $F \cap (Z \times \{z''\})$. Then $\mathbf{z}$ is an accumulation point of $F \backslash [(\{z'\} \times Z) \cup (Z \times \{z''\})]$ and it follows from Lemma 3.16 that (a) holds. The modification to the argument in the case in which $\mathbf{z} = (z', z'')$ is a vertex of $D$ results from observing that if $\mathbf{z}$ is not an accumulation point of $F \backslash ((\{z'\} \times Z) \cup (Z \times \{z''\}))$, then $\mathbf{z}$ is an accumulation point of the subset of $F \cap (\{z'\} \times Z)$, say, which lies outside $D$. $\square$

Now we can obtain the information we require about separating sets in $Z^2$.

**3.18 Proposition.** *Let $E$ be a closed set of $Z^2$ such that $\operatorname{Ind} E = 1$, which separates $E_0 = Z \times \{u_0\}$ and $E_1 = Z \times \{u_1\}$. Then $E$ contains either (a) a member of $\mathscr{Y}$, or (b) two non-trivial segments intersecting in a common end-point, precisely one of which lies in the boundary of some member of $\mathscr{D}$, or (c) a horizontal segment not lying on the boundary of any member of $\mathscr{D}$ but crossing $\{v\} \times Z$, or (d) a segment $\{v\} \times J$, or (e) a segment $\{z\} \times J$ or $J \times \{z\}$, where $z \in Z_0$.*

*Proof.* By Lemma 3.10 there exists a closed connected subset $F$ of $E$ which separates $E_0$ and $E_1$ and is a minimal separating set in the sense that no proper subset of $F$ separates $E_0$ and $E_1$. Clearly $F \cap (\{v\} \times Z)$ is non-empty. Since $F$ is connected, there exists $\mathbf{z} = (v, z'')$ on $\{v\} \times Z$ which is an accumulation point of $F \setminus (\{v\} \times Z)$. Suppose first that $\mathbf{z}$ lies on the boundary of some member $D$ of $\mathscr{D}$. If $F$ contains some non-trivial segment in the boundary of $D$, then by Lemma 3.17, either $(a)$, $(b)$ or $(e)$ holds. If $F$ does not contain a non-trivial segment in the boundary of $D$, then either $\mathbf{z}$ is an accumulation point of

$$ F \setminus \big( (\{v\} \times Z) \cup (Z \times \{z''\}) \big) $$

in which case $(a)$ holds by Lemma 3.16, or $F$ contains a non-trivial segment in $\{v\} \times Z$ so that $(d)$ holds.

Now suppose that $\mathbf{z} \in N$ so that $\mathbf{z}$ is not on the boundary of any member of $\mathscr{D}$. There is no open rectangle $O$ containing $\mathbf{z}$ such that $F \cap O$ is a segment ending at $\mathbf{z}$, for if this were the case then $F \setminus O$ would be a proper closed subset of $F$ separating $E_0$ and $E_1$. Hence either $F$ contains a horizontal segment crossing $\{v\} \times Z$ at $\mathbf{z}$ so that $(c)$ holds, or $F$ contains a segment lying on the line $\{v\} \times Z$ so that $(d)$ holds, or $\mathbf{z}$ is an accumulation point of $F \setminus \big( (\{v\} \times Z) \cup (Z \times \{z''\}) \big)$. In this last case, if $z'' \in Z_1$ then $(a)$ holds by Lemma 3.16 and if $z'' \in Z_0$, then by Lemma 3.15, $\mathbf{z}$ is an accumulation point of $F \cap (Z \times \{z''\})$ and hence $(e)$ holds by Lemma 3.14. $\square$

**3.19** *Filippov's Example.* The example is a space of the form (see Definition 3.4)

$$ X = F(Z^2, \xi, \Pi), $$

where $\xi$ is an infinite limit ordinal and $\Pi = (\Pi_0, \Pi_1, \Pi_2)$, where $\Pi_0$, $\Pi_1$, and $\Pi_2$ are disjoint subsets of $Z^2 \times T_\xi$. The ordinal $\xi$ is the first ordinal of sufficiently large cardinality that we can define a function

$$ \sigma : T_\xi \to \mathscr{Y} \cup \big( Z^2 \setminus (Q^2 \cup S) \big) $$

such that the inverse image under $\sigma$ of each member of

$$ \mathscr{Y} \cup \big( Z^2 \setminus (Q^2 \cup S) \big) $$

is cofinal in $T_\xi$. The subsets $\Pi_0$, $\Pi_1$ and $\Pi_2$ will now be defined. Let us partition the non-limit ordinals in $P$ into two disjoint sets $\Sigma_1$ and $\Sigma_2$, each of which is cofinal in $P$. The reader should refer to p. 321 for the definitions of the subsets $S$, $V$ and $N$ of $Z^2$. There are four cases.

$(a)$ If $\alpha \in T_\xi$ and $\sigma(\alpha) = Y \in \mathscr{Y}$, where $Y = \bigcup_{\gamma \in P} Y_\gamma$, then $(\mathbf{z}, \alpha) \in \Pi_0$ if $\mathbf{z} \in Y_\gamma$, where $\gamma$ is a limit ordinal, and for $i = 1, 2$, $(\mathbf{z}, \alpha) \in \Pi_i$ if $\mathbf{z} \in Y_\gamma$, where $\gamma$ is a non-limit ordinal and $\gamma \in \Sigma_i$.

(b) If $\alpha \in T_\xi$ and $\sigma(\alpha) = \mathbf{z}_0$, where $\mathbf{z}_0 \in Z^2 \backslash (V \cup N)$, then $(\mathbf{z}_0, \alpha) \in \Pi_0$, $(\mathbf{z}, \alpha) \in \Pi_1$ if $\mathbf{z}$ lies on the vertical line through $\mathbf{z}_0$, and $(\mathbf{z}, \alpha) \in \Pi_2$ if $\mathbf{z}$ lies on the horizontal line through $\mathbf{z}_0$.

(c) If $\alpha \in T_\xi$ and $\sigma(\alpha) = \mathbf{z}_0$, where $\mathbf{z}_0 \in V$ and $\mathbf{z}_0$ is a vertex of the rectangle $D$ in $\mathscr{D}$, then $(\mathbf{z}_0, \alpha) \in \Pi_0$, and if $\mathbf{z}$ lies on the horizontal or vertical line through $\mathbf{z}_0$, then $(\mathbf{z}, \alpha) \in \Pi_1$ if some non-degenerate segment lies in the intersection of the boundary of $D$ with the segment with end-points $\mathbf{z}_0$ and $\mathbf{z}$, and $(\mathbf{z}, \alpha) \in \Pi_2$ otherwise.

(d) If $\alpha \in T_\xi$ and $\sigma(\alpha) = \mathbf{z}_0$, where $\mathbf{z}_0 \in N$, then $(\mathbf{z}_0, \alpha) \in \Pi_0$, and if $\mathbf{z}$ lies on the horizontal line through $\mathbf{z}_0$, then $(\mathbf{z}, \alpha) \in \Pi_1$ if $\mathbf{z}$ lies to the left of $\mathbf{z}_0$, and $(\mathbf{z}, \alpha) \in \Pi_2$ if $\mathbf{z}$ lies to the right of $\mathbf{z}_0$.

From the discussion following Definition 3.4 we can see that

$$X = F(Z^2, \xi, \Pi)$$

is a Hausdorff space and thus is a bicompact space. The bicompact space $X$ has distinct inductive dimensions.

**3.20** *Proposition.* $\operatorname{Ind} X \geqslant 3$.

*Proof.* We recall that there is an identification mapping

$$\pi : Z^2 \times W_\xi \to X.$$

For $i = 0, 1$, let $F_i = \pi(E_i \times \{\xi\})$, where $E_i = Z \times \{u_i\}$, and let $K$ be a closed set of $X$ which separates $F_0$ and $F_1$. Then $K$ contains a cut $F$ which separates $F_0$ and $F_1$. We recall that there is a special irreducible mapping

$$\bar{\pi} : X \to Z^2 \times L_\xi,$$

and $\bar{\pi}$ maps $X_0 = \pi(Z^2 \times \{\xi\})$ homeomorphically onto $Z^2 \times \{\xi\}$ in $Z^2 \times L_\xi$. The conditions of Lemma 3.6 are satisfied. If case (a) of Lemma 3.6 arises, then $F$ contains a non-empty open set of $X_0$ and since $X_0$ is homeomorphic with $Z^2$ we see that $F$ contains a closed set homeomorphic with a non-empty closed rectangle in $Z^2$. It follows from Proposition 3.11 that $\operatorname{Ind} F \geqslant 2$. If case (b) of Lemma 3.6 arises, then $F$ contains a set of the form $\pi(E \times (\beta, \xi])$, where $\beta \in T_\xi$ and $E$ separates $E_0$ and $E_1$ in $Z^2$ and satisfies $\operatorname{Ind} E = 1$. In Proposition 3.18 the nature of the set $E$ was elucidated and we now consider the five cases. If case (a) arises, then $F$ contains a closed subspace of the form $\pi(Y \times W_\xi^\alpha)$, where $Y \in \mathscr{Y}$ and $\sigma(\alpha) = Y$; this is a space of the type considered in 3.9. If case (b) arises, then $F$ contains a closed subspace of the form $\pi((J_1 \cup J_2) \times W_\xi^\alpha)$, where $J_1$ and $J_2$ are segments intersecting in a common end-point $z_0$, precisely one of $J_1$ and $J_2$ lying in the boundary of some member of $\mathcal{D}$, and $\sigma(\alpha) = z_0$; this is a space of the type considered in 3.8. If case (c) arises, then $F$ contains a closed subspace of the form $\pi(J \times W_\xi^z)$, where $J$ is a horizontal segment intersecting $\{v\} \times Z$ in $z_0$ and $\sigma(\alpha) = z_0$; this is a space of the type considered in 3.8. If case (d) arises, then $F$ contains a closed subspace of the form $\pi(J \times [\beta + 1, \xi])$, where $J$ is a segment lying in $\{v\} \times Z$; this is a space of the type considered in 3.7. If case (e) arises, then $F$ contains a closed subspace of the form $\pi(J \times [\beta + 1, \xi])$, where $J$ is a segment lying in $Z^2 \backslash (Q^2 \cup S)$; this is a space of the type considered in 3.7. Thus $\operatorname{Ind} F \geqslant 2$ in every case. It follows that

$$\operatorname{Ind} X \geqslant 3. \ \square$$

We turn now to the determination of the small inductive dimension of the space $X$. It follows from Proposition 4.2.4 that $\operatorname{ind} X \geqslant 2$. We shall need two more lemmas.

**3.21  Lemma.** *If $z \in Q^2 \cup S$, then the subspace $\pi(\{z\} \times W_\xi)$ of $X$ is homeomorphic with $W_\xi$.*

*Proof.* If $z \in Q^2 \cup S$ and $\alpha \in T_\xi$, then $(z, \alpha) \notin \Pi_0$. If $\sigma(\alpha) \in Z^2 \backslash (Q^2 \cup S)$, this is obvious. Suppose that $\sigma(\alpha) = Y \in \mathcal{Y}$, where $Y = \bigcup_{\gamma \in P} Y_\gamma$. If $z \in Q^2$, then $z$ is not the unique point of $Y_{\omega_1}$ by definition, and $z$ cannot be the unique point of $Y_\gamma$ for some limit ordinal $\gamma$ less than $\omega_1$, since such a point must belong to $Z_0^2$. Thus $(z, \alpha) \notin \Pi_0$. If $z \in S$, then one coordinate of $z$ belongs to $Z_0$ and the other belongs to $Z_1$, so that $z$ cannot be the unique point of $Y_\gamma$ for a limit ordinal $\gamma$, since such a point must belong to $Z_0^2 \cup Z_1^2$, and hence $(z, \alpha) \notin \Pi_0$. It is now clear that $\pi(\{z\} \times W_\xi)$ is homeomorphic with $W_\xi$. $\square$

**3.22  Lemma.** *If $D \in \mathcal{D} \cup \mathcal{E}$, then $\operatorname{ind} \pi\big(\operatorname{bd}(D) \times W_\xi\big) \leqslant 1$.*

*Proof.* Let $A = \pi\big(\operatorname{bd}(D) \times W_\xi\big)$ and let $a$ be a point of $A$. Suppose first that $a = \pi(z_0, \alpha)$, where $z_0 \in \operatorname{bd}(D)$ and $\alpha \in P_\xi$. We suppose that $z_0$ is not a vertex of $D$, the necessary modification to the argument in that case being left to the reader. Each neighbourhood of $a$ in $A$ contains a neighbourhood of the form $\pi(J \times (\gamma, \alpha])$, where $J$ is a segment lying in the boundary of $D$ with end-points in $Q^2 \cup S$, and $\gamma$ is an ordinal less than $\alpha$. The boundary of this set in $A$ consists of two disjoint closed subsets contained in sets of the form $\pi(\{z\} \times W_\xi)$, where $z \in Q^2 \cup S$. By Lemma 3.21, $\operatorname{ind} \pi(\{z\} \times W_\xi) = 0$ if $z \in Q^2 \cup S$. Thus $a$ has arbitrarily small neighbourhoods in $A$ with boundaries in $A$

which are zero-dimensional. Now suppose that $a = \pi(\mathbf{z}_0, y)$, where $\mathbf{z}_0 \in \mathrm{bd}\,(D)$ and $y \in W_\xi^\alpha$. Let us suppose again that $\mathbf{z}_0$ is not a vertex of $D$. There exists a neighbourhood $M$ of $\mathbf{z}_0$ in $Z^2$ such that at least one of the identifications $\phi_1$ and $\phi_2$ is not performed on any of the sets $\{\mathbf{z}\} \times W_\xi^\alpha$ for $\mathbf{z}$ in $M \cap \mathrm{bd}\,(D)$. Let us suppose that $(\mathbf{z}, \alpha) \notin \Pi_1$ if

$$\mathbf{z} \in M \cap \mathrm{bd}\,(D).$$

Let

$$B = \{z \in \mathrm{bd}\,(D) \mid (\mathbf{z}, \alpha) \in \Pi_0\}.$$

Then $B$ is closed and $\mathrm{ind}\,B \leqslant 0$. Let $H$ be an open neighbourhood of $a$. Then $H$ contains a neighbourhood of $a$ of the form $\pi(J \times K)$, where $J$ is a segment contained in $M \cap \mathrm{bd}\,(D)$, with end-points in the complement of $B$, and $K$ is an open-and-closed interval $[k, k']$ in $W_\xi^\alpha$, where $k$ and $k'$ are not identified with other points by $\phi_2$. If $G$ is the interior of $\pi(J \times K)$ in $A$, then $G$ is open in $A$, $a \in G \subset H$ and $\mathrm{bd}_A\,(G)$ is the union of at most four disjoint closed sets $\pi(\{\mathbf{z}_1\} \times K), \pi(\{\mathbf{z}_2\} \times K)$, $\pi\big((J \cap B) \times \{k\}\big)$ and $\pi\big((J \cap B) \times \{k'\}\big)$, where $\mathbf{z}_1$ and $\mathbf{z}_2$ are the end-points of the segment $J$. Thus $\mathrm{ind}\,\mathrm{bd}_A\,(G) \leqslant 0$. This completes the proof that $\mathrm{ind}\,A \leqslant 1.\ \square$

**3.23   Theorem.** *Filippov's bicompact space $X$ satisfies*

$$\mathrm{ind}\,X = 2 \quad and \quad \mathrm{Ind}\,X = 3.$$

*Proof.* It is not difficult to prove that $\mathrm{Ind}\,X \leqslant 3$ and hence

$$2 \leqslant \mathrm{ind}\,X \leqslant \mathrm{Ind}\,X = 3.$$

It remains to show that $\mathrm{ind}\,X \leqslant 2$. Let $a$ be a point of $X$.

If $a = \pi(\mathbf{z}_0, \alpha)$, where $\alpha \in P_\xi$, then every neighbourhood of $a$ contains a neighbourhood of the form $G = \pi\big(D \times (\gamma, \alpha]\big)$, where $D \in \mathcal{D} \cup \mathcal{E}$ and $\gamma$ is an ordinal less than $\alpha$. The boundary of $G$ is contained in $\pi\big(\mathrm{bd}\,(D) \times W_\xi^\cdot\big)$ so that $\mathrm{ind}\,\mathrm{bd}\,(G) \leqslant 1$ by Lemma 3.22.

Now suppose that $a = \pi(\mathbf{z}_0, y)$, where $y \in W_\xi^\alpha$. If $D$ is an open rectangle in $Z^2$ such that $\mathbf{z}_0 \in D$ and $K$ is an open-and-closed interval in $W_\xi^\alpha$ such that $y \in K$, let $G(D, K)$ denote the interior in $X$ of $\pi(D \times K)$. If $\sigma(\alpha) \in Z^2 \backslash (Q^2 \cup S)$, then every neighbourhood of $a$ contains an open neighbourhood of $a$ of the form $G(D, K)$, where $\sigma(\alpha) \notin \mathrm{bd}\,(D)$. Then $\mathrm{ind}\,\mathrm{bd}\,[G(D, K)] \leqslant 1$. If $\sigma(\alpha) = Y \in \mathcal{Y}$ we must consider two cases. Let $Y = \bigcup_{\gamma \in P} Y_\gamma$ and suppose first that $\mathbf{z}_0 \notin Y_\gamma$ if $\gamma$ is a limit ordinal. Every neighbourhood of $a$ contains an open neighbourhood of $a$ of the form $G(D, K)$, where $(\mathbf{z}, \alpha) \notin \Pi_0 \cup \Pi_1$, say, if $\mathbf{z}$ belongs to the closure of $D$ in $Z^2$, and the end-points of $K$ are not identified with other points by $\phi_2$. Then $\operatorname{ind} \operatorname{bd}[G(D,K)] \leqslant 1$. Now suppose that $z_0 \in Y_\gamma$, where $\gamma$ is a limit ordinal. Every neighbourhood of $a$ contains an open neighbourhood of $a$ of the form $G(D,K)$, where $\operatorname{bd}(D)$ is disjoint from $Y$. Then $\operatorname{ind} \operatorname{bd}[G(D,K)] \leqslant 1$. Thus $\operatorname{ind} X \leqslant 2.\ \square$

**3.24   Remark.** If

$$X_1 = \pi\big(([u_0,v] \times Z) \times W_\xi\big) \quad \text{and} \quad X_2 = \pi\big(([v,u_1] \times Z) \times W_\xi\big),$$

then $X_1$ and $X_2$ are closed sets of $X$ and $X = X_1 \cup X_2$. It is not too difficult to show that

$$\operatorname{ind} X_1 = \operatorname{Ind} X_1 = \operatorname{ind} X_2 = \operatorname{Ind} X_2 = 2.$$

## Notes

The necessary and sufficient condition for the inverse limit of an inverse system of non-empty bicompact spaces to have covering dimension not exceeding $n$ is due to Delinić and Mardešić [1968]. The characterization of the bicompact spaces $X$ with $\dim X \leqslant n$ given in Proposition 1.8 was first established by Mardešić [1960] by different methods. It was obtained in his study of possible generalizations of Freudenthal's theorem (Freudenthal [1937]) that every compact metrizable space $X$ is homeomorphic with the inverse limit of an inverse sequence of compact polyhedra $P_i$ such that $\dim P_i \lneqq \dim X$. Bicompact spaces can be characterized as inverse limits of inverse systems of compact polyhedra. Thus it was natural to conjecture that each bicompact space $X$ is homeomorphic with the inverse limit of an inverse system of compact polyhedra $P_\alpha$ such that $\dim P_\alpha \leqslant \dim X$. Mardešić and, independently, Pasynkov [1958, 1962] showed this conjecture to be false. They showed that if a bicompact space $X$ is homeomorphic with the inverse limit of an inverse system of compact polyhedra $P_\alpha$ such that $\dim P_\alpha \leqslant 1$, then $\operatorname{ind} X \leqslant 1$. Thus any bicompact space $X$ for which $\dim X = 1$ and $\operatorname{ind} X > 1$, for instance Example 4.3.4, is a counterexample to the conjecture. One might ask if the conjecture ought to have been made in terms of inductive dimension. Pasynkov showed that this is not the case by exhibiting a bicompact space $X$ such that $\dim X = \operatorname{ind} X = \operatorname{Ind} X = 1$, which is not the limit of an inverse system of 1-dimensional compact polyhedra. Pasynkov proved furthermore that if a bicompact space $X$ is homeomorphic with the inverse limit of an inverse system of compact polyhedra $P_\alpha$ such that $\dim P_\alpha \leqslant \dim X$, in which the connecting mappings are simplicial for fixed triangulations of the $P_\alpha$, then

$$\dim X = \operatorname{ind} X = \operatorname{Ind} X.$$

Thus the problem of representing $n$-dimensional bicompact spaces as limits of $n$-dimensional compact polyhedra is connected with the problem of determining the relations between the different dimension functions for bicompact spaces. Employing this result, Pasynkov [1960] showed that $\dim X = \operatorname{ind} X = \operatorname{Ind} X$ for the quotient $X = G/H$, where $G$ is a locally compact group and $H$ is a closed subgroup of $G$.

In 1935, at a conference in Moscow, Aleksandrov raised the question whether covering dimension, small inductive dimension and large inductive dimension all coincide for bicompact spaces (Aleksandrov [1936]). Lunc [1949] found an example showing that covering dimension differs from the inductive dimensions, and shortly afterwards Lokucievskii [1949] found the rather simpler example described in Chapter 4 (Example 4.3.4). Complete information about the relations between covering dimension and the inductive dimensions is given by the examples constructed in §2 which are due to Vopěnka [1958]. The question of the coincidence of the small and large inductive dimensions proved to be even more difficult and was open for almost 35 years. The example of a bicompact space with differing inductive dimensions described in §3 was constructed by Filippov [1970]. The construction was outlined earlier (Filippov [1969]). Proposition 3.2, employed in the construction of the example, is a simple case of a result due to Ceder [1964]. Examples which are in some respects simpler than Example 3.19 have been described (Pasynkov [1970], Lifanov and Pasynkov [1970] Filippov [1970a]). Filippov [1970a] has outlined the construction of a bicompact space $R$ which satisfies $\dim R = 1$, $\operatorname{ind} R = 2$, $\operatorname{Ind} R = 3$. By general constructions, starting from this example and, say, Example 4.3.4, he is able to obtain, for each pair of positive integers $m, n$ such that $m \leqslant n \leqslant 2m - 1$, a bicompact space $T_{mn}$ which satisfies

$$\dim T_{mn} = 1, \quad \operatorname{ind} T_{mn} = m, \quad \operatorname{Ind} T_{mn} = n.$$

Thus we now have fairly complete information about the behaviour of the three principal dimension functions on bicompact spaces.

# 9

## MAPPINGS AND PRODUCTS

### 1 Mappings into spheres

We begin with some consequences of the characterization of covering dimension in terms of mappings into spheres which was one of the fundamental results of Chapter 3. The results will be needed in the next section but are of independent interest.

Let $X$ be a topological space and let $f_0, f_1: X \to S^n$ be continuous functions. Then by Proposition 1.6.3 the set

$$C = \{x \in X \mid f_0(x) \neq f_1(x)\}$$

is an open set. It is also an $F_\sigma$-set. We need to say more. For each positive integer $j$, let

$$U_j = \{x \in X \mid \|f_0(x) - f_1(x)\| > 1/j\}.$$

Then $C = \bigcup_{j \in \mathbb{N}} \overline{U}_j$. Thus $C$ is the union of countably many closed sets, each of which is the closure of an open set.

**1.1   Lemma.** *Let $X$ be a bicompact space, let $f_0, f_1: X \to S^n$ be continuous functions and let*

$$C = \{x \in X \mid f_0(x) \neq f_1(x)\}.$$

*If $\dim C \leqslant n - 1$, then $f_0$ and $f_1$ are homotopic.*

*Proof.* We observe first that $C \times I$ is an $F_\sigma$-set in $X \times I$, so that $C \times I$ is a normal space. Since $C$ is the union of countably many bicompact sets, it follows from Theorem 3.2.5 and Proposition 3.2.6 that

$$\dim C \times I \leqslant \dim C + 1 \leqslant n.$$

Let $F = X \backslash C$ and let

$$D = (X \times \{0\}) \cup (F \times I) \cup (X \times \{1\}).$$

Then $D$ is a closed set of $X \times I$, and if $E$ is a closed set of $X \times I$ which is disjoint from $D$, then $E \subset C \times I$ so that $\dim E \leqslant n$. Now let us define a continuous function $\phi: D \to S^n$ by putting

$$\phi(x, 0) = f_0(x), \phi(x, 1) = f_1(x)$$

if $x \in X$ and $\phi(x,t) = f_0(x) = f_1(x)$ if $(x,t) \in F \times I$. By Proposition 3.2.3, $\phi$ has an extension $h \colon X \times I \to S^n$. Since $h(x,0) = f_0(x)$ and $h(x,1) = f_1(x)$ if $x \in X$, the continuous functions $f_0$ and $f_1$ are homotopic. $\square$

**1.2   Lemma.** *Two continuous mappings of a bicompact space into $S^n$ are uniformly homotopic if and only if they are homotopic.*

*Proof.* Let $X$ be a bicompact space and let $f_0, f_1 \colon X \to S^n$ be continuous functions which are homotopic. Let $h \colon X \times I \to S^n$ be a continuous function such that $h(x,i) = f_i(x)$ for $i = 0, 1$ and $x$ in $X$. Let $\epsilon$ be a given positive real number and let $\{U_1, \ldots, U_k\}$ be a finite open covering of the compact metric space $h(X \times I)$ such that $\operatorname{diam} U_i < \epsilon$ for each $i$. There exist finite open coverings $\{V_1, \ldots, V_r\}$ and $\{W_1, \ldots, W_s\}$ of $X$ and $I$ respectively such that $\{V_i \times W_j\}$ is a refinement of the covering $\{h^{-1}(U_1), \ldots, h^{-1}(U_k)\}$ of $X \times I$. There exists a positive real number $\delta$ such that $u, v \in W_j$ for some $j$ if $u, v \in I$ and $|u - v| < \delta$. It follows that if $u, v \in I$ and $|u - v| < \delta$, then

$$\| h(x,u) - h(x,v) \| < \epsilon$$

for all $x$. Thus $f_0$ and $f_1$ are uniformly homotopic. $\square$

**1.3   Proposition.** *Let $X$ be a $T_4$-space, let $f_0, f_1 \colon X \to S^n$ be continuous functions and let*

$$C = \{ x \in X \mid f_0(x) \neq f_1(x) \}.$$

*If $\dim C \leqslant n - 1$, then $f_0$ and $f_1$ are uniformly homotopic.*

*Proof.* Let $\beta X$ be the Stone–Čech bicompactification of $X$ and let $g_0, g_1$ be the unique extensions of $f_0, f_1$ to $\beta X$. Let

$$D = \{ z \in \beta X \mid g_0(z) \neq g_1(z) \}.$$

Then $D \cap X = C$. By the remark preceding Lemma 1.1, $D = \bigcup E_i$, where $E_i$ is the closure of an open set of $\beta X$ for each $i$. It follows that if $B_i = E_i \cap X$, then $E_i$ is the Stone–Čech bicompactification of $B_i$. Since $\dim B_i \leqslant n - 1$, it follows from Proposition 6.4.3 that

$$\dim E_i \leqslant n - 1$$

for each $i$ and hence by the countable sum theorem in the normal subspace $D$ of $\beta X$, we have $\dim D \leqslant n - 1$. It now follows from Lemmas 1.1 and 1.2 that $g_0$ and $g_1$ are uniformly homotopic. Hence $f_0$ and $f_1$ are uniformly homotopic. $\square$

**1.4　Proposition.** *Let $X$ be a $T_4$-space and let $B_0$ and $B_1$ be closed sets of $X$ such that $B_0 \cup B_1 = X$. For $i = 0, 1$ let $f_i: B_i \to S^n$ be a continuous function, and let*

$$C = \{x \in B_0 \cap B_1 \mid f_0(x) \neq f_1(x)\}.$$

*If $\dim C \leqslant n-1$, then each of $f_0$ and $f_1$ has an extension to $X$.*

*Proof.* Let $g_0, g_1: B_0 \cap B_1 \to S^n$ be given by restriction of $f_0, f_1$ respectively. By Proposition 1.3, $g_0$ and $g_1$ are uniformly homotopic. Since $g_1$ has an extension $f_1: B_1 \to S^n$, it follows from Lemma 3.2.1 that $g_0$ has an extension $h_0: B_1 \to S^n$. The continuous function $\phi_0: X \to S^n$, such that $\phi_0 \mid B_0 = f_0$ and $\phi_0 \mid B_1 = h_0$, is the required extension of $f_0$. $\square$

We can now prove the result required in the next section.

**1.5　Proposition.** *Let $A$ be a closed set of a $T_4$-space $X$, let*

$$\phi: A \to S^n$$

*be a continuous function and let $\{G_i\}_{i \in \mathbb{N}}$ be a countable open covering of $X$ such that $\dim \mathrm{bd}\,(G_i) \leqslant n-1$ and $\phi$ has an extension to $A \cup \bar{G}_i$ for each $i$. Then $\phi$ has an extension to $X$.*

*Proof.* Let $\psi_1$ be an extension of $\phi$ to $A \cup \bar{G}_1$. Let us assume that we have defined an extension $\psi_k$ of $\phi$ to $A \cup \bigcup_{i=1}^{k} \bar{G}_i$. Let

$$B_0 = A \cup \bigcup_{i=1}^{k} \bar{G}_i$$

and let $B_1 = (A \cup \bar{G}_{k+1}) \backslash \bigcup_{i=1}^{k} G_i$. Let $\theta$ be the restriction to $B_1$ of an extension of $\phi$ to $A \cup \bar{G}_{k+1}$ and let

$$C = \{x \in B_0 \cap B_1 \mid \theta(x) \neq \psi_k(x)\}.$$

Then $C \subset (B_0 \cap B_1) \backslash A \subset \bigcup_{i=1}^{k} \mathrm{bd}\,(G_i)$. Since $\dim \mathrm{bd}\,(G_i) \leqslant n-1$ for each $i$, and $C$ is an $F_\sigma$-set, it follows that $\dim C \leqslant n-1$. It follows from Proposition 1.4 that $\psi_k$ has an extension $\psi_{k+1}$ to

$$B_0 \cup B_1 = A \cup \bigcup_{i=1}^{k+1} \bar{G}_i.$$

Now let $x$ be a point of $X$ and let $m$ be the first integer such that $x \in G_m$. Putting $\psi(x) = \psi_m(x)$ we obtain a continuous function

$$\psi: X \to S^n$$

which is an extension of $\phi$. $\square$

We conclude this section with a topic of considerable historical interest. Let $n$ be a positive integer. Let us say that a bicompact space $C$ such that $\dim C = n$ is an *n-dimensional Cantor space* if $C \backslash A$ is connected for each subset $A$ of $C$ such that $\dim A \leqslant n-2$.

**1.6   Lemma.** *If $n$ is a positive integer and a bicompact space $C$ satisfies* $\dim C = n$, *then $C$ is an n-dimensional Cantor space if and only if $C$ is not the union of proper closed sets $C_1$ and $C_2$ such that*

$$\dim C_1 \cap C_2 \leqslant n-2.$$

*Proof.* The condition is clearly necessary. Suppose that $C$ is not an $n$-dimensional Cantor space. Then there exists a subset $A$ of $C$ such that $\dim A \leqslant n-2$ and $C \backslash A = U_1 \cup U_2$, where $U_1$ and $U_2$ are non-empty disjoint open sets of $C \backslash A$. If $C \backslash A$ is not dense in $C$, then there exists a non-empty open set $V$ such that $\overline{V} \subset A$. If $C_1 = \overline{V}$ and $C_2 = X \backslash V$, then $C_1$ and $C_2$ are proper closed sets of $C$ such that

$$C = C_1 \cup C_2 \quad \text{and} \quad \dim C_1 \cap C_2 \leqslant n-2$$

since $C_1 \cap C_2$ is a closed set of $A$. If $C \backslash A$ is dense in $C$, then

$$C = \overline{U}_1 \cup \overline{U}_2.$$

Moreover $\overline{U}_1$ and $\overline{U}_2$ are proper closed sets of $C$, and

$$\dim \overline{U}_1 \cap \overline{U}_2 \leqslant n-2$$

since $\overline{U}_1 \cap \overline{U}_2$ is a closed set of $A$. Thus if the condition is satisfied then $C$ is an $n$-dimensional Cantor space. $\square$

For the proof of an existence theorem for Cantor spaces the following lemma is needed.

**1.7   Lemma.** *Let $X$ be a bicompact space, let $A$ be a closed subspace of $X$ and let $f : A \to S^n$ be a continuous function which has no extension to $X$. Then there exists a closed set $F$ of $X$ such that $f$ has no extension to $A \cup F$ but $f$ has an extension to $A \cup K$ if $K$ is a proper closed subset of $F$.*

*Proof.* Let $\mathscr{E}$ be the set of closed sets $E$ of $X$ such that $f$ has no extension to $A \cup E$ and let $\mathscr{E}$ be partially ordered by inclusion. By the Kuratowski lemma, $\mathscr{E}$ contains a maximal chain $\mathscr{F}$. Let $F = \bigcap_{E \in \mathscr{F}} E$. Suppose that $f$ has an extension to $A \cup F$. Then by Proposition 1.3.8 there exists an open set $U$ such that $A \cup F \subset U$ and $f$ has an extension to $U$. Since $X$ is compact, there exists some member $E$ of $\mathscr{F}$ such that $F \subset E \subset U$ and it follows that $f$ has an extension to $A \cup E$, which is absurd. Thus $f$ has no extension to $A \cup F$. In view of the maximality of $\mathscr{F}$, if $K$ is a proper closed subset of $F$ then $f$ has an extension to $A \cup K$.☐

We can now prove the main result about Cantor spaces.

**1.8 Proposition.** *If $n$ is a positive integer and $X$ is a bicompact space such that* $\dim X = n$, *then $X$ contains an $n$-dimensional Cantor space.*

*Proof.* Since $\dim X = n$, there exist a closed set $A$ of $X$ and a mapping $f: A \to S^{n-1}$ which has no extension to $X$. By Lemma 1.7 there exists a closed set $F$ of $X$ such that (i) $f$ has no extension to $A \cup F$ and (ii) $f$ has an extension to $A \cup K$ for each proper closed subset $K$ of $F$. The subspace $F$ of $X$ is bicompact and satisfies $\dim F = n$, for a contradiction of (i) follows from Lemma 3.2.4 under the assumption that $\dim F \leqslant n-1$. Suppose that $F = F_1 \cup F_2$, where $F_1$ and $F_2$ are proper closed subsets of $F$ and $\dim F_1 \cap F_2 \leqslant n-2$. By (ii), $f$ has an extension $f_1$ to $A \cup F_1$ and an extension $f_2$ to $A \cup F_2$. It follows from Proposition 1.4 that each of $f_1$ and $f_2$ has an extension to $A \cup F$. Thus $f$ has an extension to $A \cup F$ which is a contradiction of (i). It follows from Lemma 1.6 that $F$ is an $n$-dimensional Cantor space.☐

## 2  Mappings and dimension

In this section we shall seek relations between the dimensions of the domain and range of a continuous surjection. Most of the results will be about closed mappings. Since the set of functions with domain $X$ and range $Y$ consists of a single element if $X$ is empty, and is empty if $Y$ is empty and $X$ is non-empty, we shall always assume that the domain and range of mappings under consideration are non-empty.

If $X$ and $Y$ are topological spaces and $f: X \to Y$ is a continuous surjection, we define
$$\dim f = \sup_{y \in Y} f^{-1}(y).$$

**2.1 Proposition.** *If $X$ is a $T_4$-space, $Y$ is a weakly paracompact space and $f: X \to Y$ is a continuous closed surjection, then*
$$\dim X \leqslant \operatorname{Ind} Y + \dim f.$$

*Proof.* We must prove that if $\operatorname{Ind} Y = m$ then $\dim X \leqslant m + \dim f$. We suppose first that $m > 0$ and that the result is true for mappings onto weakly paracompact spaces with large inductive dimension less than $m$. Let $n = m + \dim f$, let $A$ be a closed set of $X$ and let $\phi: A \to S^n$ be a continuous function.

Since $Y$ is a $T_1$-space, $f^{-1}(y)$ is a closed set of $X$ and by hypothesis $\dim f^{-1}(y) \leqslant n$ for each $y$ in $Y$. By Lemma 3.2.4 there exists an open set $U_y$ containing $A \cup f^{-1}(y)$ to which $\phi$ extends. Since $f$ is a closed mapping, there exists $V_y$ open in $Y$ such that $y \in V_y$ and $f^{-1}(V_y) \subset U_y$. Let $\{W_\lambda\}_{\lambda \in \Lambda}$ be a point-finite open refinement of the covering $\{V_y\}_{y \in Y}$ of $Y$. For each non-negative integer $k$, let $E_k$ be the set of points $y$ of $Y$ such that the set $\{\lambda \in \Lambda \mid y \in W_\lambda\}$ has at most $k$ members. Then by Lemma 3.5.9, $\{E_k\}_{k \geqslant 0}$ is a closed covering of $Y$. For each $k$, let

$$D_k = f^{-1}(E_k)$$

so that $\{D_k\}_{k \geqslant 0}$ is a closed covering of $X$. For $i = 0, 1, \ldots$ we shall construct open sets $G_i$ such that

(i) $\bigcup_{i \leqslant k} D_i \subset \bigcup_{i \leqslant k} G_i$ for all $k$;
(ii) $\dim \operatorname{bd}(G_i) \leqslant n - 1$;
(iii) $f(G_i)$ is open in $Y$ and $f^{-1} f(G_i) = G_i$;
(iv) $\phi$ has an extension to $A \cup \bar{G}_i$.

Since $D_0$ is empty, we can take $G_0$ to be the empty set. Suppose that $k$ is a positive integer and that for $i < k$, open sets $G_i$ have been constructed so that (i)–(iv) are satisfied. We shall construct $G_k$. Let $F = D_k \backslash \bigcup_{i < k} G_i$. Then by (iii), $F = f^{-1}\big(E_k \backslash \bigcup_{i < k} f(G_i)\big)$. Since

$$E_{k-1} \subset \bigcup_{i < k} f(G_i)$$

it follows from Lemma 3.5.9 that $E_k \backslash \bigcup_{i < k} f(G_i) = \bigcup_{\gamma \in \Gamma} C_\gamma$, where $\{C_\gamma\}_{\gamma \in \Gamma}$ is a discrete family of closed sets and each $C_\gamma$ is contained in some $W_\lambda$. If $B_\gamma = f^{-1}(C_\gamma)$, then $\{B_\gamma\}_{\gamma \in \Gamma}$ is a discrete closed covering of $F$ and each $B_\gamma$ is contained in $U_y$ for some $y$ in $Y$. Thus $\phi$ has an extension to $A \cup B_\gamma$ for each $\gamma$. Since $\{B_\gamma\}_{\gamma \in \Gamma}$ is discrete, it follows that $\phi$ has an extension to $A \cup F$. Hence there exists an open set $P$ such that $A \cup F \subset P$ and $\phi$ has an extension to $P$. Now $f(F)$ is closed in $Y$, $f^*(P)$ is open and $f(F) \subset f^*(P)$. It follows that there exists an open set $H_k$ of $Y$ such that

$$f(F) \subset H_k \subset \bar{H}_k \subset f^*(P) \quad \text{and} \quad \operatorname{Ind} \operatorname{bd}(H_k) \leqslant m - 1.$$

Let $G_k = f^{-1}(H_k)$. Since $F \subset G_k$, condition (i) is satisfied. Since $f\big(\operatorname{bd}(G_k)\big) \subset \operatorname{bd}(H_k)$ it follows from the induction hypothesis that $\dim \operatorname{bd}(G_k) \leqslant n - 1$. Conditions (iii) and (iv) are evidently satisfied. By induction we can find a countable open covering $\{G_i\}_{i \in \mathbb{N}}$ of $X$ such that $\dim \mathrm{bd}\,(G_i) \leqslant n-1$ and $\phi$ has an extension to $A \cup \bar{G}_i$ for each $i$. Thus by Proposition 1.5, $\phi$ has an extension to $X$. It follows from Theorem 3.2.2 that $\dim X \leqslant n$. In the case $m = 0$, the same argument with $n = \dim f$ and no induction hypothesis gives a countable covering $\{G_i\}_{i \in \mathbf{N}}$ of $X$ such that $G_i$ is an open-and-closed set for each $i$ and $\phi$ has an extension to $A \cup G_i$. Again $\dim X \leqslant n$. The formula is thus established by induction. $\square$

**2.2** *Remark.* In the case of a continuous closed surjection $f\colon X \to Y$ such that $\dim f = 0$, Proposition 6.4.11 gives the stronger result that $\dim X \leqslant \dim Y$ provided that $Y$ is paracompact.

If either $X$ or $Y$ is completely paracompact we can sharpen the formula given in Proposition 2.1, replacing $\mathrm{Ind}\, Y$ by $\mathrm{ind}\, Y$. Two lemmas are required.

**2.3** **Lemma.** *Let $f\colon X \to Y$ be a continuous closed surjection, where $Y$ is a completely paracompact space. If $\mathscr{U}$ is an open covering of $X$ with the property that for each point $y$ of $Y$ there exists a member $U_y$ of $\mathscr{U}$ such that $f^{-1}(y) \subset U_y$, then $\mathscr{U}$ has a weak refinement which is the union of countably many star-finite open coverings of $X$.*

*Proof.* Since $f$ is a continuous closed surjection, for each point $y$ of $Y$ there exists an open set $V_y$ such that $y \in V_y$ and $f^{-1}(V_y) \subset U_y$. Let $\mathscr{W}$ be a weak refinement of the covering $\{V_y\}_{y \in Y}$ of $Y$ such that $\mathscr{W}$ is the union of countably many star-finite open coverings of $Y$. Then $\{f^{-1}(W)\}_{W \in \mathscr{W}}$ is a weak refinement of $\mathscr{U}$ which is the union of countably many star-finite open coverings of $X$. $\square$

**2.4** **Lemma.** *Let $X$ be a $T_4$-space, let $n$ be a positive integer and let $f\colon X \to Y$ be a continuous closed surjection such that $\dim f \leqslant n$, which has the property that for each point $y$ of $Y$ and each open set $V$ of $X$ such that $f^{-1}(y) \subset V$ there exists an open set $U$ such that $f^{-1}(y) \subset U \subset V$ and $\dim \mathrm{bd}\,(U) \leqslant n-1$. If either $X$ or $Y$ is completely paracompact then $\dim X \leqslant n$.*

*Proof.* Let $A$ be a closed set of $X$ and let $\phi\colon A \to S^n$ be a continuous function. Since $Y$ is a $T_1$-space, $f^{-1}(y)$ is a closed set of $X$ if $y \in Y$ and by hypothesis $\dim f^{-1}(y) \leqslant n$. It follows from Lemma 3.2.4 that there exists an open set $V$ such that $A \cup f^{-1}(y) \subset V$ and $\phi$ extends to $V$. By hypothesis there exists an open set $U_y$ such that

$$f^{-1}(y) \subset U_y \subset \bar{U}_y \subset V$$

and $\dim \mathrm{bd}\,(U_y) \leqslant n-1$. If $X$ is a completely paracompact space, then the open covering $\{U_y\}_{y \in Y}$ of $X$ has a weak refinement which is the union of countably many star-finite open coverings and it follows from Lemma 2.3 that the same conclusion holds in the case that $Y$ is a completely paracompact space. It now follows from Lemma 4.5.6 that the covering $\{U_y\}_{y \in Y}$ of $X$ has an open refinement $\{G_{i\lambda}\}_{i \in \mathbf{N}, \lambda \in \Lambda}$ such that $\{G_{i\lambda}\}_{\lambda \in \Lambda}$ is a discrete family for each $i$, and for each $i$ and $\lambda$, $\mathrm{bd}\,(G_{i\lambda}) \subset \mathrm{bd}\,(U_y)$ for some $y$. Thus $\dim \mathrm{bd}\,(G_{i\lambda}) \leqslant n-1$ for each $i$ and $\lambda$. Now let $G_i = \bigcup_{\lambda \in \Lambda} G_{i\lambda}$. Then $\bar{G}_i = \bigcup_{\lambda \in \Lambda} \bar{G}_{i\lambda}$ and $\mathrm{bd}\,(G_i) = \bigcup_{\lambda \in \Lambda} \mathrm{bd}(G_{i\lambda})$. Since $\{\mathrm{bd}\,(G_{i\lambda})\}_{\lambda \in \Lambda}$ is a discrete family, it follows that $\dim \mathrm{bd}\,(G_i) \leqslant n-1$. And since $\phi$ extends to $A \cup \bar{G}_{i\lambda}$ for each $\lambda$ and $\{\bar{G}_{i\lambda}\}$ is a discrete family, it follows that $\phi$ extends to $A \cup \bar{G}_i$ for each $i$. It now follows from Proposition 1.5 that $\phi$ has an extension to $X$. Thus $\dim X \leqslant n$ by Theorem 3.2.2. $\square$

**2.5  Proposition.** *Let $X$ be a $T_4$-space and let $f : X \to Y$ be a continuous closed surjection. If either $X$ or $Y$ is a completely paracompact space, then*

$$\dim X \leqslant \mathrm{ind}\,Y + \dim f.$$

*Proof.* We must prove that if $\mathrm{ind}\,Y = m$ then $\dim X \leqslant m + \dim f$. We suppose first that $m > 0$ and that the result is true for range spaces of small inductive dimension less than $m$. Let $y$ be a point of $Y$ and let $V$ be an open set of $X$ such that $f^{-1}(y) \subset V$. Since $f$ is a closed mapping there exists an open set $W$ of $Y$ such that $y \in W$ and

$$f^{-1}(W) \subset V.$$

Since $\mathrm{ind}\,Y = m$ there exists an open set $G$ such that $y \in G \subset W$ and $\mathrm{ind}\,\mathrm{bd}\,(G) \leqslant m-1$. Restriction of $f$ gives a continuous closed surjection $f^{-1}\mathrm{bd}\,(G) \to \mathrm{bd}\,(G)$. From the induction hypothesis it follows that $\dim f^{-1}\mathrm{bd}\,(G) \leqslant \dim f + m - 1$. If $U = f^{-1}(G)$, then $U$ is an open set of $X$ such that $f^{-1}(y) \subset U \subset V$ and since $\mathrm{bd}\,(U) \subset f^{-1}\mathrm{bd}\,(G)$ it follows that

$$\dim \mathrm{bd}\,(U) \leqslant \dim f + m - 1.$$

It now follows from Lemma 2.4 that $\dim X \leqslant \dim f + m$. The same argument with no induction hypothesis shows that if $\mathrm{ind}\,Y = 0$ then $\dim X \leqslant \dim f$. The result is thus established by induction. $\square$

We return now to Proposition 2.1. For range spaces for which covering dimension coincides with large inductive dimension the formula can be put in a more symmetrical form. For example:

**2.6   Proposition.** *Let $X$ be a $T_4$-space, let $Y$ be a metrizable space and let $f: X \to Y$ be a continuous closed surjection. Then*

$$\dim X \leqslant \dim Y + \dim f. \;\square$$

We can interpret this as a result about 'dimension-lowering' mappings as follows:

**2.7   Proposition.** *Let $X$ be a $T_4$-space, let $Y$ be a metrizable space and let $f: X \to Y$ be a continuous closed surjection. If $\dim X - \dim Y = k$, where $k$ is a positive integer, then there exists some point $y$ of $Y$ such that $\dim f^{-1}(y) \geqslant k$.* $\square$

It follows from Proposition 2.6 that if $X$ is a $T_4$-space, $Y$ is a metrizable space and $f: X \to Y$ is a continuous closed surjection such that $f^{-1}(y)$ is finite if $y \in Y$, then $\dim X \leqslant \dim Y$ so that $f$ does not lower dimension. We shall consider next continuous closed surjections $f: X \to Y$ such that $X$ is metrizable and $f^{-1}(y)$ is finite for each $y$ in $Y$. It follows from Proposition 2.5.7 that $Y$ is metrizable. Thus $\dim X \leqslant \dim Y$ as we have just observed. Such surjections can raise dimension. For example, if $\phi$ is the usual identification mapping of the Cantor set $C$ onto the unit interval $I$, then $\phi^{-1}(y)$ contains at most 2 points for each point $y$ of $I$. We shall see that if the number of elements in $f^{-1}(y)$ is bounded for all points $y$ of the range of $f$, then there is a bound on the amount by which $f$ raises dimension.

One preliminary result is needed. For its proof we note that if $X$ is a metrizable space and $f: X \to Y$ is a continuous closed surjection, then the subset $Z$ of $Y$ consisting of those points $y$ such that $f^{-1}(y)$ contains more than one point is an $F_\sigma$-set. For let $\{\mathscr{U}_i\}_{i \in \mathbb{N}}$ be a sequence of open coverings of $X$ such that for each open set $V$ of $X$ and each point $x$ of $V$ there exists some $i$ such that $\mathrm{St}\,(x, \mathscr{U}_i) \subset V$. For each positive integer $i$ let

$$F_i = Y \setminus \bigcup_{U \in \mathscr{U}_i} f^*(U).$$

Since $f$ is a closed mapping, $F_i$ is a closed set for each $i$. If $y \in F_i$, then $f^{-1}(y)$ is not contained in any member of $\mathscr{U}_i$. Since $\mathscr{U}_i$ is a covering of $X$ it follows that $f^{-1}(y)$ contains more than one point. Thus $F_i \subset Z$ for each $i$. If $y \in Z$ and $x \in f^{-1}(y)$, then there exists some $i$ such that $\mathrm{St}\,(x, \mathscr{U}_i)$ does not contain $f^{-1}(y)$. Then $y \in F_i$. Thus $Z = \bigcup_{i \in \mathbb{N}} F_i$.

**2.8  Lemma.** *Let $X$ and $Y$ be metrizable spaces, let $f\colon X \to Y$ be a continuous surjection such that $\dim f = 0$, and let $Z$ be the subset of $Y$ consisting of those points $y$ such that $f^{-1}(y)$ contains more than one point. If $\dim X \leqslant n$ and $\dim Z \leqslant n-1$, where $n$ is a non-negative integer, then $\dim Y \leqslant n$.*

*Proof.* When $n = 0$, $f$ is a continuous closed bijection so that $Y$ is homeomorphic with $X$. Let $n$ be a positive integer and let us assume the result true for mappings with domain of dimension less than $n$. The continuous surjection $f^{-1}(Z) \to Z$ given by restriction of $f$ is closed. It follows from Proposition 2.6 that $\dim f^{-1}(Z) \leqslant \dim Z \leqslant n-1$. Since $Z$ is an $F_\sigma$-set, $f^{-1}(Z)$ is an $F_\sigma$-set of $X$. It follows from Corollary 7.1.5 that there exists an $F_\sigma$-set $K$ of $X$ which is disjoint from $f^{-1}(Z)$ and satisfies

$$\dim K \leqslant 0 \quad \text{and} \quad \dim (X\backslash K) \leqslant n-1.$$

It follows from Theorem 7.1.1 that there exists a subset $H$ of $Z$ such that

$$\dim H \leqslant 0 \quad \text{and} \quad \dim (Z\backslash H) \leqslant n-2,$$

and by Theorem 7.1.4 we can suppose that $H$ is an $F_\sigma$-set in $Z$ and hence in $Y$. Now let

$$X_1 = X\backslash\big(K \cup f^{-1}(H)\big) \quad \text{and} \quad Y_1 = Y\backslash\big(f(K) \cup H\big).$$

Clearly $f^{-1}(Y_1) \subset X_1$ so that $Y_1 \subset f(X_1)$. But $f(X_1) \subset Y_1$ since $f \,|\, K$ is injective. Hence $f(X_1) = Y_1$ and $X_1 = f^{-1}(Y_1)$. It follows that if $f_1\colon X_1 \to Y_1$ is given by restriction of $f$, then $f_1$ is a continuous closed surjection. Furthermore since $X_1 \subset X\backslash K$ it follows that $\dim X_1 \leqslant n-1$. If $Z_1$ is the set of points $y$ of $Y_1$ such that $f_1^{-1}(y)$ contains more than one point, then $Z_1 = Z \cap Y_1$. Thus $Z_1 \subset Z\backslash H$ so that $\dim Z_1 \leqslant n-2$. It follows from the induction hypothesis that $\dim Y_1 \leqslant n-1$. Since $f(K)$ is homeomorphic with $K$, we see that $\dim f(K) \leqslant 0$. Since $H$ and $f(K)$ are $F_\sigma$-sets, it follows that $\dim H \cup f(K) \leqslant 0$. Thus $\dim Y \leqslant n$ as required. The result now follows by induction. $\square$

**2.9  Proposition.** *Let $X$ and $Y$ be metrizable spaces, let $f\colon X \to Y$ be a continuous closed surjection such that $\dim f = 0$, and let $Z$ be the subset of $Y$ consisting of those points $y$ such that $f^{-1}(y)$ contains more than one point. If $\dim Y > \dim X$, then*

$$\dim Z \geqslant \dim Y - 1.$$

*Proof.* First we note that

$$\dim Y \leqslant \dim(Y \backslash Z) + \dim Z + 1.$$

Since $Y \backslash Z$ is homeomorphic with $f^{-1}(Y \backslash Z)$, it follows that

$$\dim Y \leqslant \dim X + \dim Z + 1.$$

Suppose that $\dim Y = \infty$. Since $\dim X$ is finite it follows that

$$\dim Z = \infty.$$

Now suppose that $\dim Y = n + 1$, where $n$ is a non-negative integer. Then $\dim X \leqslant n$ and we cannot have $\dim Z \leqslant n - 1$ for this would give a contradiction of the result of Lemma 2.8. Hence

$$\dim Z \geqslant n = \dim Y - 1. \square$$

We now come to the main result about closed mappings between metrizable spaces such that the inverse image of each point of the range is finite.

**2.10    Proposition.** *Let $X$ and $Y$ be metrizable spaces, let $f: X \to Y$ be a continuous closed surjection such that $f^{-1}(y)$ is finite for each point $y$ of $Y$, and for each positive integer $i$ let $Z_i$ be the subset of $Y$ consisting of those points $y$ such that $f^{-1}(y)$ consists of precisely $i$ points. If $Z_i$ is empty except for $m + 1$ values of $i$, then*

$$\dim X \leqslant \dim Y \leqslant \dim X + m.$$

*Proof.* It follows from Proposition 2.6 that $\dim X \leqslant \dim Y$. Let $k$ be the least integer such that $Z_k$ is non-empty. Let $\{\mathscr{F}_i\}_{i \in \mathbf{N}}$ be a sequence of locally finite closed coverings of $X$ such that $\mathscr{F}_{i+1}$ is a refinement of $\mathscr{F}_i$ for each $i$, with the property that for each point $x$ and each open set $U$ such that $x \in U$ there exists some $i$ such that $\mathrm{St}(x, \mathscr{F}_i) \subset U$. For each $i$, let $\mathscr{F}_i = \{F_\lambda\}_{\lambda \in \Lambda_i}$. We can suppose that the indexing sets $\Lambda_i$ are disjoint. For each $i$, let $\mathscr{E}_i = \{f(F_\lambda)\}_{\lambda \in \Lambda_i}$. Since $f$ is a perfect mapping, $\mathscr{E}_i$ is a locally finite closed covering of $Y$ for each $i$. Let $\Gamma_i$ be the set of $k$-tuples $\gamma$ of elements of $\Lambda_i$ such that $\{F_\lambda\}_{\lambda \in \gamma}$ is a disjoint family and $\bigcap_{\gamma \in \lambda} f(F_\lambda) \neq \varnothing$. For each $\gamma$ in $\Gamma_i$ let

$$D_\gamma = \bigcap_{\lambda \in \gamma} f(F_\lambda).$$

Since $\mathscr{E}_i$ is a locally finite closed covering of $Y$ it follows that $\{D_\gamma\}_{\gamma \in \Gamma_i}$ is a locally finite family of closed sets of $Y$. Let $\Gamma = \bigcup_{i \in \mathbf{N}} \Gamma_i$ and let $\mathscr{D} = \{D_\gamma\}_{\gamma \in \Gamma}$. Then $\mathscr{D}$ is a $\sigma$-locally finite family of closed sets of $Y$

and a covering of $Y$. For if $y \in Y$, then there exist distinct points $x_1, \ldots, x_k$ of $X$ such that $\{x_1, \ldots, x_k\} \subset f^{-1}(y)$, and there exist disjoint open sets $U_1, \ldots, U_k$ such that $x_j \in U_j$ for each $j$. It follows that there exists an integer $i$ and elements $\lambda_1, \ldots, \lambda_k$ of $\Lambda_i$ such that

$$x_j \in F_{\lambda_j} \subset U_j \quad \text{for} \quad j = 1, \ldots, k.$$

If $\gamma = (\lambda_1, \ldots, \lambda_k)$, then $\gamma \in \Gamma_i$ and $y \in D_\gamma$. Thus $\mathscr{D}$ is a covering of $Y$.

Now suppose that $m = 0$ so that $f^{-1}(y)$ consists of precisely $k$ points for every $y$. If $\gamma \in \Gamma$ and $y \in D_\gamma$, it follows that $f^{-1}(y) \cap F_\lambda$ is a one-point set for each $\lambda$ in $\gamma$. Thus if $\lambda \in \gamma$, then the mapping

$$F_\lambda \cap f^{-1}(D_\gamma) \to D_\gamma$$

given by restriction of $f$ is a continuous closed bijection and thus a homeomorphism. Since $\dim F_\lambda \cap f^{-1}(D_\gamma) \leqslant \dim X$, it follows that

$$\dim D_\gamma \leqslant \dim X.$$

Thus $\{D_\gamma\}_{\gamma \in \Gamma}$ is a $\sigma$-locally finite closed covering of $Y$ such that $\dim D_\gamma \leqslant \dim X$ for each $\gamma$, and it follows from Proposition 3.5.4 that $\dim Y \leqslant \dim X$. Hence if $m = 0$, $\dim Y = \dim X$ and the theorem is proved.

Now suppose that $m > 0$ and that the result is true for continuous closed surjections such that the cardinality of the inverse image of each point of the range is finite and takes one of at most $m$ values. If $\dim Y = \dim X$ there is nothing to prove. Thus let us suppose that $\dim Y > \dim X$. Since $\{D_\gamma\}_{\gamma \in \Gamma}$ is a $\sigma$-locally finite closed covering of $Y$, there exists $\gamma_0$ in $\Gamma$ such that

$$\dim D_{\gamma_0} = \dim Y.$$

Let $\lambda_0$ be an element of $\gamma_0$, let $X_1 = F_{\lambda_0} \cap f^{-1}(D_{\gamma_0})$, let $Y_1 = D_{\gamma_0}$ and let $f_1 : X_1 \to Y_1$ be given by restriction of $f$. Then $f_1$ is a continuous closed surjection, and $\dim Y_1 = \dim Y > \dim X \geqslant \dim X_1$. If $Z$ is the subset of $Y_1$ consisting of those points $y$ such that $f_1^{-1}(y)$ contains more than one point, then by Proposition 2.9

$$\dim Z \geqslant \dim Y_1 - 1 = \dim Y - 1.$$

But if $Y_2$ is the subset of $Y$ consisting of those points $y$ such that $f^{-1}(y)$ contains more than $k$ points, then $Z$ is contained in $Y_2$. It follows that $\dim Y_2 \geqslant \dim Z$ so that

$$\dim Y_2 \geqslant \dim Y - 1.$$

But $f_2 : f^{-1}(Y_2) \to Y_2$ given by restriction of $f$ is a continuous closed surjection such that $|f_2^{-1}(y)|$ is finite for each $y$ in $Y$ and takes one of at most $m$ distinct values. Thus by the induction hypothesis

$$\dim Y_2 \leqslant \dim f^{-1}(Y_2) + m - 1.$$

Since $\dim f^{-1}(Y_2) \leqslant \dim X$, it follows that

$$\dim Y \leqslant \dim X + m$$

as required. The proof is now completed by induction. $\square$

Proposition 2.10 can be restated as a result about dimension-raising mappings.

**2.11   Proposition.** *Let $X$ and $Y$ be metrizable spaces and let $f: X \to Y$ be a continuous closed surjection. If $\dim Y - \dim X = k$, where $k$ is a positive integer, then there exists some point $y$ of $Y$ such that $f^{-1}(y)$ contains at least $k + 1$ points.*

*Proof.* Suppose that $f^{-1}(y)$ contains at most $k$ points if $y \in Y$. Proposition 2.10 can be applied and $Z_i = \varnothing$ if $i \geqslant k + 1$. It follows that $\dim Y \leqslant \dim X + k - 1$ which is a contradiction. $\square$

Let $f: X \to Y$ be a continuous closed surjection and let $Z$ be the set of points $y$ in $Y$ such that $f^{-1}(y)$ contains more than one point. In Proposition 2.9 we obtained a lower bound for the dimension of $Z$ in the case that $X$ and $Y$ are metrizable and $\dim f = 0$. We can also obtain a lower bound for the dimension of $Z$ under less restrictive hypotheses on $f$. Consideration of the usual identification mapping of the Cantor set onto the unit interval shows that the inequality we are about to obtain cannot be strengthened.

**2.12   Proposition.** *Let $X$ and $Y$ be finite-dimensional normal spaces such that $\dim X < \dim Y$ and let $f: X \to Y$ be a continuous closed surjection. If $Z$ is the subset of $Y$ consisting of those points $y$ such that $f^{-1}(y)$ contains more than one point, then*

$$\dim Z \geqslant \dim Y - \dim X - 1.$$

*Proof.* Let us suppose that $\dim X = m$ and $\dim Y = n$, where

$$0 \leqslant m < n.$$

Since $\dim Y = n$ there exists a finite open covering $\mathscr{U} = \{U_1, \ldots, U_k\}$ of $Y$, such that no open refinement of $\mathscr{U}$ has order less than $n$. Since $\{f^{-1}(U_1), \ldots, f^{-1}(U_k)\}$ is an open covering of $X$ and $\dim X = m$, there exists an open covering $\{V_1, \ldots, V_k\}$ of $X$ of order not exceeding $m$ such that $V_i \subset f^{-1}(U_i)$ for each $i$. Then $f^*(V_i) \subset U_i$. Furthermore if $\mu$ is a subset of $\{1, \ldots, k\}$ and $y \in \bigcap_{i \in \mu} f^*(V_i)$, then

$$f^{-1}(y) \subset f^{-1}\Big(\bigcap_{i \in \mu} f^*(V_i)\Big) = \bigcap_{i \in \mu} f^{-1}\big(f^*(V_i)\big) \subset \bigcap_{i \in \mu} V_i.$$

Thus the order of the family $\{f^*(V_i), \ldots, f^*(V_k)\}$ of open sets of $Y$ does not exceed $m$. Let

$$B = Y \backslash \bigcup_{i=1}^{k} f^*(V_i).$$

Then $B$ is a closed set of $Y$, and if $y \in B$ then $f^{-1}(y)$ is not contained in any $V_j$ so that $f^{-1}(y)$ contains more than one point. Thus $B \subset Z$. If $\dim Z \leqslant n - m - 2$, then $\dim B \leqslant n - m - 2$. Let $\{D_1, \ldots, D_k\}$ be a closed covering of $B$ of order not exceeding $n - m - 2$ such that $D_i \subset U_i \cap B$ for each $i$. There exist open sets $W_1, \ldots, W_k$ of $Y$ such that $D_i \subset W_i \subset U_i$ for each $i$ and the families $\{D_1, \ldots, D_k\}$ and $\{W_1, \ldots, W_k\}$ are similar. If $\mathscr{V} = \{f^*(V_1), \ldots, f^*(V_k), W_1, \ldots, W_k\}$, then $\mathscr{V}$ is an open covering of $Y$ and a refinement of $\mathscr{U}$. Since the order of $\{f^*(V_1), \ldots, f^*(V_k)\}$ does not exceed $m$ and the order of $\{W_1, \ldots, W_k\}$ does not exceed $n - m - 2$, it follows that the order of $\mathscr{V}$ does not exceed $n - 1$. This is a contradiction of the definition of $\mathscr{U}$. Hence $\dim Z \geqslant n - m - 1$ as required. $\square$

For non-metrizable spaces $X$ and $Y$, there are some results about continuous surjections $f : X \to Y$ such that $f^{-1}(y)$ is finite for each point $y$ of $Y$.

**2.13  Proposition.** *If $X$ is a normal space and $f : X \to Y$ is a continuous closed surjection such that $f^{-1}(y)$ contains at most $k + 1$ points for each point $y$ of $Y$, then $Y$ is a normal space and*

$$\dim Y \leqslant \operatorname{Ind} X + k.$$

*Proof.* It follows from Proposition 1.3.10 that $Y$ is a normal space. The proof of the inequality is by induction over $\operatorname{Ind} X$. If $\operatorname{Ind} X = 0$, then the result follows from Proposition 6.3.11. Now suppose that $\operatorname{Ind} X = m$, where $m$ is a positive integer and that the result is true for mappings with domain of large inductive dimension less than $m$. We shall prove by induction over $k$ that if $f : X \to Y$ is a continuous closed surjection such that $f^{-1}(y)$ contains at most $k + 1$ points for each point $y$ of $Y$, then $\dim Y \leqslant m + k$. If $k = 0$, then $f$ is a homeomorphism so that $\dim Y = \dim X \leqslant \operatorname{Ind} X = m$ as required. Now suppose that $k$ is a positive integer and that the result is true for surjections such that the inverse image of each point of the range consists of at most $k$ points. Let $f:X \to Y$ be a continuous closed surjection such that $f^{-1}(y)$ contains at most $k+1$ points if $y \in Y$. If $E$ is a closed set of $X$ and $G$ is an open set such that $E \subset G$, then $f^{-1}(E)$ is a closed set of $X$, and $f^{-1}(G)$ is an open set which contains $f^{-1}(E)$. Since $\operatorname{Ind} X = m$, there exists an open set $H$ such that

$$f^{-1}(E) \subset H \subset f^{-1}(G) \quad \text{and} \quad \operatorname{Ind} \operatorname{bd}(H) \leqslant m-1.$$

If $V = f^*(H)$, then $V$ is an open set such that $E \subset V \subset G$, and since $\bar{V} \subset f(\bar{H})$ it follows that $\operatorname{bd}(V) \subset K = f(\bar{H}) \backslash V$. Let

$$B = f\big(\operatorname{bd}(H)\big) \cap K$$

and let $A = \operatorname{bd}(H) \cap f^{-1}(K)$. If $g:A \to B$ is given by restriction of $f$, then $g$ is a continuous closed surjection and $g^{-1}(y)$ contains at most $k+1$ points for each point $y$ of $B$. Since $A$ is a closed subset of $\operatorname{bd}(H)$ it follows that $\operatorname{Ind} A \leqslant m-1$ so that by the first induction hypothesis $\dim B \leqslant m+k-1$. Now let $D$ be a closed set of $K$ which is disjoint from $B$. Then $D$ is closed in $Y$ and $D \subset f(H) \backslash V = f(H) \backslash f^*(H)$. Thus if $C = f^{-1}(D) \cap (X \backslash H)$ then by Lemma 6.3.9, the function $h:C \to D$ given by restriction of $f$ is a continuous closed surjection such that $h^{-1}(y)$ contains at most $k$ points if $y \in D$. Since $C$ is a closed set of $X$, $\operatorname{Ind} C \leqslant m$ so that by the second induction hypothesis,

$$\dim D \leqslant m+k-1.$$

It now follows from Proposition 3.5.6 that $\dim K \leqslant m+k-1$. Hence $\dim \operatorname{bd}(V) \leqslant m+k-1$. Thus by Lemma 4.2.9, $\dim Y \leqslant m+k$ as required. The proof is now completed by induction. $\square$

If $Y$ is a totally normal space then, using Proposition 4.4.11 and the definition of large inductive dimension instead of Proposition 3.5.6 and Lemma 4.2.9, we can adapt the above proof and obtain the following result.

**2.14 Proposition.** *Let $X$ be a normal space, let $Y$ be a totally normal space and let $f:X \to Y$ be a continuous closed surjection such that $f^{-1}(y)$ contains at most $k+1$ points for each point $y$ of $Y$. Then*

$$\operatorname{Ind} Y \leqslant \operatorname{Ind} X + k. \square$$

A continuous closed surjection $f:X \to Y$ such that $f^{-1}(y)$ is finite for each $y$ in $Y$ can raise dimension even if $X$ and $Y$ are compact metrizable spaces. We shall see that open surjections with this property preserve dimension for a wide class of domain and range spaces.

**2.15**   **Lemma.** *If $X$ and $Y$ are $T_3$-spaces and $f: X \to Y$ is a continuous open surjection such that $|f^{-1}(y)|$ is finite and constant for all points $y$ of $Y$, then*

$$\operatorname{loc dim} X = \operatorname{loc dim} Y \quad and \quad \operatorname{loc Ind} X = \operatorname{loc Ind} Y.$$

*Proof.* If $f$ is a bijection then the statement is obviously true. Suppose that $f^{-1}(y)$ contains precisely $k + 1$ points for each $y$ in $Y$, where $k$ is a positive integer. We shall show that $f$ is a local homeomorphism. Let $x$ be a point of $X$ and suppose that $f^{-1}f(x) = \{x, x_1, \ldots, x_k\}$. Since $X$ is a Hausdorff space, there exist disjoint open sets $V, V_1, \ldots, V_k$ of $X$ such that $x \in V$ and $x_j \in V_j$ for $j = 1, \ldots, k$. Then

$$G = f(V) \cap \bigcap_{j=1}^{k} f(V_j)$$

is an open neighbourhood of $f(x)$ and if $y \in G$ then $f^{-1}(y) \cap V$ is a one-point set. Thus the function $f^{-1}(G) \cap V \to G$ given by restriction of $f$ is a continuous open bijection and so is a homeomorphism. Thus $f$ is a local homeomorphism. The result now follows from Proposition 5.1.7.☐

**2.16**   **Proposition.** *If $X$ and $Y$ are weakly paracompact $T_4$-spaces and $f: X \to Y$ is a continuous open surjection such that $f^{-1}(y)$ is finite for each point $y$ of $Y$, then*

$$\dim X = \dim Y.$$

*If additionally $X$ and $Y$ are totally normal spaces, then*

$$\operatorname{Ind} X = \operatorname{Ind} Y.$$

*Proof.* Let $B_k$ be the set of points $y$ of $Y$ such that $f^{-1}(y)$ contains at most $k$ points. If $y \in Y \backslash B_k$, then $f^{-1}(y) = \{x_1, \ldots, x_s\}$, where $s > k$. There exists a disjoint family $\{V_1, \ldots, V_s\}$ of open sets of $X$ such that $x_j \in V_j$ for each $j$. Since $f$ is an open mapping, $W = \bigcap_{j=1}^{s} f(V_j)$ is an open neighbourhood of $y$ in $Y$. If $z \in W$, then $f^{-1}(z)$ contains at least $s$ points. Hence $W \cap B_k = \varnothing$. Thus $B_k$ is a closed set for each $k$ and $\{B_k\}$ is a countable closed covering of $Y$. Let $A_k = f^{-1}(B_k)$ so that $\{A_k\}$ is a countable closed covering of $X$. The mapping $A_1 \to B_1$, given by restriction of $f$, is a continuous open bijection. Thus $A_1$ and $B_1$ are homeomorphic and hence $\dim A_1 = \dim B_1$. Suppose that $\dim A_k = \dim B_k$ and let $X_k = A_{k+1} \backslash A_k$ and $Y_k = B_{k+1} \backslash B_k$. Since $X_k = f^{-1}(Y_k)$, the mapping $g: X_k \to Y_k$, given by restriction of $f$, is open. Furthermore $g^{-1}(y)$ contains precisely $k+1$ points if $y \in Y_k$. It follows from Lemma 2.15 that

$$\text{loc dim } X_k = \text{loc dim } Y_k.$$

Since $Y_k$ is open in $B_{k+1}$, it follows from Proposition 5.2.2 that

$$\text{loc dim } X_k \leqslant \text{loc dim } B_{k+1}.$$

But since $B_{k+1}$ is a weakly paracompact normal space, it follows from Proposition 5.3.4 that $\text{loc dim } B_{k+1} = \dim B_{k+1}$. If $F$ is a closed set of $A_{k+1}$ disjoint from $A_k$ then $\text{loc dim } F \leqslant \text{loc dim } X_k$ by Proposition 5.2.1. But $F$ is a weakly paracompact normal space so that

$$\dim F = \text{loc dim } F.$$

Thus if $F$ is a closed set of $A_{k+1}$ which is disjoint from $A_k$, then $\dim F \leqslant \dim B_{k+1}$. Since $\dim A_k = \dim B_k \leqslant \dim B_{k+1}$ it follows from Proposition 3.5.6 that $\dim A_{k+1} \leqslant \dim B_{k+1}$. The same argument applied to closed sets of $B_{k+1}$ disjoint from $B_k$ shows that

$$\dim B_{k+1} \leqslant \dim A_{k+1}$$

and hence $\dim A_{k+1} = \dim B_{k+1}$. Thus $\dim A_k = \dim B_k$ for all $k$ by induction. It follows from the countable sum theorem that

$$\dim X = \sup \dim A_k = \sup \dim B_k = \dim Y.$$

The same argument, employing Proposition 4.4.11 instead of Proposition 3.5.6, shows that $\text{Ind } X = \text{Ind } Y$. □

This section concludes with a famous example which is used here to show that a continuous open surjection $f$ with $\dim f = 0$ can lower dimension.

**2.17** *Example.* There exists a totally disconnected separable metric space $X$ such that $\dim X = 1$, and a continuous open surjection $f$ of $X$ onto the Cantor set such that $\dim f = 0$.

Let $L$ be the subset of the Cantor set $C$ consisting of the points $\Sigma_{n=1}^{\infty} a_n/3^n$ such that $a_n = 2$ for all sufficiently large $n$. Then $L$ consists of 1 and the left-hand end-points of intervals deleted from $I$ in forming $C$. Similarly if $R$ is the subset of $C$ consisting of the points $\Sigma_{n=1}^{\infty} a_n/3^n$ such that $a_n = 0$ for all sufficiently large $n$, then $R$ consists of 0 and the right-hand end-points of intervals deleted from $I$ in forming $C$. If $D = L \cup R$, then $D$ is a countable dense subset of $C$.

Let $a$ be the point $(\frac{1}{2}, \frac{1}{2})$ in $\mathbf{R}^2$ and let $h: \mathbf{R}^2 \to \mathbf{R}$ be projection given by $h(x, y) = y$. If $t \in D$, let $Y_t$ be the subset of $\mathbf{R}^2$ consisting of those points $z$ on the line segment joining $(t, 0)$ to $a$ such that $h(z)$ is rational, and if $t \in C \backslash D$ let $Y_t$ be the set of points $z$ of the line segment joining $(t, 0)$ to $a$ such that $h(z)$ is irrational. Now let

$$Y = \bigcup_{t \in C} Y_t$$

and let $Y$ have its topology as a subspace of the plane. Then $Y$ is a connected space. For suppose that $Y$ is not connected and let $U$ be a proper open-and-closed subset of $Y$ such that $a \in U$. For each point $t$ of $C \backslash D$ and each real number $r$ such that $0 < r < \frac{1}{2}$ let

$$Y_t(r) = \{z \in Y_t \mid h(z) \geqslant r\}$$

and let $\theta(t)$ be the greatest lower bound of the non-empty set of real numbers $r$ such that $Y_t(r) \subset U$. If $\theta(t)$ were irrational then, since $U$ is closed, the point $z$ of $Y_t$ such $h(z) = \theta(t)$ would belong to $U$. But since $U$ is open, this contradicts the definition of $\theta(t)$. Thus $\theta(t)$ is rational if $t \in C \backslash D$. Furthermore since $U$ is a proper closed subset of $Y$, the set

$$\{t \in C \backslash D \mid \theta(t) > 0\}$$

is non-empty and open in $C \backslash D$. It follows that there exists an open-and-closed interval $K$ in $C$ such that $\theta(t) > 0$ if $t \in K \backslash D$. Let the rationals in $(0, \frac{1}{2})$ be enumerated as $r_1, r_2, \ldots$ and for each positive integer $i$ let $E_i$ be the closure in $C$ of the set

$$\{t \in K \backslash D \mid \theta(t) = r_i\}.$$

The sets $E_i$, $i \in \mathbb{N}$, together with the points of $D \cap K$ form a countable closed covering of the complete metric space $K$. It follows from Corollary 2.4.6 that there exists an integer $k$ such that $E_k$ contains an open interval $J$ of $C$. Let $t_0 \in D \cap J$ and let $z_0$ be the point of $Y_{t_0}$ such that $h(z_0) = r_k$. Since $r_k > 0$, every neighbourhood of $z_0$ in $Y$ contains a point of $U$ and a point of $Y \backslash U$. Since $U$ is an open-and-closed set, this is absurd. Thus $Y$ is a connected space. It follows from Proposition 3.1.3 that $\dim Y > 0$.

Now let $X = Y \backslash \{a\}$ and for each $t$ in $C$ let $X_t = X \cap Y_t$. Then $X$ is a totally disconnected space. But $\dim X = \dim Y > 0$ by Proposition 3.5.7. Let $d_1 \in R$, $d_2 \in L$, where $d_1 < d_2$, and let $r_1, r_2$ be rationals such that $r_1 < r_2 < \frac{1}{2}$. Let us define $V(d_1, d_2; r_1, r_2)$ to be the subset of $X$ consisting of all points $z$ such that $z \in X_t$ for some $t$ satisfying

$$d_1 \leqslant t \leqslant d_2 \quad \text{and} \quad r_1 < h(z) < r_2.$$

The collection of all subsets of $X$ of this form is a base for the topology of $X$. Since the boundary of $V(d_1, d_2; r_1, r_2)$ consists of countably many points, it follows that $\operatorname{ind} X \leqslant 1$. Thus since $X$ is a separable metric space it follows that $\dim X = \operatorname{ind} X = 1$.

Finally consider the continuous surjection $f \colon X \to C$ given by putting $f(z) = t$ if $z \in X_t$. Since $f\big(V(d_1, d_2; r_1, r_2)\big)$ is the open-and-closed interval $[d_1, d_2]$ in $C$, it follows that $f$ is a continuous open surjection, and $\dim f = 0$ since $f^{-1}(t) = X_t$ for each $t$ in $C$.

## 3 The product theorem for covering dimension

In this section we shall find conditions under which the dimension of the topological product $X \times Y$ of spaces $X$ and $Y$ satisfies the inequality

$$\dim X \times Y \leqslant \dim X + \dim Y.$$

We shall call such results product theorems for dimension. Since product theorems for dimension are trivially true if precisely one factor is empty, and trivially false if both factors are empty, we shall assume throughout this section and the next that all spaces under consideration are non-empty. We already have Proposition 3.2.6 which is the product theorem for the covering dimension of bicompact spaces, and Proposition 4.5.5 which is the product theorem for the dimension of pseudo-metrizable spaces. Each of these theorems will be generalized.

One might expect to have product theorems asserting that

$$\dim X \times Y = \dim X + \dim Y$$

for the topological product $X \times Y$ of spaces $X$ and $Y$. This is not the case even for the product of separable metric spaces. The subspace $Q$ of Hilbert space consisting of the points, all of whose coordinates are rational, is homeomorphic with $Q \times Q$. But it was established in Example 4.1.8 that $\dim Q = 1$ so that

$$\dim Q \times Q < \dim Q + \dim Q.$$

We begin with generalizations of the product theorem for bicompact spaces.

**3.1  Lemma.** *If $X$ is a polyhedron and $Y$ is a bicompact space, then*

$$\dim X \times Y \leqslant \dim X + \dim Y.$$

*Proof.* The proof is by induction on $\dim X$. We note first that $X \times Y$ is normal by Proposition 2.1.13. If $\dim X = 0$ then $X$ is a discrete space so that $X \times Y$ has a discrete covering by the closed sets $\{x\} \times Y$ for $x$ in $X$. It follows from Proposition 3.5.1 that $\dim X \times Y = \dim Y$ as required. Now suppose that the result is true for $(n-1)$-dimensional polyhedra and that $X = |K|$, where $\dim K = n$. Let $J$ be the set of $n$-simplexes of $K$ and let $K_{n-1}$ be the $(n-1)$-skeleton of $K$. If $s \in J$, let

$$V_s = |s| \backslash |K_{n-1}| = |K| \backslash \bigcup \{|t| \mid t \in K, t \neq s\}.$$

Then $V_s$ is an open subset of $|K|$ and $V_s \subset |s|$. Since $|K|$ is perfectly normal, $V_s = \bigcup_{i \in \mathbf{N}} B_{si}$, where $B_{si}$ is closed for each $i$. If $s, t \in J$ and $s \neq t$, then $V_s$ and $V_t$ are disjoint. Thus for each $i$, the family $\{B_{si}\}_{s \in J}$ is discrete and hence $\{B_{si} \times Y\}_{s \in J}$ is a discrete family of closed sets of $X \times Y$. Since $B_{si} \subset |s|$, it follows that $B_{si}$ is compact and that

$$\dim B_{si} \leqslant n.$$

Hence by Proposition 3.2.6,

$$\dim (B_{si} \times Y) \leqslant n + \dim Y.$$

Thus if $A_i = \bigcup_{s \in J} (B_{si} \times Y)$ it follows from Proposition 3.5.1 that $\dim A_i \leqslant n + \dim Y$ for each $i$. Now

$$|K| \times Y = (|K_{n-1}| \times Y) \cup \bigcup_{i \in \mathbf{N}} A_i.$$

By the induction hypothesis, $\dim (|K_{n-1}| \times Y) \leqslant n - 1 + \dim Y$. Each $A_i$ is closed in $|K| \times Y$ and $|K_{n-1}| \times Y$ is closed in $|K| \times Y$, so that it follows from the countable sum theorem (Theorem 3.2.5) that

$$\dim (|K| \times Y) \leqslant n + \dim Y.$$

The proof is completed by induction. $\square$

**3.2 Proposition.** *If $X$ is a paracompact normal space and $Y$ is a bicompact space, then*

$$\dim X \times Y \leqslant \dim X + \dim Y.$$

*Proof.* Suppose that $\dim X \leqslant n$ and let $\mathscr{U}$ be an open covering of $X \times Y$. Then by Lemma 2.1.12 there exists a locally finite open covering $\mathscr{V} = \{V_\lambda\}_{\lambda \in \Lambda}$ of $X$, and for each $\lambda$ in $\Lambda$ there exists a finite open covering $\{G_{i\lambda} \mid i = 1, \ldots, n_\lambda\}$ of $Y$ such that $\{V_\lambda \times G_{i\lambda} \mid i = 1, \ldots, n_\lambda, \lambda \in \Lambda\}$ is a refinement of $\mathscr{U}$. By Proposition 3.7.4 there exist a polyhedron $Z$ such that $\dim Z \leqslant n$, a continuous function $\phi : X \to Z$ and an open covering $\{W_\lambda\}_{\lambda \in \Lambda}$ of $Z$ such that $\phi^{-1}(W_\lambda) \subset V_\lambda$ for each $\lambda$. Let

$$f : X \times Y \to Z \times Y$$

be the continuous function defined by

$$f(x, y) = \big(\phi(x), y\big) \quad \text{if} \quad (x, y) \in X \times Y.$$

By Lemma 3.1, $\dim Z \times Y \leqslant n + \dim Y$. Thus the open covering $\{W_\lambda \times G_{i\lambda} \mid i = 1, \dots, n_\lambda, \lambda \in \Lambda\}$ of $Z \times Y$ has an open refinement $\mathscr{H}$ of order not exceeding $n + \dim Y$. Hence $\{f^{-1}(H)\}_{H \in \mathscr{H}}$ is an open covering of $X \times Y$ of order not exceeding $n + \dim Y$ and a refinement of $\mathscr{U}$. Thus $\dim X \times Y \leqslant n + \dim Y$. $\square$

The following result also generalizes Proposition 3.2.6.

**3.3 Proposition.** *If $X$ and $Y$ are Hausdorff spaces and $X \times Y$ is strongly paracompact, then*

$$\dim X \times Y \leqslant \dim X + \dim Y.$$

*Proof.* We observe first that $X$ and $Y$ are $T_4$-spaces. Since $X \times Y$ is a strongly paracompact subspace of the bicompact space $\beta X \times \beta Y$, it follows from Proposition 3.6.7 that

$$\dim X \times Y \leqslant \dim \beta X \times \beta Y.$$

But by Proposition 3.2.6,

$$\dim \beta X \times \beta Y \leqslant \dim \beta X + \dim \beta Y,$$

and by Proposition 6.4.3 we have $\dim \beta X = \dim X$ and

$$\dim \beta Y = \dim Y$$

so that

$$\dim X \times Y \leqslant \dim X + \dim Y. \square$$

We turn next to a generalization of the product theorem for metrizable spaces.

**3.4 Lemma.** *Let $X$ be a Hausdorff space and let $Y$ be a metrizable space such that $X \times Y$ is countably paracompact and normal. If*

$$\{U_1, \dots, U_k\}$$

*is a finite open covering of $X \times Y$, then there exist open $F_\sigma$-sets $W_1, \dots, W_k$ in $\beta X \times Y$ such that $X \times Y \subset \bigcup_{i=1}^{k} W_i$ and $(X \times Y) \cap W_i \subset U_i$ for $i = 1, \dots, k$.*

*Proof.* Let $\{U_\lambda\}_{\lambda \in \Lambda}$ be a $\sigma$-locally finite base for the topology of $Y$ such that $U_\lambda$ is non-empty for every $\lambda$. For each $i = 1, \dots, k$ and each $\lambda$ in $\Lambda$ let $V_{\lambda i}$ be the union of the open sets $V$ of $X$ such that $V \times U_\lambda \subset U_i$.

Then $V_{\lambda i}$ is open in $X$ and $V_{\lambda i} \times U_\lambda \subset U_i$, and $\{V_{\lambda i} \times U_\lambda\}_{i=1,\ldots,k,\lambda \in \Lambda}$ is a $\sigma$-locally finite open covering of the countably paracompact normal space $X \times Y$. By Lemma 2.1.21 and Proposition 1.3.9 there exists a locally finite open covering $\{H_{\lambda i}\}_{i=1,\ldots,k,\lambda \in \Lambda}$ of $X \times Y$ such that

$$\bar{H}_{\lambda i} \subset V_{\lambda i} \times U_\lambda \quad \text{for} \quad i = 1, \ldots, k$$

and $\lambda$ in $\Lambda$. Now for $\lambda, \mu$ in $\Lambda$ and $i = 1, \ldots, k$ let $H_{\lambda \mu i}$ be the union of the open sets $H$ of $X$ such that $H \times U_\mu \subset H_{\lambda i}$. Then $H_{\lambda \mu i}$ is an open set of $X$ and $H_{\lambda \mu i} \times U_\mu \subset H_{\lambda i}$. Furthermore for fixed $\mu$, the family

$$\{H_{\lambda \mu i}\}_{i=1,\ldots,k,\lambda \in \Lambda}$$

is locally finite. For let $x$ be a point of $X$ and let $P$ be a neighbourhood of $x$ which meets $H_{\lambda \mu i}$ for infinitely many pairs $(\lambda, i)$. Let $y$ be a point of $U_\mu$ and let $Q$ be a neighbourhood of $y$. Then $P \times Q$ is a neighbourhood of $(x, y)$ which meets $H_{\lambda \mu i} \times U_\mu$ for infinitely many pairs $(\lambda, i)$. But $H_{\lambda \mu i} \times U_\mu \subset H_{\lambda i}$ and the family $\{H_{\lambda i}\}$ is locally finite. It follows that there exists some neighbourhood of $x$ which meets $H_{\lambda \mu i}$ for only finitely many pairs $(\lambda, i)$, and we have the asserted local finiteness. It follows that if $F_{\mu i} = \bigcup_{\lambda \in \Lambda} \bar{H}_{\lambda \mu i}$, then $F_{\mu i}$ is closed in $X$. Since

$$H_{\lambda \mu i} \times U_\mu \subset H_{\lambda i} \subset \bar{H}_{\lambda i} \subset V_{\lambda i} \times U_\lambda$$

it follows that $\bar{H}_{\lambda \mu i} \subset V_{\lambda i}$ and if $H_{\lambda \mu i}$ is non-empty then $U_\mu \subset U_\lambda$ so that $V_{\lambda i} \subset V_{\mu i}$. Thus for each $i = 1, \ldots, k$ and each $\mu, F_{\mu i} \subset V_{\mu i}$. Since $X$ is a $T_4$-space, the closure of $F_{\mu i}$ in $\beta X$ is contained in

$$\hat{V}_{\mu i} = \beta X \backslash \mathrm{cl}_{\beta X}(X \backslash V_{\mu i}).$$

It follows that there exists an open $F_\sigma$-set $W_{\mu i}$ of $\beta X$ such that

$$F_{\mu i} \subset X \cap W_{\mu i} \subset V_{\mu i}.$$

Then $\{W_{\mu i} \times U_\mu\}_{i=1,\ldots,k,\mu \in \Lambda}$ is a $\sigma$-locally finite family of open $F_\sigma$-sets of $\beta X \times Y$. It follows that if for each $i = 1, \ldots, k$ we define

$$W_i = \bigcup_{\mu \in \Lambda}(W_{\mu i} \times U_\mu)$$

then each $W_i$ is an open $F_\sigma$-set of $\beta X \times Y$. For each $i = 1, \ldots, k$ and each $\mu$

$$F_{\mu i} \times U_\mu \subset (X \times Y) \cap (W_{\mu i} \times U_\mu) \subset V_{\mu i} \times U_\mu.$$

Thus $(X \times Y) \cap W_i \subset U_i$ for each $i$. If $(x, y) \in X \times Y$, then $(x, y) \in H_{\lambda i}$ for some $i, \lambda$ so that there exists $\mu$ such that $(x, y) \in H_{\lambda \mu i} \times U_\mu$. Hence $(x, y) \in F_{\mu i} \times U_\mu \subset W_i$. Thus $X \times Y \subset \bigcup_{i=1}^k W_i. \square$

**3.5   Proposition.** *If $X$ is a Hausdorff space and $Y$ is a metrizable space such that $X \times Y$ is countably paracompact and normal, then*

$$\dim X \times Y \leqslant \dim X + \dim Y.$$

*Proof.* Let $\dim X + \dim Y = n$. Since $X$ is a $T_4$-space, $\dim X = \dim \beta X$ by Proposition 6.4.3. Thus it follows from Proposition 3.2 that

$$\dim \beta X \times Y \leqslant \dim \beta X + \dim Y = \dim X + \dim Y = n.$$

Now let $\{U_1, \ldots, U_k\}$ be an open covering of $X \times Y$. By Lemma 3.4 there exist open $F_\sigma$-sets $W_1, \ldots, W_k$ of $\beta X \times Y$ such that

$$(X \times Y) \cap W_i \subset U_i \quad \text{and} \quad X \times Y \subset \bigcup_{i=1}^k W_i.$$

If $W = \bigcup_{i=1}^k W_i$, then $W$ is an open $F_\sigma$-set of the normal space $\beta X \times Y$ so that $\dim W \leqslant n$ by Corollary 3.6.3. Since $\{W_1, \ldots, W_k\}$ is an open covering of $W$, there exists an open covering $\{H_1, \ldots, H_k\}$ of $W$ of order not exceeding $n$ such that $H_i \subset W_i$ for each $i$. If $G_i = H_i \cap (X \times Y)$, then $\{G_1, \ldots, G_k\}$ is an open covering of $X \times Y$ of order not exceeding $n$ such that $G_i \subset U_i$ for each $i$. Thus $\dim X \times Y \leqslant n.\square$

**3.6   Corollary.** *If $X$ is a perfectly normal Hausdorff space and $Y$ is a metrizable space, then*

$$\dim X \times Y \leqslant \dim X + \dim Y.$$

*Proof.* This follows from Proposition 3.5 since $X \times Y$ is a perfectly normal space by Proposition 7.5.11, and by Proposition 2.1.19 a perfectly normal space is countably paracompact and normal.$\square$

**3.7   Example.** The product theorem for covering dimension need not be satisfied for the product of a hereditarily paracompact Hausdorff space and a separable metric space.

Let $X$ be the space with underlying set the unit interval $I$ and topology consisting of all sets of the form $U \cup S$, where $U$ is open in the usual topology for $I$, and $S$ is a subset of the set of irrationals. It is easily verified that $X$ is a hereditarily paracompact Hausdorff space. Furthermore $\dim X = 0$. For let $\mathcal{U} = \{U_1, \ldots, U_k\}$ be an open covering of $X$. Let $Q$ be the set of rationals in $I$ and let $Q$ be enumerated as $r_1, r_2, \ldots$. For each $i$ there exists an open interval $(a_i, b_i)$, where $a_i$ and $b_i$ are irrational, such that $r_i \in (a_i, b_i)$ and $(a_i, b_i)$ is contained in some member of $\mathcal{U}$. For each $i$, let

$$V_i = (a_i, b_i) \setminus \bigcup_{t<i} [a_t, b_t].$$

Then $V_t$ is an open set which is contained in some member of $\mathscr{U}$, and $Q \subset \bigcup_{i \in \mathbf{N}} V_i$. Let

$$S = X \setminus \bigcup_{i \in \mathbf{N}} V_i.$$

Then $S$ is a set of irrationals. If $s \in S$, choose $j(s)$ such that $s \in U_{j(s)}$ and for $j = 1, \ldots, k$ let $S_j = \{s \in S \mid j(s) = j\}$. Then $S = \bigcup_{j=1}^{k} S_j$, the sets $S_j$ are disjoint and $S_j \subset U_j$ for each $j$. The sets $V_i$, $i \in \mathbf{N}$ together with the sets $S_1, \ldots, S_k$ form a disjoint open covering of $X$ which is a refinement of $\mathscr{U}$.

Now let $Y$ be the set of irrationals in the unit interval $I$ with its usual topology. Then $Y$ is a separable metric space and $\dim Y = 0$. But $X \times Y$ is not normal. For let $A = Q \times Y$ and let

$$B = \{(x, y) \in X \times Y \mid x = y\},$$

so that $B$ consists of all pairs $(x, x)$, where $x$ is an irrational in $I$. Then $A$ and $B$ are disjoint closed sets in $X \times Y$. Let $W$ be an open set of $X \times Y$ such that $B \subset W$, and for each positive integer $n$ let

$$W_n = \{x \in X \setminus Q \mid \{x\} \times B_{1/n}(x) \subset W\},$$

where $B_{1/n}(x)$ denotes the set of irrational numbers $y$ in $I$ such that $|x - y| < 1/n$. Then $X \setminus Q = \bigcup_{n \in \mathbf{N}} W_n$ so that $Q \supset \bigcap_{n \in \mathbf{N}} (X \setminus \overline{W}_n)$. Since $Q$ is not a $G_\delta$-set in the usual topology for $I$, it follows that $Q$ is not a $G_\delta$-set in $X$. Thus $Q \neq \bigcap_{n \in \mathbf{N}} (X \setminus \overline{W}_n)$, so that there exists $k$ such that $\overline{W}_k \cap Q \neq \varnothing$. Let $x \in \overline{W}_k \cap Q$ and choose $y$ in $Y$ such that $|x - y| < 1/2k$. Then $(x, y) \in A$ and we shall prove that $(x, y) \in \overline{W}$. Suppose that $(x, y) \in G \times H$, where $G$ and $H$ are open in $X$ and $Y$ respectively. Choose $x'$ in $G \cap W_k$ such that $|x - x'| < 1/2k$. Then $(x', y) \in G \times H$ and

$$|x' - y| \leqslant |x' - x| + |x - y| < 1/k,$$

so that $(x', y) \in W$ since $x' \in \overline{W}_k$. Thus $(x, y) \in \overline{W}$ as required. It follows that $A$ and $B$ cannot be included in disjoint open sets. Hence $X \times Y$ is not normal and it follows that $\dim X \times Y > 0$ so that

$$\dim X \times Y > \dim X + \dim Y.$$

## 4  The product theorem for large inductive dimension

In this section we shall find sufficient conditions to ensure that

$$\operatorname{Ind} X \times Y \leqslant \operatorname{Ind} X + \operatorname{Ind} Y,$$

where $X$ and $Y$ are topological spaces. Proposition 4.5.5 gives this result for pseudo-metrizable spaces $X$ and $Y$. We shall find analogues of the product theorems for covering dimension under the additional strong assumption that the product is totally normal.

**4.1  Proposition.** *If $X$ and $Y$ are topological spaces such that $X \times Y$ is a completely paracompact totally normal space, then*

$$\operatorname{Ind} X \times Y \leqslant \operatorname{Ind} X + \operatorname{Ind} Y.$$

*Proof.* The proof is by induction on $\operatorname{Ind} X + \operatorname{Ind} Y$. If

$$\operatorname{Ind} X + \operatorname{Ind} Y = 0,$$

then $\operatorname{Ind} X = 0$ and $\operatorname{Ind} Y = 0$. Since $X \times Y$ is totally normal, $X$ and $Y$ are regular spaces so that $\operatorname{ind} X = 0$ and $\operatorname{ind} Y = 0$. Thus $X$ and $Y$ have bases for their topologies which consist of open-and-closed sets. Thus $\operatorname{ind} X \times Y = 0$ so that $\operatorname{Ind} X \times Y = 0$ by Proposition 4.5.8 since $X \times Y$ is a completely paracompact totally normal space. Now suppose that the result is true for products in which the sum of the large inductive dimensions of the factors is less than $n$, and let $\operatorname{Ind} X = p$, $\operatorname{Ind} Y = q$, where $p$ and $q$ are non-negative integers such that $p + q = n$. Let $(x, y) \in X \times Y$ and let $G$ be an open neighbourhood of $(x, y)$ in $X \times Y$. Since $X$ and $Y$ are regular, there exist closed sets $A, B$ of $X$, $Y$ respectively and open sets $P, Q$ of $X$, $Y$ respectively such that

$$(x, y) \in A \times B \subset P \times Q \subset G.$$

It follows that there exist $U, V$ open in $X$, $Y$ respectively such that $(x, y) \in U \times V \subset G$, $\operatorname{Ind} \operatorname{bd}(U) \leqslant p - 1$ and $\operatorname{Ind} \operatorname{bd}(V) \leqslant q - 1$. Now

$$\operatorname{bd}(U \times V) = \big(\operatorname{bd}(U) \times \overline{V}\big) \cup \big(\overline{U} \times \operatorname{bd}(V)\big)$$

so that $\operatorname{Ind} \operatorname{bd}(U \times V) \leqslant n - 1$ by Theorem 4.4.8 since by the induction hypothesis $\operatorname{Ind} \operatorname{bd}(U) \times \overline{V} \leqslant n - 1$ and $\operatorname{Ind} \overline{U} \times \operatorname{bd}(V) \leqslant n - 1$. It follows that $\operatorname{ind} \operatorname{bd}(U \times V) \leqslant n - 1$. Hence $\operatorname{ind} X \times Y \leqslant n$. Since $X \times Y$ is a completely paracompact totally normal space it follows from Proposition 4.5.8 that $\operatorname{Ind} X \times Y \leqslant n$. The proof is completed by induction.□

The other product theorems for large inductive dimension obtained here will follow from a general result. For the formulation of this result a new concept must be introduced. Let us call a subset of a topological product $X \times Y$ of the form $A \times B$ an open rectangle if $A$ and $B$ are open in $X$ and $Y$ respectively and a closed rectangle if $A$ and $B$ are closed.

**4.2**   **Definition.** *If $X$ and $Y$ are topological spaces, then $X \times Y$ is an F-product if given a pair $K, L$ of disjoint closed sets of $X \times Y$ there exist a $\sigma$-locally finite covering $\{U_\lambda\}_{\lambda \in \Lambda}$ of $X \times Y$ by open rectangles which is a refinement of the covering $\{(X \times Y)\backslash K, (X \times Y)\backslash L\}$ of $X \times Y$, and a covering $\{F_\lambda\}_{\lambda \in \Lambda}$ of $X \times Y$ by closed rectangles such that $F_\lambda \subset U_\lambda$ for every $\lambda$.*

We observe that a closed rectangle in an $F$-product is an $F$-product. We shall prove a product theorem for the large inductive dimension of $F$-products. The remainder of the section will be devoted to examples of $F$-products.

**4.3**   **Proposition.** *If $X$ and $Y$ are topological spaces such that $X \times Y$ is a totally normal space and an F-product, then*

$$\operatorname{Ind} X \times Y \leqslant \operatorname{Ind} X + \operatorname{Ind} Y.$$

*Proof.* The proof is by induction on $\operatorname{Ind} X + \operatorname{Ind} Y$. Suppose that $\operatorname{Ind} X \leqslant n, \operatorname{Ind} Y \leqslant m$, where $n, m$ are non-negative integers, and that the result holds for products in which the sum of the large inductive dimensions of the two factors is less than $n + m$. Let $K, L$ be disjoint closed sets of $X \times Y$. Then since $X \times Y$ is an $F$-product, there exists a $\sigma$-locally finite open covering $\{U_\lambda\}_{\lambda \in \Lambda}$ of $X \times Y$ such that $U_\lambda = V_\lambda \times W_\lambda$ for each $\lambda$, where $V_\lambda$ and $W_\lambda$ are open in $X$ and $Y$ respectively and $U_\lambda$ is disjoint either from $K$ or from $L$ and there exists a closed covering $\{F_\lambda\}_{\lambda \in \Lambda}$ of $X \times Y$ such that $F_\lambda = D_\lambda \times E_\lambda$ for each $\lambda$, where $D_\lambda$ and $E_\lambda$ are closed in $X$ and $Y$ respectively and $F_\lambda \subset U_\lambda$. Thus $D_\lambda \subset V_\lambda$ and $E_\lambda \subset W_\lambda$ for each $\lambda$. Since $X \times Y$ is totally normal, $X$ and $Y$ are normal spaces. It follows that there exist open sets $G_\lambda$ and $H_\lambda$ of $X$ and $Y$ respectively such that

$$D_\lambda \subset G_\lambda \subset \bar{G}_\lambda \subset V_\lambda \quad \text{and} \quad \operatorname{Ind} \operatorname{bd}(G_\lambda) \leqslant n-1$$

and

$$E_\lambda \subset H_\lambda \subset \bar{H}_\lambda \subset W_\lambda \quad \text{and} \quad \operatorname{Ind} \operatorname{bd}(H_\lambda) \leqslant m-1.$$

For each $\lambda$,

$$F_\lambda \subset G_\lambda \times H_\lambda \subset \bar{G}_\lambda \times \bar{H}_\lambda \subset U_\lambda,$$

and

$$\operatorname{bd}(G_\lambda \times H_\lambda) = \left(\operatorname{bd}(G_\lambda) \times \bar{H}_\lambda\right) \cup \left(\bar{G}_\lambda \times \operatorname{bd}(H_\lambda)\right).$$

Since $\operatorname{bd}(G_\lambda) \times \bar{H}_\lambda$ and $\bar{G}_\lambda \times \operatorname{bd}(H_\lambda)$ are totally normal $F$-products, it follows from the induction hypothesis that

$$\operatorname{Ind} \operatorname{bd}(G_\lambda) \times \bar{H}_\lambda \leqslant n+m-1 \quad \text{and} \quad \operatorname{Ind} \bar{G}_\lambda \times \operatorname{bd}(H_\lambda) \leqslant n+m-1.$$

Thus by Theorem 4.4.8,

$$\text{Ind bd}\,(G_\lambda \times H_\lambda) \leqslant n+m-1.$$

Now let $\Lambda = \bigcup_{i \in \mathbf{N}} \Lambda_i$, where for each $j, \Lambda_j \subset \Lambda_{j+1}$ and the family $\{U_\lambda\}_{\lambda \in \Lambda_j}$ is locally finite. For each positive integer $j$ let

$$N_j = X \times Y \backslash \bigcup \{\bar{G}_\lambda \times \bar{H}_\lambda \mid \lambda \in \Lambda_j \quad \text{and} \quad (\bar{G}_\lambda \times \bar{H}_\lambda) \cap K = \varnothing\},$$

so that $N_j$ is an open set of $X \times Y$ and $N_j \supset N_{j+1} \supset K$. Then $\bigcap_{j \in \mathbf{N}} \bar{N}_j$ is disjoint from $L$. For if $z \in L$ then for some $j$ there exists $\lambda$ in $\Lambda_j$ such that $z \in F_\lambda \subset U_\lambda \subset (X \times Y) \backslash K$. Since $\bar{G}_\lambda \times \bar{H}_\lambda \subset U_\lambda$, it follows that $(\bar{G}_\lambda \times \bar{H}_\lambda) \cap K = \varnothing$. Thus $z \in G_\lambda \times H_\lambda$ and $(G_\lambda \times H_\lambda) \cap N_j = \varnothing$ so that $z \notin \bar{N}_j$. Also

$$\text{bd}\,(N_j) \subset A_j = \bigcup_{\lambda \in \Lambda_j} \text{bd}\,(G_\lambda \times H_\lambda).$$

For if $z \in \text{bd}\,(N_j)$, then since $z \notin N_j$ there exists $\lambda$ in $\Lambda_j$ such that $z \in \bar{G}_\lambda \times \bar{H}_\lambda$ and $(\bar{G}_\lambda \times \bar{H}_\lambda) \cap K = \varnothing$. Since $z \in N_j$ and $(G_\lambda \times H_\lambda) \cap N_j = \varnothing$ we see that $z \notin G_\lambda \times H_\lambda$. Hence $z \in \text{bd}\,(G_\lambda \times H_\lambda)$. Since $\text{bd}\,(G_\lambda \times H_\lambda) \subset U_\lambda$ and $\{U_\lambda\}_{\lambda \in \Lambda_j}$ is locally finite it follows that $\{\text{bd}\,(G_\lambda \times H_\lambda)\}_{\lambda \in \Lambda_j}$ is a locally finite closed covering of $A_j$. Since $\text{Ind bd}\,(G_\lambda \times H_\lambda) \leqslant n+m-1$ for each $\lambda$, it follows from Corollary 4.4.10 that $\text{Ind}\,A_j \leqslant n+m-1$ and hence

$$\text{Ind bd}\,(N_j) \leqslant n+m-1$$

for each positive integer $j$. Similarly for each positive integer $j$ let

$$M_j = X \times Y \backslash \bigcup \{\bar{G}_\lambda \times \bar{H}_\lambda \mid \lambda \in \Lambda_j \quad \text{and} \quad (\bar{G}_\lambda \times \bar{H}_\lambda) \cap L = \varnothing\}.$$

Then for each $j$, $M_j$ is an open set of $X \times Y$ such that $M_j \supset M_{j+1} \supset L$. Furthermore $\bigcap_{j \in \mathbf{N}} \bar{M}_j$ is disjoint from $K$ and $\text{Ind bd}\,(M_j) \leqslant n+m-1$ for each $j$.

Now let 
$$P = \bigcup_{j \in \mathbf{N}} N_j \cap \big((X \times Y) \backslash \bar{M}_j\big).$$

Then $P$ is an open set of $X \times Y$, and

$$K \subset P \subset (X \times Y) \backslash L.$$

For if $z \in K$, there exists $k$ such that $z \notin \bar{M}_k$. Since $z \in N_j$ for all $j$ we see that $z \in N_k \cap \big((X \times Y) \backslash \bar{M}_k\big)$ and hence $z \in P$. Also

$$L \subset \bigcap_{j \in \mathbf{N}} \bar{M}_j \subset (X \times Y) \backslash P$$

so that $P \subset (X \times Y) \backslash L$, and

$$\text{bd}\,(P) \subset Z = \bigcup_{j \in \mathbf{N}} \text{bd}\,(N_j) \cup \bigcup_{j \in \mathbf{N}} \text{bd}\,(M_j).$$

For suppose that $z \notin Z$. If $z \notin N_1$, then since $z \notin \mathrm{bd}\,(N_1)$ we see that $z \notin \overline{N}_1$. Since $P \subset N_1$, we see that $z \notin \overline{P}$. Thus $z \notin \mathrm{bd}\,(P)$. If $z \in N_1$, then either (i) there exists $j$ such that $z \in N_j$ and $z \notin N_{j+1}$, or (ii) $z \in \bigcap_{j \in \mathbf{N}} N_j$. In case (i), since $z \notin \mathrm{bd}\,(N_{j+1})$ it follows that $z \notin \overline{N}_{j+1}$ and we must consider two possibilities. If $z \notin M_j$, since $z \notin \mathrm{bd}\,(M_j)$ it follows that $z \notin \overline{M}_j$. Hence $z \in N_j \cap \big((X \times Y) \backslash \overline{M}_j\big) \subset P$ so that $z \notin \mathrm{bd}\,(P)$. If $z \in M_j$ then $M_j \cap \big((X \times Y) \backslash \overline{N}_{j+1}\big)$ is an open set containing $z$. If $i \leqslant j$ then $M_j$ is disjoint from $\big((X \times Y) \backslash \overline{M}_i\big)$ and if $i \geqslant j+1$ then $(X \times Y) \backslash \overline{N}_{j+1}$ is disjoint from $N_i$. It follows that

$$P \cap M_j \cap \big((X \times Y) \backslash \overline{N}_{j+1}\big) = \varnothing$$

so that $z \notin \overline{P}$ and hence $z \notin \mathrm{bd}\,(P)$. Finally, suppose that case (ii) arises so that $z \in \bigcap_{j \in \mathbf{N}} N_j$. There exists $j$ such that $z \in F_\lambda$ for some $\lambda$ in $\Lambda_j$. Since $z \in N_j$ and $z \in G_\lambda \times H_\lambda$ it follows that $(\overline{G}_\lambda \times \overline{H}_\lambda) \cap K \neq \varnothing$. Since $\overline{G}_\lambda \times \overline{H}_\lambda \subset U_\lambda$ we must have $(\overline{G}_\lambda \times \overline{H}_\lambda) \cap L = \varnothing$ so that $z \notin M_j$. Since $z \notin \mathrm{bd}\,(M_j)$, it follows that $z \notin \overline{M}_j$ so that $z \in N_j \cap \big((X \times Y) \backslash \overline{M}_j\big) \subset P$ and hence $z \notin \mathrm{bd}\,(P)$. This completes the proof that $\mathrm{bd}\,(P) \subset Z$. But $Z$ is a totally normal space and it follows from Theorem 4.4.8 that $\mathrm{Ind}\,Z \leqslant n+m-1$. Thus

$$\mathrm{Ind}\,\mathrm{bd}\,(P) \leqslant n+m-1,$$

from which it follows that $\mathrm{Ind}\,X \times Y \leqslant n+m$. If $\mathrm{Ind}\,X = 0$ and $\mathrm{Ind}\,Y = 0$, then a simpler version of the above argument with no induction hypothesis gives $\mathrm{Ind}\,X \times Y = 0$. The result follows by induction. $\square$

**4.4  Proposition.** *If $X$ is a perfectly normal space and $Y$ is a pseudo-metrizable space, then*

$$\mathrm{Ind}\,X \times Y \leqslant \mathrm{Ind}\,X + \mathrm{Ind}\,Y.$$

*Proof.* By Proposition 7.5.11, $X \times Y$ is a perfectly normal space. Thus the result will follow from Proposition 4.3 if we show that $X \times Y$ is an $F$-product. Let $K_1$ and $K_2$ be disjoint closed sets of $X \times Y$. Let $\{V_\lambda\}_{\lambda \in \Lambda}$ be a $\sigma$-locally finite base for the topology of $Y$, and for $i = 1, 2$ and $\lambda$ in $\Lambda$ let $G_{i\lambda}$ be the union of the open sets $G$ of $X$ such that $G \times V_\lambda \subset (X \times Y) \backslash K_i$. Then $G_{i\lambda}$ is open in $X$ and

$$(X \times Y) \backslash K_i = \bigcup_{\lambda \in \Lambda} (G_{i\lambda} \times V_\lambda).$$

Since $X$ is a perfectly normal space, for each $\lambda$ in $\Lambda$ and $i = 1, 2$, $G_{i\lambda} = \bigcup_{k \in \mathbf{N}} D_{i\lambda k}$, where $D_{i\lambda k}$ is a closed set of $X$ and $D_{i\lambda k} \subset D_{i\lambda, k+1}$ for each $k$. Since $Y$ is perfectly normal, $V_\lambda = \bigcup_{k \in \mathbf{N}} E_{\lambda k}$, where $E_{\lambda k}$ is a

closed set of $Y$ and $E_{\lambda k} \subset E_{\lambda, k+1}$ for each $k$. Let $\Omega = \{1, 2\} \times \Lambda \times \mathbf{N}$.
If $\omega = (i, \lambda, k) \in \Omega$, let

$$U_\omega = G_{i\lambda} \times V_\lambda \quad \text{and} \quad F_\omega = D_{i\lambda k} \times E_{\lambda k}.$$

Then $\{U_\omega\}_{\omega \in \Omega}$ is a $\sigma$-locally finite covering of $X \times Y$ by open rectangles,
each of which is disjoint either from $K_1$ or from $K_2$, and $\{F_\omega\}_{\omega \in \Omega}$ is a
covering of $X \times Y$ by closed rectangles such that $F_\omega \subset U_\omega$ for each $\omega$. $\square$

**4.5   Proposition.** *If $X$ is a paracompact space and $Y$ is a compact
space such that $X \times Y$ is totally normal, then*

$$\operatorname{Ind} X \times Y \leqslant \operatorname{Ind} X + \operatorname{Ind} Y.$$

*Proof.* It is enough to show that $X \times Y$ is an $F$-product. Let $K$ and $L$
be disjoint closed sets of $X \times Y$. By Lemma 2.1.12 the open covering
$\{(X \times Y)\backslash K, (X \times Y)\backslash L\}$ of $X \times Y$ has a refinement of the form

$$\mathcal{U} = \{V_\lambda \times G_{i\lambda} \mid i = 1, \ldots, n_\lambda, \lambda \in \Lambda\},$$

where $\{V_\lambda\}_{\lambda \in \Lambda}$ is a locally finite open covering of $X$ and

$$\{G_{i\lambda} \mid i = 1, \ldots, n_\lambda\}$$

is a finite open covering of $Y$ for each $\lambda$. Then $\mathcal{U}$ is a locally finite
open covering of $X \times Y$ by open rectangles each of which is disjoint
either from $K$ or from $L$. Since $X$ and $Y$ are normal spaces, there
exists a locally finite closed covering $\{D_\lambda\}_{\lambda \in \Lambda}$ of $X$ such that $D_\lambda \subset U_\lambda$
for each $\lambda$, and for each $\lambda$ there exists a closed covering $\{E_{i\lambda} \mid i = 1, \ldots, n_\lambda\}$ of $Y$ such that $E_{i\lambda} \subset G_{i\lambda}$ for each $i$. Then $\{D_\lambda \times E_{i\lambda} \mid i = 1, \ldots, n_\lambda, \lambda \in \Lambda\}$ is a covering of $X \times Y$ by closed rectangles and
$D_\lambda \times E_{i\lambda} \subset V_\lambda \times G_{i\lambda}$ for each $i, \lambda$. $\square$

**4.6   Proposition.** *If $X$ and $Y$ are paracompact $M$-spaces and $X \times Y$
is a totally normal space, then*

$$\operatorname{Ind} X \times Y \leqslant \operatorname{Ind} X + \operatorname{Ind} Y.$$

*Proof.* It is enough to show that $X \times Y$ is an $F$-product. Let $f: X \to A$
and $g: Y \to B$ be perfect mappings, where $A$ and $B$ are metric spaces.
Let $\{\mathcal{G}_i\}_{i \in \mathbf{N}}$ be a sequence of locally finite coverings of $A \times B$ by open
rectangles such that if $G$ is open in $A \times B$ and $(a, b) \in G$, then

$$\operatorname{St}\big((a, b), \mathcal{G}_i\big) \subset G$$

for some $i$. For each positive integer $i$, let $\mathcal{G}_i = \{G_\lambda\}_{\lambda \in \Lambda_i}$ and let

$\{E_\lambda\}_{\lambda \in \Lambda_i}$ be a covering of $A \times B$ by closed rectangles such that $E_\lambda \subset G_\lambda$ if $\lambda \in \Lambda_i$. Let $f \times g = h \colon X \times Y \to A \times B$ and for each $\lambda$ in $\Lambda_i$ let

$$U_\lambda = h^{-1}(G_\lambda) \quad \text{and} \quad F_\lambda = h^{-1}(E_\lambda).$$

Then for each $i$, $\{U_\lambda\}_{\lambda \in \Lambda_i}$ is a locally finite covering of $X \times Y$ by open rectangles and $\{F_\lambda\}_{\lambda \in \Lambda_i}$ is a covering of $X \times Y$ by closed rectangles such that $F_\lambda \subset U_\lambda$ for each $\lambda$.

Let $K$ and $L$ be disjoint closed sets of $X \times Y$. If $(a, b) \in A \times B$, then $h^{-1}(a, b) = f^{-1}(a) \times g^{-1}(b)$ is a compact subset of $X \times Y$. It follows that there exist open rectangles $P_1, \ldots, P_k$ and $Q_1, \ldots, Q_k$ such that $h^{-1}(a, b) \subset \bigcup_{j=1}^k P_j$ and for each $j$, $\overline{P}_j \subset Q_j$ and $Q_j$ is disjoint either from $K$ or from $L$. The mapping $h$ is closed so that there exists an open neighbourhood $N$ of $(a, b)$ in $A \times B$ such that $h^{-1}(N) \subset \bigcup_{j=1}^k P_j$. There exists an integer $i$ and $\lambda$ in $\Lambda_i$ such that $(a, b) \in E_\lambda \subset G_\lambda \subset N$, from which it follows that

$$h^{-1}(a, b) \subset F_\lambda \subset U_\lambda \subset \bigcup_{j=1}^k P_j.$$

Now for $j = 1, \ldots, k$ let

$$W_j = U_\lambda \cap Q_j \quad \text{and} \quad D_j = F_\lambda \cap P_j.$$

Then $W_1, \ldots, W_k$ are open rectangles such that $U_\lambda = \bigcup_{j=1}^k W_j$, and $D_1, \ldots, D_k$ are closed rectangles such that $D_j \subset W_j$ for each $j$ and

$$F_\lambda = \bigcup_{j=1}^k D_j.$$

For each point $(a, b)$ let us choose $\lambda$ in $\Lambda_i$ for some $i$ such that:

(i) $h^{-1}(a, b) \subset F_\lambda \subset U_\lambda$;

(ii) there is a finite family $\{W_\gamma\}_{\gamma \in \Gamma(\lambda)}$ of open rectangles of $X \times Y$, each of which is disjoint either from $K$ or from $L$, such that

$$U_\lambda = \bigcup_{\gamma \in \Gamma(\lambda)} W_\gamma;$$

(iii) for each $\gamma$ in $\Gamma(\lambda)$ there is a closed rectangle $D_\gamma$ such that $D_\gamma \subset W_\gamma$ and $F_\lambda = \bigcup_{\gamma \in \Gamma(\lambda)} D_\gamma$.

Let $\Omega$ be the subset of $\bigcup_{i \in \mathbb{N}} \Lambda_i$ consisting of those elements $\lambda$ thus selected. We can suppose that if $\lambda, \mu \in \Omega$ and $\lambda \ne \mu$, then $\Gamma(\lambda)$ and $\Gamma(\mu)$ are disjoint. Let $\Gamma = \bigcup_{\lambda \in \Omega} \Gamma(\lambda)$. Then $\{D_\gamma\}_{\gamma \in \Gamma}$ is a covering of $X \times Y$ by closed rectangles. For each $\gamma$, $D_\gamma \subset W_\gamma$, where $W_\gamma$ is an open rectangle which is disjoint either from $K$ or from $L$. Furthermore the family $\{W_\gamma\}_{\gamma \in \Gamma}$ is $\sigma$-locally finite, for if $\Gamma_i = \bigcup_{\lambda \in \Omega \cap \Lambda_i} \Gamma(\lambda)$,' then $\{W_\gamma\}_{\gamma \in \Gamma_i}$ is a locally finite family for each positive integer $i$. Thus $X \times Y$ is an $F$-product. $\square$

Notes

Propositions 1.3, 1.4 and 1.5, exploiting the characterization of dimension in terms of extension of mappings into spheres were given by Hurewicz and Wallman [1941] for separable metric spaces. The extension to normal spaces is due to Smirnov [1951]. The $n$-dimensional Cantor spaces defined here have been called *Cantor manifolds* but such an inappropriate name should not be perpetuated. Hurewicz and Tumarkin independently proved that every $n$-dimensional compact metric space contains an $n$-dimensional Cantor space, solving a problem posed by Urysohn. The existence theorem for Cantor spaces in bicompact spaces (Proposition 1.8) is due to Aleksandrov [1947].

Hurewicz proved that if $X$ and $Y$ are separable metric spaces and $f: X \to Y$ is a continuous closed surjection, then

$$\dim X \leqslant \dim Y + \dim f.$$

The Hurewicz formula given in Proposition 2.1 is due to Zarelua [1963]. In the case that $Y$ is paracompact, the formula was established in this form by Morita [1956]. Employing sheaf cohomology and the Leray spectral sequence of the mapping $f$, Skljarenko [1962] obtained the Hurewicz formula in the symmetrical form above for paracompact Hausdorff spaces $X$ and $Y$. Pasynkov [1965a] has generalized Skljarenko's result for the case of a closed mapping of a $T_4$-space onto a paracompact Hausdorff space and announced [1972] that the formula also holds when $Y$ is a weakly paracompact Hausdorff space, which is a strengthening of Proposition 2.1. Proposition 2.5 is due to Zarelua [1963].

Hurewicz [1933] proved that if $X$ and $Y$ are separable metric spaces and $f: X \to Y$ is a continuous closed surjection such that $f^{-1}(y)$ contains at most $k+1$ points if $y \in Y$, then

$$\dim Y \leqslant \dim X + k.$$

Roberts [1941] showed that the Hurewicz formula is the best possible result for separable metric spaces by proving that if $m$ and $n$ are integers such that $0 < m \leqslant n$, then there exist separable metric spaces $X$ and $Y$ such that $\dim X = n - m$, $\dim Y = n$, and a continuous closed surjection $f: X \to Y$ such that $f^{-1}(y)$ contains at most $m+1$ points if $y \in Y$. The corresponding result for general metric spaces was proved by Nagami [1960]. Proposition 2.10 and the results leading to it were obtained by Freudenthal [1932] for separable metric spaces and Nagami [1970] for metrizable spaces. Keesling [1968] has proved a more general result for metrizable spaces. The forms of the Hurewicz theorem for general spaces given in Propositions 2.13 and 2.14 are due to Morita [1956]. Zarelua [1969] has proved the final result of this type, showing that if $X$ and $Y$ are $T_4$-spaces and $f:X \to Y$ is a continuous closed surjection such that $f^{-1}(y)$ contains at most $k+1$ points if $y \in Y$, then

$$\dim Y \leqslant \dim X + k.$$

Zarelua's method of proof is to associate with $f$ a resolution of the constant sheaf over $Y$ and to study the spectral sequence of this resolution. Skordev [1970a] has given a simpler construction of the resolution associated with $f$. By a study of the Leray spectral sequence of the mapping, Skordev [1970] has proved a stronger result than Proposition 2.9 from which the Hurewicz–Zarelua theorem can also be deduced. Proposition 2.12 was obtained by Arhangel'skiĭ [1968].

Proposition 2.16 is due essentially to Nagami [1960]. Aleksandrov [1936a] proved that if $X$ and $Y$ are compact metric spaces and $f:X \to Y$ is a continuous open surjection such that $f^{-1}(y)$ is countable if $y \in Y$, then $\dim X = \dim Y$. Pasynkov [1967a] has announced that this result remains true for bicompact spaces $X$ and $Y$. Kolmogorov [1937] gave an example of a continuous open surjection between separable metric spaces which raises dimension. Roberts [1947] deduced from this result the existence of a continuous open surjection which raises dimension and has the property that the inverse image of each point of the range is countable. Example 2.17 is due to Knaster and Kuratowski [1921]. Roberts [1947] pointed out that the mapping of the totally disconnected space of positive dimension onto the Cantor set is open.

Morita [1953] proved Propositions 3.2 and 3.3. Lemma 3.4 and Proposition 3.5 were first proved by Kodama [1969]. The proof of Lemma 3.4 given here is due to Engelking [1973]. Since

$$\dim X \times Y \leqslant \dim X + \dim Y$$

if $X$ is either metrizable or bicompact and $Y$ is either metrizable or bicompact, the following conjecture due to Nagata [1966] seems natural: $\dim X \times Y \leqslant \dim X + \dim Y$ if $X$ and $Y$ are paracompact $M$-spaces. Example 3.7 is due to Michael [1963]. Kimura [1964] pointed out that it provides a counterexample to the product theorem for covering dimension. It should be noted that the non-normality of $X \times Y$ forces the failure of the product theorem. We shall return to this point in the notes on Chapter 10.

The example due to Erdös showed that the logarithmic law

$$\dim X \times Y = \dim X + \dim Y$$

need not hold for classes of spaces for which the product theorem is true. Anderson and Keisler [1967] have shown that if $n$ is a non-negative integer, then there exists a subspace $X$ of the Euclidean space $\mathbf{R}^{n+1}$ such that $\dim X = n$, and the dimension of the countable product of copies of $X$ is equal to $n$, so that $\dim X^k = n$ for every positive integer $k$. Pontrjagin [1930] found compact metric spaces $X$ and $Y$ such that $\dim X = \dim Y = 2$ and $\dim X \times Y = 3$. Two problems have been considered: (i) to characterize the compact metric spaces $X$ for which the logarithmic law holds for all compact metric spaces $Y$ and (ii) to characterize the paracompact Hausdorff spaces $X$ for which the logarithmic law holds for all bicompact spaces $Y$. Boltyanskiĭ [1949] gave a solution to (i) in terms of homological dimension. The methods of cohomological dimension theory have been used for problem (ii). Kuz'minov [1968] gave a survey of this field and described the status of problem (ii).

Proposition 4.1 is due essentially to Katsuta [1966]. The concept of an $F$-product was defined by Nagata [1967] who proved all the remaining results in §4. He showed more generally that $X \times Y$ is an $F$-product if (i) $X$ is a normal $P$-space (Morita [1964]) and $Y$ is a metrizable space, or (ii) $X$ is a locally compact paracompact Hausdorff space and $Y$ is a paracompact Hausdorff space. Nagami [1969] extended the product theorem for metrizable spaces, showing that if $X$ is a Hausdorff $P$-space and $Y$ is a $\Sigma$-space (Nagami [1969a]) such that $X \times Y$ is hereditarily paracompact, then

$$\operatorname{Ind} X \times Y \leqslant \operatorname{Ind} X + \operatorname{Ind} Y.$$

The assumption of total normality of the product in the propositions of §4 is quite restrictive. The product of perfectly normal bicompact spaces need not be completely normal. There are some results involving no such assumption. Lifanov [1968] proved that if $X$ and $Y$ are bicompact spaces such that $\operatorname{Ind} X = \operatorname{Ind} Y = 1$, then

$$\dim X \times Y = \operatorname{Ind} X \times Y = 2.$$

He also stated [1969] that if $X$ is a paracompact Hausdorff space and $Y$ is a bicompact space, then

$$\operatorname{Ind} X \times Y \leqslant \Delta(X) + \Delta(Y).$$

Pasynkov [1969] has announced that if $X$ and $Y$ are totally normal spaces such that $X \times Y$ is paracompact and an $F$-product, then

$$\operatorname{Ind} X \times Y \leqslant \operatorname{Ind} X + \operatorname{Ind} Y.$$

It follows that this product theorem holds if $X$ and $Y$ are totally normal bicompact spaces. This was also announced by Lifanov [1968a]. That the product theorem does not hold in general for bicompact spaces has been established by Filippov [1972]. He constructed bicompact spaces $X$ and $Y$ such that $\operatorname{Ind} X = 1$, $\operatorname{Ind} Y = 2$ and $\operatorname{Ind} X \times Y \geqslant 4$.

# 10

## ALGEBRAS OF CONTINUOUS REAL FUNCTIONS AND DIMENSION

### 1 A modification of covering dimension

This section begins with the definition of a concept of dimension for distributive lattices. Let $L$ be a distributive lattice with universal bounds 0 and 1. A covering of $L$ is a finite family $\{u_1, \ldots, u_k\}$ of members of $L$ such that $u_1 \vee \ldots \vee u_k = 1$. The order of the covering $\{u_1, \ldots, u_k\}$ of $L$ is the largest integer $n$ for which there exists a subset $\{r_0, \ldots, r_n\}$ of $\{1, \ldots, k\}$ with $n+1$ members such that $u_{r_0} \wedge \ldots \wedge u_{r_n} \neq 0$. If $\mathscr{U} = \{u_1, \ldots, u_k\}$ and $\mathscr{V} = \{v_1, \ldots, v_m\}$ are coverings of $L$, then $\mathscr{U}$ is a refinement of $\mathscr{V}$ if for each $i = 1, \ldots, k$ there exists some $j$ such that $u_i \leqslant v_j$. We say that $\dim L \leqslant n$ if each covering of $L$ has a refinement of order not exceeding $n$. We say that $\dim L = n$ if it is true that $\dim L \leqslant n$ but false that $\dim L \leqslant n-1$. If $\dim L \leqslant n$ is false for every $n$ then $\dim L = \infty$.

If $X$ is a topological space and $\mathscr{L}$ is a base for the topology of $X$ which is closed under finite unions and finite intersections, then $\mathscr{L}$ is a distributive lattice with respect to inclusion with least element $\varnothing$ and greatest element $X$. We define

$$\dim_{\mathscr{L}} X = \dim \mathscr{L}.$$

If $\mathscr{L}$ is the set of all open sets of $X$, then $\dim_{\mathscr{L}} X = \dim X$, the usual covering dimension. We shall see that if $X$ is a compact space then $\dim_{\mathscr{L}} X = \dim X$, whatever $\mathscr{L}$. For the proof we require:

**1.1 Lemma.** *Let $X$ be a compact space and let $\mathscr{L}$ be a base for the topology of $X$ which is closed under finite unions and finite intersections. If $\{U_1, \ldots, U_k\}$ is an open covering of $X$, then there exists a covering $\{W_1, \ldots, W_k\}$ of $X$ by elements of $\mathscr{L}$ such that $W_i \subset U_i$ for every $i$.*

*Proof.* For each point $x$ of $X$, there exists $V_x$ in $\mathscr{L}$ such that $x \in V_x \subset U_i$ for some $i$. Let $\{V_1, \ldots, V_s\}$ be a finite subcovering of the open covering $\{V_x\}_{x \in X}$ of $X$. For each $j = 1, \ldots, s$, choose $t(j)$ such that $V_j \subset U_{t(j)}$ and let $W_i = \bigcup_{t(j)=i} V_j$. Then $W_i \in \mathscr{L}$ and $W_i \subset U_i$ for every $i$, and $\{W_1, \ldots, W_k\}$ is a covering of $X$. $\square$

**1.2 Proposition.** *If $X$ is a compact space and $\mathscr{L}$ is a base for the topology of $X$ which is closed under finite unions and finite intersections, then* $\dim_{\mathscr{L}} X = \dim X$.

*Proof.* Let $\mathscr{U}$ be a finite open covering of $X$. By Lemma 1.1 there exists a covering $\mathscr{W}$ of $X$ by elements of $\mathscr{L}$ which refines $\mathscr{U}$. Then $\mathscr{W}$ is a covering of the lattice $\mathscr{L}$, so that if $\dim \mathscr{L} \leqslant n$, there is a refinement $\mathscr{V}$ of $\mathscr{W}$ of order not exceeding $n$. Moreover $\mathscr{V}$ is a finite open covering of $X$ of order not exceeding $n$ and refines $\mathscr{U}$ so that $\dim X \leqslant n$. Thus $\dim X \leqslant \dim \mathscr{L}$. Now let $\mathscr{U}$ be a covering of the lattice $\mathscr{L}$ so that $\mathscr{U}$ is a finite covering of $X$ by members of $\mathscr{L}$. If $\dim X \leqslant n$, there exists an open covering $\mathscr{V} = \{V_1, \ldots, V_k\}$ of $X$ of order not exceeding $n$ which is a refinement of $\mathscr{U}$. By Lemma 1.1 there exists a covering $\mathscr{W} = \{W_1, \ldots, W_k\}$ of $X$ by members of $\mathscr{L}$ such that $W_i \subset V_i$ for each $i$. Then $\mathscr{W}$ is a covering of the lattice $\mathscr{L}$ of order not exceeding $n$ which is a refinement of $\mathscr{U}$ so that $\dim \mathscr{L} \leqslant n$. Thus $\dim \mathscr{L} \leqslant \dim X$. Hence $\dim \mathscr{L} = \dim X$.□

Now let $X$ be a Tihonov space and let $\mathscr{L}$ be the set of cozero-sets of $X$. By Proposition 1.5.11, $\mathscr{L}$ is a base for the topology of $X$. Moreover $\mathscr{L}$ is closed under finite unions and intersections. We shall find $\dim_{\mathscr{L}} X$. By Proposition 1.6.13, the maximal ideal space $\mathscr{M}$ of the distributive lattice $\mathscr{L}$ is the Stone–Čech bicompactification $\beta X$ of $X$. The space $\mathscr{M}$ is compact and has a base $\mathbf{L}$ for its topology which consists of the sets

$$\mathscr{M}_U = \{M \in \mathscr{M} \mid U \notin M\}$$

for $U$ in $\mathscr{L}$. If $U, V \in \mathscr{L}$, then

$$\mathscr{M}_{U \cup V} = \mathscr{M}_U \cup \mathscr{M}_V \quad \text{and} \quad \mathscr{M}_{U \cap V} = \mathscr{M}_U \cap \mathscr{M}_V.$$

Thus $\mathbf{L}$ is closed under finite unions and finite intersections so that by Proposition 1.2

$$\dim \mathbf{L} = \dim \mathscr{M}.$$

Furthermore there is a surjective lattice homomorphism $\mathscr{L} \to \mathbf{L}$ given by $U \mapsto \mathscr{M}_U$. This homomorphism is in fact an isomorphism of lattices. For if $U, V \in \mathscr{L}$ and $U \neq V$, then there exists $x$ such that, say, $x \in U$ and $x \notin V$. If $M = \{W \in \mathscr{L} \mid x \notin W\}$, then $M \in \mathscr{M}$ and $U \notin M$, but $V \in M$ so that $\mathscr{M}_U \neq \mathscr{M}_V$. Since $\mathscr{L}$ and $\mathbf{L}$ are isomorphic it follows that

$$\dim \mathscr{L} = \dim \mathbf{L}.$$

But by definition $\dim_{\mathscr{L}} X = \dim \mathscr{L}$ so that we have

$$\dim_{\mathscr{L}} X = \dim \mathscr{M} = \dim \beta X.$$

In view of this result it seems natural to study a new dimension function.

**1.3 Definition.** *If $X$ is a topological space, then $\partial\mathrm{im}\, X$ is the least integer $n$ such that every finite covering of $X$ by cozero-sets has a refinement which is a finite covering, of order not exceeding $n$, by cozero-sets, or $\partial\mathrm{im}\, X = \infty$ if there is no such integer.*

**1.4 Theorem.** *If $X$ is a Tihonov space, then*

$$\partial\mathrm{im}\, X = \dim \beta X.$$

*Proof.* If $X$ is a Tihonov space and $\mathscr{L}$ is the base of cozero-sets then $\partial\mathrm{im}\, X = \dim_{\mathscr{L}} X$. But it was established above that

$$\dim'_{\mathscr{L}} X = \dim \beta X. \square$$

**1.5 Example.** Let $X$ be the Tihonov plank which is the topological product $P \times N$, where $P, N$ are the linearly ordered spaces consisting of the ordinals not exceeding the first uncountable ordinal $\omega_1$ and the first infinite ordinal $\omega_0$ respectively. It was established in Example 4.3.1 that $\dim X = 0$, but if $Y = X \setminus \{(\omega_1, \omega_0)\}$ then $\dim Y = 1$. It is easy to establish that $\beta Y = X$ and hence $\partial\mathrm{im}\, Y = \dim X = 0$. Thus $Y$ is an example of a Tihonov space for which the dimension $\partial\mathrm{im}$ differs from $\dim$. This cannot arise for normal spaces.

**1.6 Proposition.** *If $X$ is a normal space, then*

$$\partial\mathrm{im}\, X = \dim X.$$

*Proof.* Let $X$ be a normal space, and let $\mathscr{U} = \{U_1, \ldots, U_k\}$ be an open covering of $X$. Since $X$ is a normal space, by Proposition 1.3.12 there exist continuous functions $f_i : X \to I$ such that $f_i(x) > 0$ only if $x \in U_i$ and $\Sigma_{i=1}^{k} f_i(x) = 1$ for every $x$ in $X$. If $V_i = \{x \in X \mid f_i(x) > 0\}$, then $\mathscr{V} = \{V_1, \ldots, V_k\}$ is a covering of $X$ by cozero-sets such that $V_i \subset U_i$ for every $i$. Furthermore the order of $\mathscr{V}$ does not exceed $n$ if the order of $\mathscr{U}$ does not exceed $n$. It is now easy to see that if $X$ is a normal space then $\partial\mathrm{im}\, X \leqslant n$ if and only if $\dim X \leqslant n$. $\square$

From Proposition 1.6 and Theorem 1.4 we recover a result proved earlier.

**1.7   Corollary.** *If $X$ is a $T_4$-space, then* $\dim X = \dim \beta X$. $\square$

We recall that a subset $A$ of a Tihonov space $X$ is called $\beta$-closed if each continuous function $f:A \to I$ has an extension $g:X \to I$. Proposition 1.6.11 states that if $A$ is a $\beta$-closed subspace of a Tihonov space $X$, then the closure of $A$ in $\beta X$ is $\beta A$. It should be noted that each closed set in a $T_4$-space is $\beta$-closed. We have a subset theorem and a countable sum theorem for the dimension $\partial$im.

**1.8   Proposition.** *If $A$ is a $\beta$-closed set in a Tihonov space $X$, then* $\partial$im $A \leqslant \partial$im $X$.

*Proof.* The closure of $A$ in $\beta X$ is $\beta A$. Thus $\beta A$ is a closed subset of $\beta X$ and we have

$$\partial\text{im } A = \dim \beta A \leqslant \dim \beta X = \partial\text{im } X. \square$$

**1.9   Proposition.** *If $X$ is a Tihonov space and $X = \bigcup_{i \in \mathbf{N}} A_i$, where $A_i$ is a $\beta$-closed subset of $X$ and $\partial$im $A_i \leqslant n$ for each $i$, then $\partial$im $X \leqslant n$.*

*Proof.* The closure of $A_i$ in $\beta X$ is $\beta A_i$. Let $Y = \bigcup_{i \in \mathbf{N}} \beta A_i$. Then $Y$ is a regular space which is the union of countably many compact subspaces. Thus $Y$ is a regular Lindelöf space and hence $Y$ is normal by Proposition 2.2.9. For each $i$, $\beta A_i$ is closed in $Y$ and $\dim \beta A_i \leqslant n$. Thus $\dim Y \leqslant n$ by Theorem 3.2.5 (the countable sum theorem). But $X \subset Y \subset \beta X$, so that $\beta Y = \beta X$ by Proposition 1.6.8. Thus

$$\partial\text{im } X = \dim \beta X = \dim \beta Y = \dim Y.$$

Hence $\partial$im $X \leqslant n$. $\square$

## 2   Normed algebras

In the remainder of this chapter we shall exploit the fact that the set $C(X)$ of continuous bounded real-valued functions on a topological space $X$ has the structure of a complete normed algebra. This section begins with a review of the elementary properties of complete normed algebras. We shall then find a construction for the Stone–Čech bicompactification and characterize the dimension of a Tihonov space $X$ in terms of the algebraic structure of $C(X)$.

An *algebra* is a real vector space $A$ on which is defined a binary operation of multiplication which associates with each ordered pair $(x, y)$ of elements of $A$ an element $xy$ of $A$ such that:

(*a*) multiplication is associative and commutative and there exists $1 \in A$ such that $1x = x$ for all $x$ in $A$;

(*b*) if $x, y, z \in A$ and $\lambda, \mu \in \mathbf{R}$, then

$$x(y+z) = xy+xz \quad \text{and} \quad (\lambda x)(\mu y) = (\lambda \mu)(xy).$$

If $A$ and $B$ are algebras, a function $\phi: A \to B$ is said to be a homomorphism of algebras if $\phi$ is a linear transformation such that $\phi(1) = 1$ and
$$\phi(xy) = \phi(x)\phi(y) \quad \text{if} \quad x, y \in A.$$

An isomorphism of algebras is a bijective homomorphism.

If $A$ is an algebra, a subset $K$ of $A$ is called a *subalgebra* of $A$ if $K$ is a vector subspace of $A$ such that $1 \in K$ and $xy \in K$ if $x, y \in K$. Trivially $A$ is a subalgebra of $A$ and it is clear that any intersection of subalgebras of an algebra $A$ is a subalgebra of $A$. Thus if $H$ is a subset of $A$, there exists a subalgebra $K[H]$ of $A$ such that $H \subset K[H]$ and $K[H]$ is the smallest subalgebra containing $H$ in the sense that if $K$ is a subalgebra of $A$ and $H \subset K$ then $K[H] \subset K$. The subalgebra $K[H]$ consists of all elements of the form $\lambda_0 + \lambda_1 u_1 + \ldots + \lambda_m u_m$, where each $u_i$ is a product of finitely many members of $H$ and $\lambda_0, \ldots, \lambda_m$ are real numbers. If $\phi: A \to B$ is a homomorphism of algebras we define the *image* of $\phi$ to be

$$\operatorname{Im} \phi = \{y \in B \mid y = \phi(x) \text{ for some } x \text{ in } A\}.$$

It is easy to verify that $\operatorname{Im} \phi$ is a subalgebra of $B$.

If $A$ is an algebra, a vector subspace $J$ of $A$ is called an *ideal* if $x \in J$ implies that $xz \in J$ for all $z$ in $A$. Clearly $A$ is an ideal of $A$. An ideal of $A$ which is a proper subset of $A$ is called a proper ideal. Obviously an ideal $J$ of $A$ is proper if and only if $1 \notin J$. Clearly any union or intersection of ideals is an ideal. It follows that if $H$ is a subset of $A$, then there exists a smallest ideal containing $H$ which is called the ideal generated by $H$. If $\phi: A \to B$ is a homomorphism of algebras, the *kernel* of $\phi$ is defined to be

$$\operatorname{Ker} \phi = \{x \in A \mid \phi(x) = 0\}.$$

It is easy to verify that $\operatorname{Ker} \phi$ is an ideal of $A$ and since $\phi(1) = 1$ we see that $\operatorname{Ker} \phi$ is a proper ideal of $A$. A proper ideal of an algebra $A$ is said to be a *maximal ideal* if it is not contained in any distinct proper ideal of $A$. It follows from Zorn's lemma that each proper ideal of $A$ is contained in a maximal ideal. A proper ideal $J$ of $A$ is a maximal ideal if and only if for each $a$ such that $a \notin J$, the ideal generated by $J \cup \{a\}$ is not proper, or equivalently, there exist $x$ in $A$ and $b$ in $J$ such that $xa+b = 1$.

An algebra $A$ is called a *normed algebra* if there is a real number $\|x\|$, called the norm of $x$, associated with each element $x$ of $A$ with the properties that for all $x, y$ in $A$:

(1) $\|x\| \geqslant 0$ and $\|x\| = 0$ if and only if $x = 0$;

(2) $\|x + y\| \leqslant \|x\| + \|y\|$;

(3) $\|\lambda x\| = |\lambda| \, \|x\|$ if $\lambda \in \mathbf{R}$;

(4) $\|xy\| \leqslant \|x\| \, \|y\|$;

(5) $\|1\| = 1$.

If $A$ is a normed algebra then we can define a metric $d$ on $A$ by putting

$$d(x, y) = \|x - y\| \quad \text{if} \quad x, y \in A.$$

We shall regard a normed algebra as a metric space with the metric thus associated with its norm. In a normed algebra the closure of each subalgebra is a subalgebra. It follows that if $A$ is a normed algebra and $H$ is a subset of $A$, then the smallest closed subalgebra of $A$ which contains $H$, which we shall denote by $A[H]$, is the closure in $A$ of $K[H]$. If $A$ and $B$ are normed algebras, a homomorphism $\phi: A \to B$ of algebras is said to be a homomorphism of normed algebras if it is continuous. An algebra homomorphism $\phi: A \to B$ is a homomorphism of normed algebras if and only if there exists a real number $\lambda$ such that $\|\phi(x)\| \leqslant \lambda \|x\|$ for all $x$ in $A$. A homomorphism $\phi: A \to B$ of normed algebras is called an *isometry* if $\|\phi(x)\| = \|x\|$ for all $x$ in $A$. It is clear that if $\phi$ is an isometry then $\phi$ is injective. In fact $\phi$ is an embedding. For if $U$ is open in $A$ and $x_0 \in U$, then there exists a positive real number $\delta$ such that $x \in U$ if $\|x - x_0\| < \delta$. Suppose that $x \in A$ and $\|\phi(x) - \phi(x_0)\| < \delta$. Then $\|\phi(x - x_0)\| < \delta$ so that $\|x - x_0\| < \delta$. Hence $x \in U$ so that $\phi(x) \in \phi(U)$. Thus $\phi$ is a continuous injection such that $\phi(U)$ is open in $\phi(A)$ for each open set $U$ of $A$ and hence $\phi$ is an embedding.

A normed algebra is said to be a *complete normed algebra*, or a Banach algebra, if it is complete as a metric space (see Definition 2.4.1).

**2.1 Proposition.** *If $A$ is a complete normed algebra, $B$ is a normed algebra and $\phi: A \to B$ is an isometry, then* $\text{Im } \phi$, *the image of $\phi$, is a closed subalgebra of $B$.*

*Proof.* Suppose that $u \in (\text{Im } \phi)^-$. Then by Proposition 2.3.11 there exists a sequence $\{\phi(x_n)\}_{n \in \mathbf{N}}$ in $\text{Im } \phi$ which converges to $u$. Thus given $\epsilon > 0$ there exists an integer $N$ such that $\|\phi(x_n) - \phi(x_m)\| < \epsilon$ if $m, n \geqslant N$. Since

$$\|\phi(x_n) - \phi(x_m)\| = \|\phi(x_n - x_m)\| = \|x_n - x_m\|,$$

it follows that $\{x_n\}_{n\in\mathbb{N}}$ is a Cauchy sequence in $A$. By hypothesis, the sequence $\{x_n\}_{n\in\mathbb{N}}$ converges to $x$, say, in $A$. Given $\delta > 0$ there exists an integer $k$ such that $\|x - x_k\| < \delta/2$ and $\|u - \phi(x_k)\| < \delta/2$ so that we have

$$\|u - \phi(x)\| \leqslant \|u - \phi(x_k)\| + \|\phi(x_k) - \phi(x)\| < \delta.$$

Hence $\|u - \phi(x)\| = 0$ so that $u = \phi(x)$. Thus $u \in \operatorname{Im}\phi.\square$

**2.2   Remark.** If $A$ and $B$ are complete normed algebras and the homomorphism $\phi: A \to B$ is an isometry, then $\operatorname{Im}\phi$ is a closed subalgebra of $B$ and the complete normed algebras $A$ and $\operatorname{Im}\phi$ are homeomorphic as spaces and isomorphic as algebras.

Let $X$ be a topological space and let $C(X)$ denote the set of continuous bounded real-valued functions on $X$. Then $C(X)$ is a complete normed algebra. The zero element is the constant function taking the value $0$, and if $f, g \in C(X)$ and $\lambda \in \mathbb{R}$, then the members $f + g$ and $\lambda f$ of $C(X)$ are defined by

$$(f + g)(x) = f(x) + g(x) \quad \text{and} \quad (\lambda f)(x) = \lambda f(x)$$

for $x$ in $X$. The unit $1$ is the constant function $\iota$ on $X$ given by $\iota(x) = 1$ for all $x$ in $X$, and if $f, g \in C(X)$, then the member $fg$ of $C(X)$ is defined by

$$(fg)(x) = f(x) g(x) \quad \text{if} \quad x \in X.$$

The norm $\|f\|$ of $f$ in $C(X)$ is defined by

$$\|f\| = \sup_{x\in X} |f(x)|.$$

It is easily verified that with these definitions $C(X)$ has the structure of a normed algebra. Furthermore the metric on $C(X)$ associated with its norm is the metric on $C(X)$ considered in §4 of Chapter 2, and it was established there that $C(X)$ is a complete metric space. Thus $C(X)$ is a complete normed algebra. We remark that a subalgebra of $C(X)$ is a subset $K$ of $C(X)$ which contains all constant functions and has the property that if $f, g \in K$, then $f + g \in K$ and $fg \in K$.

If $X$ and $Y$ are topological spaces and $h: X \to Y$ is a continuous function, then the composite $f \circ h$ is a member of $C(X)$ if $f \in C(Y)$. We define

$$h^*: C(Y) \to C(X)$$

by $h^*(f) = f \circ h$ for $f$ in $C(Y)$. It is easily verified that $h^*$ is an algebra homomorphism. If $f \in C(Y)$, then

$$\|h^*(f)\| = \sup_{x \in X} |f(h(x))| \leqslant \|f\|.$$

Thus $h^*$ is continuous and hence is a homomorphism of normed algebras.

**2.3 Proposition.** *If $X$ and $Y$ are topological spaces and $h: X \to Y$ is a continuous function such that $h(X)$ is dense in $Y$, then the homomorphism*

$$h^*: C(Y) \to C(X)$$

*is an isometry, and $h^*C(Y)$ is a closed subalgebra of $C(X)$.*

*Proof.* Suppose that $f \in C(Y)$. Let $\epsilon$ be a positive real number. Then

$$V = \{y \in Y \mid |f(y)| > \|f\| - \epsilon\}$$

is a non-empty open set. Since $h(X)$ is dense in $Y$, there exists a point $x_0$ of $X$ such that $h(x_0) \in V$. Thus $|h^*(f)(x_0)| > \|f\| - \epsilon$ so that

$$\|f\| - \epsilon < \|h^*(f)\| \leqslant \|f\|.$$

Since $\epsilon$ is arbitrary, it follows that $\|h^*(f)\| = \|f\|$ as required. It follows from Proposition 2.1 that $h^*C(Y)$ is a closed subalgebra of $C(X)$. $\square$

**2.4 Corollary.** *If $X$ is a Tihonov space, then $C(X)$ and $C(\beta X)$ are isomorphic normed algebras.*

*Proof.* By applying Proposition 2.3 to the inclusion $i$ of $X$ in its Stone–Čech bicompactification $\beta X$ we obtain an isometry $i^*: C(\beta X) \to C(X)$ and $i^*(f) = f|X$ if $f \in C(\beta X)$. But if $g \in C(X)$ there exists some $f$ in $C(\beta X)$ such that $f|X = g$. Thus $i^*$ is surjective and hence is an isomorphism. $\square$

We shall exploit the last result to give a new construction of the Stone–Čech bicompactification of a Tihonov space.

**2.5 Proposition.** *Let $X$ be a bicompact space. For each point $x$ of $X$ the set*

$$M_x = \{f \in C(X) \mid f(x) = 0\}$$

*is a maximal ideal of the algebra $C(X)$. Furthermore every maximal ideal of $C(X)$ is of the form $M_x$ for some $x$, and if $x$ and $y$ are distinct points of $X$, then $M_x \neq M_y$.*

*Proof.* Let $x$ be a point of $X$. If $\mathbf{R}$ has the obvious algebra structure then the function $\phi:C(X)\to\mathbf{R}$ given by $\phi(f)=f(x)$ is an algebra homomorphism. Since $M_x$ is the kernel of $\phi$ it follows that $M_x$ is a proper ideal of $C(X)$. Furthermore $M_x$ is maximal. For if $g\notin M_x$, then $g(x)\neq 0$. Let $h$ be the constant function on $X$ with value $1/g(x)$ and let $f$ be the function defined by

$$f(y) = 1-g(y)/g(x) \quad \text{if} \quad y\in X.$$

Then $f\in M_x$ and $hg+f=\iota$ so that $M_x$ is a maximal ideal. If $x,y\in X$ and $x\neq y$, then since $X$ is bicompact there exists $f$ in $C(X)$ such that $f(x)=0$ and $f(y)=1$. Thus $M_x\neq M_y$. Finally let $M$ be a maximal ideal of $C(X)$. Suppose that $f\in M$ and that the zero-set $Z(f)$ of $f$ is empty. Then $1/f$ is continuous so that $1/f\in C(X)$ since $X$ is compact. But $f(1/f)=\iota$ so that $\iota\in M$, which contradicts the hypothesis that $M$ is a proper ideal of $C(X)$. Thus if $f\in M$, then $Z(f)\neq\varnothing$. Now if $f_1,\dots,f_k\in M$, then $\Sigma_{i=1}^k f_i^2\in M$ so that

$$\bigcap_{i=1}^{k} Z(f_i) = Z\left(\sum_{i=1}^{k} f_i^2\right) \neq \varnothing.$$

Thus the family $\{Z(f)\}_{f\in M}$ of closed sets has the finite intersection property. Since $X$ is a compact space it follows that $\bigcap_{f\in M}Z(f)$ is non-empty. If $x\in\bigcap_{f\in M}Z(f)$, then $f(x)=0$ for every $f$ in $M$ so that $M\subset M_x$. Since $M$ is a maximal ideal it follows that $M=M_x$. $\square$

For a topological space $X$, let $\mathscr{M}(X)$ be the set of maximal ideals of $C(X)$. If $X$ is a bicompact space, then it follows from Proposition 2.5 that there is a bijection between $X$ and $\mathscr{M}(X)$ which associates with each point $x$ of $X$ the maximal ideal

$$M_x = \{f\in C(X)\,|\,f(x)=0\}.$$

Under this bijection, the image of the cozero-set $X\backslash Z(f)$, where $f\in C(X)$, is the set

$$\{M\in\mathscr{M}(X)\,|\,f\notin M\}.$$

Since the cozero-sets form a base for the topology of $X$, it follows that we can take the sets $\{M\in\mathscr{M}(X)\,|\,f\notin M\}$ for $f$ in $C(X)$ as the base for a topology, and if $\mathscr{M}(X)$ has this topology, then $X$ and $\mathscr{M}(X)$ are homeomorphic.

Now let $X$ be a Tihonov space. Then the Stone–Čech bicompactification $\beta X$ of $X$ is homeomorphic with $\mathscr{M}(\beta X)$. But by Corollary 2.4 there is an isomorphism between $C(\beta X)$ and $C(X)$ under which each member of $C(\beta X)$ corresponds to its restriction to $X$. Thus there is a bijection between $\mathcal{M}(X)$ and $\mathcal{M}(\beta X)$ and it follows that the maximal ideals of $C(X)$ are precisely the sets

$$M_z = \{f \in C(X) \,|\, \tilde{f}(z) = 0\}$$

for $z$ in $\beta X$, where $\tilde{f}$ denotes the unique extension of $f$ to $\beta X$. For each $f$ in $C(X)$ let

$$\mathcal{M}_f = \{M \in \mathcal{M}(X) \,|\, f \notin M\}.$$

Under the bijection between $\mathcal{M}(X)$ and $\mathcal{M}(\beta X)$, the sets $\mathcal{M}_f$, for $f$ in $C(X)$, correspond to the members of the base for the topology of $\mathcal{M}(\beta X)$ described above. Thus if we take the sets $\mathcal{M}_f$ for $f$ in $C(X)$ as the base for a topology on $\mathcal{M}(X)$, then $\mathcal{M}(X)$ and $\mathcal{M}(\beta X)$ are homeomorphic. Thus $\mathcal{M}(X)$, together with the embedding of $X$ in $\mathcal{M}(X)$ given by mapping each $x$ in $X$ to the maximal ideal $M_x$ of $C(X)$, is a bicompactification of $X$ equivalent to the Stone–Čech bicompactification $\beta X$ of $X$. Hence the Stone–Čech bicompactification of the Tihonov space $X$ is completely determined by the algebraic structure of $C(X)$. The base $\mathcal{L}$ for the topology of $\mathcal{M}(X)$ which consists of the sets $\mathcal{M}_f$, for $f$ in $C(X)$ is a lattice with respect to inclusion. From the discussion of §1, it follows that

$$\partial\mathrm{im}\, X = \dim \beta X = \dim \mathcal{M}(X) = \dim \mathcal{L},$$

where $\dim \mathcal{L}$ denotes the dimension of the lattice $\mathcal{L}$ in the sense of §1. Thus $\partial\mathrm{im}\, X$ is the least integer $n$ for which every finite covering of $\mathcal{M}(X)$ by members of $\mathcal{L}$ has a refinement of order not exceeding $n$, consisting of members of $\mathcal{L}$, and $\partial\mathrm{im}\, X = \infty$ if and only if there is no such integer. It follows that the dimension $\partial\mathrm{im}\, X$ of a Tihonov space $X$ is determined by the algebraic structure of $C(X)$. In §4 we shall obtain a characterization of the dimension $\partial\mathrm{im}\, X$ in terms of $C(X)$ which will be more interesting since it is an algebraic characterization of dimension which is not simply a translation of the definition of $\partial\mathrm{im}\, X$ into a statement about maximal ideals of $C(X)$.

To conclude this section we obtain a form of the Stone–Weierstrass theorem. If $f_1, \ldots, f_k \in C(X)$ let us write $K[f_1, \ldots, f_k]$ and $A[f_1, \ldots, f_k]$ to denote the smallest subalgebra and the smallest closed subalgebra respectively of $C(X)$ containing $f_1, \ldots, f_k$.

**2.6  Lemma.** *If $X$ is a topological space and $f \in C(X)$, then $|f| \in A[f]$.*

*Proof.* Clearly it is enough to prove the result in the case $\|f\| = 1$. Thus we suppose that $\|f\| = 1$. Suppose given $\epsilon > 0$. Since the binomial expansion of $(1-t)^{\frac{1}{2}}$ converges uniformly for $|t| \leqslant 1$, there exists a polynomial $p$ such that

$$|(1-t)^{\frac{1}{2}} - p(t)| < \epsilon/2 \quad \text{if} \quad |t| \leqslant 1.$$

Now $|f(x)| = \left(1 - (1 - f(x)^2)\right)^{\frac{1}{2}}$ and since $|1 - f(x)^2| \leqslant 1$ we have

$$|\,|f(x)| - p(1 - f(x)^2)\,| < \epsilon/2$$

for all $x$ in $X$. Thus if $g = p(\iota - f^2)$, then $g \in K[f]$ and if $x \in X$, then $|\,|f(x)| - g(x)\,| < \epsilon/2$ so that

$$\|\,|f| - g\,\| < \epsilon.$$

Since $A[f]$ is the closure of $K[f]$ it follows that $|f| \in A[f]$. $\square$

**2.7   Corollary.** *If* $X$ *is a topological space and* $f, g \in C(X)$, *then* $\max\{f, g\} \in A[f, g]$ *and* $\min\{f, g\} \in A[f, g]$.

*Proof.* By Lemma 2.6, $|f - g| \in A[f - g] \subset A[f, g]$. But

$$\max\{f, g\} = \tfrac{1}{2}(f + g + |f - g|),$$

$$\min\{f, g\} = \tfrac{1}{2}(f + g - |f - g|). \square$$

**2.8   Lemma.** *If* $X$ *is a topological space and* $M$ *is a subset of* $X$, *then the subset of* $C(X)$ *which consists of all functions which are constant on* $M$ *is a closed subalgebra of* $C(X)$.

*Proof.* Let $K$ be the subset of $C(X)$ consisting of all functions which are constant on $M$. It is clear that $K$ is a subalgebra of $C(X)$. Suppose that $g \in C(X)$ and $g \notin K$. Then there exist points $x_0, x_1$ of $M$ such that $g(x_0) < g(x_1)$. Let $\epsilon = \tfrac{1}{2}\left(g(x_1) - g(x_0)\right)$. If $f \in C(X)$ and $\|f - g\| < \epsilon$, then $|f(x_0) - g(x_0)| < \epsilon$ and $|f(x_1) - g(x_1)| < \epsilon$. Thus

$$f(x_0) < \tfrac{1}{2}\left(g(x_0) + g(x_1)\right) < f(x_1).$$

It follows that $f \notin K$. Hence $K$ is a closed subalgebra of $C(X)$. $\square$

Let $X$ be a topological space and let $H$ be a subset of $C(X)$. We can define an equivalence relation $\equiv$ on $X$ as follows: if $x, y \in X$, then $x \equiv y$ if $f(x) = f(y)$ for every $f$ in $H$. The equivalence classes with respect to this relation are called *constancy sets* of $H$. Clearly each constancy set of $H$ is a closed set of $X$.

**2.9   Proposition.** *Let* $X$ *be a compact space and let* $K$ *be a subalgebra of* $C(X)$. *Then the closure* $\bar{K}$ *of* $K$ *consists of all functions in* $C(X)$ *which are constant on every constancy set of* $K$.

*Proof.* Let $\tilde{K}$ be the subset of $C(X)$ consisting of all functions which are constant on the constancy sets of $K$. It follows from Lemma 2.8 that $\tilde{K}$ is the intersection of closed subalgebras of $C(X)$ so that $\tilde{K}$ is a closed subalgebra of $C(X)$. Thus $\bar{K} \subset \tilde{K}$. Now suppose that $f \in \tilde{K}$ and that $a, b \in X$. If $f(a) = f(b)$, then there exists a constant function in $K$ which agrees with $f$ at $a$ and $b$. If $f(a) \neq f(b)$, then since $f$ is constant on each constancy set of $K$, there exists $h$ in $K$ such that $h(a) \neq h(b)$. Let

$$\lambda = \big(f(b) - f(a)\big)/\big(h(b) - h(a)\big)$$

and let $g = \lambda\big(h - h(a)\iota\big) + f(a)\iota$. Then $g \in K$ and $g$ agrees with $f$ at $a$ and $b$. Thus if $f \in \tilde{K}$, then for each pair $a, b$ of points of $X$ there exists $g_{ab}$ in $K$ such that $g_{ab}$ agrees with $f$ at $a$ and at $b$. Suppose given $\epsilon > 0$. For fixed $b$, and each $a$ in $A$ let

$$V_a = \{x \in X \mid g_{ab}(x) < f(x) + \epsilon\}.$$

Then $\{V_a\}_{a \in X}$ is an open covering of the compact space $X$, and hence there exist $a_1, \ldots, a_n$ such that $X = \bigcup_{i=1}^{n} V_{a_i}$. If

$$g_b = \min\{g_{a_1 b}, \ldots, g_{a_n b}\},$$

then $g_b \in \bar{K}$ by Corollary 2.7. Clearly $g_b(x) < f(x) + \epsilon$ for all $x$, and $g_b(x) > f(x) - \epsilon$ if $x \in W_b$, where

$$W_b = \bigcap_{i=1}^{n} \{x \in X \mid g_{a_i b}(x) > f(x) - \epsilon\}.$$

Since $b \in W_b$ for every $b$ in $X$ we see that $\{W_b\}_{b \in X}$ is an open covering of the compact space $X$. Hence there exist $b_1, \ldots, b_m$ such that

$$\bigcup_{i=1}^{m} W_{b_i} = X.$$

If
$$h = \max\{g_{b_1}, \ldots, g_{b_m}\},$$

then $h \in \bar{K}$ by Corollary 2.7, and if $x \in X$, then $|f(x) - h(x)| < \epsilon$ so that $\|f - h\| \leqslant \epsilon$. It follows that $f \in \bar{K}$. Thus $\tilde{K} = \bar{K}$. $\square$

Now we have a form of the Stone–Weierstrass theorem. It should be noted that if a subset of $C(X)$ separates the points of $X$, then $X$ is a Hausdorff space.

**2.10  Proposition.** *Let $X$ be a bicompact space and let $K$ be a subalgebra of $C(X)$ which separates the points of $X$. Then $K$ is dense in $C(X)$.*

*Proof.* Since $K$ separates the points of $X$, the constancy sets of $K$ are the one-point subsets of $X$. Thus each member of $C(X)$ is constant on every constancy set of $K$. Hence $\bar{K} = C(X)$ by Proposition 2.9. $\square$

## 3   Dimension and bicompactification

It is clear that a Tihonov space $X$ has a bicompactification of weight equal to $w(X)$, and the Stone–Čech bicompactification $\beta X$ has covering dimension equal to $\partial \text{im}\, X$. In this section it will be shown that $X$ has a bicompactification of weight equal to $w(X)$ and covering dimension equal to $\partial \text{im}\, X$. We begin with a reformulation of the characterization of covering dimension in terms of inessential mappings into cubes.

**3.1   Definition.** *Let $X$ be a topological space and let $H$ and $G$ be subsets of $C(X)$ such that $H \subset G$. The dimensional rank of the set $H$ relative to the set $G$, denoted by $\mathrm{dr}\,(H, G)$, is the least non-negative integer $n$ such that for each $(n+1)$-tuple $(h_0, \ldots, h_n)$ of members of $H$ and each positive real number $\epsilon$, there exist some $(n+1)$-tuple $(g_0, \ldots, g_n)$ of members of $G$ and some positive real number $\delta$ such that $\|h_i - g_i\| < \epsilon$ for all $i$ and $\bigcap_{i=0}^{n} g_i^{-1}(-\delta, \delta) = \varnothing$. If there exists no such integer then $\mathrm{dr}\,(H, G) = \infty$.*

We note that if $\mathrm{dr}\,(H, G) = m$, where $m$ is a non-negative integer, then the condition of Definition 3.1 is satisfied for each $(n+1)$-tuple of members of $H$, where $n \geqslant m$. We put $\mathrm{dr}\,(H, H) = \mathrm{dr}\, H$ and call $\mathrm{dr}\, H$ the dimensional rank of the subset $H$ of $C(X)$. The dimensional rank of $C(X)$ determines the covering dimension of a normal space $X$.

**3.2   Proposition.** *If $X$ is a normal space, then*

$$\dim X = \mathrm{dr}\, C(X).$$

*Proof.* Let $X$ be a normal space such that $\dim X = n$, where $n$ is a non-negative integer. Let $(f_0, \ldots, f_n)$ be an $(n+1)$-tuple of members of $C(X)$ and let $\epsilon$ be a positive real number. Let $f \colon X \to \mathbf{R}^{n+1}$ be given by

$$f(x) = \big(f_0(x), \ldots, f_n(x)\big) \quad \text{if} \quad x \in X.$$

Let $E = \{y \in \mathbf{R}^{n+1} \mid \|y\| < \epsilon/3\}$ and let $B = f^{-1}(E)$. If $B$ is non-empty let $f' \colon B \to E$ be given by restriction of $f$. Since $B$ is a closed set of $X$ it follows that $\dim B \leqslant n$, so that $f'$ must be an inessential mapping by Proposition 3.3.1. Hence there exists $g:B\to S$, where $S$ is the boundary sphere of $E$, such that $g(x) = f(x)$ if $x\in f^{-1}(S)$. Since

$$\mathrm{bd}\,(B) \subset f^{-1}(S),$$

it follows that we can define a continuous function $h:X\to \mathbf{R}^{n+1}$ by putting $h(x) = g(x)$ if $x\in B$ and $h(x) = f(x)$ if $x\in (X\backslash B)^-$. If for each $x$ in $X$

$$h(x) = \big(h_0(x),\dots,h_n(x)\big),$$

then $(h_0,\dots,h_n)$ is an $(n+1)$-tuple of members of $C(X)$. If $x\in X$ and $f(x) \neq h(x)$, then $x\in B, f(x)\in E$ and $h(x)\in S$ so that $\|f(x)-h(x)\| < 2\epsilon/3$. It follows that $\|f_i-h_i\| < \epsilon$ for $i = 0,\dots,n$. Finally let $\delta = \epsilon/3\sqrt{(n+1)}$. If $x\in \bigcap_{i=0}^n h_i^{-1}(-\delta,\delta)$, then $\|h(x)\| < \epsilon/3$ which is absurd since $h(x)\in E\backslash S$ for all $x$. Therefore $\bigcap_{i=0}^n h_i^{-1}(-\delta,\delta) = \varnothing$. Thus $\mathrm{dr}\,C(X) \leqslant n$.

Now we show that $\mathrm{dr}\,C(X)$ is not less than $n$. Since $\dim X = n$, it follows from Proposition 3.3.2 that there exists a continuous mapping $f:X\to E^n$ such that $\mathbf{0}$ is not an unstable value of $f$. Thus there exists a positive number $\epsilon'$ with the property that $\mathbf{0}\in g(X)$ for each continuous function $g:X\to \mathbf{R}^n$ such that $\|f(x)-g(x)\| < \epsilon'$ for all $x$ in $X$. Let

$$f(x) = \big(f_1(x),\dots,f_n(x)\big) \quad\text{if}\quad x\in X.$$

Then $(f_1,\dots,f_n)$ is an $n$-tuple of members of $C(X)$. Let us put $\epsilon = \epsilon'/\sqrt{n}$. If $(g_1,\dots,g_n)$ is an $n$-tuple of members of $C(X)$ such that $\|f_i-g_i\| < \epsilon$ for $i = 1,\dots,n$ and the function $g:X\to \mathbf{R}^n$ is given by

$$g(x) = \big(g_1(x),\dots,g_n(x)\big),$$

then $\|f(x)-g(x)\| < \epsilon'$ for all $x$ in $X$. It follows that there exists $x_0\in X$ such that $g(x_0) = \mathbf{0}$ and hence $g_i(x_0) = 0$ for $i = 1,\dots,n$. Thus $x_0\in \bigcap_{i=1}^n g_i^{-1}(-\delta,\delta)$ for each positive real number $\delta$. It follows that $\mathrm{dr}\,C(X) \geqslant n$. Thus we have $\mathrm{dr}\,C(X) = n$ as required. $\square$

**3.3 Lemma.** *If $X$ is a topological space and $H$ and $G$ are subsets of $C(X)$ such that $H \subset G$, then*

$$\mathrm{dr}\,(H,G) = \mathrm{dr}\,(\bar{H},\bar{G}).$$

*Proof.* It is easily seen from the definition that

$$\mathrm{dr}\,(H,\bar{G}) \leqslant \mathrm{dr}\,(H,G) \quad\text{and}\quad \mathrm{dr}\,(H,\bar{G}) \leqslant \mathrm{dr}\,(\bar{H},\bar{G}).$$

Now suppose that $\mathrm{dr}\,(H,\bar{G}) \leqslant n$. Let $(f_0,\dots,f_n)$ be an $(n+1)$-tuple of members of $\bar{H}$ and let $\epsilon$ be a positive real number. Then there exists an $(n+1)$-tuple $(h_0,\dots,h_n)$ of members of $H$ such that

$$\|f_i-h_i\| < \epsilon/3$$

for $i = 0, \ldots, n$. Since $\mathrm{dr}\,(H, \bar{G}) \leqslant n$, there exists an $(n+1)$-tuple $(k_0, \ldots, k_n)$ of members of $G$ and a positive real number $\delta$ such that $\|h_i - k_i\| < \epsilon/3$ for $i = 0, \ldots, n$ and $\bigcap_{i=0}^{n} k_i^{-1}(-\delta, \delta) = \varnothing$. Also there exists an $(n+1)$-tuple $(g_0, \ldots, g_n)$ of members of $G$ such that

$$\|k_i - g_i\| < \min\{\epsilon/3, \delta/2\}$$

for $i = 0, \ldots, n$. Thus given an $(n+1)$-tuple $(f_0, \ldots, f_n)$ of members of $H$ and $\epsilon > 0$ we have found an $(n+1)$-tuple $(g_0, \ldots, g_n)$ of members of $G$ and $\delta > 0$ such that $\|f_i - g_i\| < \epsilon$ for $i = 0, \ldots, n$ and

$$\bigcap_{i=0}^{n} g_i^{-1}(-\delta, \delta) = \varnothing.$$

Hence $\mathrm{dr}\,(H, G) \leqslant n$ and $\mathrm{dr}\,(\bar{H}, \bar{G}) \leqslant n$. Thus

$$\mathrm{dr}\,(H, G) = \mathrm{dr}\,(H, \bar{G}) = \mathrm{dr}\,(\bar{H}, \bar{G}). \square$$

We recall from §2 that if $H$ is a subset of $C(X)$, then the smallest subalgebra $K[H]$ of $C(X)$ containing $H$ consists of all functions of the form

$$f = \lambda_0 \iota + \sum_{i=1}^{m} \lambda_i u_i,$$

where $\lambda_0, \ldots, \lambda_m$ are real numbers, $\iota$ is the constant function on $X$ taking the value 1, and each of the functions $u_1, \ldots, u_m$ is the product of finitely many members of $H$. The smallest closed subalgebra $A[H]$ of $C(X)$ containing $H$ is the closure in $C(X)$ of $K[H]$. Now we define $Q[H]$ to be the subset of $C(X)$ consisting of all functions of the form

$$f = \lambda_0 \iota + \sum_{i=1}^{m} \lambda_i u_i,$$

where $\lambda_0, \ldots, \lambda_m$ are rational numbers and each of the functions $u_1, \ldots, u_m$ is the product of finitely many members of $H$.

**3.4  Lemma.** *If $H$ is a subset of $C(X)$, then the smallest closed subalgebra $A[H]$ of $C(X)$ containing $H$ is the closure in $C(X)$ of $Q[H]$.*

*Proof.* Let $f = \lambda_0 \iota + \sum_{i=1}^{m} \lambda_i u_i$, where $\lambda_0, \ldots, \lambda_m$ are real numbers and the functions $u_1, \ldots, u_m$ are finite products of members of $H$. Given $\epsilon > 0$, choose rational numbers $q_0, \ldots, q_m$ such that for $i = 0, \ldots, m$

$$|\lambda_i - q_i| < \epsilon/(m+1) M_i,$$

where $M_i = \|u_i\|$ if $i > 0$ and $M_0 = 1$. If $g = q_0\iota + \Sigma_{i=1}^m q_i u_i$, then $g \in Q[H]$ and

$$\|f - g\| = \left\| (\lambda_0 - q_0) + \sum_{i=1}^m (\lambda_i - q_i) u_i \right\|$$

$$\leqslant |\lambda_0 - q_0| + \sum_{i=1}^m |\lambda_i - q_i| \, \|u_i\| < \epsilon.$$

Thus $f \in (Q[H])^-$. Hence $K[H] \subset (Q[H])^-$ so that $A[H] \subset (Q[H])^-$. The reverse inclusion evidently holds.$\square$

The preceding result is used in the proof of Proposition 3.6. One more lemma is needed to ease that proof.

**3.5 Lemma.** *Let $X$ be a topological space, let $n$ be a non-negative integer and let $\{H_k\}_{k\in\mathbf{N}}$ be a sequence of subsets of $C(X)$ such that*

$$Q[H_k] \subset H_{k+1} \quad and \quad \mathrm{dr}\,(H_k, H_{k+1}) \leqslant n$$

*for all $k$. Then $H = \bigcup_{k\in\mathbf{N}} H_k$ satisfies $H = Q[H]$ and $\mathrm{dr}\,H \leqslant n$.*

*Proof.* Let $f \in Q[H]$. Then $f$ is a multinomial with rational coefficients in members $h_0, \ldots, h_m$ of $H$. Since $\{H_k\}$ is an increasing sequence of sets there exists $k$ such that $h_i \in H_k$ for all $i$. Thus $f \in Q[H_k] \subset H_{k+1}$. Hence $Q[H] = H$.

Now let $(h_0, \ldots, h_n)$ be an $(n+1)$-tuple of members of $H$. Then as we have just noted there exists some $k$ such that $h_i \in H_k$ for $i = 0, \ldots, n$. Since $\mathrm{dr}\,(H_k, H_{k+1}) \leqslant n$, it follows that $\mathrm{dr}\,(H_k, H) \leqslant n$. It is now easily seen that the condition of Definition 3.1 is satisfied for the $(n+1)$-tuple $(h_0, \ldots, h_n)$. Thus $\mathrm{dr}\,H \leqslant n$, as asserted.$\square$

We note that if $H$ is an infinite subset of $C(X)$, then the cardinality of $Q[H]$ is equal to $|H|$.

**3.6 Proposition.** *If $X$ is a normal space and $G$ is a subset of $C(X)$ such that $|G| \leqslant \tau$, where $\tau$ is an infinite cardinal number, then there exists a subset $H$ of $C(X)$ such that $G \subset H = Q[H]$, $|H| \leqslant \tau$ and*

$$\mathrm{dr}\,H = \dim X.$$

*Proof.* Suppose first that $\dim X = n$, where $n$ is a non-negative integer. By Proposition 3.2, $\mathrm{dr}\,C(X) = n$, so that there exist $n$ members $h_1, \ldots, h_n$ of $C(X)$ and some $\epsilon > 0$ for which it is impossible to find members $g_1, \ldots, g_n$ of $C(X)$ and $\delta > 0$ to satisfy the conditions

$$\|h_i - g_i\| < \epsilon$$

for $i = 1, \ldots, n$ and $\bigcap_{i=1}^{n} g_i^{-1}(-\delta, \delta) = \varnothing$. Let $H_1 = G \cup \{h_1, \ldots, h_n\}$. Clearly $|H_1| \leqslant \tau$. Now suppose that we have found subsets $H_1, \ldots, H_k$ of $C(X)$ each of cardinality not exceeding $\tau$ such that $Q[H_i] \subset H_{i+1}$ and $\mathrm{dr}\,(H_i, H_{i+1}) \leqslant n$ for $i = 1, \ldots, k-1$. Let $\Gamma$ be the set of $(n+1)$-tuples of members of $H_k$. Since $\mathrm{dr}\,C(X) = n$, if $\gamma = (h_0, \ldots, h_n)$ is a member of $\Gamma$ and $m$ is a positive integer, then there exist an $(n+1)$-tuple $(g_0, \ldots, g_n)$ of members of $C(X)$ and $\delta > 0$ such that

$$\|g_i - h_i\| < 1/m$$

and $\bigcap_{i=0}^{n} g_i^{-1}(-\delta, \delta) = \varnothing$. Let

$$A_{\gamma m} = \{g \in C(X) \,|\, g = g_i \quad \text{for some} \quad i = 0, \ldots, n\}$$

and let $A_k = \bigcup_{\gamma, m} A_{\gamma m}$. Evidently $|A_k| \leqslant \tau$. Thus if we define

$$H_{k+1} = Q[H_k] \cup A_k,$$

then $|H_{k+1}| \leqslant \tau$, $Q[H_k] \subset H_{k+1}$ and $\mathrm{dr}\,(H_k, H_{k+1}) \leqslant n$ by the construction of $A_k$. It follows by induction that we have a sequence $\{H_k\}$ of subsets of $X$ such that $G \subset H_1$, and for each $k$, $|H_k| \leqslant \tau$, $Q[H_k] \subset H_{k+1}$ and $\mathrm{dr}\,(H_k, H_{k+1}) \leqslant n$. If $H = \bigcup H_k$, then $|H| \leqslant \tau$ and it follows from Lemma 3.5 that $H = Q[H]$ and $\mathrm{dr}\,H \leqslant n$. But it is clear from the construction of $H$ that $\mathrm{dr}\,(H_1, H) \geqslant n$ and hence $\mathrm{dr}\,H \geqslant n$. Thus $\mathrm{dr}\,H = n$.

Now suppose that $\dim X = \infty$. Since $\mathrm{dr}\,C(X) = \infty$, for each non-negative integer $n$ there exist a positive number $\epsilon_n$ and an $(n+1)$-tuple $(h_{n0}, \ldots, h_{nn})$ of members of $C(X)$ such that it is impossible to find an $(n+1)$-tuple $(g_0, \ldots, g_n)$ of members of $C(X)$ and a positive number $\delta$ such that $\|h_i - g_i\| < \epsilon_n$ for each $i$ and $\bigcap_{i=0}^{n} g_i^{-1}(-\delta, \delta) = \varnothing$. Let

$$B = \{h \in C(X) \,|\, h = h_{ni} \text{ for some } n, i\}.$$

If $H = Q[G \cup B]$, then $|H| \leqslant \tau$, $H = Q[H]$ and from the construction of $B$ it is clear that $\mathrm{dr}\,H = \infty$. $\square$

Let $\mathscr{H} = \{h_\lambda\}_{\lambda \in \Lambda}$ be an indexed family of members of $C(X)$. For each $\lambda$ in $\Lambda$, let $P_\lambda$ be the smallest closed interval in $\mathbf{R}$ which contains $h_\lambda(X)$. We shall call the topological product $P = \Pi_{\lambda \in \Lambda} P_\lambda$ the cube associated with $\mathscr{H}$. For each $\lambda$, let $\pi_\lambda : P \to P_\lambda$ be the projection. The unique continuous function $p : X \to P$, such that

$$(\pi_\lambda \circ p)(x) = h_\lambda(x) \quad \text{if} \quad x \in X \quad \text{and} \quad \lambda \in \Lambda,$$

will be called the natural mapping of $X$ into the cube associated with $\mathscr{H}$. Furthermore, let

$$\mathscr{H}^* = \{h \in C(X) \mid h = h_\lambda \text{ for some } \lambda \text{ in } \Lambda\}.$$

We shall employ these definitions and notations throughout the rest of the chapter.

**3.7　Proposition.** *Let $X$ be a topological space, let $\mathscr{H}$ be an indexed family of members of $C(X)$ and let $p: X \to P$ be the natural mapping of $X$ into the cube associated with $\mathscr{H}$. Then*

$$\dim\big(p(X)\big)^- = \mathrm{dr}\, Q[\mathscr{H}^*].$$

*Proof.* Let $Y = \big(p(X)\big)^-$ and let $q: X \to Y$ be given by restriction of $p$. It follows from Proposition 2.3 that

$$q^*: C(Y) \to C(X)$$

is an isometry so that $C(Y)$ is isomorphic with the closed subalgebra $q^*C(Y)$ of $C(X)$. Let $P = \Pi_{\lambda \in \Lambda} P_\lambda$ and for each $\lambda$ let $\pi_\lambda: P \to P_\lambda$ be the projection. Let us regard $\pi_\lambda$ as a member of $C(P)$. Let

$$R = \{\rho \in C(Y) \mid \rho = \pi_\lambda | Y \text{ for some } \lambda\}.$$

Clearly $R$ separates the points of $Y$. Since $Y$ is bicompact it follows from Proposition 2.10 that $A[R] = C(Y)$. Since $q^*(R) = \mathscr{H}^*$ it follows that $q^*C(Y) = A[\mathscr{H}^*]$. By Proposition 3.2, we have $\dim Y = \mathrm{dr}\, C(Y)$. Clearly $\mathrm{dr}\, C(Y) = \mathrm{dr}\, q^*C(Y) = \mathrm{dr}\, A[\mathscr{H}^*]$. But $A[\mathscr{H}^*] = (Q[\mathscr{H}^*])^-$ by Lemma 3.4. Thus $\mathrm{dr}\, A[\mathscr{H}^*] = \mathrm{dr}\, Q[\mathscr{H}^*]$ by Lemma 3.3, so that $\dim Y = \mathrm{dr}\, Q[\mathscr{H}^*]$. $\square$

We come now to the main result of this section. Its statement is rather elaborate in order to include a number of interesting special cases. It follows from Proposition 1.5.16, that if $X$ is a Tihonov space of infinite weight, then there exists a subset $H$ of $C(X)$ such that $|H| = w(X)$ which separates points of $X$ from closed sets.

**3.8　Proposition.** *Let $X$ be a $T_4$-space of infinite weight, let $Y$ be a bicompact space of infinite weight and let $f: X \to Y$ be a continuous function. Then there exist*

　　*(a) a bicompactification $\tilde{X}$ of $X$ such that $w(\tilde{X}) \leqslant \max\{w(X), w(Y)\}$ and an extension $\tilde{f}: \tilde{X} \to Y$ of $f$, and*

　　*(b) a bicompact space $Z$ such that $w(Z) \leqslant w(Y)$, and continuous functions $\phi: \tilde{X} \to Z$ and $\psi: Z \to Y$ with $\tilde{f} = \psi \circ \phi$ such that furthermore*

$$\dim X = \dim \tilde{X} = \dim Z.$$

*Proof.* Let $w(Y) = \tau$ and let $\sigma = \max\{w(X), w(Y)\}$. Let $\Lambda_0$ be a set of cardinality $\sigma$ and let $\Lambda_1, \Lambda_2$ be subsets of $\Lambda_0$ each of cardinality $\tau$ such that $\Lambda_2 \subset \Lambda_1$, the set $\Lambda_1 \backslash \Lambda_2$ is of cardinality $\tau$, and $\Lambda_0 \backslash \Lambda_1$ is of cardinality $\sigma$. Since $w(Y) = \tau$, we can find a family $\{g_\lambda\}_{\lambda \in \Lambda_2}$ of members of $C(Y)$ which separates points of $Y$ from closed sets. For each $\lambda$ in $\Lambda_2$ let $h_\lambda = g_\lambda \circ f \in C(X)$ and let $\mathscr{H}_2 = \{h_\lambda\}_{\lambda \in \Lambda_2}$. If

$$\mathscr{H}_2^* = \{h \in C(X) \mid h = h_\lambda \text{ for some } \lambda \text{ in } \Lambda_2\},$$

then the cardinality of $\mathscr{H}_2^*$ does not exceed $\tau$, and by Proposition 3.6 there exists a subset $H_1$ of $C(X)$ of cardinality not exceeding $\tau$ such that $\mathscr{H}_2^* \subset H_1 = Q[H_1]$ and $\mathrm{dr}\, H_1 = \dim X$. Let $B$ be a subset of $C(X)$ which separates points of $X$ from closed sets and is of cardinality equal to $w(X)$. Then $H_1 \cup B$ is a subset of $C(X)$ of cardinality not exceeding $\sigma$ so that by Proposition 3.6 there exists a subset $H_0$ of $C(X)$ of cardinality not exceeding $\sigma$ such that $H_1 \cup B \subset H_0 = Q[H_0]$ and $\mathrm{dr}\, H_0 = \dim X$. Let us index the members of $H_1 \backslash \mathscr{H}_2^*$ by $\Lambda_1 \backslash \Lambda_2$ and the members of $H_0 \backslash H_1$ by $\Lambda_0 \backslash \Lambda_1$ so that we obtain families $\mathscr{H}_i = \{h_\lambda\}_{\lambda \in \Lambda_i}$ for $i = 0, 1, 2$ such that for $i = 0, 1$,

$$H_i = \mathscr{H}_i^* = \{h \in C(X) \mid h = h_\lambda \text{ for some } \lambda \text{ in } \Lambda_i\}.$$

For $i = 0, 1, 2$ let $P_i$ be the cube associated with $\mathscr{H}_i$ and let $p_i: X \to P_i$ be the natural mapping of $X$ into the cube associated with $\mathscr{H}_i$. Evidently there are projections $P_0 \overset{q_0}{\to} P_1 \overset{q_1}{\to} P_2$ such that the diagram

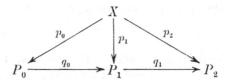

is commutative. Since $B$ separates points of $X$ from closed sets, the family $\mathscr{H}_0$ separates points of $X$ from closed sets so that $p_0$ is an embedding. Thus if $\tilde{X} = (p_0(X))^-$, then $\tilde{X}$ is a bicompactification of $X$. Since $X \subset P_0$, it follows that $w(\tilde{X}) \leqslant \sigma$. By Proposition 3.7,

$$\dim X = \mathrm{dr}\, Q[\mathscr{H}_0^*] = \mathrm{dr}\, Q[H_0] = \mathrm{dr}\, H_0 = \dim X.$$

The family $\mathscr{G} = \{g_\lambda\}_{\lambda \in \Lambda_2}$ separates points of $Y$ from closed sets. Thus if $R$ is the cube associated with $\mathscr{G}$, then the natural mapping $r: Y \to R$ is an embedding. Since $h_\lambda = g_\lambda \circ f$ for each $\lambda$ in $\Lambda_2$, it follows that $P_2$ is a subspace of $R$ and $rf(x) = p_2(x)$ for each $x$ in $X$. Since $r(Y)$ is closed in $R$ and $q_1 q_0 p_0(X) = p_2(X) \subset r(Y)$, it follows that $q_1 q_0(\tilde{X}) \subset r(Y)$. Let $\theta: r(Y) \to Y$ be the homeomorphism which is inverse to $r$. Then we can define $\tilde{f}\colon \tilde{X} \to Y$ by putting $\tilde{f}(w) = \theta q_1 q_0(w)$ if $w \in \tilde{X}$. Clearly $\tilde{f}$ is an extension of $f$. Finally let $Z = \left(p_1(X)\right)^-$. Then $Z$ is a bicompact space and since $Z \subset P_1$, it follows that $w(Z) \leqslant \tau$. By Proposition 3.7,

$$\dim Z = \operatorname{dr} Q[\mathscr{H}_1^*] = \operatorname{dr} Q[H_1] = \operatorname{dr} H_1 = \dim X.$$

Moreover $q_0(\tilde{X}) \subset Z$ and $q_1(Z) \subset r(Y)$. Thus we can define continuous functions $\phi\colon \tilde{X} \to Z$ and $\psi\colon Z \to Y$ by putting $\phi(w) = q_0(w)$ if $w \in \tilde{X}$ and $\psi(z) = \theta q_1(z)$ if $z \in Z$. Evidently $\tilde{f} = \psi \circ \phi$. $\square$

Ignoring some of the information contained in Proposition 3.8 we obtain a result which will have an important application later.

**3.9   Corollary.** *If $X$ is a $T_4$-space of infinite weight, $Y$ is a bicompact space of infinite weight and $f\colon X \to Y$ is a continuous function, then there exist a bicompact space $Z$ such that*

$$\dim Z = \dim X \quad and \quad w(Z) \leqslant w(Y),$$

*and continuous functions $\theta\colon X \to Z$ and $\psi\colon Z \to Y$ such that $\psi \circ \theta = f$.*

*Proof.* This follows immediately from Proposition 3.8, taking $\theta$ to be the composite of the inclusion of $X$ in $\tilde{X}$ and $\phi$. $\square$

**3.10   Remark.** If $X$ is a $T_4$-space of infinite weight, $Y$ is a bicompact space of infinite weight and $f\colon X \to Y$ is a continuous function such that $f(X)$ is dense in $Y$, then the bicompactification $\tilde{X}$ of $X$ given by Proposition 3.8 satisfies

$$w(\tilde{X}) = \max\{w(X), w(Y)\}.$$

For the extension $\tilde{f}\colon \tilde{X} \to Y$ must be a surjection so that it follows from Proposition 1.5.6 that $w(Y) \leqslant w(\tilde{X})$. Since $w(X) \leqslant w(\tilde{X})$ it follows that $w(\tilde{X}) \geqslant \max\{w(X), w(Y)\}$ so that the asserted equality follows from Proposition 3.8. Furthermore the continuous function $\psi\colon Z \to Y$ must also be a surjection. It follows from Proposition 1.5.6 that $w(Y) \leqslant w(Z)$ and hence $w(Y) = w(Z)$. We note that by construction the continuous function $\phi\colon \tilde{X} \to Z$ is a surjection.

As a special case we have the following factorization theorem.

**3.11   Proposition.** *If $X$ and $Y$ are bicompact spaces of infinite weight and $f\colon X \to Y$ is a continuous surjection, then there exist a bicompact space $Z$ such that*

$$\dim Z = \dim X \quad and \quad w(Z) = w(Y),$$

*and continuous surjections $g\colon X \to Z$ and $h\colon Z \to Y$ such that $f = h \circ g$.* $\square$

**3.12   Proposition.** *If $X$ is a Tihonov space and $Y$ is a bicompactification of $X$, then there exists a bicompactification $\tilde{X}$ of $X$ which follows $Y$ such that*

$$\dim \tilde{X} = \partial\mathrm{im}\, X \quad and \quad w(\tilde{X}) = w(Y).$$

*Proof.* Let us apply the factorization theorem to the natural identification mapping $f:\beta X \to Y$. By Proposition 3.11 there exist a bicompact space $\tilde{X}$ such that

$$\dim \tilde{X} = \dim \beta X = \partial\mathrm{im}\, X$$

and $w(\tilde{X}) = w(Y)$, and continuous surjections $g:\beta X \to \tilde{X}$ and $h:\tilde{X}\to Y$ such that $f = h \circ g$. It follows from Proposition 1.6.5 that $\tilde{X}$ is a bicompactification of $X$ which follows $Y$.□

**3.13   Corollary.** *If $X$ is a Tihonov space, then there exists a bicompactification $\tilde{X}$ of $X$ such that*

$$\dim \tilde{X} = \partial\mathrm{im}\, X \quad and \quad w(\tilde{X}) = w(X).$$

*Proof.* This follows at once from Proposition 3.12 on taking $Y$ to be a bicompactification of $X$ such that $w(Y) = w(X)$. It follows from Proposition 1.5.16 that such a bicompactification exists.□

This section concludes with another consequence of the factorization theorem which is complementary to Proposition 3.12. For the proof, a remark about families of functions which separate points from closed sets is needed. Let $X$ be a Tihonov space and let $\mathscr{H} = \{h_\lambda\}_{\lambda\in\Lambda}$ be a family of members of $C(X)$ which separates points of $X$ from closed sets. Since the natural mapping of $X$ into the cube associated with $\mathscr{H}$ is an embedding, it follows that $|\Lambda| \geqslant w(X)$. We shall now show that we can choose a subset $\Lambda_1$ of $\Lambda$ such that $|\Lambda_1| = w(X)$ and the family $\{h_\lambda\}_{\lambda\in\Lambda_1}$ separates points of $X$ from closed sets. Let $\mathscr{B}$ be a base for the topology of $X$ such that $|\mathscr{B}| = w(X)$, and consider the set of ordered pairs $(B_1, B_2)$ of members of $\mathscr{B}$ for which there exists some $\lambda$ in $\Lambda$ such that $(h_\lambda(B_2))^-$ is disjoint from $(h_\lambda(X\backslash B_1))^-$. For each such pair, choose one such $\lambda$ and let $\Lambda_1$ be the subset of $\Lambda$ thus obtained. Evidently $|\Lambda_1| \leqslant w(X)$. Now let $x$ be a point of $X$ and let $F$ be a closed set of $X$ such that $x \notin F$. There exists a member $B_1$ of $\mathscr{B}$ such that $x\in B_1$ and $B_1$ is disjoint from $F$. There exists $\mu$ in $\Lambda$ such that $h_\mu(x) \notin (h_\mu(X\backslash B_1))^-$, and there exists $B_2$ in $\mathscr{B}$ such that $x\in B_2$ and $(h_\mu(B_2))^-$ is disjoint from $(h_\mu(X\backslash B_1))^-$. If $\lambda$ is the member of $\Lambda_1$

associated with $(B_1, B_2)$, then $\left(h_\lambda(B_2)\right)^-$ is disjoint from $\left(h_\lambda(X\backslash B_1)\right)^-$. But $h_\lambda(x) \in h_\lambda(B_2)$ and $\left(h_\lambda(F)\right)^- \subset \left(h_\lambda(X\backslash B_1)\right)^-$. Thus the family

$$\mathscr{H}_1 = \{h_\lambda\}_{\lambda \in \Lambda_1}$$

separates points of $X$ from closed sets. Since $|\Lambda_1| \leqslant w(X)$ and the natural mapping of $X$ into the cube associated with $\mathscr{H}_1$ is an embedding, it follows that $|\Lambda_1| = w(X)$.

**3.14 Proposition.** *If $X$ is a Tihonov space and $Y$ is a bicompactification of $X$, then there exists a bicompactification $\tilde{X}$ of $X$ which precedes $Y$ such that*

$$\dim \tilde{X} = \dim Y \quad and \quad w(\tilde{X}) = w(X).$$

*Proof.* Let $\Lambda$ be a set of cardinality equal to $w(Y)$ and let $\{f_\lambda\}_{\lambda \in \Lambda}$ be a family of members of $C(Y)$ which separates points of $Y$ from closed sets. For each $\lambda$, let $h_\lambda = f_\lambda \circ j$, where $j : X \to Y$ is the natural embedding. Then $\mathscr{H} = \{h_\lambda\}_{\lambda \in \Lambda}$ is a family of members of $C(X)$ which separates points of $X$ from closed sets. It follows from the remark above that there exists a subset $\Lambda_1$ of $\Lambda$ such that $|\Lambda_1| = w(X)$ and the family $\mathscr{H}_1 = \{h_\lambda\}_{\lambda \in \Lambda_1}$ separates points of $X$ from closed sets. Let $P$ and $P_1$ be the cubes associated with $\mathscr{H}$ and $\mathscr{H}_1$ respectively, let $p : X \to P$ and $p_1 : X \to P_1$ be the natural mappings and let $q : P \to P_1$ be the projection. Since $\mathscr{H}_1$ separates points of $X$ from closed sets, the subspace $Z = \left(p_1(X)\right)^-$ of $P_1$ is a bicompactification of $X$ such that $w(Z) = w(X)$. Since $Y$ is a bicompactification of $X$, the smallest closed interval in $\mathbf{R}$ which contains $h_\lambda(X)$ is the smallest closed interval which contains $f_\lambda(Y)$ for $\lambda$ in $\Lambda$. Thus there exists an embedding $f : Y \to P$ such that $f \circ j = p$. The function $\theta : Y \to Z$ given by $\theta(y) = qf(y)$ if $y \in Y$ is a continuous surjection such that $\theta j(x) = p_1(x)$ if $x \in X$. Thus the bicompactification $Z$ precedes $Y$ and $\theta : Y \to Z$ is the natural identification mapping. It follows from Proposition 3.11 that there exist a bicompact space $\tilde{X}$ such that $\dim \tilde{X} = \dim Y$ and $w(\tilde{X}) = w(Z) = w(X)$ and continuous surjections $\phi : Y \to \tilde{X}$ and $\psi : \tilde{X} \to Z$ such that $\psi \circ \phi = \theta$. By Proposition 1.6.5, $\tilde{X}$ is a bicompactification of $X$ which precedes $Y$. $\square$

## 4   Analytic dimension

A closed subalgebra $L$ of a normed algebra $A$ is an *analytic subalgebra* if each $x$ in $A$ which satisfies an equation of the form

$$x^n + a_1 x^{n-1} + \ldots + a_n = 0,$$

where $a_1, \ldots, a_n \in L$, is a member of $L$.

Trivially $A$ is an analytic subalgebra of $A$ and it is clear that the intersection of analytic subalgebras of $A$ is an analytic subalgebra of $A$. Thus if $H$ is a subset of $A$ there exists a smallest analytic subalgebra $L[H]$ of $A$ which contains $H$. If $A$ and $B$ are complete normed algebras and $\phi:A\to B$ is an isometry, then $\phi(L)$ is an analytic subalgebra of the closed subalgebra $\phi(A)$ of $B$ for each analytic subalgebra $L$ of $A$. Furthermore for each subset $H$ of $A$, $\phi(L[H])$ is the smallest analytic subalgebra of $\phi(A)$ which contains $\phi(H)$.

We shall find a simpler description of the analytic subalgebras of $C(X)$ for a bicompact space $X$.

**4.1 Lemma.** *Let $X$ be a bicompact space and let $L$ be a closed subalgebra of $C(X)$ with the property that if $f\in C(X)$ and $f^2\in L$ then $f\in L$. Then every constancy set of $L$ is connected.*

*Proof.* Suppose that $Y$ is a constancy set of $L$ which is not connected. Then $Y = Y_1 \cup Y_2$, where $Y_1$ and $Y_2$ are disjoint non-empty sets which are open-and-closed in $Y$. Since $Y$ is a closed set, $Y_1$ and $Y_2$ are closed in $X$. Since $X$ is a normal space there exist $U_1, U_2$ open in $X$ such that $Y_i \subset U_i$ for $i = 1, 2$ and $\overline{U}_1, \overline{U}_2$ are disjoint. Let $x_1 \in Y_1$ and let $M$ be the subset of $L$ consisting of those functions $f$ such that $f(x_1) = 0$. If $f\in L$ and $g$ is defined by

$$g(x) = f(x)-f(x_1) \quad \text{if} \quad x\in X,$$

then $g\in L$ and $g(x_1) = 0$. Thus $M$ is a non-empty subset of $L$. Also

$$Y = \bigcap_{g\in M} Z(g),$$

where as usual $Z(g)$ denotes the zero-set of $g$. For it is clear that $Y \subset \bigcap_{g\in M}Z(g)$, and the argument which shows that $M$ is non-empty also shows that each member of $L$ is constant on $Z(g)$ for some $g$ in $M$, so that $\bigcap_{g\in M}Z(g) \subset Y$. Since each set $Z(g)$ is closed in the compact space $X$ and $\bigcap_{g\in M}Z(g) \subset U_1 \cup U_2$, there exist members $g_1,\ldots,g_m$ of $M$ such that

$$\bigcap_{i=1}^{m} Z(g_i) \subset U_1 \cup U_2.$$

If $h = \Sigma_{i=1}^m g_i^2$, then $h\in M$ and $Z(h) = \bigcap_{i=1}^m Z(g_i)$ so that

$$Y \subset Z(h) \subset U_1 \cup U_2.$$

For each positive integer $n$, let

$$G_n = \{x\in U_1 \mid h(x) < 1/n\}.$$

Then $\bigcap_{n\in\mathbf{N}}G_n \subset \overline{U}_1 \cap Z(h) \subset U_1$ so that there exists $k$ such that $\overline{G}_k \subset U_1$. Now let $f = \iota - kh$. Then $f \in L$ and $f(x) = 0$ if $x \in \overline{G}_k\backslash G_k$. Thus we can define $\phi$ in $C(X)$ by

$$\phi(x) = \begin{cases} f(x) & \text{if} \quad x \in \overline{G}_k, \\ -f(x) & \text{if} \quad x \in X\backslash G_k. \end{cases}$$

Since $\phi^2 = f^2$ it follows that $\phi^2 \in L$ so that by hypothesis $\phi \in L$. Thus $\phi$ is constant on $Y$. Since $h(x_1) = 0$, it follows that $f(x_1) = 1$ and hence $\phi(x_1) = 1$. But if $x_2 \in Y_2$ then $f(x_2) = 1$ since $f$ is constant on $Y$. Since $x_2 \notin G_k$ we must have $\phi(x_2) = -1$ which is absurd. Hence every constancy set of $L$ is connected.$\square$

**4.2  Proposition.** *If $X$ is a bicompact space, then a closed subalgebra $L$ of $C(X)$ is an analytic subalgebra if and only if $L$ has the property that $f \in L$ if $f$ is a member of $C(X)$ such that $f^2 \in L$.*

*Proof.* Clearly an analytic subalgebra has this property. Now let $L$ be a closed subalgebra of $C(X)$ with the property, and let $f$ in $C(X)$ satisfy the equation

$$f^n + h_1 f^{n-1} + \dots + h_{n-1}f + h_n = 0,$$

where $h_1, \dots, h_n \in L$. Let $Y$ be a constancy set of $L$ and let $h_i$ take the value $a_i$ on $Y$. Then if $x \in Y$,

$$f(x)^n + a_1 f(x)^{n-1} + \dots + a_{n-1}f(x) + a_n = 0.$$

Since a polynomial equation with real coefficients has finitely many roots, $f$ takes only finitely many values on $Y$. But $Y$ is connected by Lemma 4.1. Hence $f$ must be constant on $Y$. Since $L$ is a closed subalgebra of $C(X)$, it follows from Proposition 2.9 that $f \in L$. Thus $L$ is an analytic subalgebra of $C(X)$.$\square$

We shall next determine $L[H]$ for a subset $H$ of $C(X)$ where $X$ is a bicompact space.

**4.3  Lemma.** *If $X$ is a bicompact space and $M$ is a connected subset of $X$, then the subset of $C(X)$ which consists of all functions which are constant on $M$ is an analytic subalgebra of $C(X)$.*

*Proof.* Let $L$ be the subset of $C(X)$ consisting of all functions which are constant on $M$. By Lemma 2.8, $L$ is a closed subalgebra of $C(X)$. Suppose that $f \in C(X)$ and $f^2 \in L$ so that $f^2$ is constant on $M$. Then $f$

can assume at most two values on $M$, so that since $M$ is connected, $f$ must be constant on $M$. Hence $f \in L$. It follows from Proposition 4.2 that $L$ is an analytic subalgebra of $C(X)$. $\square$

**4.4   Proposition.** *Let $X$ be a bicompact space and let $H$ be a subset of $C(X)$. Then $L[H]$ is the subset of $C(X)$ consisting of all functions which are constant on each connected subset of each constancy set of $H$.*

*Proof.* Let $B$ be the subset of $C(X)$ consisting of all functions which are constant on each connected subset of each constancy set of $H$. It follows from Lemma 4.3 that $B$ is an intersection of analytic subalgebras of $C(X)$, and hence $B$ is an analytic subalgebra of $C(X)$. Furthermore $H \subset B$ so that $L[H] \subset B$. Now each constancy set of $L[H]$ is connected by Lemma 4.1, and is contained in a constancy set of $L[H]$. Since $L[H]$ is a closed subalgebra of $C(X)$, it follows from Proposition 2.10 that $B \subset L[H]$ and hence $B = L[H]$. $\square$

Let $H$ be a subset of a normed algebra $A$. We define the *rank*, $\operatorname{rk} H$, of $H$ to be the least cardinal number $\tau$ for which there exists a subset $K$ of $A$ such that $|K| = \tau$ and $H \subset L[K]$. In particular $\operatorname{rk} A$ is the least cardinal number $\tau$ for which there exists a subset $K$ of $A$ such that $|K| = \tau$ and $A = L[K]$. A subset $K$ of $A$ such that $L[K] = A$ will be called an *analytic base* of $A$. Thus $\operatorname{rk} A$ is the least of the cardinal numbers of analytic bases of $A$.

The *analytic dimension*, $\operatorname{ad} A$, of a normed algebra $A$ is the least cardinal number $\tau$ such that each finite subset $H$ of $A$ has rank not exceeding $\tau$. Thus if $\operatorname{ad} A = \tau$, and $H$ is a finite subset of $A$, then $\operatorname{rk} H \leqslant \tau$ so that there exists a subset $K$ of $A$ such that $|K| \leqslant \tau$ and $H \subset L[K]$. We observe that $\operatorname{ad} A = 0$ and only if every finite subset of $A$ is contained in $L[\varnothing]$ and hence $L[\varnothing] = A$. It follows that $A$ has no proper analytic subalgebra.

If $H$ is a finite subset of $A$, then $H \subset L[H]$ and it follows that $\operatorname{rk} H < \aleph_0$. Hence $0 \leqslant \operatorname{ad} A \leqslant \aleph_0$. It is clear that $\operatorname{ad} A \leqslant \operatorname{rk} A$. Suppose that $\operatorname{rk} A$ is finite, and let $H$ be an analytic base of $A$ of cardinality equal to $\operatorname{rk} A$. Then $\operatorname{rk} H = \operatorname{rk} A$ and since $H$ is finite it follows that $\operatorname{rk} A \leqslant \operatorname{ad} A$. Thus if $\operatorname{rk} A$ is finite then $\operatorname{rk} A = \operatorname{ad} A$. We note that if $\operatorname{ad} A = 0$ then $A$ has no proper analytic subalgebra and it follows that $\operatorname{rk} A = 0$.

Clearly $\operatorname{rk} A = \operatorname{rk} B$ and $\operatorname{ad} A = \operatorname{ad} B$ if there exists an isometry of $A$ onto $B$. It follows that if $X$ is a Tihonov space, then

$$\operatorname{rk} C(X) = \operatorname{rk} C(\beta X)$$

and $\operatorname{ad} C(X) = \operatorname{ad} C(\beta X)$. We shall prove that if $X$ is a Tihonov space then

$$\partial \operatorname{im} X = \operatorname{ad} C(X).$$

In view of the identities $\partial \operatorname{im} X = \dim \beta X$ and $\operatorname{ad} C(X) = \operatorname{ad} C(\beta X)$ it is enough to establish that if $Y$ is a bicompact space then

$$\dim Y = \operatorname{ad} C(Y).$$

We begin with the result in dimension zero.

**4.5  Proposition.** *If $X$ is a bicompact space, then $\operatorname{ad} C(X) = 0$ if and only if $\dim X = 0$.*

*Proof.* We noted above that $\operatorname{ad} C(X) = 0$ if and only if $C(X)$ has no proper analytic subalgebra. It follows from Lemma 4.3 that if $X$ is bicompact and $C(X)$ has no proper analytic subalgebra, then $X$ has no connected subset containing more than one point. If the bicompact space $X$ is totally disconnected, it follows from Proposition 4.4 that $X$ has no proper analytic subalgebra. Thus if $X$ is a bicompact space, $\operatorname{ad} C(X) = 0$ if and only if $X$ is totally disconnected. But by Proposition 3.1.3, a bicompact space $X$ is totally disconnected if and only if $\dim X = 0$. $\square$

For the proof that $\dim X = \operatorname{ad} C(X)$ for a bicompact space $X$ such that $\dim X > 0$, a great deal of machinery must be developed. We begin with some concepts of independent interest.

An open covering $\mathscr{U}$ of a space $X$ is said to *envelop* a subset $S$ of $X$ if $S = \bigcup_{\lambda \in \Lambda} S_\lambda$, where $\{S_\lambda\}_{\lambda \in \Lambda}$ is a disjoint family of sets such that each $S_\lambda$ is open in $S$ and is contained in some member of $\mathscr{U}$.

  In the case of a bicompact subset there is a simpler characterization of envelopment.

**4.6  Proposition.** *Let $\mathscr{U}$ be an open covering of a space $X$ and let $K$ be a bicompact subset of $X$. Then $\mathscr{U}$ envelopes $K$ if and only if each connected subset of $K$ is contained in some member of $\mathscr{U}$.*

*Proof.* If $\mathscr{U}$ envelops $K$ then since $K$ is compact, $K = \bigcup_{i=1}^{m} K_i$, where the $K_i$ are disjoint sets open in $K$ and each $K_i$ is contained in some member of $\mathscr{U}$. Clearly each connected subset of $K$ must be contained in some set $K_i$. Conversely, suppose that each connected subset of the bicompact set $K$ is contained in some member of $\mathscr{U}$. For each point $x$ of $K$ let $S_x$ be the intersection of all the open-and-closed subsets of $K$ which contain $x$. Since $K$ is bicompact, it follows from Proposition 1.5.5 that $S_x$ is connected so that, by hypothesis, there exists some member $U_x$ of $\mathcal{U}$ which contains $S_x$, and it follows from Remark 1.5.2 (3) that there exists an open-and-closed set $H_x$ of $K$ such that $S_x \subset H_x \subset U_x$. Since $K$ is compact, the covering $\{H_x\}_{x \in K}$ contains a finite subcovering $\{H_1, \ldots, H_m\}$. Let $G_1 = H_1$ and for $1 < j \leqslant m$ let

$$G_j = H_j \setminus \bigcup_{i<j} H_i.$$

Since each $H_i$ is an open-and-closed set, $\{G_1, \ldots, G_m\}$ is a disjoint covering of $K$ by open sets of $K$ and each $G_i$ is contained in some member of $\mathcal{U}$. $\square$

We recall that if $X$ is a topological space and $n$ is a positive integer, then $C_n(X)$ is the complete metric space of continuous functions $f: X \to \mathbf{R}^n$ such that $f(X)$ is a bounded subset of $\mathbf{R}^n$. In particular $C_1(X) = C(X)$. If $f \in C_n(X)$, then

$$\|f\| = \sup_{x \in X} \|f(x)\|$$

and the metric $d$ in $C_n(X)$ is defined by

$$d(f, g) = \|f - g\| \quad \text{if} \quad f, g \in C_n(X).$$

We define $C_0(X)$ to be the one-point subspace of $C(X)$ which consists of the constant function taking the value 0 on $X$.

If $\mathcal{U}$ is an open covering of a space $X$, $n$ is a non-negative integer and $f \in C_n(X)$, then $f$ is said to be *zero-dimensional with respect to* $\mathcal{U}$ if there exists a positive real number $\epsilon$ such that $\mathcal{U}$ envelops every subset $S$ of $X$ such that $\operatorname{diam} f(S) < \epsilon$. We shall see that if $\mathcal{U}$ is a locally finite open covering of a normal space $X$ such that $\dim X \leqslant n$, then there is a plentiful supply of mappings in $C_n(X)$ which are zero-dimensional with respect to $\mathcal{U}$. A lemma is needed.

**4.7 Lemma.** *Let $\epsilon$ be a positive real number, let $n$ be a positive integer, let $\mathcal{H} = \{H_\lambda\}_{\lambda \in \Lambda}$ be a locally finite open covering of order not exceeding $n$ of a normal space $X$ and let $f$ be a member of $C_n(X)$ such that*

$$\operatorname{diam} f(H_\lambda) < \epsilon/2n$$

*for all $\lambda$. Then there exists $g$ in $C_n(X)$ which is zero-dimensional with respect to $\mathcal{H}$ and satisfies $\|g - f\| < \epsilon$.*

*Proof.* Let $\alpha, \beta$ be positive real numbers and for each positive integer $n$ consider the statement $(a_n)$: if $\mathscr{H} = \{H_\lambda\}_{\lambda \in \Lambda}$ is a locally finite open covering of order not exceeding $n$ of a normal space $X$, and

$$f_1, \ldots, f_n \in C(X)$$

and satisfy $\operatorname{diam} f_k(H_\lambda) < \alpha$ for all $k, \lambda$, then there exist $g_1, \ldots, g_n$ in $C(X)$ such that $\|f_k - g_k\| \leqslant \alpha + \beta$ for $k = 1, \ldots, n$ and $\mathscr{H}$ envelops each subset $S$ of $X$ such that $\operatorname{diam} g_k(S) < \beta$ for $k = 1, \ldots, n$.

We shall prove by induction that $(a_n)$ is true for all $n$. We suppose that $(a_{n-1})$ is true and prove $(a_n)$. Let $\{G_\lambda\}_{\lambda \in \Lambda}$ be a locally finite open covering of $X$ such that $\bar{G}_\lambda \subset H_\lambda$ for each $\lambda$. Let $\Gamma$ be the set of subsets of $\Lambda$ with precisely $n+1$ elements and if $\gamma \in \Gamma$ let

$$U_\gamma = \bigcap_{\lambda \in \gamma} H_\gamma \quad \text{and} \quad V_\gamma = \bigcap_{\lambda \in \gamma} G_\lambda.$$

Then $\{U_\gamma\}_{\gamma \in \Gamma}$ is a disjoint family of open sets of $X$. Let $Y = X \backslash \bigcup_{\gamma \in \Gamma} V_\gamma$. Then $Y$ is a normal space and $\{G_\lambda \cap Y\}_{\lambda \in \Lambda}$ is a locally finite open covering of $Y$ of order not exceeding $n-1$. For each $k = 1, \ldots, n-1$ and each $\lambda$, $\operatorname{diam} f_k(G_\lambda \cap Y) < \alpha$. Thus there exist $h_1, \ldots, h_{n-1}$ in $C(Y)$ such that $|f_k(y) - h_k(y)| \leqslant \alpha + \beta$ if $y \in Y$ and $\{G_\lambda \cap Y\}_{\lambda \in \Lambda}$ envelops each subset $T$ of $Y$ such that $\operatorname{diam} h_k(T) < \beta$ for $k = 1, \ldots, n-1$. By the Tietze–Urysohn extension theorem, for each $k$ there exists $\phi_k$ in $C(X)$ such that $\|\phi_k\| \leqslant \alpha + \beta$ and $\phi_k(y) = h_k(y) - f_k(y)$ if $y \in Y$. Let $g_k = f_k + \phi_k$. Then we have found $g_1, \ldots, g_{n-1}$ in $C(X)$ such that $\|f_k - g_k\| \leqslant \alpha + \beta$ for each $k$ and $\mathscr{H}$ envelops each subset $T$ of $Y$ such that $\operatorname{diam} g_k(T) < \beta$ for each $k$.

Now let

$$A = \bigcup_{\gamma \in \Gamma} \bar{V}_\gamma \quad \text{and} \quad B = X \backslash \bigcup_{\gamma \in \Gamma} U_\gamma.$$

Then $A$ and $B$ are disjoint closed sets of $X$. Hence there exists $\phi_n$ in $C(X)$ such that $0 \leqslant \phi_n(x) \leqslant \alpha + \beta$ for all $x$, $\phi_n(x) = 0$ if $x \in A$ and $\phi_n(x) = \alpha + \beta$ if $x \in B$. If we put $g_n = f_n + \phi_n$, then it is clear that $\|f_n - g_n\| \leqslant \alpha + \beta$.

Let $S$ be a subset of $X$ such that $\operatorname{diam} g_k(S) < \beta$ for $k = 1, \ldots, n$. Suppose that $\gamma_0 \in \Gamma$ and $S \cap \bar{V}_{\gamma_0} \neq \varnothing$. If $x \in S \cap \bar{V}_{\gamma_0}$ and $y \in S \cap \bar{U}_{\gamma_0}$, then $|f_n(x) - f_n(y)| \leqslant \alpha$ and $|g_n(x) - g_n(y)| < \beta$ so that

$$|\phi_n(x) - \phi_n(y)| < \alpha + \beta.$$

Since $x \in A$, it follows that $\phi_n(x) = 0$ so that $\phi_n(y) < \alpha + \beta$ and we see that $y \notin B$. Thus $y \in U_\gamma$ for some $\gamma$ and since $\{U_\gamma\}$ is a disjoint family of open sets it follows that $y \in U_{\gamma_0}$. Let $\Delta = \{\gamma \in \Gamma \mid S \cap \bar{V}_\gamma \neq \varnothing\}$. If

$\gamma \in \Delta$, then $S \cap \overline{U}_\gamma = S \cap U_\gamma$, so that $S \cap U_\gamma$ is open-and-closed in $S$. Let $W = \bigcup_{\gamma \in \Delta} S \cap U_\gamma$. Since $\{U_\gamma\}$ is a discrete family, it follows that $W$ is an open-and-closed set in $S$ and that $\mathscr{H}$ envelops $W$. Clearly $S \backslash W$ is disjoint from $\overline{V}_\gamma$ for each $\gamma$ in $\Gamma$ so that $S \backslash W \subset X \backslash \bigcup_{\gamma \in \Gamma} \overline{V}_\gamma \subset Y$. It follows that $\mathscr{H}$ envelops $S \backslash W$. Thus $\mathscr{H}$ envelops $S$ which completes the proof of $(a_n)$. The construction of $g_n$ above supplies the method of proof of $(a_1)$. Thus $(a_n)$ is true for all $n$.

Let us return to the hypotheses of the lemma. For each $x$ in $X$ let

$$f(x) = \big(f_1(x), \ldots, f_n(x)\big).$$

Then $f_1, \ldots, f_n \in C(X)$ and $\operatorname{diam} f_k(H_\lambda) < \epsilon/2n$ for all $k, \lambda$. Applying $(a_n)$ with $\alpha = \epsilon/2n$ and $\beta = \epsilon/4n$ we obtain $g_1, \ldots, g_n$ in $C(X)$ such that $\|f_k - g_k\| \leqslant 3\epsilon/4n$ for $k = 1, \ldots, n$ and $\mathscr{H}$ envelops each subset $S$ of $X$ such that $\operatorname{diam} g_k(S) < \epsilon/4n$ for $k = 1, \ldots, n$. Let $g$ be the member of $C_n(X)$ defined by

$$g(x) = \big(g_1(x), \ldots, g_n(x)\big) \quad \text{if} \quad x \in X.$$

Then $\|f - g\| \leqslant 3\epsilon/4 < \epsilon$, and if $S$ is a subset of $X$ such that

$$\operatorname{diam} g(S) < \epsilon/4n,$$

then $\operatorname{diam} g_k(S) < \epsilon/4n$ for $k = 1, \ldots, n$ so that $\mathscr{H}$ envelops $S$. Thus $g$ is zero-dimensional with respect to $\mathscr{H}$. $\square$

**4.8  Proposition.** *If $\mathscr{U}$ is a locally finite open covering of a normal space $X$ such that $\dim X \leqslant n$, then the set*

$$\{g \in C_n(X) \mid g \text{ is zero-dimensional w.r.t. } \mathscr{U}\}$$

*is a dense open subset of $C_n(X)$.*

*Proof.* If $\dim X = 0$, then $\mathscr{U}$ has a disjoint open refinement so that $\mathscr{U}$ envelops every subset of $X$. Thus the unique member of $C_0(X)$ is zero-dimensional with respect to $\mathscr{U}$. Now let $n$ be a positive integer and let

$$G = \{g \in C_n(X) \mid g \text{ is zero-dimensional w.r.t. } \mathscr{U}\}.$$

If $g \in G$, then there exists $\epsilon > 0$ such that $\mathscr{U}$ envelops each subset $S$ of $X$ such that $\operatorname{diam} g(S) < \epsilon$. Suppose that $h \in C_n(X)$ and $\|h - g\| < \epsilon/3$. If $S$ is a subset of $X$ such that $\operatorname{diam} h(S) < \epsilon/3$, then $\operatorname{diam} g(S) < \epsilon$ so that $\mathscr{U}$ envelops $S$. Thus $h \in G$. Hence $G$ is open in $C_n(X)$.

It remains to show that $G$ is dense in $C_n(X)$. Let $f$ be a member of $C_n(X)$ and let $\epsilon$ be a positive real number. We shall find $g$ in $G$ such that $\|g - f\| < \epsilon$. Since $f(X)$ is a subspace of $\mathbf{R}^n$, there exists a locally finite open covering $\{V_\gamma\}_{\gamma \in \Gamma}$ of $f(X)$ such that $\operatorname{diam} V_\gamma < \epsilon/2n$ for each $\gamma$.

Since dim $X \leqslant n$ there exists a locally finite open covering $\mathcal{H} = \{H_\lambda\}_{\lambda \in \Lambda}$ of $X$ of order not exceeding $n$ which is a common refinement of $\mathcal{U}$ and the locally finite open covering $\{f^{-1}(V_\gamma)\}_{\gamma \in \Gamma}$ of $X$. Clearly

$$\mathrm{diam}\, f(H_\lambda) < \epsilon/2n$$

for each $\lambda$. It follows from Lemma 4.7 that there exists $g$ in $C_n(X)$ which is zero-dimensional with respect to $\mathcal{H}$ such that $\|g - f\| < \epsilon$. Since $\mathcal{H}$ is a refinement of $\mathcal{U}$ it follows that $g \in G$. $\square$

Three lemmas now lead to the algebraic characterization of dimension.

**4.9   Lemma.** *Let $X$ be a compact space and let $f_1, \ldots, f_m$ be members of $C(X)$. For each positive real number $\epsilon$ there exists a finite open covering $\{V_1, \ldots, V_s\}$ of $X$ such that $\mathrm{diam}\, f_k(V_i) < \epsilon$ for $k = 1, \ldots, m$ and $i = 1, \ldots, s$.*

*Proof.* For each $k = 1, \ldots, m$ let $\{W_1^k, \ldots, W_{r(k)}^k\}$ be a finite open covering of the compact metric space $f_k(X)$ by sets of diameter less than $\epsilon$. For the required finite open covering of $X$ we can take all sets of the form $\bigcap_{k=1}^m f_k^{-1}(W_{i(k)}^k)$, where $1 \leqslant i(k) \leqslant r(k)$ for each $k$. $\square$

**4.10   Lemma.** *Let $X$ be a bicompact space, let $g$ be a member of $C_n(X)$ and let $\mathcal{U}$ be a finite open covering of $X$ which envelops $g^{-1}(y)$ for each $y$ in $g(X)$. Then there exists a finite open covering $\mathcal{W}$ of $g(X)$ such that $\mathcal{U}$ envelops $g^{-1}(W)$ for each member $W$ of $\mathcal{W}$.*

*Proof.* Suppose that $y \in g(X)$. Since $\mathcal{U}$ envelops $g^{-1}(y)$ and $g^{-1}(y)$ is compact, it follows that $g^{-1}(y) = \bigcup_{i=1}^m S_i$, where the $S_i$ are disjoint sets which are open-and-closed in $g^{-1}(y)$ and each set $S_i$ is contained in some member of $\mathcal{U}$. Since $g^{-1}(y)$ is a closed set, the sets $S_1, \ldots, S_m$ are disjoint closed sets of $X$. Since $X$ is a normal space there exist disjoint open sets $G_1, \ldots, G_m$ of $X$ such that $S_i \subset G_i$ for each $i$ and we can suppose that each $G_i$ is contained in some member of $\mathcal{U}$. Thus if $G = \bigcup_{i=1}^m G_i$, then $G$ is open in $X$, $g^{-1}(y) \subset G$ and $\mathcal{U}$ envelops $G$. Since $g$ is a closed mapping there exists an open neighbourhood $V_y$ of $y$ in $g(X)$ such that $g^{-1}(V_y) \subset G$. Clearly $\mathcal{U}$ envelops $g^{-1}(V_y)$. Thus if $\mathcal{W}$ is a finite subcovering of the open covering $\{V_y\}_{y \in g(X)}$ of the compact space $g(X)$, then $\mathcal{W}$ is the required covering of $g(X)$. $\square$

**4.11   Lemma.** *Let $X$ be a bicompact space such that $\dim X \leqslant n$, where $n$ is a positive integer. If $\mathcal{V}$ is a finite open covering of $X$ then the set*

$$\{g \in C_n(X) \mid \mathcal{V} \text{ envelops } g^{-1}(y) \text{ if } y \in g(X)\}$$

*is a dense open subset of $C_n(X)$.*

*Proof.* Let

$$G = \{g \in C_n(X) \mid \mathscr{V} \text{ envelops } g^{-1}(y) \text{ if } y \in g(X)\}$$

and let $G_0$ be the subset of $C_n(X)$ consisting of those functions which are zero-dimensional with respect to $\mathscr{V}$. Clearly $G_0 \subset G$. But if $f \in G$, by Lemma 4.10 there exists a finite open covering $\mathscr{W}$ of the compact metric space $f(X)$ such that $\mathscr{V}$ envelops $f^{-1}(W)$ for each $W$ in $\mathscr{W}$. By Proposition 2.3.22 there exists a positive real number $\epsilon$ such that $B$ is contained in some member of $\mathscr{W}$ if $B$ is a subset of $f(X)$ such that $\mathrm{diam}\, B < \epsilon$. Thus if $S$ is a subset of $X$ such that $\mathrm{diam}\, f(S) < \epsilon$, then $f(S) \subset W$ for some member $W$ of $\mathscr{W}$ and hence $S \subset f^{-1}(W)$. Thus $\mathscr{V}$ envelops $S$. It follows that $f \in G_0$. Thus $G = G_0$. Now it follows from Proposition 4.8 that $G$ is a dense open subset of $C_n(X)$. $\square$

Now we have the main result of this section.

**4.12   Proposition.** *If $X$ is a Tihonov space, then for each non-negative integer $n$ the following statements about $X$ are equivalent:*
  (a) *$\partial \mathrm{im}\, X \leqslant n$;*
  (b) *each countable subset of $C(X)$ has rank not exceeding $n$;*
  (c) *$\mathrm{ad}\, C(X) \leqslant n$.*

*Proof.* As we observed earlier, it is enough to prove this result in the case of a bicompact space $X$. The result is true if $n = 0$. For if $X$ is a bicompact space then by Proposition 4.5, $\dim X = 0$ if and only if $\mathrm{ad}\, C(X) = 0$, and $\mathrm{ad}\, C(X) = 0$ if and only if $C(X)$ has no proper analytic subalgebra, which is equivalent to condition (b) in the case $n = 0$. For the remainder of the proof we suppose that $X$ is a bicompact space and that $n$ is a positive integer.

(a) $\Rightarrow$ (b). Let $\{f_k\}_{k \in \mathbf{N}}$ be a countable set in $C(X)$ and suppose that $\dim X \leqslant n$. By Lemma 4.9, for each positive integer $m$ there exists a finite open covering $\mathscr{V}_m$ of $X$ such that $\mathrm{diam}\, f_k(V) < 1/m$ for

$$k = 1, \ldots, m$$

and each $V$ in $\mathscr{V}_m$. Let

$$G_m = \{g \in C_n(X) \mid \mathscr{V}_m \text{ envelops } g^{-1}(y) \text{ if } y \in g(X)\}.$$

By Lemma 4.11, $G_m$ is a dense open subset of the complete metric space $C_n(X)$. By Baire's theorem (Proposition 2.4.5), $G = \bigcap_{m \in \mathbf{N}} G_m$ is a non-empty (in fact dense) subset of $C_n(X)$. Suppose that $h \in G$ and let

$$h(x) = \big(h_1(x), \ldots, h_n(x)\big) \quad \text{if} \quad x \in X.$$

We shall show that $f_k \in L[H]$ for every $k$, where $H$ is the subset $\{h_1, \ldots, h_n\}$ of $C(X)$. By Proposition 4.4, $L[H]$ consists of all functions

$f$ in $C(X)$ such that $f$ is constant on each connected subset of a constancy set of $H$. But the constancy sets of $H$ are the sets $h^{-1}(y)$ for $y$ in $h(X)$. Suppose that $y \in h(X)$ and let $S$ be a connected subset of $h^{-1}(y)$. For each positive integer $m$, $\mathscr{V}_m$ envelops $h^{-1}(y)$ so that since $S$ is connected, $S$ is contained in some member of $\mathscr{V}_m$. It follows that if $k \in \mathbf{N}$, then $\operatorname{diam} f_k(S) < 1/m$ for every $m$ such that $m \geqslant k$, so that $\operatorname{diam} f_k(S) = 0$ from which we conclude that $f_k(S)$ is a one-point set and hence $f_k$ is constant on $S$. Thus $f_k \in L[H]$.

(b) $\Rightarrow$ (c) is clear.

(c) $\Rightarrow$ (a). Let $X$ be a bicompact space such that $\operatorname{ad} C(X) \leqslant n$ and let $\mathscr{U} = \{U_1, \ldots, U_k\}$ be a finite open covering of $X$. Since $X$ is a normal space there exists a closed covering $\{F_1, \ldots, F_k\}$ of $X$ such that $F_i \subset U_i$ for each $i$. Hence by Urysohn's lemma, there exist $f_1, \ldots, f_k$ in $C(X)$ such that $F_i \subset Z(f_i) \subset U_i$ for each $i$, where $Z(f_i)$ is the zero-set of $f_i$. By hypothesis there exists a subset $G = \{g_1, \ldots, g_n\}$ of $C(X)$ such that $f_i \in L[G]$ for $i = 1, \ldots, k$. Let $g$ in $C_n(X)$ be defined by

$$g(x) = \big(g_1(x), \ldots, g_n(x)\big) \quad \text{if} \quad x \in X.$$

Then by Proposition 4.4, each function $f_i$ is constant on each connected subset of each set $g^{-1}(y)$, where $y \in g(X)$. Suppose that $y \in g(X)$ and let $S$ be a connected subset of $g^{-1}(y)$. Since $\{Z(f_1), \ldots, Z(f_k)\}$ is a covering of $X$, there exists $i$ such that $Z(f_i)$ meets $S$. Since $f_i$ is constant on $S$, it follows that $S \subset Z(f_i) \subset U_i$. Thus by Proposition 4.6, $\mathscr{U}$ envelops $g^{-1}(y)$ if $y \in g(X)$. It follows from Lemma 4.10 that there exists a finite open covering $\mathscr{W}$ of $g(X)$ such that $\mathscr{U}$ envelops $g^{-1}(W)$ for each $W$ in $\mathscr{W}$. Since $\dim g(X) \leqslant n$, there exists a finite open refinement $\{G_1, \ldots, G_m\}$ of $\mathscr{W}$ of order not exceeding $n$. For each $i$, $\mathscr{U}$ envelops $g^{-1}(G_i)$. Thus $g^{-1}(G_i) = \bigcup_{\lambda \in \Lambda_i} H_{i\lambda}$, where the disjoint sets $H_{i\lambda}$ are open in $g^{-1}(G_i)$, so open in $X$, and each $H_{i\lambda}$ is contained in some member of $\mathscr{U}$. If $\mathscr{H} = \{H_{i\lambda} \mid \lambda \in \Lambda_i, i = 1, \ldots, m\}$, then $\mathscr{H}$ is an open covering of $X$ which is a refinement of $\mathscr{U}$ and the order of $\mathscr{H}$ does not exceed $n$. Thus $\dim X \leqslant n$. $\square$

In the statement of the next theorem we interpret $\partial\mathrm{im}\, X = \operatorname{ad} C(X)$ in case either term is infinite to imply that $\operatorname{ad} C(X) = \aleph_0$ if and only if $\partial\mathrm{im}\, X = \infty$. The reader is asked to accept without further mention similar conventions henceforth.

**4.13    Theorem.** *If $X$ is a Tihonov space, then*

$$\partial\mathrm{im}\, X = \operatorname{ad} C(X).$$

*Proof.* This is an immediate consequence of Proposition 4.12. $\square$

We observed earlier that $\operatorname{ad} C(X) = 0$ if and only if $\operatorname{rk} C(X) = 0$. Thus for a bicompact space $X$ we have $\dim X = \operatorname{ind} X = 0$ if and only if $\operatorname{rk} C(X) = 0$. Furthermore if $\operatorname{rk} C(X)$ is finite then $\operatorname{ad} C(X) = \operatorname{rk} C(X)$. There exist bicompact spaces $X$ for which $\operatorname{ad} C(X)$ is finite but $\operatorname{rk} C(X) > \aleph_0$. For example the 'long line' $L$ considered in §3 of Chapter 4 is a bicompact space such that $\dim L = 1$ and hence $\operatorname{ad} C(L) = 1$ but $\operatorname{rk} C(L) > \aleph_0$. For if $H = \{h_i\}$ is a countable subset of $C(L)$, it is easy to see that for each $i$ there exist $z_i$ in $L \backslash \{\omega_i\}$ such that $h_i$ is constant on the interval $[z_i, \omega_1]$. Thus taking $z_0 = \sup z_i$ we see that the connected set $[z_0, \omega_1]$ is contained in a constancy set of $H$. It follows from Proposition 4.4 that $L[H] \neq C(L)$. Thus $\operatorname{rk} C(L) > \aleph_0$. In the remainder of this section we shall obtain a connection between $\operatorname{rk} C(X)$ and $\operatorname{ind} X$ for a bicompact space $X$.

First we establish a relation between analytic bases for $C(X)$ and mappings of $X$ into cubes. If $X$ and $Y$ are topological spaces, a continuous function $f: X \to Y$ is said to be a *light mapping* if $f^{-1}(y)$ is totally disconnected for each $y$ in $f(X)$.

**4.14  Proposition.** *If $X$ is a bicompact space and $\mathscr{H} = \{h_\lambda\}_{\lambda \in \Lambda}$ is a family of members of $C(X)$ such that*

$$\mathscr{H}^* = \{h \in C(X) \mid h = h_\lambda \text{ for some } \lambda \text{ in } \Lambda\}$$

*is an analytic base for $C(X)$, then the natural mapping of $X$ into the cube associated with $\mathscr{H}$ is a light mapping.*

*Proof.* Let $P$ be the cube associated with $\mathscr{H}$ and let $p: X \to P$ be the natural mapping so that $p(x) = \{h_\lambda(x)\}_{\lambda \in \Lambda}$ if $x \in X$. If $y \in p(X)$ then $p^{-1}(y)$ is a constancy set of $\mathscr{H}^*$. Since $L[\mathscr{H}^*] = C(X)$ it follows from Proposition 4.4 that each connected subset of a constancy set of $\mathscr{H}^*$ is a one-point set. Thus $p^{-1}(y)$ is totally disconnected. $\square$

We next introduce a class of mappings which includes the light mappings between bicompact spaces and which has useful properties in dimension theory. The study of these mappings will be continued in the next section.

**4.15  Definition.** *If $X$ and $Y$ are topological spaces, a continuous function $f: X \to Y$ is called a* decomposing mapping *if for each point $x$ of $X$ and each open set $V$ such that $x \in V$ there exists an open set $W$ of $Y$ such that $f(x) \in W$ and $f^{-1}(W) = G \cup H$, where $G$ and $H$ are disjoint open sets and $x \in G \subset V$.*

If $f: X \to Y$ is a decomposing mapping, then for each $y$ in $f(X)$ we have $\operatorname{ind} f^{-1}(y) = 0$. For if $x \in f^{-1}(y)$ and $U$ is an open neighbourhood of $x$ in $f^{-1}(y)$, there exists $V$ open in $X$ such that $U = V \cap f^{-1}(y)$, and there exists an open neighbourhood $W$ of $y$ such that $f^{-1}(W) = G \cup H$, where $G$ and $H$ are disjoint open sets of $X$ and $x \in G \subset V$. Since $f^{-1}(y) \subset G \cup H$ and $G$ is open-and-closed in $G \cup H$, we see that

$$G_1 = G \cap f^{-1}(y)$$

is open-and-closed in $f^{-1}(y)$, and $x \in G_1 \subset U$. Thus $\operatorname{ind} f^{-1}(y) = 0$ for $y$ in $f(X)$. There is a partial converse.

**4.16 Proposition.** *If $X$ is a normal space, $Y$ is a $T_1$-space and $f: X \to Y$ is a closed mapping such that $\operatorname{ind} f^{-1}(y) = 0$ for each $y$ in $f(X)$, then $f$ is a decomposing mapping.*

*Proof.* Let $x$ be a point of $X$, let $V$ be an open neighbourhood of $x$ in $X$ and let $y = f(x)$. If $E = f^{-1}(y)$ then $\operatorname{ind} E = 0$ and hence there exists an open-and-closed set $C$ of $E$ such that $x \in C \subset V \cap E$. Since $Y$ is a $T_1$-space, $E$ is closed in $X$ so that $C$ and $E \backslash C$ are disjoint closed sets in $X$. Since $X$ is a normal space there exist disjoint open sets $P$ and $Q$ such that $C \subset P \subset V$ and $E \backslash C \subset Q$. Since $f$ is a closed mapping and $f^{-1}(y) \subset P \cup Q$, there exists an open neighbourhood $W$ of $y$ in $Y$ such that $f^{-1}(W) \subset P \cup Q$. If $G = P \cap f^{-1}(W)$ and $H = Q \cap f^{-1}(W)$ then $G$ and $H$ are disjoint open sets of $X$ such that $f^{-1}(W) = G \cup H$ and $x \in G \subset V$. Thus $f$ is a decomposing mapping. $\square$

**4.17 Corollary.** *A mapping of a bicompact space into a Hausdorff space is decomposing if and only if it is a light mapping.*

*Proof.* Let $X$ be a bicompact space, let $Y$ be a Hausdorff space and let $f: X \to Y$ be a continuous function. Then $f$ is a closed mapping and for each $y$ in $f(X)$, $f^{-1}(y)$ is totally disconnected if and only if

$$\operatorname{ind} f^{-1}(y) = 0.$$

It follows from Proposition 4.16 and the remark which precedes it that $f$ is a decomposing mapping if and only if it is a light mapping. $\square$

For our immediate needs the most important property of decomposing mappings is stated in the next proposition. In the proof we shall use the obvious fact that the restriction of a decomposing mapping is a decomposing mapping.

**4.18 Proposition.** *If there exists a decomposing mapping of a space $X$ into a space $Y$, then $\operatorname{ind} X \leqslant \operatorname{ind} Y$.*

*Proof.* Let $f: X \to Y$ be a decomposing mapping and suppose first that ind $Y = 0$. Let $x$ be a point of $X$ and let $V$ be an open neighbourhood of $x$. Then there exists an open neighbourhood $W$ of $f(x)$ such that $f^{-1}(W) = G \cup H$, where $G$ and $H$ are disjoint open sets and $x \in G \subset V$. Since ind $Y = 0$ there exists an open-and-closed set $W_1$ such that $y \in W_1 \subset W$. If $G_1 = f^{-1}(W_1) \cap G, H_1 = f^{-1}(W_1) \cap H$, then

$$f^{-1}(W_1) = G_1 \cup H_1$$

and $G_1, H_1$ are disjoint open sets. Since $f^{-1}(W_1)$ is an open-and-closed set, it follows that $G_1$ and $H_1$ are open-and-closed sets. Since $x \in G_1 \subset V$ it follows that ind $X = 0$. Now suppose that ind $Y = n$, where $n$ is a positive integer, and that the result is established for decomposing mappings into spaces of small inductive dimension less than $n$. Let $x$ be a point of $X$ and let $V$ be an open neighbourhood of $x$ in $X$. Then there exists an open set $W$ of $Y$ such that $f(x) \in W$ and $f^{-1}(W) = G \cup H$, where $H$ and $G$ are disjoint open sets and $x \in G \subset V$. Since ind $Y = n$ we can suppose that ind bd $(W) \leqslant n-1$. The mapping

$$f^{-1}\big(\mathrm{bd}\,(W)\big) \to \mathrm{bd}\,(W)$$

given by restriction of $f$ is decomposing and hence by the induction hypothesis ind $f^{-1}\big(\mathrm{bd}\,(W)\big) \leqslant n-1$. But $\bar{G} \subset X \backslash H$ so that

$$\mathrm{bd}\,(G) = \bar{G} \backslash f^{-1}(W) \subset \mathrm{bd}\big(f^{-1}(W)\big) \subset f^{-1}\big(\mathrm{bd}\,(W)\big).$$

Thus ind bd $(G) \leqslant n-1$. Hence ind $X \leqslant n =$ ind $Y$. The result now follows by induction. $\square$

**4.19   Proposition.** *If $X$ is a bicompact space such that* $\mathrm{rk}\,C(X) \leqslant \aleph_0$, *then*
$$\dim X = \mathrm{ind}\,X = \mathrm{rk}\,C(X) = \mathrm{ad}\,C(X).$$

*Proof.* If $\mathrm{rk}\,C(X) = \aleph_0$ then

$$\mathrm{ad}\,C(X) = \dim X \leqslant \mathrm{ind}\,X \leqslant \mathrm{rk}\,C(X),$$

and there exists a countable subset $H$ of $C(X)$ such that $L[H] = C(X)$ and no finite subset with this property. Hence $\mathrm{rk}\,H = \aleph_0$. It follows from Proposition 4.12 that $\mathrm{ad}\,C(X) = \aleph_0$ and we have the asserted equality. If $\mathrm{rk}\,C(X) = 0$, then $\mathrm{ad}\,C(X) = 0$ so that $\dim X = 0$ and hence ind $X = 0$. Finally suppose that $\mathrm{rk}\,C(X) = n$, where $n$ is a non-negative integer. Then $\dim X = \mathrm{ad}\,C(X) = n$, and there exists an analytic base for $C(X)$ with $n$ elements. It follows from Proposition 4.14 and Corollary 4.17 that there exists a decomposing mapping of $X$ into an $n$-dimensional cube. It follows from Proposition 4.18 that ind $X \leqslant n$. But $\dim X \leqslant \mathrm{ind}\,X$ so that ind $X = n$. $\square$

**4.20** *Remark.* For a bicompact space $X$, from $\operatorname{rk} C(X) > \aleph_0$ we can only deduce that $\operatorname{ind} X > 0$. In the next section we shall see that a bicompact space $X$ such that $\operatorname{rk} C(X) \leqslant \aleph_0$ satisfies

$$\dim X = \operatorname{ind} X = \operatorname{Ind} X.$$

Thus any bicompact space $X$ for which covering dimension differs from inductive dimension or the inductive dimensions differ (such as the spaces studied in Chapter 8) must satisfy $\operatorname{rk} C(X) > \aleph_0$.

**4.21** *Proposition.* The dimension (in any sense) of a compact metrizable space $X$ is equal to $\operatorname{rk} C(X)$.

*Proof.* If $X$ is a compact metrizable space, then $X$ is of countable weight, so that there exists a countable subset $H$ of $C(X)$ which separates points of $X$ from closed sets. It follows from Proposition 2.10 that $A[H] = C(X)$ and hence $L[H] = C(X)$. Thus $\operatorname{rk} C(X) \leqslant \aleph_0$. $\square$

## 5 Z-spaces

In this section we shall continue the study of decomposing mappings. We begin with another special case. This case belongs more properly to the dimension theory of metric spaces but it is included because of its historical interest.

**5.1** *Definition.* If $X$ and $Y$ are metric spaces, then a continuous function $f: X \to Y$ is said to be a uniformly zero-dimensional mapping if given a positive real number $\epsilon$, there exists a positive real number $\delta$ such that the covering of $X$ by all open sets $G$ such that $\operatorname{diam} G < \epsilon$ envelops every subset $S$ of $X$ such that $\operatorname{diam} f(S) < \delta$.

**5.2** *Proposition.* If $X$ is a metric space such that $\dim X \leqslant n$, then $C_n(X)$ contains a uniformly zero-dimensional mapping.

*Proof.* For each positive integer $k$, let $\mathcal{V}_k$ be a locally finite open covering of $X$ by sets $V$ such that $\operatorname{diam} V < 1/k$. By Proposition 4.8, for each $k$ the set

$$G_k = \{g \in C_n(X) \mid g \text{ is zero-dimensional w.r.t. } \mathcal{V}_k\}$$

is a dense open subset of the complete metric space $C_n(X)$. It follows from Baire's theorem that $G = \bigcap_{k \in \mathbf{N}} G_k$ is non-empty. It is clear that if $g \in G$ then $g$ is a uniformly zero-dimensional mapping. $\square$

402 ALGEBRAS OF FUNCTIONS [CH. 10

It is clear that the restriction of a uniformly zero-dimensional mapping is uniformly zero-dimensional and it is easy to verify that the composite of uniformly zero-dimensional mappings is uniformly zero-dimensional.

**5.3  Proposition.** *If $X$ and $Y$ are metric spaces and there exists a uniformly zero-dimensional mapping of $X$ into $Y$, then*

$$\dim X \leqslant \dim Y.$$

*Proof.* Let $f: X \to Y$ be a uniformly zero-dimensional mapping. Suppose first that $\dim Y = 0$. For each positive integer $k$ there exists $\delta_k > 0$ such that for each subset $T$ of $Y$ with $\operatorname{diam} T < \delta_k$, the covering of $X$ by all open sets $G$ such that $\operatorname{diam} G < 1/k$ envelops $f^{-1}(T)$. Since $\dim Y = 0$, there exists a disjoint open covering $\{G_{k\lambda}\}_{\lambda \in \Lambda}$ of $Y$ such that $\operatorname{diam} G_{k\lambda} < \delta_k$ for each $\lambda$. If $\mathscr{H}_k = \{f^{-1}(G_{k\lambda})\}_{\lambda \in \Lambda}$, then $\mathscr{H}_k$ is a disjoint open covering of $X$ and for each $\lambda$, $f^{-1}(G_{k\lambda})$ is the disjoint union of open sets each of which is of diameter less than $1/k$. Thus we obtain from $\mathscr{H}_k$ a disjoint open covering of $X$ by sets each of which is of diameter less than $1/k$. It follows from Lemma 4.5.1 that $\dim X \leqslant 0$.

Now suppose that $\dim Y = n$, where $n$ is a positive integer. By Proposition 5.2 there exists a uniformly zero-dimensional mapping $g: Y \to \mathbf{R}^n$. Let $h = g \circ f$ so that $h: X \to \mathbf{R}^n$ is a uniformly zero-dimensional mapping. Let us put $\mathbf{R}^n = \bigcup_{k=0}^n A_k$, where $\dim A_k = 0$ for each $k$. Then $X = \bigcup_{k=0}^n h^{-1}(A_k)$ and for each $k$ we have a uniformly zero-dimensional mapping $h^{-1}(A_k) \to A_k$, given by restriction of $h$, so that $\dim h^{-1}(A_k) \leqslant 0$. It follows from Proposition 3.5.11 that $\dim X \leqslant n$. $\square$

**5.4  Theorem.** *If $X$ is a metrizable space then the following statements about $X$ are equivalent*:

(a) $\dim X \leqslant n$;

(b) *for each metric on $X$ which induces its topology, there exists a uniformly zero-dimensional bounded mapping of $X$ into $\mathbf{R}^n$*;

(c) *for some metric on $X$ which induces its topology, there exists a uniformly zero-dimensional bounded mapping of $X$ into $\mathbf{R}^n$*.

*Proof.* (a) $\Rightarrow$ (b). This is immediate from Proposition 5.2.

(b) $\Rightarrow$ (c). This is obvious.

(c) $\Rightarrow$ (a). This is immediate from Proposition 5.3. $\square$

The connection with decomposing mappings will now be established.

**5.5  Proposition.** *A uniformly zero-dimensional mapping between metric spaces is a decomposing mapping.*

*Proof.* Let $X$ and $Y$ be metric spaces and let $f: X \to Y$ be a uniformly zero-dimensional mapping. Let $x$ be a point of $X$ and let $V$ be an open set of $X$ such that $x \in V$. There exists $\epsilon > 0$ such that $B_\epsilon(x) \subset V$, and there exists $\delta > 0$ such that each subset $S$ of $X$ such that

$$\operatorname{diam} f(S) < \delta$$

is enveloped by the covering of $X$ by open sets $U$ such that $\operatorname{diam} U < \epsilon$. Let $W = B_{\delta/2}\big(f(x)\big)$. Then $f^{-1}(W)$ is the union of a disjoint family $\mathscr{G}$ of open sets each of which is of diameter less than $\epsilon$. Let $G$ be the member of $\mathscr{G}$ which contains $x$ and let $H$ be the union of the remaining members of $\mathscr{G}$. Then $f^{-1}(W) = G \cup H$, where $G$ and $H$ are disjoint open sets and $x \in G \subset V$. $\square$

It follows from Propositions 5.2 and 5.5 that each finite-dimensional metrizable space admits a decomposing mapping into a compact metric space. Similarly from Proposition 4.14 and Corollary 4.17 it follows that each bicompact space $X$ such that $\operatorname{rk} C(X) \leqslant \aleph_0$ admits a decomposing mapping into a compact metric space. These observations prompt the next definition.

**5.6  Definition.** *A Hausdorff space $X$ is called a $Z$-space if there exists a decomposing mapping of $X$ into a compact metric space.*

Since the restriction of a decomposing mapping is a decomposing mapping, each subspace of a $Z$-space is a $Z$-space. The next result shows that each $Z$-space is a Tihonov space.

**5.7  Proposition.** *If $X$ is a topological space and there exists a decomposing mapping of $X$ into a completely regular space, then $X$ is a completely regular space.*

*Proof.* Let $Y$ be a completely regular space and let $f: X \to Y$ be a decomposing mapping. Let $x_0$ be a point of $X$ and let $F$ be a closed set of $X$ such that $x_0 \notin F$. There exists $W$ open in $Y$ such that $f(x_0) \in W$ and $f^{-1}(W) = G \cup H$, where $G$ and $H$ are disjoint open sets and $x_0 \in G \subset X \backslash F$. There exists a continuous function $\phi: Y \to I$ such that $\phi\big(f(x_0)\big) = 0$ and $\phi(y) = 1$ if $y \notin W$. Now let us define $\psi: X \to I$ by putting

$$\psi(x) = \begin{cases} \phi\big(f(x)\big) & \text{if} \quad x \in \bar{G}, \\ 1 & \text{if} \quad x \in X \backslash G. \end{cases}$$

The definition is meaningful, for if $x \in \bar{G} \backslash G$ then $x \notin G \cup H$ so that $f(x) \notin W$ and hence $\phi(f(x)) = 1$. Since the restrictions of $\psi$ to $\bar{G}$ and $X \backslash G$ are continuous, $\psi$ is a continuous function, and $\psi(x_0) = 0$ whilst $\psi(x) = 1$ if $x \in F$. Thus $X$ is a completely regular space. $\square$

**5.8 Definition.** *If $X$ is a Z-space, then the* dimensional weight $\mathrm{dw}\,(X)$ *of $X$ is the least integer $n$ for which $C_n(X)$ contains a decomposing mapping or* $\mathrm{dw}\,(X) = \infty$ *if there exists no such integer.*

Since the composite of decomposing mappings is a decomposing mapping and a closed embedding is a decomposing mapping, it follows that if $X$ is a Z-space such that $\mathrm{dw}\,(X) = n$, then $C_m(X)$ contains a decomposing mapping for every integer $m$ such that $m \geqslant n$. Since the restriction of a decomposing mapping is a decomposing mapping, it follows that if $A$ is a subspace of a Z-space $X$, then $\mathrm{dw}\,(A) \leqslant \mathrm{dw}\,(X)$.

**5.9 Proposition.** *If $X$ is a Z-space, then*

$$\mathrm{ind}\,X \leqslant \mathrm{dw}\,(X)$$

*and* $\mathrm{ind}\,X = 0$ *if and only if* $\mathrm{dw}\,(X) = 0$.

*Proof.* To prove the first assertion it is enough to show that $\mathrm{ind}\,X \leqslant n$ if $\mathrm{dw}\,(X) \leqslant n$. But if $\mathrm{dw}\,(X) \leqslant n$, then there exists a decomposing mapping $f \colon X \to P$, where $P$ is a compact subset of $\mathbf{R}^n$ and it follows from Proposition 4.18 that $\mathrm{ind}\,X \leqslant n$. We establish the second assertion by showing that every mapping with domain a space $X$ such that $\mathrm{ind}\,X = 0$ is a decomposing mapping, so that in particular the unique member of $C_0(X)$ is a decomposing mapping. Thus let $X$ be a space such that $\mathrm{ind}\,X = 0$ and let $f \colon X \to Y$ be a continuous function. Let $x$ be a point of $X$, let $V$ be an open neighbourhood of $x$ in $X$ and let $W$ be any open neighbourhood of $f(x)$ in $Y$. Since $\mathrm{ind}\,X = 0$, there exists an open-and-closed set $G$ of $X$ such that $x \in G \subset V \cap f^{-1}(W)$. If $H = f^{-1}(W) \backslash G$, then $G$ and $H$ are disjoint open sets of $X$ such that $f^{-1}(W) = G \cup H$ and $x \in G \subset V$. Thus $f$ is a decomposing mapping. $\square$

**5.10 Proposition.** *If $X$ is a normal Z-space, then*

$$\mathrm{ind}\,X \leqslant \mathrm{dw}\,(X) \leqslant \dim X \leqslant \mathrm{Ind}\,X.$$

*Proof.* It is enough to show that if $X$ is a normal Z-space such that $\dim X \leqslant n$, then $\mathrm{dw}\,(X) \leqslant n$. There exists a decomposing mapping $h \colon X \to Y$, where $Y$ is a compact metric space. For each positive integer $k$ there exists a finite covering $\mathcal{U}_k$ of $Y$ by open sets of diameter less than $1/k$. Let $\mathcal{V}_k$ be the open covering of $X$ by the sets $h^{-1}(U)$ for $U$ in $\mathcal{U}_k$. By Proposition 4.8 the set $P_k$ of functions $g$ in $C_n(X)$ which are zero-dimensional with respect to $\mathcal{V}_k$ is a dense open subset of $C_n(X)$. Since $C_n(X)$ is a complete metric space, the set $P = \bigcap_{k\in\mathbf{N}} P_k$ is dense in $C_n(X)$ and therefore non-empty. We shall show that if $g \in P$ then $g$ is a decomposing mapping. Let $x$ be a point of $X$ and let $V$ be an open neighbourhood of $x$ in $X$. Since $h$ is a decomposing mapping there exists an open neighbourhood $W$ of $h(x)$ in $Y$ such that $h^{-1}(W) = G \cup H$, where $G$ and $H$ are disjoint open sets in $X$ and $x \in G \subset V$. There exists a positive integer $m$ such that $B_{1/m}\big(h(x)\big) \subset W$. Since $g$ is zero-dimensional with respect to $\mathcal{V}_m$ there exists $\epsilon > 0$ such that $\mathcal{V}_m$ envelops $g^{-1}(T)$ for every subset $T$ of $\mathbf{R}^n$ such that $\operatorname{diam} T < \epsilon$. Thus there exists an open neighbourhood $W_1$ of $g(x)$ in $\mathbf{R}^n$ such that $g^{-1}(W_1)$ is the union of a disjoint family of open sets each of which is contained in some member of $\mathcal{V}_m$. Let $Q$ be the member of this family which contains $x$. Since $\operatorname{diam} h(Q) < 1/m$ and $h(x) \in h(Q)$, it follows that $h(Q) \subset B_{1/m}\big(h(x)\big) \subset W$, so that

$$x \in Q \subset h^{-1}(W).$$

Now let $G_1 = Q \cap G$ and $H_1 = g^{-1}(W_1)\backslash G_1$. Then $G_1$ and $H_1$ are disjoint sets, $g^{-1}(W_1) = G_1 \cup H_1$ and $x \in G_1 \subset V$. It is clear that $G_1$ is an open set. Since $Q \subset h^{-1}(W)$ and $G$ is closed in $h^{-1}(W)$, it follows that $G_1 = Q \cap G$ is closed in $Q$. But $Q$ is closed in $g^{-1}(W_1)$. Thus $G_1$ is a closed set in $g^{-1}(W_1)$ and hence $H_1$ is an open set. Thus $g$ is a decomposing mapping. It follows that $\operatorname{dw}(X) \leqslant n$. $\square$

**5.11**   *Remarks.* (1) The proof of Proposition 5.10 gives a stronger result: if $X$ is a normal Z-space such that $\dim X \leqslant n$, then the decomposing mappings form a dense $G_\delta$-set in $C_n(X)$.

    (2) For Roy's metrizable space $P$ studied in §4 of Chapter 7 we have

$$\operatorname{ind} P = \operatorname{dw}(P) = 0 \quad \text{and} \quad \dim P = \operatorname{Ind} P = 1.$$

We have the following condition for the coincidence of dimensional weight and dimension.

**5.12**   *Theorem. If $X$ is a completely paracompact Z-space, then*

$$\operatorname{ind} X = \operatorname{dw}(X) = \dim X = \operatorname{Ind} X.$$

*Proof.* Since $X$ is completely paracompact and regular, $X$ is a completely paracompact normal space. Thus $\dim X \leqslant \operatorname{ind} X$ by Proposition 4.5.8. It follows from Proposition 5.10 that

$$\operatorname{ind} X = \operatorname{dw}(X) = \dim X.$$

We have $\dim X = \operatorname{Ind} X$ if $\dim X = 0$. Let us suppose that covering dimension and large inductive dimension coincide for completely paracompact $Z$-spaces of covering dimension less than $n$, where $n$ is a positive integer. Suppose that $X$ is a completely paracompact $Z$-space such that $\dim X = \operatorname{ind} X = n$, let $E$ be a closed set of $X$ and let $G$ be an open set such that $E \subset G$. Then by Lemma 4.5.7 there exists an open set $U$ such that $E \subset U \subset G$ and $\operatorname{bd}(U)$ is the union of countably many closed sets $F_i$ such that $\operatorname{ind} F_i \leqslant n-1$. For each $i$, $F_i$ is completely paracompact so that $\dim F_i \leqslant n-1$. Since $\operatorname{bd}(U)$ is normal, the countable sum theorem for covering dimension holds in $\operatorname{bd}(U)$. Thus $\dim \operatorname{bd}(U) \leqslant n-1$ and hence by the induction hypothesis $\operatorname{Ind} \operatorname{bd}(U) \leqslant n-1$. Thus $\operatorname{Ind} X \leqslant n$ so that

$$\dim X = \operatorname{Ind} X = n.$$

The result is now established by induction. $\square$

We conclude this section with characterizations of decomposing mappings and $Z$-spaces. Three lemmas precede these characterization theorems.

**5.13   Lemma.** *Let $X$ and $Y$ be topological spaces and let $f: X \to Y$ be a decomposing mapping. Let $\{g_\lambda\}_{\lambda \in \Lambda}$ be a family of members of $C(Y)$ which separates points of $Y$ from closed sets and let*

$$G = \{g \in C(X) \mid g = g_\lambda \circ f \text{ for some } \lambda \text{ in } \Lambda\}.$$

*Then the subset*

$$H = \{h \in C(X) \mid h^2 = hg \text{ for some } g \text{ in } K[G]\}$$

*of $C(X)$ separates points of $X$ from closed sets.*

*Proof.* Let $F$ be a closed set of $X$ and let $x_0$ be a point of $X \backslash F$. Since $f$ is a decomposing mapping, there exists an open neighbourhood $U$ of $y_0 = f(x_0)$ in $Y$ such that $f^{-1}(U) = G \cup H$, where $G$ and $H$ are disjoint open sets of $X$ and $x_0 \in G \subset X \backslash F$. Since the family $\{g_\lambda\}_{\lambda \in \Lambda}$ separates points of $Y$ from closed sets, there exists $\lambda$ in $\Lambda$ such that $z_0 = g_\lambda(y_0) \notin (g_\lambda(Y \backslash U))^-$. There exists $\epsilon > 0$ such that the open interval $W = (z_0 - \epsilon, z_0 + \epsilon)$ is disjoint from $(g_\lambda(Y \backslash U))^-$. Hence $g_\lambda^{-1}(W) \subset U$ so that if

$$G_1 = G \cap (g_\lambda \circ f)^{-1}(W), \quad H_1 = H \cap (g_\lambda \circ f)^{-1}(W),$$

then $G_1$ and $H_1$ are disjoint open sets of $X$, $G_1 \cup H_1 = (g_\lambda \circ f)^{-1}(W)$
and $x_0 \in G_1 \subset X \backslash F$. Now let $g$ in $C(X)$ be defined by

$$g(x) = \big(g_\lambda(f(x)) - z_0 - \epsilon\big)\big(g_\lambda(f(x)) - z_0 + \epsilon\big) \quad \text{if} \quad x \in X,$$

and let $h$ be the function on $X$ defined by

$$h(x) = \begin{cases} g(x) & \text{if} \quad x \in \bar{G}_1, \\ 0 & \text{if} \quad x \in X \backslash G_1. \end{cases}$$

If $x \in \bar{G}_1 \backslash G_1$, then $g_\lambda\big(f(x)\big) \in \overline{W} \backslash W$ so that $g(x) = 0$. It follows that $h$ is
meaningfully defined and hence $h \in C(X)$. Since it is clear that
$g \in K[G]$ and

$$h^2 - hg = 0,$$

it follows that $h \in H$. Since $h(x_0) = -\epsilon^2 \neq 0$ and $h(x) = 0$ if $x \in F$,
we see that $h(x_0) \notin \big(h(F)\big)^-$. Thus $H$ separates points of $X$ from closed
sets. $\square$

**5.14  Lemma.** *Let $G$ and $H$ be subsets of a normed algebra $A$ such that
for each $h$ in $H$ there exists some $g$ in $K[G]$ such that $h^2 = hg$. Then $G$
is an analytic base for $A[G \cup H]$.*

*Proof.* Let $K$ be an analytic subalgebra of $A[G \cup H]$ which contains $G$.
Then $K$ is a closed subalgebra of $A$ which contains $G$ so that

$$K[G] \subset K.$$

If $h \in H$ there exists $g$ in $K[G]$ such that $h^2 - hg = 0$. Then

$$h \in A[G \cup H]$$

and $g \in K$ so that $h \in K$ since $K$ is an analytic subalgebra of $A[G \cup H]$.
Thus $K$ is a closed subalgebra of $A$ such that $G \cup H \subset K \subset A[G \cup H]$.
Hence $K = A[G \cup H]$. $\square$

**5.15  Lemma.** *Let $X$ be a Hausdorff space, let $Y$ be a bicompact space
of infinite weight and let $f \colon X \to Y$ be a decomposing mapping. Then there
exist a bicompactification $\tilde{X}$ of $X$ such that $w(\tilde{X}) \leqslant \max\{w(X), w(Y)\}$
and an extension $\tilde{f} \colon \tilde{X} \to Y$ of $f$ which is a light mapping.*

*Proof.* Let $\Lambda$ be a set such that $|\Lambda| = \max\{w(X), w(Y)\}$ and let $\Lambda_1$ be
a subset of $\Lambda$ such that $|\Lambda_1| = w(Y)$ and $|\Lambda \backslash \Lambda_1| = w(X)$. Let

$$\mathcal{G} = \{g_\lambda\}_{\lambda \in \Lambda_1}$$

be a family of members of $C(Y)$ which separates points of $Y$ from
closed sets. If $P_1$ is the cube associated with $\mathcal{G}$, then the natural

mapping of $Y$ into $P_1$ is a closed embedding and we identify $Y$ with its image under this embedding. Let

$$G = \{g \in C(X) \mid g = g_\lambda \circ f \text{ for some } \lambda \text{ in } \Lambda_1\},$$

and let $H$ be the subset of $C(X)$ consisting of those functions $h$ such that $h^2 = hg$ for some $g \in K[G]$. By Lemma 5.13, $H$ separates points of $X$ from closed sets. As noted in §3, we can select a subset of $H$ of cardinality equal to $w(X)$ which separates points of $X$ from closed sets. Let this subset be indexed by $\Lambda \backslash \Lambda_1$ to give a family $\{h_\lambda\}_{\lambda \in \Lambda \backslash \Lambda_1}$ of members of $C(X)$ which separates points of $X$ from closed sets. If $\lambda \in \Lambda_1$ let $h_\lambda = g_\lambda \circ f$. Then $\mathscr{H} = \{h_\lambda\}_{\lambda \in \Lambda}$ is a family of members of $C(X)$ which separates points of $X$ from closed sets. Let $P$ be the cube associated with $\mathscr{H}$ and let $p : X \to P$ be the natural mapping. Then $p$ is an embedding so that $\tilde{X} = \big( p(X) \big)^-$ is a bicompactification of $X$ and $w(\tilde{X}) \leqslant |\Lambda| = \max \{w(X), w(Y)\}$.

For each $\lambda$ let $\pi_\lambda$ be the projection of $P$ onto the $\lambda$th factor and let us regard $\pi_\lambda$ as a member of $C(P)$. For $\lambda$ in $\Lambda_1$ let $\rho_\lambda = \pi_\lambda | \tilde{X}$, and let $\mathscr{R}_1 = \{\rho_\lambda\}_{\lambda \in \Lambda_1}$. Let $R_1$ be the cube associated with $\mathscr{R}_1$ and let $r : \tilde{X} \to R_1$ be the natural mapping. For each $\lambda$ in $\Lambda_1$, $h_\lambda(X) \subset g_\lambda(Y)$ so that $R_1$ is a subspace of $P_1$. Also $r(\tilde{X}) \subset Y$, for if $x \in X$, then $p(x) = \{h_\lambda(x)\}_{\lambda \in \Lambda}$ so that $rp(x) = \{h_\lambda(x)\}_{\lambda \in \Lambda_1} = \{g_\lambda f(x)\}_{\lambda \in \Lambda_1} = f(x)$, taking note of the identification of $Y$ with its image in $P_1$. Thus $r(\tilde{X}) \subset Y$ since $Y$ is closed. If $\tilde{f} : \tilde{X} \to Y$ is defined by restriction of $r$, then $\tilde{f}$ is an extension of $f$.

If $q : X \to \tilde{X}$ is the dense embedding given by restriction of $p$, then as established in the proof of Proposition 3.7 the homomorphism

$$q^* : C(\tilde{X}) \to C(X)$$

maps $C(\tilde{X})$ isometrically and isomorphically onto $A[\mathscr{H}^*]$. By Lemma 5.14, $G$ is an analytic base for $A[\mathscr{H}^*]$. Since $\rho_\lambda \circ q = g_\lambda \circ f$ for each $\lambda$ in $\Lambda_1$, it follows that $q^*(\mathscr{R}_1^*) = G$ and hence $\mathscr{R}_1^*$ is an analytic base for $C(X)$. It follows from Proposition 4.14 that the natural mapping $r : \tilde{X} \to R_1$ is a light mapping. Thus $\tilde{f}$ is a light mapping. $\square$

We can now obtain a characterization of the decomposing mappings.

**5.16   Proposition.** *If $X$ and $Y$ are Tihonov spaces, a continuous function $f : X \to Y$ is a decomposing mapping if and only if there exist bicompactifications $\tilde{X}$ and $\tilde{Y}$ of $X$ and $Y$ respectively and a light mapping $\tilde{f} : \tilde{X} \to \tilde{Y}$ which is an extension of $f$.*

*Proof.* Since by Corollary 4.17 a light mapping between bicompact spaces is a decomposing mapping and the restriction of a decomposing mapping is a decomposing mapping, the condition is sufficient. Now let $X$ and $Y$ be Tihonov spaces and let $f: X \to Y$ be a decomposing mapping. If $\tilde{Y}$ is any bicompactification of $Y$ and we apply Lemma 5.15 to the composite of $f$ with the inclusion of $Y$ in $\tilde{Y}$, we obtain a bicompactification $\tilde{X}$ of $X$ and a light mapping $\tilde{f}: \tilde{X} \to \tilde{Y}$ which is an extension of $f$. $\square$

We can also characterize $Z$-spaces.

**5.17   Proposition.** *A topological space is a $Z$-space if and only if it is a Tihonov space which has a bicompactification which is a $Z$-space.*

*Proof.* The condition is clearly sufficient since each subspace of a $Z$-space is a $Z$-space. If $X$ is a $Z$-space then $X$ is a Tihonov space and there exists a decomposing mapping $f: X \to Y$, where $Y$ is a compact metric space. By Lemma 5.15, there exists a bicompactification $\tilde{X}$ of $X$ and a light mapping $\tilde{f}: \tilde{X} \to Y$ which extends $f$. Since by Corollary 4.17 a light mapping between bicompact spaces is decomposing, it follows that $X$ is a $Z$-space. $\square$

Finally we have another result about the existence of a bicompactification with certain weight and dimension.

**5.18   Proposition.** *If $X$ is a $Z$-space of infinite weight, then there exists a bicompactification $\tilde{X}$ of $X$ which is a $Z$-space and satisfies $w(\tilde{X}) = w(X)$ and*

$$\dim \tilde{X} = \operatorname{ind} \tilde{X} = \operatorname{Ind} \tilde{Y} = \operatorname{dw}(X).$$

*Proof.* If $X$ is a $Z$-space, then there exists a decomposing mapping $g: X \to Y$, where $Y$ is a compact metric space and $Y \subset \mathbf{R}^n$ if $\operatorname{dw}(X) = n$, where $n$ is a non-negative integer. By Lemma 5.15, there exists a bicompactification $\tilde{X}$ of $X$ such that $w(X) \leqslant w(\tilde{X}) \leqslant \max\{w(X), w(Y)\}$ and a decomposing mapping $\tilde{f}: \tilde{X} \to Y$. Thus $\tilde{X}$ is a $Z$-space and since $w(Y) \leqslant \aleph_0$ and $w(X) \geqslant \aleph_0$ we see that $w(\tilde{X}) = w(X)$. Furthermore $\operatorname{dw}(\tilde{X}) = \operatorname{dw}(X)$. Since $\operatorname{dw}(X) \leqslant \operatorname{dw}(\tilde{X})$, to see this it is enough to note that if $\operatorname{dw}(X) = n$, where $n$ is a non-negative integer, then the existence of the decomposing mapping $\tilde{f}: \tilde{X} \to Y \subset \mathbf{R}^n$ implies that $\operatorname{dw}(\tilde{X}) \leqslant n$. Since $\tilde{X}$ is a compact $Z$-space by Theorem 5.12 we have

$$\operatorname{ind} \tilde{X} = \dim \tilde{X} = \operatorname{Ind} \tilde{X} = \operatorname{dw}(\tilde{X}) = \operatorname{dw}(X). \square$$

## 6  A universal space

We conclude by finding a universal space for the class of bicompact spaces of given weight and covering dimension and hence for the class of Tihonov spaces of given weight and dimension $\partial\text{im}$.

**6.1  Proposition.** *If $\tau$ is an infinite cardinal number and $n$ is a non-negative integer, then there exists a universal space for the class of bicompact spaces of weight equal to $\tau$ and covering dimension equal to $n$.*

*Proof.* Let $\Lambda$ be the set of homeomorphism classes of bicompact spaces of weight equal to $\tau$ and covering dimension equal to $n$. For each $\lambda$ in $\Lambda$ select a subspace $X_\lambda$ of the Tihonov cube $I^\tau$ of weight $\tau$ such that $X_\lambda$ is a member of $\lambda$. Let $X$ be the topological sum of the family of spaces $\{X_\lambda\}_{\lambda \in \Lambda}$. Then $X$ is a $T_4$-space and since $\dim X_\lambda = n$ for each $\lambda$, it follows that $\dim X = n$. Let $f: X \to I^\tau$ be the continuous function such that $f \,|\, X_\lambda$ is the inclusion of $X_\lambda$ in $I^\tau$ for each $\lambda$. It follows from Corollary 3.9 that there exist a bicompact space $P_{n\tau}$ such that $\dim P_{n\tau} = \dim X = n$ and $w(P_{n\tau}) \leqslant w(I^\tau) = \tau$, and continuous functions $g: X \to P_{n\tau}$ and $h: P_{n\tau} \to I^\tau$ such that $h \circ g = f$. Since $f \,|\, X_\lambda$ is an embedding, it is easily seen that $g \,|\, X_\lambda$ is an embedding for each $\lambda$. Thus each bicompact space of weight equal to $\tau$ and covering dimension equal to $n$ can be embedded in $P_{n\tau}$. $\square$

**6.2  Remark.** Similarly there exists a bicompact space $P_{\infty\tau}$ which is a universal space for the class of bicompact spaces of weight equal to $\tau$ and infinite covering dimension.

**6.3  Theorem.** *If $\tau$ is an infinite cardinal number and $n$ is a non-negative integer or $n = \infty$, then the bicompact space $P_{n\tau}$ is a universal space for the class of Tihonov spaces $X$ such that $\partial\text{im}\, X = n$ and $w(X) = \tau$.*

*Proof.* If $X$ is a Tihonov space such that $\partial\text{im}\, X = n$ and $w(X) = \tau$, then by Corollary 3.13, $X$ can be embedded in a bicompact space $\tilde{X}$ such that $\dim \tilde{X} = n$ and $w(\tilde{X}) = \tau$, and $\tilde{X}$ can be embedded in $P_{n\tau}$. $\square$

## Notes

The first applications of the algebra of continuous bounded real-valued functions in dimension theory are due to Katětov. He proved (Katětov [1950]) that if $X$ is a bicompact space such that $\text{rk}\, C(X) \leqslant \aleph_0$, then ind $X = \operatorname{rk} C(X)$. He introduced (Katětov [1950a]) the concept of analytic dimension and proved, using the machinery developed in the earlier paper, that dim $X = \operatorname{ad} C(X)$ if $X$ is a bicompact space. It should be noted that the norm of $C(X)$, and hence its metric, is determined by the algebra structure of $C(X)$ so that Katětov's theorem gives an algebraic characterization of dimension. Katětov defined the dimension $\partial$im and extended his theorem to Tihonov spaces. The first use of coverings by cozero-sets in dimension theory seems to be by Aleksandrov [1940]. Katětov proved subset and countable sum theorems for $\partial$im. Smirnov [1956] developed the theory of the dimension $\partial$im as an application of his dimension theory for proximity spaces. He proved analogues of Theorem 3.2.2 and Propositions 3.3.1 and 3.7.4 for the dimension $\partial$im in Tihonov spaces. Gard and Johnson [1968] characterized the dimension $\partial$im for arbitrary topological spaces in terms of essential mappings into the closed $n$-dimensional ball. The dimension theory for lattices touched on in § 1 is due to Kist [1967]. He associated with each base $\mathscr{L}$, closed with respect to finite unions and intersections, for the topology of a space $X$, a Wallman-type compactification of $X$ which has covering dimension equal to dim $\mathscr{L}$.

In connection with a modification of covering dimension for non-normal spaces, Morita has raised an interesting question. The known cases of failure of the product theorem for covering dimension result from non-normality of the product (Terasawa [1973]). Morita [1970] has asked if the product theorem

$$\partial\operatorname{im} X \times Y \leqslant \partial\operatorname{im} X + \partial\operatorname{im} Y$$

holds for all Tihonov spaces $X$ and $Y$. He has made some progress towards verifying this conjecture (Morita [1973]).

The form of the Stone–Weierstrass theorem given in §2 was obtained by M. H. Stone [1937]. The exposition here follows that of Gillman and Jerison [1960] who presented a self-contained account of Katětov's theorem.

The concept of dimensional rank and the results of §3 are due to Zarelua [1964]. His form of Proposition 3.8 is more general. He showed that if $\{A_\alpha\}_{\alpha \in \Omega}$ is a family of closed sets of $X$ such that the cardinality of $\Omega$ does not exceed $w(Y)$, then it can be supposed additionally that for each $\alpha$

$$\dim A_\alpha = \dim \operatorname{cl}_{\tilde{X}}(A_\alpha) = \dim \operatorname{cl}_Z\big(\phi(A_\alpha)\big).$$

The existence for a $T_4$-space of a bicompactification of the same weight

and dimension was proved by Skljarenko [1958a]. The result for
separable metric spaces was obtained by Hurewicz [1927a, 1930a].
Zarelua's results also include those of Forge [1961]. Proposition 3.11
is due to Mardešić. It was established for his work (Mardešić [1960])
mentioned in the notes on Chapter 8.

In §3 we were concerned with the existence of bicompactifications
of a space $X$ which have the same dimension as $X$. Isbell [1964]
defined mindim $X$ for a Tihonov space $X$ to be the minimum of dim $Y$
over all bicompactifications $Y$ of $X$. Clearly mindim $X \leqslant$ dim $X$. It
can happen that mindim $X <$ dim $X$. Since by Proposition 4.1.3 a
Tihonov space $X$ such that ind $X = 0$ can be embedded in a zero-
dimensional (in all senses) bicompact space it follows that

$$\text{mindim } X = 0$$

if and only if ind $X = 0$. Thus the space $M$ considered in Remark
5.4.6 for which ind $M = 0$ and dim $M = n > 0$ satisfies mindim $M = 0$.
Roy's metrizable space $P$ studied in §4 of Chapter 7 satisfies dim $P = 1$
and mindim $P = 0$. It is clear from Proposition 3.6.7 that if the
Tihonov space $X$ is strongly paracompact, then dim $X =$ mindim $X$.

Another topic in the dimension theory of bicompactifications which
has attracted much interest concerns the dimension of remainders.
If $X$ is a Tihonov space and $Y$ is a bicompactification of $X$, then
$Y\backslash X$ is called the remainder of $X$ in $Y$. For a Tihonov space $X$ and a
non-negative integer $n$, two statements about remainders are of
interest: (1) there exists a bicompactification $Y$ of $X$ such that
$\partial \text{im}\,(Y\backslash X) \leqslant n$; (2) there exists a bicompactification $Y$ of $X$ such
that $w(Y) = w(X)$ and $\partial \text{im}\,(Y\backslash X) \leqslant n$. Freudenthal [1942] found a
necessary and sufficient condition for the validity of (2) for separable
metrizable spaces in the case $n = 0$. He showed (Freudenthal [1951])
that it is a sufficient condition for the validity of (1) for all Tihonov
spaces in the case $n = 0$. Skljarenko [1958] isolated a class of Tihonov
spaces for which, in the case $n = 0$, Freudenthal's condition is necessary and sufficient for the validity of (1) and (2). For this class of
spaces, Smirnov [1966] found necessary and sufficient conditions for
the validity of (1) and (2) for every positive integer $n$. Aarts [1966]
found necessary and sufficient conditions for the validity of (2) for
each $n$ for separable metrizable spaces. In connection with (2), for
separable metrizable spaces, there is a long-standing conjecture (de
Groot [1942]). For a separable metrizable space $X$, de Groot defined
def $X$, the deficiency of $X$, to be the least integer $n$ for which $X$ has a
bicompactification $Y$ with a countable base such that $Y\backslash X$ has

dimension (in all senses) equal to $n$. It is clear that for a separable metrizable space $X$, validity of (2) is equivalent to the condition $\operatorname{def} X \leqslant n$. A concept of inductive compactness, analogous with small inductive dimension, was also introduced by de Groot. A space $X$ satisfies $\operatorname{cmp} X = -1$ if and only if $X$ is compact; if $n$ is a non-negative integer, then $\operatorname{cmp} X \leqslant n$ if for each $x$ in $X$ and each open neighbourhood $G$ of $x$ there exists an open set $U$ such that $x \in U \subset G$ and $\operatorname{cmp} \operatorname{bd}(U) \leqslant n-1$; $\operatorname{cmp} X = n$ if $\operatorname{cmp} X \leqslant n$ and it is not true that $\operatorname{cmp} X \leqslant n-1$. The conjecture is that if $X$ is a separable metric space then $\operatorname{def} X = \operatorname{cmp} X$. It is known that $\operatorname{def} X = n$ if and only if $\operatorname{cmp} X = n$ for $n \leqslant 0$. Information about this conjecture and the properties of deficiency and inductive compactness was given by de Groot and Nishiura [1966].

Katětov [1952] defined the concept of envelopment, but the name seems to be due to Gillman and Jerison. In the same work of Katětov, mappings zero-dimensional with respect to a covering and uniformly zero-dimensional mappings between metric spaces were introduced and Proposition 4.8 was proved. Theorem 5.4 is the basis of Katětov's proof of the equality of covering dimension and large inductive dimension for metrizable spaces.

Katětov proved that for a bicompact space $X$ such that

$$\operatorname{rk} C(X) \leqslant \aleph_0$$

there exists a light mapping of $X$ into the Hilbert cube and

$$\dim X = \operatorname{ind} X = n,$$

where $n$ is the least integer such that exists a light mapping of $X$ into the cube $I^n$. This led Zarelua [1963a] to introduce the concept of decomposing mappings, the class of $Z$-spaces and to define the dimensional weight of a $Z$-space. He proved that every metrizable space is a $Z$-space (in §5 it is established only that every finite-dimensional metrizable space is a $Z$-space). He also showed that the class of $Z$-spaces includes all finite-dimensional quotients of locally compact groups. The characterizations of decomposing mappings and $Z$-spaces are due to Zarelua [1964]. He also showed that if $X$ is a metric space, then $\operatorname{ind} X \leqslant \operatorname{rk} U(X)$, where $U(X)$ is the subset of $C(X)$ consisting of the uniformly continuous functions. Since (Nagata [1960]) $\dim X \geqslant \operatorname{rk} U(X)$, it follows that if $X$ is a metric space which is strongly metrizable, then its dimension (in all senses) is equal to $\operatorname{rk} U(X)$.

The universal space described in §6 was obtained by Pasynkov

[1964]. Zarelua [1964a] also proved the existence of such a universal space employing methods similar to those used in §3. Pasynkov's method is also suitable for a proof of the existence (Theorem 7.3.4) of a universal metrizable space of given weight and dimension. A factorization theorem for mappings into metrizable spaces is employed (Pasynkov [1967]) in place of Mardešić's factorization theorem. Pasynkov [1971] has announced that Mardešić's factorization theorem can be strengthened as follows: if $X$ and $Y$ are bicompact spaces and $f: X \to Y$ is a continuous function, then there exist a bicompact space $Z$ and continuous functions $g: X \to Z$ and $h: Z \to Y$ such that

$$f = h \circ g, \quad w(Y) \leqslant w(Z),$$

$\dim Z \leqslant \dim X$ and $\operatorname{Ind} Z \leqslant \operatorname{Ind} X$. From this theorem the existence of a universal space can be deduced: if $m$ and $n$ are integers such that $0 < n \leqslant m$ and $\tau$ is an infinite cardinal number, then there exists bicompact space $Q_{nm\tau}$ of weight $\tau$ such that

$$\dim Q_{nm\tau} = n \quad \text{and} \quad \operatorname{Ind} Q_{nm\tau} = m,$$

which is a universal space for the class of $T_4$-spaces $X$ such that $w(X) \leqslant \tau$, $\dim X \leqslant n$ and $\operatorname{Ind} X \leqslant m$.

# BIBLIOGRAPHY

Aarts, J. M. [1966]. *Dimension and deficiency in general topology.* Thesis, Amsterdam.
— [1971]. A characterization of strong inductive dimension. *Fund. Math.* **70**, 147–55.
Aleksandrov, P. S. [1928]. Untersuchungen über Gestalt und Lage abgeschlossener Mengen beliebiger Dimension. *Ann. Math.* **30**, 101–87.
— [1928a]. Über den allgemeinen Dimensionsbegriff und seine Beziehungen zur elementaren geometrischen Anschauung. *Math. Ann.* **98**, 617–35.
— [1932]. Dimensionstheorie. *Math. Ann.* **106**, 161–238.
— [1936]. Einige Problemstellungen in der mengentheoretische Topologie. *Mat. Sb.* **1**, 3–80.
— [1936a]. Über abzählbar-fache offene Abbildungen. *Dokl. Akad. Nauk SSSR* **4**, 283.
— [1940]. Über die Dimension der bikompakten Räume. *Dokl. Akad. Nauk SSSR* **26**, 619–22.
— [1941]. Der endliche dimensionstheoretische Summensatz für bikompakte Räume. *Soobšč. Akad. Nauk Gruzin SSR* **2**, 1–6.
— [1947]. On the dimension of normal spaces. *Proc. Roy. Soc. London Ser. A* **189**, 11–39.
— [1951]. The present status of the theory of dimension. *Uspehi Mat. Nauk* **6**, no. 4, 43–68 (Russian). English translation: *Amer. Math. Soc. Transl. Ser.* 2, **1**, 1–26.
— [1960]. On some results in the theory of topological spaces obtained during the last twenty-five years. *Uspehi Mat. Nauk* **15**, no. 2, 25–95 (Russian). English translation: *Russian Math. Surveys* **15** (1960) no. 2, 23–83.
— [1964]. Some basic directions in general topology. *Uspehi Mat. Nauk* **19**, no. 6, 3–46 (Russian). English translation: *Russian Math. Surveys* **19** (1964) no. 6, 1–39.
Aleksandrov, P. S. and Ponomarev, V. I. [1960]. On some classes of $n$-dimensional spaces. *Sibirsk. Mat. Ž.* **1**, 3–13.
Anderson, R. D. and Keisler, J. E. [1967]. An example in dimension theory. *Proc. Amer. Math. Soc.* **18**, 709–13.
Arhangel'skiĭ, A. V. [1961]. New criteria for paracompactness and metrizability of an arbitrary $T_1$-space. *Dokl. Akad. Nauk SSSR* **141**, 13–15 (Russian). English translation: *Soviet Math. Dokl.* **2** (1961), 1367–9.
— [1963]. On the rank of families of sets and dimension of spaces. *Fund. Math.* **52**, 257–75 (Russian).
— [1968]. On closed mappings increasing dimension. *Czechoslovak Math. J.* **18**, 389–91.
— [1969]. A criterion for $n$-dimensionality and an approach to the proof of the equality dim = Ind for metric spaces. *Dokl. Akad. Nauk* **187**, 490–3 (Russian). English translation: *Soviet Math. Dokl.* **10** (1969), 869–72.
Bing, R. H. [1951]. Metrization of topological spaces. *Canad. J. Math.* **3**, 175–86.

[ 415 ]

**416** BIBLIOGRAPHY

Boltyanskiĭ, V. G. [1949]. On dimensional full-valuedness of compacta. *Dokl. Akad. Nauk SSSR* **67**, 773–7 (Russian). English translation: *Amer. Math. Soc. Transl. Ser.* 1 no. 48 (1951), 7–11.

Borsuk, K. [1937]. Sur les prolongements des transformations continues. *Fund. Math.* **28**, 99–110.

Brouwer, L. E. J. [1911]. Beweis der Invarianz der Dimensionenzahl. *Math. Ann.* **70**, 161–5.

— [1913]. Über den natürlichen Dimensionsbegriff. *J. Reine Angew. Math.* **142**, 146–52.

Čech, E. [1932]. Sur la dimension des espaces parfaitement normaux. *Bull. Intern. Acad. Tcheque Sci.* **33**, 38–55.

— [1933]. Contribution to the theory of dimension. *Časopis Pěst. Mat. Fys.* **62**, 277–91 (Czech).

— [1948]. Problem P53. *Colloq. Math.* **1**, 332.

Ceder, J. G. [1964]. On maximally resolvable spaces. *Fund. Math.* **55**, 87–93.

Delinić, K. and Mardešić, S. [1968]. A necessary and sufficient condition for $n$-dimensionality of inverse limits. *Proc. Int. Symp. on topology and its applications (Herceg-Novi)*, 124–9.

Dowker, C. H. [1947]. Mapping theorems for non-compact spaces. *Amer. J. Math.* **69**, 200–42.

— [1948]. An extension of Aleksandrov's mapping theorem. *Bull. Amer. Math. Soc.* **54**, 386–91.

— [1953]. Inductive dimension of completely normal spaces. *Quart. J. Math. Oxford Ser.* 2 **4**, 267–81.

— [1955]. Local dimension of normal spaces. *Quart. J. Math. Ser.* 2 **6**, 101–20.

Dowker, C. H. and Hurewicz, W. [1956]. Dimension of metric spaces. *Fund. Math.* **43**, 83–7.

Dugundji, J. [1966]. *Topology*. Boston.

Eilenberg, S. and Otto, E. [1938]. Quelques propriétés caractéristiques de la dimension. *Fund. Math.* **31**, 149–53.

Engelking, R. [1968]. *Outline of General Topology*. Elsevier, Amsterdam.

— [1973]. Some new proofs in dimension theory. To appear.

Erdös, P. [1940]. The dimension of the rational points in Hilbert space. *Ann. Math.* **41**, 734–6.

Fedorčuk, V. V. [1968]. Bicompacta with non-coinciding dimensionalities. *Dokl. Akad. Nauk SSSR* **182**, 275–7 (Russian). English translation: *Soviet Math. Dokl.* **9** (1968), 1148–50.

Filippov, V. V. [1969]. A bicompactum with non-coinciding inductive dimensionalities. *Dokl. Akad. Nauk SSSR* **184**, 1050–3 (Russian). English translation: *Soviet Math. Dokl.* **10** (1969), 208–11.

— [1969a]. A bicompactum satisfying the first axiom of countability with non-coinciding ind and dim dimensions. *Dokl. Akad. Nauk SSSR* **186**, 1020–2 (Russian). English translation: *Soviet Math. Dokl.* **10** (1969), 713–15.

— [1970]. Solution of a problem of P. S. Aleksandrov (a bicompact space with distinct inductive dimensions). *Mat. Sb.* **83**, 42–60 (Russian). English translation: *Math. USSR Sbornik* **12** (1970), 41–57.

— [1970a]. On bicompacta with non-coinciding inductive dimensions. *Dokl. Akad. Nauk SSSR* **192**, 289–93 (Russian). English translation: *Soviet Math. Dokl.* **11** (1970), 635–8.

— [1972]. On the inductive dimension of the product of bicompacta. *Dokl. Akad. Nauk SSSR* **202**, 1016–19 (Russian). English translation: *Soviet Math. Dokl.* **13** (1972), 250–4.

Forge, A. B. [1961]. Dimension preserving compactifications allowing extensions of continuous functions. *Duke Math. J.* 28, 625–7.

Freudenthal, H. [1932]. Über dimensionserhöhende stetige Abbildungen. *S.-B. Preuss. Akad. Wiss. H.* 5, 34–8.

[1937]. Entwicklungen von Räumen und ihren Gruppen. *Compositio Math.* 4, 145–234.

[1942]. Neuaufbau der Endentheorie. *Ann. Math.* 43, 261–79.

[1951]. Kompaktifizierungen und Bikompaktifizierungen. *Nederl. Akad. Wetensch. Proc. Ser. A* 54, 184–92.

Gard, J. R. and Johnson, R. D. [1968]. Four dimension equivalences. *Canad. J. Math.* 20, 48–50.

Gillman, L. and Jerison, M. [1960]. *Rings of Continuous Functions.* Van Nostrand, New York.

de Groot, J. [1942]. *Topologische Studien.* Thesis, Groningen.

de Groot, J. and Nishiura, T. [1966]. Inductive compactness as a generalization of semicompactness. *Fund. Math.* 58, 201–18.

Hanai, S. [1954]. On closed mappings. *Proc. Japan Acad.* 30, 285–8.

Hemmingsen, E. [1946]. Some theorems in dimension theory for normal Hausdorff spaces. *Duke Math. J.* 13, 495–504.

Henderson, D. W. [1967]. An infinite-dimensional compactum with no positive-dimensional compact subsets – a simpler construction. *Amer. J. Math.* 89, 105–21.

[1968]. A lower bound for transfinite dimension. *Fund. Math.* 63, 167–73.

Henriksen, M. and Isbell, J. R. [1958]. Some properties of compactifications. *Duke Math. J.* 25, 83–106.

Hurewicz, W. [1927]. Normalbereiche und Dimensionstheorie. *Math. Ann.* 96, 736–64.

[1927a]. Über das Verhältnis separabler Räume zu kompakten Räumen. *Proc. Akad. Wetensch. Amst.* 30, 425–30.

[1930]. Ein Theorem der Dimensionstheorie. *Ann. Math.* 31, 176–80.

[1930a]. Einbettung separabler Räume in gleich dimensional kompakte Räume. *Monatsh. Math.* 37, 199–208.

[1933]. Über dimensionserhöhende stetige Abbildungen. *J. Reine Angew. Math.* 169, 71–8.

[1933a]. Über Abbildungen von endlichdimensionalen Räumen auf Teilmengen Cartesischer Räume. *S.-B. Preuss. Akad. Wiss H.* 24/25, 754–68.

[1935]. Über Abbildung der topologischer Räume auf die $n$-dimensionalen Sphäre. *Fund. Math.* 24, 144–50.

Hurewicz, W. and Wallman, H. [1941]. *Dimension Theory.* Princeton University Press.

Isbell, J. R. [1964]. Uniform Spaces. *Amer. Math. Soc. Math. Surveys* 12.

Katětov, M. [1950]. On rings of continuous functions and the dimension of bicompact spaces. *Časopis Pěst. Mat. Fys.* 75, 1–16 (Russian).

[1950a]. A theorem on the Lebesgue dimension. *Časopis Pěst. Mat. Fys.* 75, 79–87.

[1951]. On the dimension of metric spaces. *Dokl. Akad. Nauk SSSR* 79, 189–91 (Russian).

[1952]. On the dimension of non-separable spaces, I. *Czechoslovak Math. J.* 2, 333–68 (Russian).

Katsuta, Y. [1966]. A note on the inductive dimension of product spaces. *Proc. Japan Acad.* 42, 1011–5.

**418** BIBLIOGRAPHY

Keesling, J. [1968]. Mappings and dimension in general metric spaces. *Pacific J. Math.* **25**, 277–88.

Kelley, J. L. [1955]. *General Topology.* van Nostrand, Princeton.

Kimura, N. [1964]. On the covering dimension of product spaces. *Proc. Japan Acad.* **40**, 267–71.

Kist, J. [1967]. On the dimension of Wallman compactifications. *Proc. London Math. Soc.* **17**, 761–7.

Kljušin, V. L. [1964]. Perfect mappings of paracompact spaces. *Dokl. Akad. Nauk SSSR* **159**, 734–7 (Russian). English translation: *Soviet Math. Dokl.* **5** (1964), 1583–6.

Knaster, B. and Kuratowski, C. [1921]. Sur les ensembles connexes. *Fund. Math.* **2**, 206–55.

Kodama, Y. [1969]. On subset theorems and the dimension of products. *Amer. J. Math.* **91**, 486–97.

Kolmogorov, A. [1937]. Über offene Abbildungen. *Ann. Math.* **38**, 36–8.

Kowalsky, H. J. [1957]. Einbettung metrischer Räume. *Arch. Math.* **8**, 336–9.

Kuratowski, C. [1932]. Sur l'application des espaces fonctionnels à la théorie de la dimension. *Fund. Math.* **18**, 285–92.

  [1937]. Sur les théorèmes de 'plongement' dans la théorie de la dimension. *Fund. Math.* **28**, 336–42.

Kuz'minov, V. I. [1968]. Homological dimension theory. *Uspehi Mat. Nauk* **23**, no. 5, 3–50 (Russian). English translation: *Russian Math. Surveys* **23** (1968) no. 5, 1–45.

Landau, M. [1969]. Strong transfinite ordinal dimension. *Proc. Amer. Math. Soc.* **21**, 591–6.

Lebesgue, H. [1911]. Sur la non applicabilité de deux domaines appartenant à des espaces de $n$ et $n+p$ dimensions. *Math. Ann.* **70**, 166–8.

Lefschetz, S. [1931]. On compact spaces. *Ann. Math.* **32**, 521–38.

Levšenko, B. T. [1965]. Spaces of transfinite dimension. *Mat. Sb.* **67**, 255–66 (Russian).

Lifanov, I. K. [1967]. Some theorems concerning the Ind dimension. *Dokl. Akad. Nauk SSSR* **177**, 33–6 (Russian). English translation: *Soviet Math. Dokl.* **8** (1967), 1362–5.

  [1968]. The dimensionality of the product of unidimensional bicompacta. *Dokl. Akad. Nauk SSSR* **180**, 534–7 (Russian). English translation: *Soviet Math. Dokl.* **9** (1968), 648–51.

  [1968a]. Dimensionality of products of bicompact spaces. *Dokl. Akad. Nauk SSSR* **182**, 1274–7 (Russian). English translation: *Soviet Math. Dokl.* **9** (1968), 1280–3.

  [1969]. On the large inductive dimension. *Dokl. Akad. Nauk SSSR* **184**, 1288–91. (Russian). English translation: *Soviet Math. Dokl.* **10** (1969), 247–50.

Lifanov, I. K. and Pasynkov, B. A. [1970]. Examples of bicompacta with non-coinciding inductive dimensions. *Dokl. Akad. Nauk. SSSR* **192**, 276–8 (Russian). English translation: *Soviet Math. Dokl.* **11** (1970), 619–21.

Lokucievskii, O. V. [1949]. On the dimension of bicompacta. *Dokl. Akad. Nauk. SSSR* **67**, 217–19 (Russian).

Lunc, A. L. [1949]. A bicompactum whose inductive dimension is greater than its dimension defined by means of coverings. *Dokl. Akad. Nauk. SSSR* **66**, 801–3 (Russian).

Mack, J. E. and Pears, A. R. [1974]. Closed covers, dimension and quasi-order spaces. *Proc. London Math. Soc.* (3) **29**, 289–316.

Mardešić, S. [1960]. On covering dimension and inverse limits of compact spaces. *Illinois J. Math.* **4**, 278–91.

Menger, K. [1923]. Über die Dimensionalität von Punktmengen, I. *Monatsh. Math.* **33**, 148–60.

[1924]. Über die Dimensionalität von Punktmengen, II. *Monatsh. Math.* **34**, 137–61.

[1926]. Über umfassendste *n*-dimensionale Mengen. *Proc. Akad. Wetensch. Amst.* **29**, 1125–8.

[1929]. Zur Begründung einer axiomatischen Theorie der Dimension. *Monatsh. Math.* **36**, 193–218.

Michael, E. A. [1953]. A note on paracompact spaces. *Proc. Amer. Math. Soc.* **4**, 831–8.

[1957]. Another note on paracompact spaces. *Proc. Amer. Math. Soc.* **8**, 822–8.

[1959]. Yet another note on paracompact spaces. *Proc. Amer. Math. Soc.* **10**, 309–14.

[1963]. The product of a normal space and a metric space need not be normal. *Bull. Amer. Math. Soc.* **69**, 375–6.

Moore, R. L. [1935]. A set of axioms for plane analysis situs. *Fund. Math.* **25**, 13–28.

Morita, K. [1948]. Star-finite coverings and the star-finite property. *Math. Japon.* **1**, 60–8.

[1950]. On the dimension of normal spaces, I. *Japan. J. Math.* **20**, 5–36.

[1950a]. On the dimension of normal spaces, II. *J. Math. Soc. Japan.* **2**, 16–33.

[1953]. On the dimension of product spaces. *Amer. J. Math.* **75**, 205–23.

[1953a]. On spaces having the weak topology with respect to a closed covering. *Proc. Japan Acad.* **29**, 537–43.

[1954]. Normal families and dimension theory for metric spaces. *Math. Ann.* **128**, 350–62.

[1955]. A condition for the metrizability of topological spaces and for *n*-dimensionality. *Sci. Rep. Tokyo Kyoiku Daigaku Sect. A.* **5**, 33–6.

[1956]. On closed mappings and dimension. *Proc. Japan Acad.* **32**, 161–5.

[1962]. Paracompactness and product spaces. *Fund. Math.* **50**, 223–36.

[1964]. Products of normal spaces with metric spaces. *Math. Ann.* **154**, 365–82.

[1970]. Topological completions and *M*-spaces. *Sci. Rep. Tokyo Kyoiku Daigaku*, **10**, 271–88.

[1973]. On the dimension of the product of Tychonoff spaces. *General Topology and Appl.* **3**, 125–33.

Nagami, K. [1960]. Mappings of finite order and dimension theory. *Japan. J. Math.* **30**, 25–54.

[1962]. Mappings defined on 0-dimensional spaces and dimension. *J. Math. Soc. Japan* **14**, 101–18.

[1966]. A normal space $Z$ with ind $Z = 0$, dim $Z = 1$, Ind $Z = 2$. *J. Math. Soc. Japan* **18**, 158–65.

[1969]. A note on the large inductive dimension of totally normal spaces. *J. Math. Soc. Japan* **21**, 282–90.

[1969a]. $\Sigma$-spaces. *Fund. Math.* **65**, 168–92.

[1970]. *Dimension Theory.* Academic Press, New York.

[1971]. Dimension for $\sigma$-metric spaces. *J. Math. Soc. Japan* **23**, 123–9.

[1972]. Perfect classes of spaces. *Proc. Japan Acad.* **48**, 21–4.

# 420 BIBLIOGRAPHY

Nagami, K. and Roberts, J. H. [1967]. A study of metric-dependent dimension functions. *Trans. Amer. Math. Soc.* **129**, 414–35.

Nagata, J. [1950]. On a necessary and sufficient condition of metrizability. *J. Inst. Polytech. Osaka City Univ. Ser. A.* **1**, 93–100.

[1957]. On imbedding theorem for non-separable metric spaces. *J. Inst. Polytech. Osaka City Univ. Ser. A.* **8**, 9–14.

[1960]. On rings of continuous functions and the dimension of metric spaces. *Proc. Japan Acad.* **36**, 49–52.

[1961]. On dimension and metrization. *Proc. 1st. Prague Topo. Symp.* 282–5.

[1963]. A remark on general imbedding theorems in dimension theory. *Proc. Japan Acad.* **39**, 197–9.

[1965]. *Modern Dimension Theory.* Elsevier, Amsterdam.

[1966]. A survey of dimension theory. *Proc. 2nd. Prague Topo. Symp.* 259–70.

[1967]. Product theorems in dimension theory. *Bull. Acad. Polon. Sci. Sér. Sci. Math. Astronom. Phys.* **15**, 439–48.

[1971]. A survey of dimension theory, II. *General Topology and Appl.* **1**, 65–77.

Nishiura, T. [1966]. Inductive invariants and dimension theory. *Fund. Math.* **59**, 243–62.

Nöbeling, G. [1930]. Über eine $n$-dimensionale Universalmenge im $R_{2n+1}$. *Math. Ann.* **104**, 71–80.

Nyikos, P. [1973]. Prabir Roy's space $\Delta$ is not N-compact. *General Topology and Appl.* **3**, 197–210.

Ostrand, P. A. [1971]. Covering dimension in general spaces. *General Topology and Appl.* **1**, 209–21.

Pasynkov, B. A. [1958]. On polyhedral spectra and the dimension of bicompacta, in particular bicompact groups. *Dokl. Akad. Nauk. SSSR* **121**, 45–8 (Russian).

[1960]. The coincidence of various definitions of dimension for locally compact groups. *Dokl. Akad. Nauk SSSR* **132**, 1035–7 (Russian). English translation: *Soviet Math. Dokl.* **1** (1960), 720–2.

[1962]. On spectra and the dimension of topological spaces. *Mat. Sb.* **57**, 449–76 (Russian).

[1963]. $\omega$-mappings and inverse spectra. *Dokl. Akad. Nauk SSSR* **150**, 488–91 (Russian). English translation: *Soviet Math. Dokl.* **4** (1963), 706–9.

[1964]. On universal bicompacta of given weight and dimension. *Dokl. Akad. Nauk SSSR* **154**, 1042–3 (Russian). English translation: *Soviet Math. Dokl.* **5** (1964), 245–6.

[1965]. On the spectral decomposition of topological spaces. *Mat. Sb.* **66**, 35–79 (Russian).

[1965a]. On a formula of W. Hurewicz. *Vestnik Moskov. Univ. Ser. I Mat. Meh.* **20**, no. 4, 3–5 (Russian).

[1967]. Universal bicompact and metric spaces of given dimension. *Fund. Math.* **60**, 285–308 (Russian).

[1967a]. On open mappings. *Dokl. Akad. Nauk SSSR* **175**, 292–5 (Russian). English translation: *Soviet Math. Dokl.* **8** (1967), 853–6.

[1969]. On inductive dimensions. *Dokl. Akad. Nauk SSSR* **189**, 254–7 (Russian). English translation: *Soviet Math. Dokl.* **10** (1969), 1402–6.

[1970]. On bicompacta with non-coinciding dimensions. *Dokl. Akad. Nauk SSSR* **192**, 503–6 (Russian). English translation: *Soviet Math. Dokl.* **11** (1970), 619–21.

[1971]. On the dimension of normal spaces. *Dokl. Akad. Nauk SSSR* **201**,

1049–52 (Russian). English translation: *Soviet Math. Dokl.* 12 (1971), 1784–7.

[1972]. A factorization theorem for non-closed sets. *Dokl. Akad. Nauk SSSR* 202, 1274–6 (Russian). English translation: *Soviet Math. Dokl.* 13 (1972), 292–5.

Pears, A. R. [1970]. A sum theorem for local inductive dimension. *J. London Math. Soc.* (2) 2, 321–2.

[1971]. A note on transfinite dimension. *Fund. Math.* 71, 215–21.

[1971a]. Dimension theory of $M$-spaces. *J. London Math. Soc.* (2) 3, 109–12.

Poincaré, H. [1912]. Pourquoi l'espace a trois dimensions. *Revue de Métaphysique et de Morale* 20, 483–304.

Ponomarev, V. I. [1962]. Some applications of projective spectra to the theory of topological spaces. *Dokl. Akad. Nauk SSSR* 144, 993–6 (Russian). English translation: *Soviet Math. Dokl.* 3 (1962), 863–6.

[1963]. Projective spectra and continuous mappings of paracompacta. *Mat. Sb.* 60, 89–119 (Russian). English translation: *Amer. Math. Soc. Transl. Ser.* 2, 39 (1964), 133–64.

Pontrjagin, L. S. [1930]. Sur une hypothèse fondamentale de la théorie de la dimension. *C. R. Acad. Sci. Paris Sér. A-B* 190, 1105–7.

Proskuryakov, I. V. [1951]. On the theory of topological spaces. *Moskov. Gos. Univ. Uč. Zapiski* 148, 219–23 (Russian).

Pupko, V. I. [1961]. The monotonicity of the dimension of a class of subsets of normal spaces. *Vestnik. Moskov. Univ. Ser. I Mat. Meh.* no. 2, 41–45 (Russian).

Roberts, J. H. [1941]. A theorem on dimension. *Duke Math. J.* 8, 565–74.

[1947]. Open transformations and dimension. *Bull. Amer. Math. Soc.* 53, 176–8.

[1948]. A problem in dimension theory. *Amer. J. Math.* 70, 126–8.

Roy, P. [1962]. Failure of equivalence of dimension concepts for metric spaces. *Bull. Amer. Math. Soc.* 68, 609–13.

[1968]. Nonequality of dimensions for metric spaces. *Trans. Amer. Math. Soc.* 134, 117–32.

Šedivá, V. [1959]. On collectionwise normal and strongly paracompact spaces. *Czechoslovak Math. J.* 9, 50–62 (Russian).

Sitnikov, K. A. [1953]. An example of a two-dimensional set in three-dimensional Euclidean space. *Dokl. Akad. Nauk SSSR* 88, 21–4 (Russian).

Skljarenko, E. G. [1958]. Bicompactifications of semibicompact spaces. *Dokl. Akad. Nauk SSSR* 120, 1200–3 (Russian).

[1958a]. On the embedding of normal spaces into bicompacta of the same weight and dimension. *Dokl. Akad. Nauk SSSR* 123, 36–9 (Russian).

[1962]. A theorem on mappings lowering dimension. *Bull. Acad. Polon. Sci. Sér. Sci. Math. Astronom. Phys.* 10, 429–32 (Russian).

Skordev, G. S. [1970]. Dimension-increasing mappings. *Mat. Zametki* 7, 697–705 (Russian).

[1970a]. On resolutions of continuous mappings. *Mat. Sb.* 82, 532–50 (Russian). English translation: *Math. USSR Sbornik* 11 (1970), 491–506.

Smirnov, Ju. M. [1951]. Some relations in the theory of dimension. *Mat. Sb.* 29, 157–72 (Russian).

[1951a]. On normally situated sets in normal spaces. *Mat. Sb.* 29, 173–6 (Russian).

Smirnov, Ju. M. [1951 b]. A necessary and sufficient condition for the metrizability of a topological space. *Dokl. Akad. Nauk SSSR* **77**, 197–206 (Russian).

[1956]. On the dimension of proximity spaces. *Mat. Sb.* **38**, 283–302 (Russian). English translation: *Amer. Math. Soc. Transl. Ser.* 2, **21** (1962), 1–20.

[1958]. An example of a zero-dimensional normal space having infinite covering dimension. *Dokl. Akad. Nauk SSSR* **123**, 40–2 (Russian).

[1959]. On universal spaces for certain classes of infinite dimensional spaces. *Izv. Akad. Nauk SSSR Ser. Mat.* **23**, 185–96 (Russian). English translation: *Amer. Math. Soc. Transl. Ser.* 2, **21** (1962), 21–33.

[1966]. On the dimension of increments of bicompact extensions of proximity spaces and topological spaces I, II. *Mat. Sb.* **69**, 141–60, **71**, 454–82 (Russian). English translation: *Amer. Math. Soc. Transl. Ser.* 2, **84** (1969), 197–217, 219–51.

Stone, A. H. [1948]. Paracompactness and product spaces. *Bull. Amer. Math. Soc.* **54**, 977–82.

[1960]. Sequences of coverings. *Pacific J. Math.* **10**, 689–91.

Stone, M. H. [1937]. Applications of the theory of Boolean rings to general topology. *Trans. Amer. Math. Soc.* **41**, 765–72.

Terasawa, J. [1973]. On the zero-dimensionality of some non-normal product spaces. *Sci. Rep. Tokyo Kyoiku Daigaku Sect. A*. To appear.

Tihonov, A. N. [1930]. Über die topologische Erweiterung von Räumen. *Math. Ann.* **102**, 544–61.

Toulmin, G. H. [1954]. Shuffling ordinals and transfinite dimension. *Proc. London Math. Soc.* **4**, 177–95.

Tumarkin, L. A. [1926]. Beitrag zur allgemeinen Dimensionstheorie. *Mat. Sb.* **33**, 57–86.

[1928]. Über die Dimension nicht abgeschlossener Mengen. *Math. Ann.* **98**, 637–56.

Tukey, J. W. [1940]. *Convergence and Uniformity in Topology*. Kraus Repr., New York.

Urysohn, P. S. [1922]. Les multiplicités Cantoriennes. *C.R. Acad. Sci. Paris Sér. A-B* **157**, 440–2.

[1925]. Memoire sur les multiplicités Cantoriennes. *Fund. Math.* **7**, 30–137.

[1926]. Memoire sur les multiplicités Cantoriennes (suite). *Fund. Math.* **8**, 225–359.

Vedenisov, N. B. [1939]. Remarks on the dimension of topological spaces. *Uč. Zapiski Moskov. Gos. Univ.* **30**, 131–40 (Russian).

[1941]. On dimension in the sense of Čech. *Izv. Akad. Nauk Ser. Mat.* **5**, 211–16 (Russian).

Vopěnka, P. [1958]. On the dimension of compact spaces. *Czechoslovak Math. J.* **8**, 319–27 (Russian).

[1959]. Remarks on the dimension of metric spaces. *Czechoslovak Math. J.* **9**, 519–22. (Russian).

Wallace, A. D. [1945]. Dimensional types. *Bull. Amer. Math. Soc.* **51**, 679–81.

Wallman, H. [1938]. Lattices and topological spaces. *Ann. Math.* **39**, 112–26.

Zarelua, A. V. [1963]. On a theorem of Hurewicz. *Mat. Sb.* **60**, 17–28 (Russian).

[1963 a]. On the equality of dimensions. *Mat. Sb.* **62**, 295–319 (Russian).

[1964]. On the extension of mappings to bicompactifications preserving some special properties. *Sibirsk. Mat. Ž.* **5**, 532–48 (Russian).

[1964a]. A universal bicompactum of given weight and dimension. *Dokl. Akad. Nauk SSSR* **154**, 1015–8 (Russian). English translation: *Soviet Math. Dokl.* **5** (1964), 214–18.

[1969]. Finite-to-one mappings of topological spaces and cohomology manifolds. *Sibirsk. Mat. Ž.* **10**, 64–92 (Russian). English translation: *Siberian Math. J.* **10** (1969), 45–63.

# INDEX

Accumulation point, 4
Algebra, 369
  complete normed, 371
  normed, 371
Almost-open mapping, 207
Analytic base, 390
Analytic dimension ad, 390
Analytic subalgebra, 387
  $L[H]$, 388

Baire space $B(\tau)$, 84
Baire's theorem, 93
Base for a topology, 3
$\beta$-closed set, 49
Bicompactification, 42
  Stone–Čech, 46
Bicompact space, 37
Bing's theorem, 80
Boundary, 12
Brouwer theorem, 10

Canonical covering, 214
Canonical mapping
  into a nerve, 108
  of an inverse system, 52
Cantor manifold, 362
Cantor space, $n$-dimensional, 335
Cantor set, 150
Cauchy sequence, 91
Chain, 2
Closed covering, 4
Closed mapping, 6
Closed set, 3
Closed subalgebra $A[H]$, 371
Closed unit ball, 10
Closure, 3
Cofinal subset of a directed set, 52
Compact space, 35
Completely normal space, 27
Completely paracompact space, 76
Completely regular space, 39
Complete normed algebra, 371
Complete space, 91

Component, 4
Connected space, 3
Connecting mapping, 52
Constancy set, 376
Continuous function, 5
Countable sum theorem
  for covering dimension, 125
  for large inductive dimension, 171
Countably paracompact space, 66
Covering, 4
  canonical, 214
  closed 4
  numerable, 22
  open, 4
  shrinkable, 20
Covering dimension dim, 111
Cozero-set, 18
Cushioned refinement, 68
Cut, 312
CW-complex, 296
$C(X)$, 92
$C^n(X)$, 92

Decomposing mapping, 398
Dense set, 4
Diameter, diam, 77
Dimension
  analytic, ad, 390
  approximation, 238
  at a point, 197
  cofinal approximation, 238
  covering, dim, 111
  $\delta$, 217
  $\Delta$, 217
  $\partial$im, 368
  large inductive, Ind, 155
  local, loc dim, 188
  local inductive, loc Ind, 188
  of a CW-complex, 297
  of a distributive lattice, 366
  of a mapping, 336
  of a simplicial complex, 105
  small inductive, ind, 150

[ 424 ]

Pseudo-metric, 9
  space, 9
  topology, 9
Pseudo-metrizable space, 9

Quasi-order, 1
Quasi-ordered set, 1
Quotient space, 8

R, set of real numbers, 5
Rank of a family of sets, 145
Rank rk of a subset of a normed
  algebra, 390
Refinement, 4
  closed, 4
  cushioned, 68
  open, 4
  star-, 70
  strong star-, 70
  weak, 76
Regular family of coverings, 216
Regular space, 8
Replica of an $N$-space, 290
Retract, 7
Retraction, 7
Roy's example, 271ff

Separable space, 86
Separation axioms, 8
Separation of sets, 119
Shrinkable covering, 20
$\sigma$-discrete family, 11
$\sigma$-locally finite family, 11
Similar families, 24
Simplex, 105
Simplicial complex, 105
Skeleton of a simplicial complex, 106
Small inductive dimension ind, 150
Space
  Baire, 84
  bicompact, 37
  Cantor, 335
  compact, 35
  complete, 91
  completely normal, 27
  completely paracompact, 76
  completely regular, 39
  countably paracompact, 66
  dominated by a covering, 14
  Euclidean, 10
  first-countable, 82

Hausdorff, 8
hereditarily paracompact, 73
Hilbert, 88
Lindelöf, 75
linearly ordered, 5
metric, 9
metrizable, 10
$N$-, 288
normal, 17
paracompact, 57
paracompact $M$-, 101
perfectly normal, 33
perfectly zero-dimensional, 217
pseudo-metrizable, 9
quotient, 8
regular, 8
separable, 86
strongly paracompact, 74
strongly pseudo-metrizable, 83
Tihonov, 39
topological, 3
$T_i$-, $i = 0,1,2$, 8
$T_3$-, 9
$T_4$-, 17
totally disconnected, 112
totally normal, 31
universal, 41
weakly paracompact, 74
$Z$-, 403
Special irreducible mapping, 311
Standard base for an inverse limit,
  298
Star-finite family, 74
Star of a vertex, 107
Star-refinement, 70
Stone–Čech bicompactification, 46
Stone–Weiestrass theorem, 377
Strongly directed family of coverings,
  221
Strongly paracompact space, 74
Strongly pseudo-metrizable space, 83
Strong star-refinement, 70
Subalgebra, 370
  $K[H]$, 370
Subbase for a topology, 3
Subcomplex, 106
Subset theorems
  for dim, 114, 138ff
  for ind, 151
  for Ind, 158, 166ff
  for local dimension, 190 ff